POLARITY CONTROL FOR SYNTHESIS

TSE-LOK HO
The NutraSweet Co.

A Wiley–Interscience Publication

JOHN WILEY & SONS, INC.

New York • Chichester • Brisbane • Toronto • Singapore

4276978

CHEMISTRY

Copyright © 1991 by John Wiley & Sons, Inc.

Library of Congress Cataloging in Publication Data:

Ho, Tse-Lok.
    Polarity control for synthesis / Tse-Lok Ho.
       p.   cm.
    Includes bibliographical references.
    ISBN 0-471-53850-7
    1. Organic compounds—Synthesis. 2. Polarization (Electricity)
    I. Title.
    QD262.H6   1991
    547.2—dc20   90-23148
                                                    CIP

Printed in the United States of America

10   9   8   7   6   5   4   3   2   1

To Professor E.J. Corey

Inspiring mentor for generations of organic chemists
and
fecund purveyor of synthetic ideas

PREFACE

"...intellectual processes such as the recognition and use of synthons require considerable ability and knowledge; here, too, genius and originality find ample opportunity for expression."   —E.J. Corey, in *Pure Appl. Chem.* (1967) **14**: 30.

The main purpose of this book is to identify the polar reactions that are frequently employed in carbogenic synthesis, and delineate the choice of their use in a contextual sense. The aspect of choice is most crucial to the success, efficiency, and elegance of a synthesis. While the construction of a complete or partial molecular skeleton of a synthetic target, or the introduction and interconversion of functional groups can be handled in myriad ways, there are certain orderings and combinations of reactions that are outstanding.

The revolution of synthetic design brought forth by Prof. E.J. Corey on his establishment of retrosynthetic analysis has transformed this chemical endeavor from an arcane art into a logical science. The synthon concept espoused therein is central to the analysis and the generation of synthetic routes, but equally important, in practical terms, is the assembly of these synthons in a polar reaction, the union of two synthons requires that they are of opposite polarity. The distribution of functionalities in a synthon or its synthetic equivalent is of major significance, as a remote functional group can often determine the suitability of a reaction.

This volume touches on the retrosynthetic analysis, but it does not dwell on it. It is not intended to compete with the excellent monographs of Corey and Cheng, and Warren. Rather, its emphasis is on the exploitation and manipulation of the electronic properties of the synthons in response to synthetic necessity. By taking interesting examples from the chemical literature and keeping the level of discussion to about the graduate level, I hope to convey a sense of excitement of synthesis to a wide audience. Practicing synthetic chemists may find some benefit from consulting it.

My thoughts on the control of reactivity by remote substituents began in the mid-to-late 1970s when I widened my outlook on the application of the hard and soft acids and bases (HSAB) principle to organic chemistry. Analysis of many reactions led me to recognize patterns relevant to reactivity in redox reactions, odd-membered ring compounds, etc. which also occurred to Prof. Dieter Seebach. The further stimulation provided to this author by the article on reactivity umpolung [(1979) *Angew. Chem. Int. Ed. Engl.* **18**: 239] must be acknowledged.

The reader should not expect full information of the total synthesis of various compounds discussed because that is not the purpose of this book. Neither does this book serve as a comprehensive reference, and in this spirit, only one or a few examples of a particular reaction with reference to complex synthesis are provided.

I am indebted to Professors E.J. Corey and Paul Helquist for reading portions of the draft and making several comments. I also thank Mrs. Karen Macali for drawing many of the formulas. As always, it is the sacrifices of my family – Honor, Jocelyn, and Daphne – that have made my writing possible.

Tse-Lok Ho

Glenview, Illinois
August 1990

CONTENTS

ABBREVIATIONS

Ac	acetyl, acetate
acac	acetylacetonate (as ligand)
Am	*n*-amyl
9-BBN	9-borabicyclo[3.3.1]nonane
Bn	benzyl
Boc	*tert*-butyloxycarbonyl
Bu	*n*-butyl
sBu	*sec*-butyl
tBu	*tert*-butyl
Bz	benzoyl
18-c-6	18-crown-6
CAN	ceric ammonium nitrate
Cbz	carbobenzyloxy
cp	cyclopentadiene
CSA	camphorsulfonic acid
DABCO	1,4-diazabicyclo[2.2.2]octane
DBN	1,5-diazabicyclo[4.3.0]non-5-ene
DBU	1,8-diazabicyclo[5.4.0]undec-7-ene
DCC	dicyclohexylcarbodiimide
DDQ	2,3-dichloro-5,6-dicyano-1,4-benzoquinone
DEAD	diethyl azodicarboxylate
DHP	dihydropyran
Dibal-H	diisobutylaluminum hydride
DMAP	4-dimethylaminopyridine
DME	1,2-dimethoxyethane (glyme)
DMF	*N*,*N*-dimethylformamide

DMSO	dimethyl sulfoxide
EG	ethylene glycol
Et	ethyl
EVE	ethyl vinyl ether
HMPA	hexamethylphosphoric triamide
HMPT	hexamethylphosphorous triamide
hv	light
Im	imidazole
K222	potassium [2.2.2]cryptate
L-selectride	lithium tri-*sec*-butylborohydride
LAH	lithium aluminum hydride
LDA	lithium diisopropylamide
MCPBA	*m*-chloroperoxybenzoic acid
Me	methyl
MEM	methoxyethoxymethyl
Mes	mesityl
MOM	methoxymethyl
Ms	methanesulfonyl
MVK	methyl vinyl ketone
NBS	*N*-bromosuccinimide
NCS	*N*-chlorosuccinimide
NIS	*N*-iodosuccinimide
NMO	*N*-methylmorpholine *N*-oxide
PCC	pyridinium chlorochromate
PDC	pyridinium dichromate
PG	prostaglandin
Ph	phenyl
PPA	polyphosphoric acid
Pr	*n*-propyl
iPr	*iso*-propyl
py	pyridine
Ni	Raney nickel
Red-Al	sodium dihydrobis(2-methoxyethoxy)aluminate
Tf	trifluoromethanesulfonyl
TFA	trifluoroacetic acid
TFAA	trifluoroacetic anhydride
THF	tetrahydrofuran
THP	tetrahydropyran
TMEDA	*N,N,N',N'*-tetramethylethylenediamine
TMS	trimethylsilyl
Tol	*p*-tolyl
Tr	trityl
Z	carbobenzyloxy
Δ	heat

1

INTRODUCTION

Chemical synthesis must be based on a sound scheme. Even in the recent past the logical formulation of synthetic routes was shrouded in mystery, and at best it was a tortuous experience.

During the early days of organic chemistry, synthesis was not an important part of research. Activities in this area were simple transformations, performed mainly on aromatic compounds. Thus, Perkin's attempt in synthesizing quinine by oxidation of toluidine was as bold as it was fanciful since the structure of the alkaloid was still unknown. Only after many of the fundamental organic reactions had been established did the pursuit of multistep synthesis become feasible. This period also saw the discovery of more powerful reactions such as the Diels–Alder reaction, and parallel developments in erecting their mechanistic frameworks and creating new techniques of chemical analysis. A milestone is the successful synthesis of quinine by R.B. Woodward and W.v.E. Doering [1945] whose complexity and control was far beyond what Perkin could have imagined.

Despite efforts by Woodward and a few other protagonists who elevated this field to prominence, their masterpieces did not leave many clues to show how they were planned. Nor did they teach a systematic attack on new and unrelated synthetic problems. In a real sense a new era of chemical synthesis was ushered in by E.J. Corey who pioneered and now promotes retrosynthetic analysis [Corey, 1968c, 1985d, 1989b] whereby a target molecule is dissected or disconnected to reveal appropriate and logical precursors. Each of these precursors undergoes renewed bond disconnection ultimately to yield building blocks which are either commercially available or easily prepared. Various reasonable pathways are then generated by reconnecting the retrosynthetic steps (called *transforms*) in the forward direction and the shortest, most efficient, or novel one can then be selected to attempt its execution. The probability of success for a synthesis is high when all the retrosynthetic steps have exact counterparts of synthetic reactions to achieve the forward progress.

Retrosynthetic analysis with its succinct guidelines is amenable to interfacing with a computer. However, even without a machine, the analytic protocol has great heuristic value and it has also demystified the thought process that must precede and guide a synthesis. Presently, retrosynthetic analysis plays an integral and indispensable role in both teaching and research.

Bond disconnection generates fragmented species which may or may not be capable of existence. These idealized predecessors which often carry cationic and anionic atomic sites are called *synthons**. Although the placement of complementary electric charges to these atoms is arbitrary, a certain way is always more logical. For example, of the two disconnections of a ketone:

$$-CH_2-\overset{\overset{\displaystyle O}{\|}}{C}- \Rightarrow -CH_2^- \quad {}^+\overset{\overset{\displaystyle O}{\|}}{C}-$$

$$\Rightarrow -CH_2^+ \quad {}^-\overset{\overset{\displaystyle O}{\|}}{C}-$$

the one leading to an acylium ion and a carbanion is preferred because experience of organic reactivity indicates the diverse stabilities of acylium ions and acyl anions. However, the less logical disconnection should not be ignored. Circumstances such as availability of reagents and synthetic equivalents may strongly argue for the adoption of the less logical recombination. In such cases umpolung (polarity inversion) manipulation is called for.

The double arrow ($\Rightarrow$) is used to indicate the retrosynthetic direction or transform. Transforms are numerous, but the most significant ones are those capable of molecular simplification, especially when stereocontrol is included. Also useful are those that modify target molecules to allow subsequent

---

* Synthons are structural units within a molecule which are related to possible synthetic operations. Over the years this term has been used by various chemists in different contexts to mean building blocks, reagents, convenient starting materials, and so forth. In the correct sense, retrosynthetic analysis should start by identifying *retrons*, which are keying elements of minimal but essential framework atoms, then derive synthons from retrons. The difference between retrons and synthons may be illustrated by considering a β-amino ketone.

$$R-\overset{\overset{\displaystyle }{\underset{\underset{\displaystyle O}{\|}}{C}}}{}-CH_2-CH_2-NR_2' \Rightarrow C-\overset{\overset{\displaystyle }{\underset{\underset{\displaystyle O}{\|}}{C}}}{}-C-N \Rightarrow C-C^-\quad C=N^+$$

retron

$$\Rightarrow \overset{\overset{\displaystyle }{\underset{\underset{\displaystyle O}{\|}}{C}}}{}-C=C\quad N$$

synthons

application of simplifying transforms. These can effect either (1) skeletal changes via connection or rearrangement, (2) functional group interchange (FGI) and transposition, or (3) stereocenter inversion or transfer.

It may be surprising to witness the value of transforms that actually increase the molecular complexity. In the corresponding synthetic reactions the added structural elements are generally removed after their service in activation, protection, or stereocontrol is terminated. Without these structural elements, useful transforms cannot be generated. For example, the acyloin condensation transform performed on a macrocyclic ketone necessitates the incorporation of an extra hydroxyl group:

A Diels–Alder transform on bicyclo[2.2.1]hept-5-en-2-one is unfruitful because ketene does not undergo [4 + 2]cycloaddition with cyclopentadiene. However, the corresponding cyanohydrin acetate fulfills the reactivity needs.

Transform selection should take great heed to the hierarchical scale of importance which is correlated with its power of structural simplification. In a complete retrosynthetic analysis the transform-based strategy is often very rewarding. A complementary structure-based strategy relies on astute recognition of a potential starting material or building block. The "chiron approach" [Hanessian 1983], which champions the application of chiral substances of nature such as carbohydrates, α-amino acids, and terpenes, belongs to this category.

These are other strategies that consider molecular topology, stereochemistry, or functional groups as the central issue in retrosynthetic analysis. However, it is crucial to keep the maximum number of concurrent independent strategies in mind. Many of the extremely efficient synthetic methods feature a combination of transforms, and among these the tandem Michael-aldol condensations, tandem Claisen–Cope rearrangements, tandem aza-Cope rearrangement–Mannich cyclization are only a few representatives.

Let us now outline, on the basis of retrosynthetic method, the critical stages of a synthetic masterpiece that was completed in the mid-1950s, so as to retrace the logic underlying the plan. This is the synthesis of reserpine [Woodward, 1958].

Reserpine is a pentacyclic indole alkaloid with six chirality centers, five of which are concentrated in the E-ring (including the D/E ring juncture). The dense functionality demands meticulous planning and execution of the synthesis.

After a straightforward disconnection of reserpine to 6-methoxytryptamine and a DE-component (Pictet–Spengler or Bischler–Napieralski transform, etc) one may consider the derivation of the latter subtarget. In this crucial intermediate the most conspicuous stereochemical feature is an all-*cis* configuration of the methine hydrogens at the nonoxygenated carbon centers, which is also integrated into a cyclohexane framework. Such a framework is a partial Diels–Alder retron that requires no further adjustment of the stereochemical outcome from the cycloaddition. Furthermore, FGI of the vicinal dioxy substituents leads to a 2-cyclohexene-1-carboxylic acid derivative, and thus a full Diels–Alder retron emerges. The availability of many methods for the functionalization of a cyclohexene and the control provided by the allylic carboxyl group make the introduction of the oxy substituents to the Diels–Alder adduct a routine task.

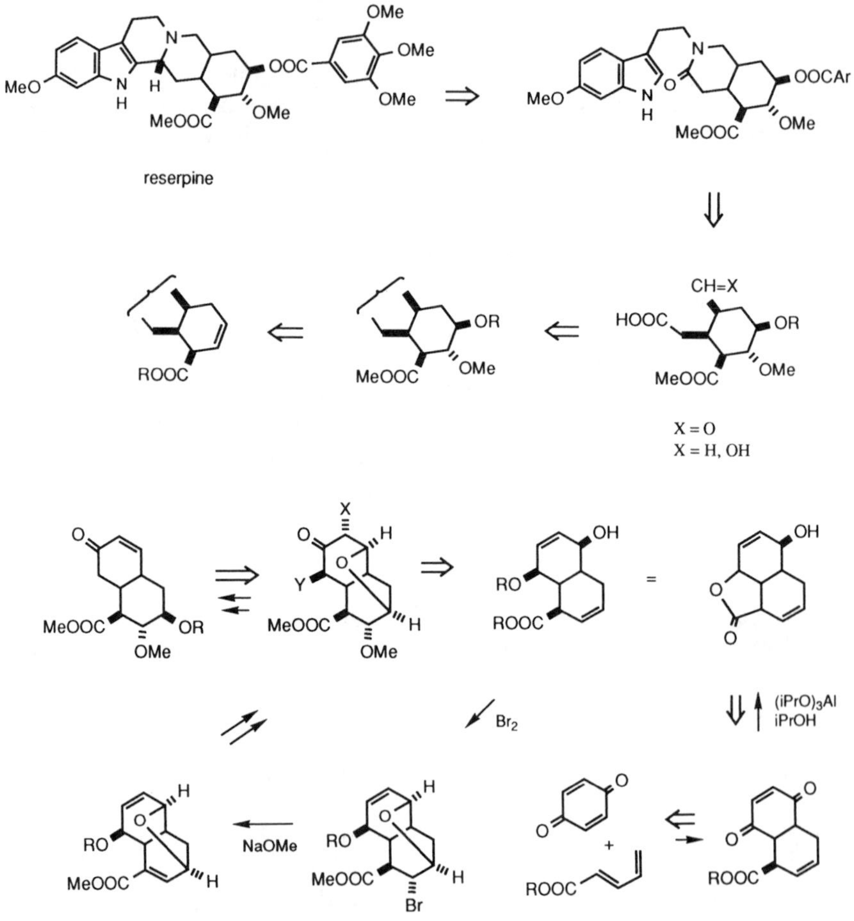

The actual synthesis started from a condensation of 2,4-pentadienecarboxylic acid or ester with *p*-benzoquinone. *A priori*, excision of a one-carbon unit from the dienophilic portion must be planned into the overall scheme. This was exquisitely executed by exploiting neighboring group participation in various stages. For example, γ-lactonization occurring upon Meerwein–Pondorff reduction of the Diels–Alder adduct rigidified the molecular conformation and permitted a bromoetherification which was instrumental to the establishment of the remaining asymmetric centers in the E-ring. Replacement of the bromine atom by a methoxy group with retention of configuration was the result of an elimination–addition process. The S_N2 option was disfavored sterically and electrostatically, as the methoxide ion donor would have to overcome repulsion of the ethereal and carbonyl oxygen atoms during its attack.

The locked conformation of lactone was also responsible for its facile elaboration into the cyclohexenone precursor of the subtarget. Thus, the regioselectivity of bromohydrin formation, with respect to opening of the bromonium ion in a diaxial fashion, directly relates to such a conformation, whether the introduction of the hydroxyl group is via relay at the lactone carbonyl (i.e., assisted bromonium ion opening by the carbonyl and entry of a water molecule via the ortho acid) is a moot point. After oxidation of the bromohydrin, zinc dust reduction liberated the two ethereal functions simultaneously with the generation of the cyclohexenone. An oxidative cleavage pathway for the cyclohexenone to reach the subtarget is clear.

Condensation of the differentiated carbon chains of the desired intermediate with 6-methoxytryptamine furnished a lactam which was submitted to a Bischler–Napieralski cyclization and *in situ* reduction with sodium borohydride. The 3-epireserpine product was probably not predicted with absolute certainty, and even nowadays a retrosynthetic analysis by most chemists would ignore this aspect. We therefore must admire the elegant solution to this stereochemical problem by effecting a conformational change upon formation of the 16,18-lactone whereby the large indolyl group was forced to adopt an axial configuration (i.e., from a comfortable equatorial state with respect to the D-ring in the pre-lactonized pentacycle). This thermodynamic shift was the driving force for epimerization of C-3.

1.1. POLARITY ALTERNATION AND CHEMICAL REACTIVITY

It may be broadly stated that chemical reactivity of carbogenic molecules is bestowed by functional groups. Saturated aliphatic hydrocarbons are called paraffins (Latin: *parum* = too little, *affinis* = attraction) with sound reasons.

Functional groups having heteroatoms whose electronegativities are different from carbon's render the attaching hydrocarbon backbone polar to various degree. Depending on the nature of the functional group, or the electronegativity of the heteroatom adjoining the carbon atom, electronic properties of the carbon atom are modified. Naturally a more electronegative atom likes to associate with a more electropositive carbon, hence there is a tendency for the heteroatom to force its carbon neighbor to assume a more electropositive state. This is achieved by withdrawing the bonding electrons away from the carbon atom.

In turn, the "electropositivated" carbon would transmit this effect on its other neighbors, resulting in an alternating polarity along a chain of atoms. This phenomenon is a manifestation of the most fundamental law of nature, and it was recognized by, *inter alia*, Lapworth [1898] but has largely been ignored in more recent decades. As will be seen throughout this book the polarity alternation rule serves magnificently in assessing chemical reactivity patterns.

Two kinds of substituents on a carbon framework need be considered. (1) The donor, *d*, groups include those with *n*- and π-electron pairs {OH, OR, OCOR, NRR', NR(COR'), SH, SR, Hal} and the $+I$ substituents such as alkyl groups, and (2) the acceptor *a*, groups contain empty orbitals {CHO, COR, COOH, COOR, CONR$_2$CN, SO$_n$R, NO$_2$, SiR$_3$} and the $-I$ substituents (e.g., R$_3$N$^+$). Although one may further differentiate each class of functional groups into σ- and π-types, the differentiation is generally unnecessary. However, care must be taken when dealing with groups capable of valence change during reaction. These include the nitro group, and sulfur- and phhosphorus-based substituents. The role change of such groups during a reaction naturally affects neighboring atoms, as reflected in the fascinating unmanipulated umpolung which has been called "contrapolarization"* [Ho, 1989a].

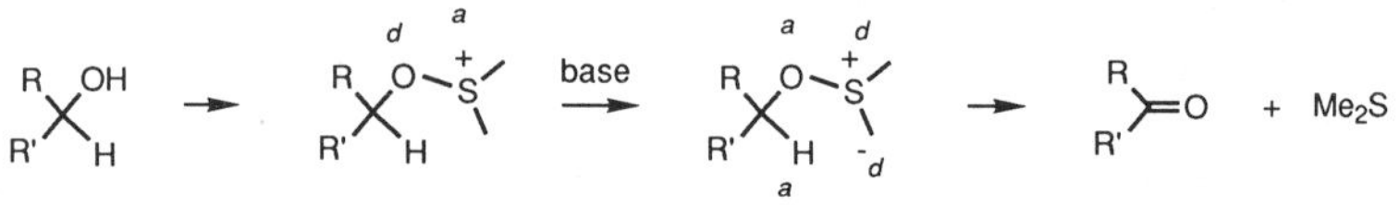

Most relevant to the majority of reactions to be discussed in this book are the mutual effects of two or more polar functionalities in a molecule and their electronic influences on the intervening atoms. The two possible situations are: the polarity alternation rule is either obeyed or violated, and the reactivity patterns are clearly very different. It can be safely assumed that the ground-state energies for fragments conforming to the rule are lower, and those violating cases must be thermodynamically less stable.

In the following diagram generalized molecular fragments with assigned polarity are presented. The boxed fragments are those with favorable electronic arrangements, and they are said to have *conjoint* electronic arrangements. Conversely, the unboxed fragments are those with a node of conflicting electronic interactions, be it at an atom or in a bond. Such arrangements are *disjoint*.†

* Perhaps it is advisable to indicate that contrapolarization is always involved in redox processes, among other reactions. To briefly illustrate the role changes of atoms the Moffatt–Pfitzner oxidation is shown here. An oxysulfonium species is formed and decomposed via sigmatropic rearrangement of the derived ylide. The oxygen and the sulfur atoms contrapolarize in a redox separation into two molecules.

† The author has used the *conjoint/disjoint* terminology for years. It seems that these terms correspond to the terms *consonant/dissonant* that originated from Prof. D.A. Evans.

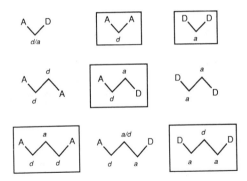

Using this simple polarity alternation protocol it is possible to assess directly situations in terms of their favorableness without going through all sorts of arguments (resonance, etc) or calculations, although this author is not advocating the abandonment of the prevalent theories. The polarity alternation canon is a useful mnemonic device whose origins are the most basic electronic interactions of bonding partners.

We realize that α-oxy- and α-amino carbenium ions are stabilized in comparison with the all-carbon analogs on the basis of ideal d–a pairings. By the same token a carbanion gains enormous stability when it is flanked by an acceptor group (a–d pair).

$$\underset{d \quad a}{R\text{-}O\text{-}CH_2{}^+} \qquad \underset{d \quad a}{R_2N\text{-}CH_2{}^+}$$

$$\underset{a \quad d}{\overset{\overset{O}{\parallel}}{{}^-C\text{-}CH_2{}^-}} \qquad \underset{a \quad d}{NC\text{-}CH_2{}^-} \qquad \underset{a \quad d}{O_2N\text{-}CH_2{}^-} \qquad \underset{a \quad d}{O_2\overset{\displaystyle |}{S}\text{-}CH_2{}^-}$$

$$\underset{a \quad d}{R_3Si\text{-}CH_2{}^-}$$

As expected, α-acceptor-substituted carbocations are destabilized. This phenomenon has been manifested in solvolytic studies of α-cyanocycloalkyl tosylates [Gassman, 1980; Kirmse, 1990].

Without resorting to laborious calculations the application of the polarity alternation rule predicts the relative stability of $XCH_2CH_2^+$ as compared with $CH_3CH_2^+$, and that of $XCH_2CH_2^-$ as compared with $CH_3CH_2^-$. When $X = Li$ the cation is stabilized and the anion destabilized, and the trend is reversed with $X = F$.

Sundry aspects of chemical properties may be correlated by considering polarity alternation. For example, the abnormally long C—C bond of the oxalate ion apparently contradicts the expectation of a conjugate system. Yet the union of two acceptor atoms should destabilize and therefore stretch the C—C bond.

A low ionization potential (IP) of the pyridazine lone pair electrons [Gleiter, 1970] indicates the favorableness of removing the d–d pairing. On the other hand, pyrimidine shows a higher IP than pyridine.

The ease of deprotonation from trihalomethane follows the trend: CHI_3 > $CHBr_3$ > $CHCl_3$ ⋙ CHF_3. Thus $CHXF_2$ is kinetically quite inert, yet the stability of difluorocarbene is quite high [Hine, 1958]. In the light of the polarity alternation influence the findings are perfectly reasonable. The fluorine atoms render the geminal hydrogen highly donor-like, yet upon α-dehydrohalogenation the resulting carbene center which is an acceptor would gain stabilization by the two powerful donor substituents. In fact, even without donor stabilization, carbene formation is a logical fate of the simple $^-CH_2$—X anion in view of the unfavorable d–d arrangement present.

Dithioacetals/ketals are hydrolytically much more stable than the simple acetals/ketals. This is due to the fact that a sulfur atom may assume acceptor character which in turn confers some donor character to the adjacent carbon atom, discouraging the reaction with water. The same effect is witnessed in the comparative stability of (trimethylsilyl)methyl triflate vs alkyl triflates [Vedejs, 1979] and the inertness of (trimethylsilyl)methyl halides toward various donors. In fact, with hard nucleophiles, reactions of the halides occur predominantly at Si.

The very selective thioketalization of 1,1,1-trifluoro-2,4-alkanediones at C-4 [Barros, 1988] is easily understood. The fluorine atoms effectively transmit their donor properties to the proximal carbonyl.

The predilection for aminolysis of 3-nitrophthalic anhydride at the C-1 carbonyl [Wieland, 1969] can be ascribed to an influence of the nitro group via a polarity alternation mechanism.

1.2. UMPOLUNG (POLARITY INVERSION)

Umpolung is one of the most fruitful concepts by which the expansion of synthetic horizon suddenly lost its stricture. In principle, the possibility of forming a X—Y bond is doubled by this technique. If the traditional pathway calls for the combination $(X^+ + Y^-)$, then an umpolung method allows the complementary $(X'^- + Y'^+)$ operation to achieve the goal.

Not only does this increased flexibility offer at times a better choice of precursors that are more readily available, the alternative assembly mode may also enable preservation of functionalities which are labile under the conventional reaction conditions. More importantly, chemists can now enjoy the latitude of disconnecting a bond in the "illogical" sense for the formulation of a synthetic scheme.

The use of 2-lithio-1,3-dithianes as acyl anion equivalents started an avalanche of activities [Seebach, 1979; Hase, 1987] that ultimately established the umpolung chemistry as an integral part of a standard organic chemistry course. It is instructive then to illustrate the power of this fundamental principle by examining a route to vermiculine [Seebach, 1977]. Three alkylation steps involving a dithiane served to construct a chain of functionalized atoms which required only a Wittig reaction to complete a protected molecular half. The efficiency of the overall process belies the intricate maneuver that would have been required to stitch together three disjoint carbonyl groups in each of the carbon chains by classical methods.

vermiculine

Another example pertaining to a synthesis of *exo*-brevicomin [Masaki, 1982] is interesting in that it describes the creation of a conjoint six-membered ring starting from a disjoint material, i.e., tartaric acid. Necessarily a polarity inversion must be involved. The contrapolarizable sulfonyl group showed its versatility.

exo-brevicomin   A = SO$_2$Ph

More of umpolung strategies for synthesis will be found in various sections of this book.

2

SOME ASPECTS OF ORGANIC REACTIVITY

2.1. CAPTODATIVE EFFECTS

The captodative situation is a special case in which an atom is bonded simultaneously to a donor and an acceptor. The opposing electronic influences necessarily create an ambiguity for the central atom such that attenuation of effect from the more powerful effector is expected. If coincidentally the two substituents are equipotent the central atom is electronically "neutral".

The captodative effect [Viehe, 1985] is best expressed by the stabilization of a free-radical center, the well-known oxidative dimerization of indoxyl [Russell, 1969] ultimately to give indigo is an expression of this situation. On the other hand, the tendency for cation or anion formation is depressed. Examples of the modified electronic character include the intervention of free-radical intermediates in the Claisen rearrangement of allyl α-fluoroacetate in the presence of a strong base [J.T. Welch, 1988], the selective silylation of a 1,3-oxathian-3,3-dioxide at C-4 [Fuji, 1986], the selective ring opening of a sulfinyl epoxide [Satoh, 1987], and the kinetic alkylation of an oxaalkyl-bridged piperazine-2,5-diones at the bridgehead away from the oxygen atom [Williams, 1983].

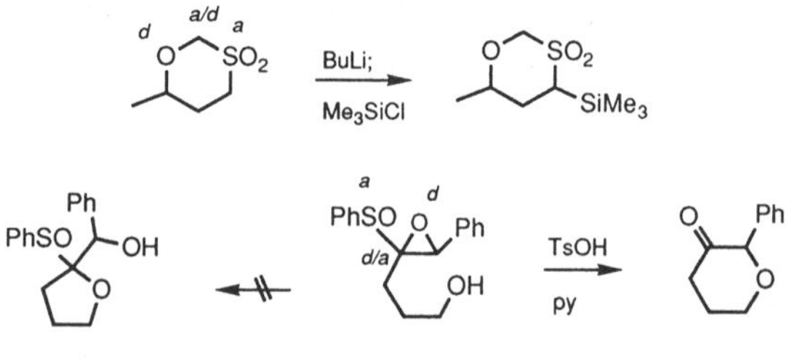

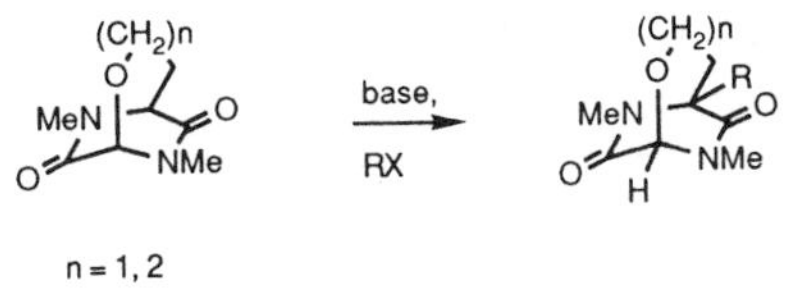

Relating to the behavior of allyl α-fluoroacetate the highly regioselective (12:1) aza-Cope rearrangement of the enamine(s) derived from the N-allyl-N-methyl-α-fluorocyclohexanone iminium ion [J.T. Welch, 1990] may also be due to an avoided carbanion formation at the captodative site. The relevance of the deprotonation to the initial step of the Favorskii rearrangement should be noted.

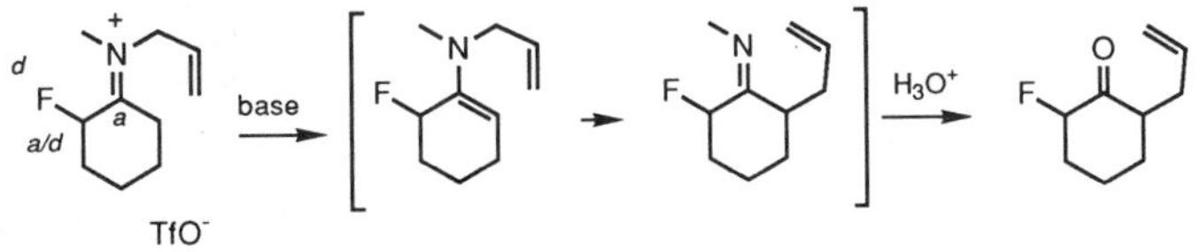

In the piperazinedione reaction the bridged structure is of no consequence to the trend of reactivity. The general belief that an alkoxy group exerts inductive electron withdrawal from the α-position and hence facilitates deprotonation at that site has no solid foundation. In α-alkoxy ketones enolization occurs preferentially at the α'-carbon. This behavior has an important implication in syntheses of hitachimycin [A.B. Smith, 1988] and pseudoyohimbine [Wenkert, 1984].

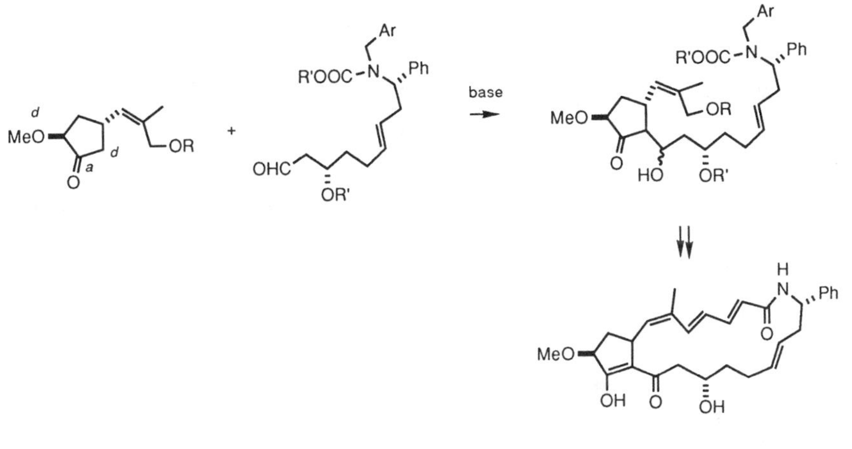

hitachimycin

The intramolecular aldolization of 8-trimethylsilyl-2,6-octanedione is regioselective [Sommer, 1954]. The silicon atom is β to the internal carbonyl group, nevertheless it has the same effect as an α-donor. In other words, the α-methylene is more reluctant to enolize.

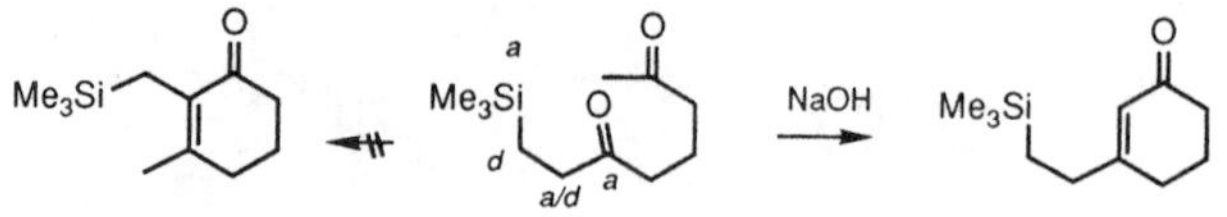

A consequence of the captodative functional group arrangement in picolinic acid is the facile decarboxylation. The result is an ylide which can also be formulated as an α-aminocarbene. Of course such $d-a$ species and the conjugate acid enjoy a much better electronic harmony.

Acyclic analogs such as the imino derivatives of benzolformic acid also undergo ready decarboxylation [Aly, 1985].

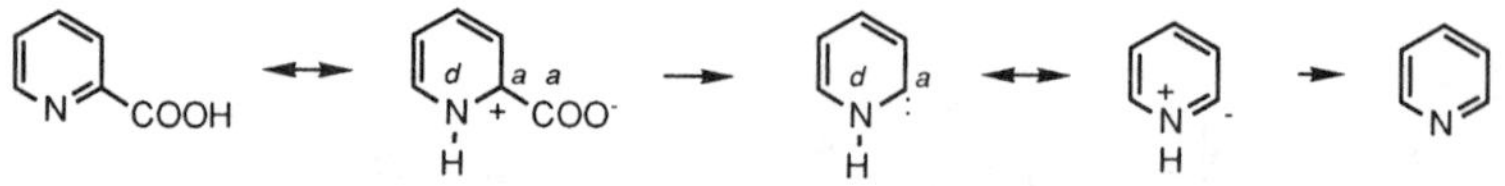

Another case in which chemical reactivity may be ascribed to a captodative effect is the allyl transfer reaction from allyltrimethylsilane to α,β-unsaturated acylsilanes [Danheiser, 1985]. Lewis acid coordination of the carbonyl group generated a reactive acceptor site (the 1-oxy-1-silylallyl cation is captodative) at the β-carbon which was captured immediately. On the other hand, the inertness of the corresponding α,β-unsaturated esters must be due to the much more highly stabilized 1,1-dioxyallyl cations.

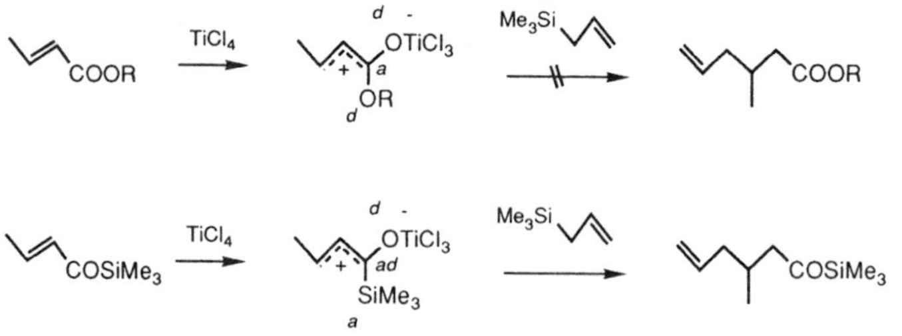

The behavior of N-benzoyl-ε-caprolactam and its α-chloro derivative toward sulfuric acid is interesting. Selective cleavage of the cyclic amide bond and the benzoyl group, respectively, have been observed [Tull, 1964]. The chlorine substituent apparently discourages the generation of a β-acylium ion which would magnify the nonalternating interaction.

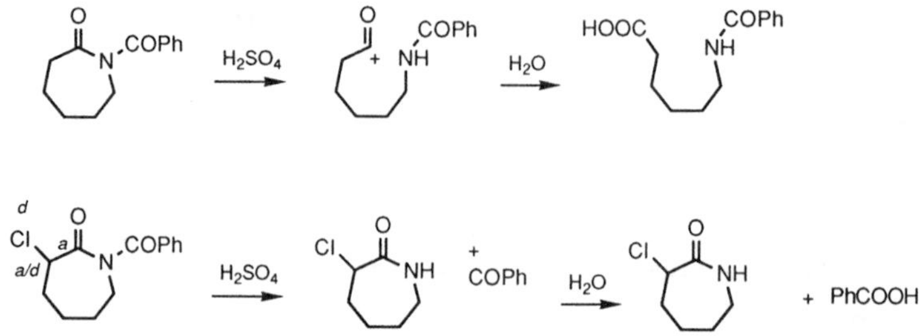

α-Chloroalkylfurans are notoriously labile. Moderation of this reactivity has been achieved by substituting the other α-position with a silyl group [Takanishi, 1987]. The rationale may be that this other α-carbon is captodative so that the assistance in ionization of the chlorine atom by the nuclear oxygen is suppressed.

Finally, a few reminders are warranted. Captodative arrays do not preclude formation of ionic species, only the tendency for the generation of such reactive intermediates is lowered. For example, α-cyanohydrin ethers, α-amino nitriles, and α-phosphonyl carbamates have been deprotonated with strong bases and alkylated. The intermediates are supposedly anionic and they definitely serve as acyl anion equivalents. Many syntheses have relied on them, e.g., that of periplanone B [T. Takahashi, 1986], isonitramine [Quirion, 1988], and conhydrine [Shono, 1983].

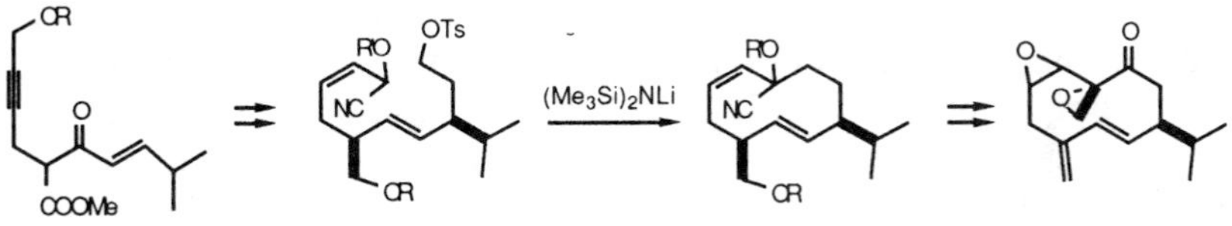

periplanone-B

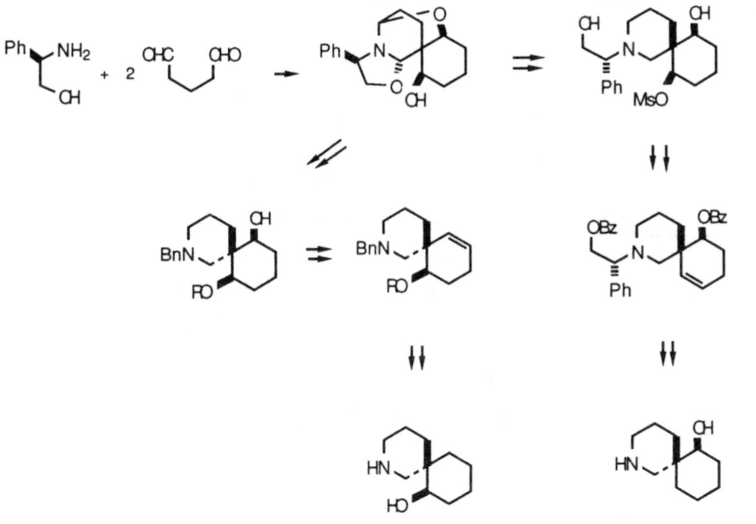

(+)-isonitramine   (-)-isonitramine

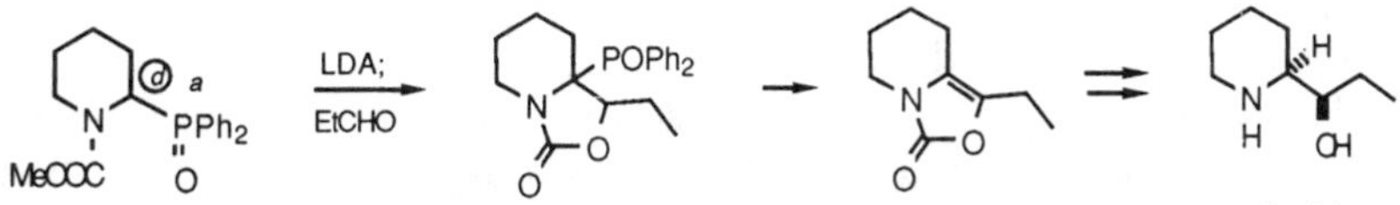

conhydrine

2.2. ACTIVATION OF ORGANIC MOLECULES

Strange as it may appear, an organic molecule may be activated for its reactions by either accentuation of polarity alternation around the reaction center or disruption of the polarity alternation. In the accentuation the electronic character of the center, be it a donor or an acceptor, is enhanced. It is true that stability of a molecule is gained by distinct and smooth polarity alternation, but such an electronic set-up also renders the molecule kinetically reactive. On the other hand, disruption of polarity alternation would increase the ground-state energy of the reactant, therefore the molecule wishes to get rid of the unfavorable arrangement, e.g., by reaction.

Accentuation of polarity alternation usually involves extension of the alternation sequence. The simplest example is the facilitation of C—X bond cleavage by protonation. In the same manner occurs activation of carboxylic acid by forming an anhydride (either homo or mixed). Nature uses this strategy in phosphorylating a variety of molecules effectively with adenosine triphosphate (ATP) which is an anhydride. Germane to the anhydride activation is the marked acylating potential of N-acylimidazoles, N-acyl-2- and 4-pyridones, and N-acyl-4-dimethylaminopyridines.

The activation of carboxylic acids by forming acyloxyphosphonium salts is the crux of a method for amide synthesis [Mukaiyama, 1976]. The most striking aspect of this transformation is the hard–soft discrimination of the reagents such that only after the two soft compounds react to afford a hard phosphonium salt can the hard donors participate in the reaction.

An oxidative acyl transfer involves treatment of monoacyl-1,3-propanedithiols with iodine in the presence of a donor (e.g., MeOH) [Takagi, 1976]. Contra-polarization of the sulfur atom is in operation.

It is noted that in the acylsulfonium intermediate, even though the sulfur atom bears a positive charge, it is designated as a donor (cf. N-acylpyridinium species). We must emphasize that *the role of a donor or an acceptor is assigned to bonding partners in the sense of bond formation*. When dealing with bond cleavage we always consider the pre-fragmentated state and retain the same assignment of donors and acceptors.

The epoxidizing capacity of peroxycarboxylic acids (and other peracids such as peroxysulfuric acid) and silyl hydroperoxides [Ho, 1979; Rebek, 1979] is due to enhanced acceptor properties of the terminal hydroxyl as imposed by the carbonyl group and the silicon atom, respectively. Alkyl hydroperoxides and hydrogen peroxide are not activated in such a fashion, and they require Lewis acid catalysts to facilitate the HO^+ transfer [Sharpless, 1979].

The propensity for amino transfer from α-heteromethyl azides varies with the α-substituent [Trost, 1983]. Generally, an acceptor favors the reaction with Grignard reagents, as one might expect from the reinforcement of the acceptor role of the terminal nitrogen atom.

As will be seen in much more detail in the section on the Mannich reaction, the Robinson tropinone synthesis, consisting of admixing succinaldehyde, methylamine and acetone, can be greatly improved by replacing acetone with acetonedicarboxylic acid. This is a clear-cut case of reactivity enhancement by the accentuation process. It should be noted that this strategy is mastered by nature in its elaboration of many biomolecules including fatty acids, polyketides, terpenes and steroids.

Aldol ring closure to provide the diquinane enone system is not easy. With additional activation of the donor carbon, e.g., with an ester or phosphonyl group, the condensation proceeds well. Thus in an approach to quadrone [Danishefsky, 1980] an "extraneous" ester group also helped the introduction of an acetic ester chain into the angular position by a Michael addition.

The activation scheme was crucial to another case [Rao, 1988] without which a condensation involving another unprotected ketone would definitely have taken precedence.

Somewhat related is the cyclization step of a route to lupinine and epilupinine [Takahata, 1986] starting from an alkylthiolactim. Of course the extra ester pendant is an essential progenitor of the hydroxymethyl chain.

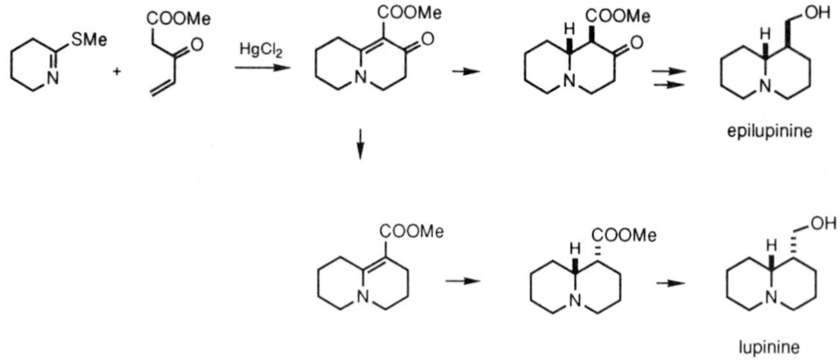

epilupinine

lupinine

Both roles of activation and regiocontrol were played by the trimethylsilyl group during the oxidative transformation of two isomeric silylallene alcohols into a γ- and a δ-lactone, respectively [Bertrand, 1980]. Epoxidation was followed by rearrangement to the cyclopropanone intermediate which was attacked intramolecularly by the hydroxyl group. The ring opening was directed by the silyl substituent.

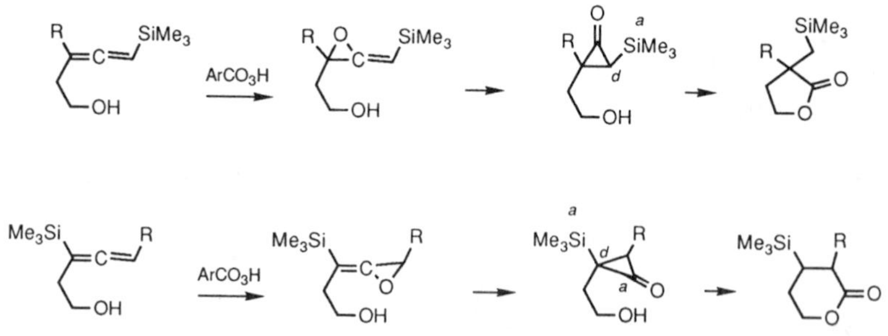

A reactive double bond can be protected in the form of a cyclopentadiene adduct. This synthetic tactic could have gained much higher popularity had it been possible to effect the retro-Diels–Alder reaction under milder conditions. A solution to this problem emerged when it was found that certain 7-trimethylsilylnorbornene derivatives may decompose at as low a temperature as 110°C [Magnus, 1987]. It can be seen that the acceptor substituent confers substantial acceptor character at either of the bridgehead carbon atoms, and should this atom also be β to an acceptor site as is usually encountered in Diels–Alder adducts, a weakening of the C—C bond which was originally formed in the forward reaction results. This process of rendering the pericyclic reaction polar to facilitate it actually has precedence in the decomposition of the cycloadducts of furan and fulvene. Note that the single-atom bridge is always a donor. (It is amusing to compare the thermal behavior of norbornene derivatives in which the 7-position is an acceptor. Loss of this bridge seems to be preferred, as in decarbonylation of 7-oxonorbornenes. In other words, the weakened bond is shifted one atom away.)

While on the topic of retro-Diels–Alder reactions some observations on the reactivity of Diels–Alder addends may be mentioned. Given a typical diene the cycloaddition is more facile with dienophiles with increasing or more pronounced polarity alternation. For example, acryloxyboranes [Furuta, 1988] are better dienophiles than acrylic acid. The dienophilic reactivity of allylidenemetal carbonyls [Wulff, 1983], vinyl metals [Minot, 1976], and alkoxyallyl cations [Gassman, 1987] can be traced to similar arrangements.

The stability of the Fischer carbene complexes is contributed by the heteroatom donor attaching to the carbenic center. Although most of these complexes contain alkoxy groups, other strong donors including the fluorine atom should also stabilize such carbene complexes.

There seems to exist a paradox concerning the electrophilic character of the Fischer carbene complexes in which the central atom is in a *low* oxidation state, and the Schrock complexes in which the metal atom is in a *high* oxidation state. Besides other factors, the presence of donor substituents appears to be very important in governing the behavior of the two classes of compounds. In simple terms, the substituent effects that accentuate polarity alternation become dominant. A significant consequence of the nature of the Fischer carbene complexes is their application in the synthesis of disjoint 1,4-hydroquinones [Dötz, 1984].

Intramolecular carbenoid insertion to a remote C—H bond leads to a cyclic product. Generally, five-membered rings are created preferentially, as illustrated in an elegant synthesis of sterpurene [Zhao, 1990].

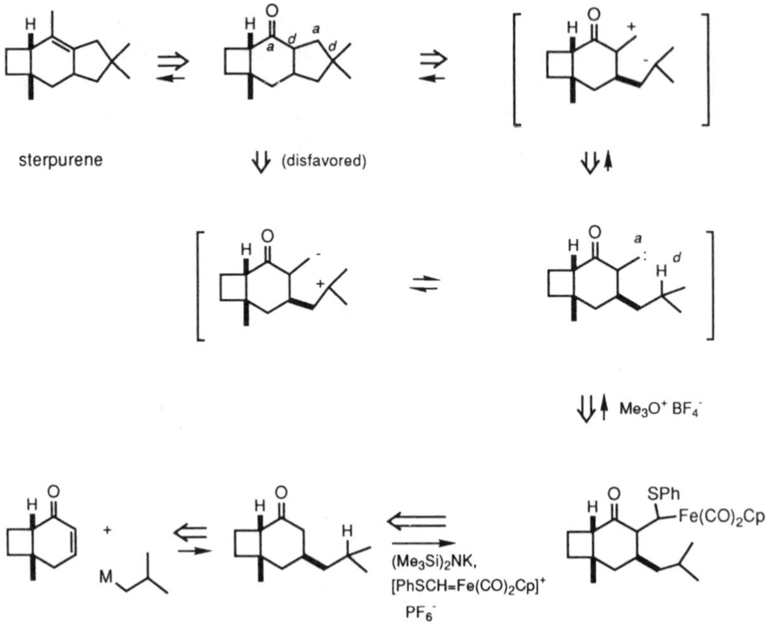

Cyclization to a cyclopentane proceeds via a six-center transition state. Despite the concertedness of these insertion reactions it might still be useful to analyze them in a polar context. (Even free-radical and pericyclic reactions show influences by polar functionalities.) We may then consider the insertion *formally* involving a hydride transfer to the carbenic center which is followed by an immediate collapse of the zwitterion.

It must be emphasized that a six-center transition state is favored not only on steric grounds, the conjoint relationship between the reaction centers should lower the activation energy, however small that lowering may be. In the sterpurene synthesis the carbenoid site is excellently selected with respect to its generation as well as introduction of its progenitor to a bicyclic ketone intermediate because it is β to the carbonyl group (*a–d–a* linkage).

Perhaps it should be mentioned that steric effects could divert the insertion–cyclization to give a six-membered ring product. A case in point is the formation of the δ-lactone ring toward synthesis of pentalenolactone E methyl ester [Cane, 1984]. Cyclopentanation would have produced a compound with the diquinane skeleton spiroannulated at two adjacent carbon atoms.

Some examples of polarity alternation disruption as a driving force for reaction are shown below. The Pummerer rearrangement intermediate from an α-sulfinyl carbonyl substance contains an *a–a* array of atoms, and such species is extremely reactive toward donor reagents. The double cyclization achieved in a quantitative yield during a synthesis of 3-demethoxyerythratidinone [Ishibashi, 1988] attests to the validity of this strategy.

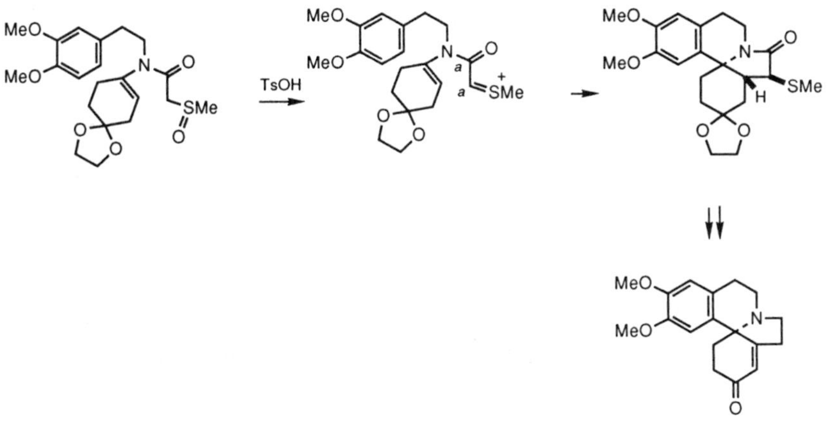

3-demethoxyerythratidinone

α-Diketones are very electrophilic owing to the presence of an a–a linkage. Rearrangement (benzilic acid type) is readily induced.

Newer developments in cationic polyene cyclization to give a steroid skeleton include the use of a tertiary, geminally silylated cyclohexenol as initiator [Bishop, 1983]. Discrimination of the two tertiary cationic sites for bond formation is most pleasing, and this discrimination must be largely due to the less favorable placement of a full positive charge next to the silicon (a–a array). Although the allyl cation would be delocalized, it is expected that the all-carbon substituted tertiary cation contributes more weight.

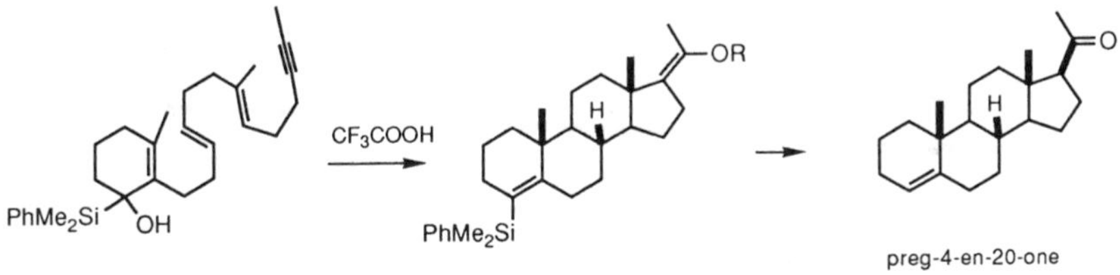

preg-4-en-20-one

The hydride reduction of 3,3,5-trimethylcyclohexanone and its 5-substituted analogs shows a rate correlation with the nature of the substituent [Agami, 1980]. An acceptor cyano group is a rate enhancer while the donor fluorine is a retarder. In the former case C-α and C-β constitute an a–a bonding situation.

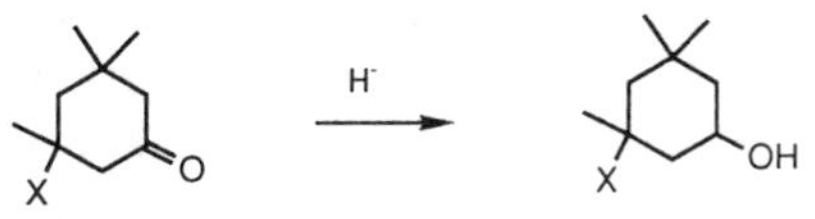

	$10^3 k_{ax}$	$10^3 k_{eq}$
X = CN	15	235
X = H	1.5	11.5
X = F	0.7	5.3

2.3. AROMATIC SUBSTITUTIONS

2.3.1. Electrophilic Substitutions

The effect of existing substituents on the orientation of electrophilic aromatic substitutions is well known. Generally, to account for the major products, resonance structures of the σ-complexes (intermediates) are considered. Favorable electronic interactions between the substituent and the charge in one or more of the canonical structures in comparison with those of the alternative σ-complexes (e.g., *ortho* vs *meta*) suggest the preferential formation of one type of product. The effect is conveniently assessed by following the member atoms around the ring with designated alternating polarity, starting from the existing substituent. For example, a donor substituent decrees the *ipso* carbon an acceptor role, the *ortho* carbon, a donor, etc. Without exception, a donor substituent is *o-/p*-directing and an acceptor substituent, *m*-directing, in electrophilic substitutions. Particularly significant is that a $-I$ substituent that contains lone-pair electrons (e.g., a halogen) is also *o-/p*-directing, although it decreases the reaction rate.

Sometimes very subtle electronic effects are revealed in electrophilic aromatic substitutions. The regioselective ring opening of benzocyclopropenes [Bee, 1980] is a case in point. Note the effect of a single methyl substituent on the reaction pathway.

m-Bromination of phenols and their methyl ethers [Jacquesy, 1980] under superacid conditions is due to the *O*-protonated species being the substrates, in which all the nuclear carbon atoms suffer polarity inversion. Electrophilic substitution of aniline in acidic media follows the same pattern.

As indicated in the formula, acceptor-substituted benzenes are also *ipso*-directing in electrophilic substitution. This reaction mode is rarely detected due to steric repulsion between the attacking electrophile and the existing group. When this steric effect is lessened by lengthening of the C—A bond in compounds with A = SiMe$_3$, SnR$_3$, HgR, *ipso* substitution becomes the preferred reaction course. In fact, 8-methoxy-*N*-methyl-1,2,3,4-tetrahydroisoquinoline has been acquired by a desilylative Pictet–Spengler cyclization [R.B. Miller, 1988].

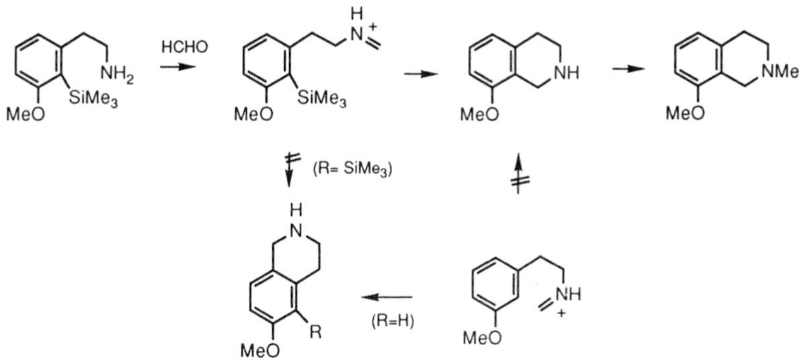

Friedel–Crafts acylation is generally irreversible. In other words, deacylation of phenones is virtually unobservable. Remarkably, treatment of 1-fluoro-fluorenone with polyphosphoric acid gives 3-fluorofluorenone (an equilibrium mixture of 82:8 in favor of the latter) [Agranat, 1977], with the result of removing peri interaction, however small that may be. It must be emphasized that the fluoro substituent stabilizes the incipient carbenium ion to enable the C—C bond cleavage, and directs the reacylation.

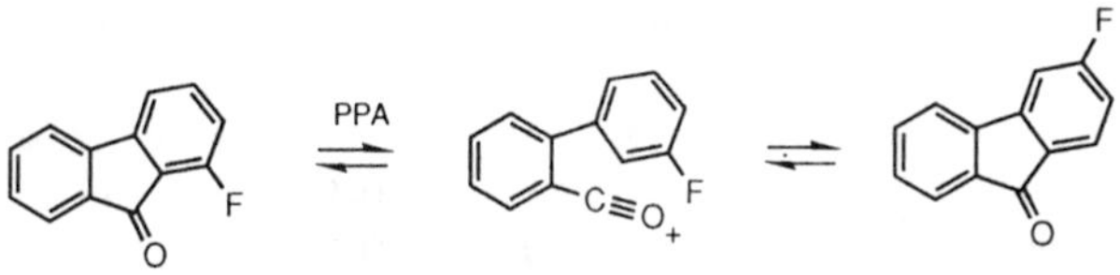

An attempt to use a dehydroabietane B-seco acid derivative to synthesize taxodione [Burnell, 1987] via an intramolecular acylation to reconstitute the tricyclic system resulted in a rearranged phenone. The powerful methoxy group directed the formation of a spirocyclic intermediate which could only return to the acylium ion of the starting material or collapse by a dealkylative pathway.

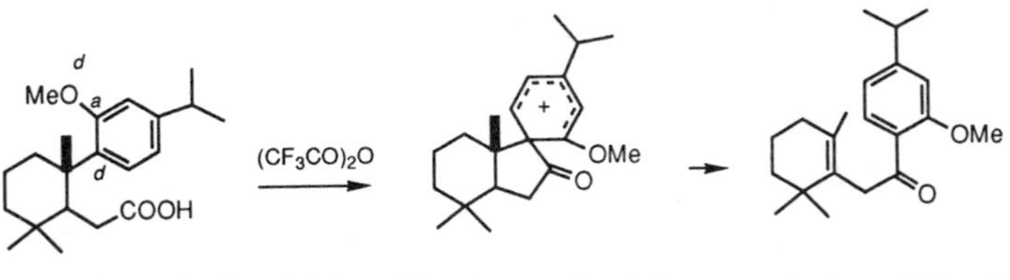

The Bischler–Napieralski cyclization embodying an electrophilic attack on the benzenoid moiety of a *β*-phenethylamine and therefore it is strongly favored by the presence of donor substituents such as methoxy groups. Interestingly, a *meta* nitro group does not totally inhibit the cyclization [McCourbery, 1949]. It should be noted that the C—C bond formation involved a donor site of the aromatic ring.

Electrophilic substitutions of pyridine, although requiring much more drastic conditions owing to inductive electron withdrawal by the nitrogen atom, take place at C-3. Its congeners show regioselectivity at the alternative carbon atoms [Ashe, 1985], as a result of the different character exhibited by phosphorus, arsenic, and antimony. Nitrogen is a donor, whereas the others have strong acceptor traits.

(M = P, As, Sb)

Interestingly, pyridine-N-oxide nitrates at C-4. The nitrogen atom spends its lone-pair electrons to bond with an oxygen atom and is thereby converted into an acceptor.

Very intricate situations arise when five-membered heterocycles such as furan and pyrrole are subject to electrophilic substitution. These nonalternant molecules undergo reaction at an unsubstituted α-carbon such that the σ-intermediates retain a cationic center next to the donor atom and the (donor) π-electron system.

The oxidation of 2-trialkylsilylfurans with peracid to afford 3-butenolides [Kuwajima, 1981] may be considered as an *ipso* substitution.

2.3.2. Nucleophilic Substitutions

Nucleophilic substitutions of simple aromatic compounds are difficult because of electron repulsion and the disfavor of hydride displacement. However, certain metal carbonyl complexes of arenes are electron deficient and liable to attack by donor reagents. Oxidative work-up of the σ-adducts regenerates the aromatic compounds with the new substituents. More importantly, opposite regio-selectivity to the corresponding electrophilic substitution (of the uncomplex substrates) has been demonstrated [Semmelhack, 1979]. The polarity alternation rule proves useful in predicting the orientation effect. (For detailed discussion of the reaction, see Kündig [1989].)

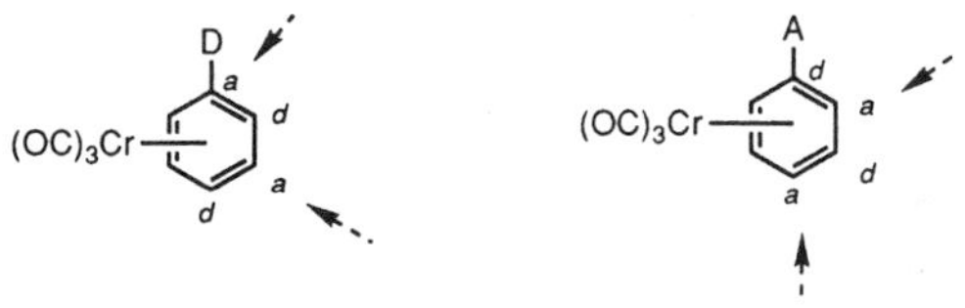

It is thus possible to effect nucleophilic reaction at the *meta* position of an anisole. A synthesis of acorenone [Semmelhack, 1980] was based on this reaction.

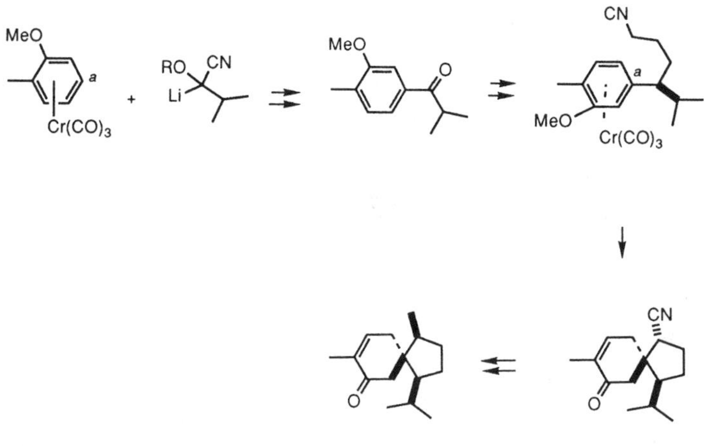

acorenone

More familiar nucleophilic aromatic substitutions involve aryl halides which contain acceptor substituents (notably nitro) in conjoint (*o*- and/ *p*-) positions. Mechanistically, such reactions are akin to those of β-halo-α,β-enones. Let us examine a strategy for synthesis of thyroxine.

Thyroxine is easily correlated with tyrosine, therefore it is most logical to use the α-amino acid to synthesize the thyroid hormone [Chalmers, 1949]. While introduction of the iodine atoms can be achieved at any convenient stage, attention must be paid to detailing the construction of the diphenyl ether linkage. In view of the sensitivity of the α-amino acid residue, even properly protected, it would be difficult to use the phenolic oxygen of tyrosine as a donor to displace a substituent (e.g., a halogen) from a *p*-oxyphenyl derivative because the leaving group is not activated (note the disjoint relationship). On the other hand, the tyrosine hydroxyl could be converted into a leaving group, such as tosyloxy, and exchanging the iodo groups with acceptors would further facilitate the etherification step.

Despite the limited choice of the surrogate acceptors, the nitro group appears ideal because of its ready conversion into the iodo group via the diazonium ion. Consequently, coupling of the dinitro-*O*-tosyltyrosine derivative with *p*-methoxyphenol would provide a useful intermediate.

(-)-thyroxine

(X = NO₂)   L-tyrosine

Pyridines, quinolines and certain other heteroaromatics are susceptible to attack by donors at C-2 and C-4 (acceptor sites) without further activation, although *N*-alkylation or *N*-acylation further increases their reactivity. Usually, harder bases react at C-2 (closer to the hard nitrogen atom) and softer bases at C-4.

The Chichibabin amination is of mechanistic significance because the C-2 hydrogen leaves as a hydride ion. The combined effect of two nitrogen atoms in the intermediate is to heighten the acceptor character of the central carbon which in turn renders the hydrogen adequately donative.

It should be noted that phosphabenzene, arsabenzene, and stibabenzene react with nucleophiles at the heteroatom (acceptor site) [Ashe, 1985].

Indole alkaloid synthesis [Wenkert, 1968a, 1981a] based on the general scheme of *N*-tryptophylpyridinium ion modification and Pictet–Spengler cyclization often calls for carbon chain attachment to C-4 of the pyridine ring. Such reactivity has also been exploited in approaches to sesbanine [Wanner, 1981; Wada, 1985]. In this case C-4 is double activated, i.e., by *N*-acylation and the presence of an ester group at C-3.

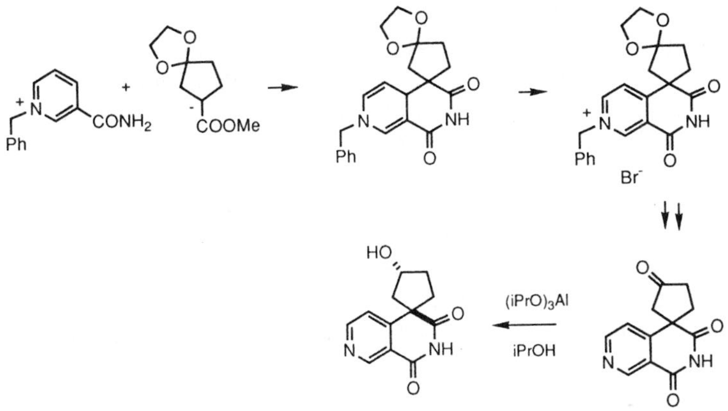

sesbanine

An intramolecular version of the C—C bond formation is embodied in a route to nauclefine [Sainsbury, 1977]. Thus, exocyclic disconnection from the γ-carbon of the pyridine moiety generates a readily available *N*-acylenamine. Compatible donor/acceptor centers are present in the intermediate.

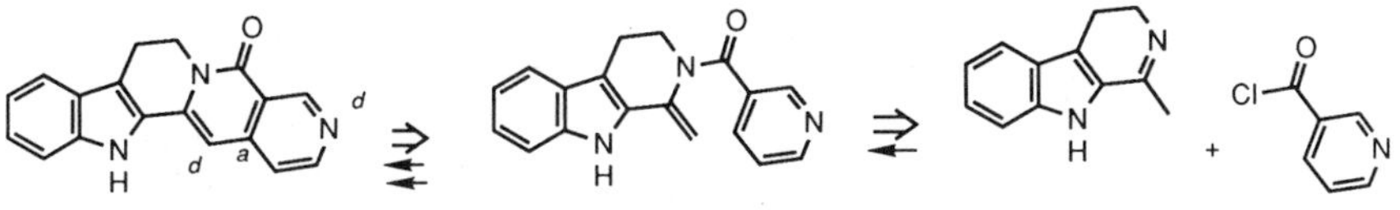

nauclefine

Alkenylation of the quinoline system is illustrated by a synthesis of quinine/quinidine [Taylor, 1974]. Here a chloride ion is displaced. (However, it must be emphasized that 1-chloronaphthalene does not undergo a similar reaction with Wittig reagents.)

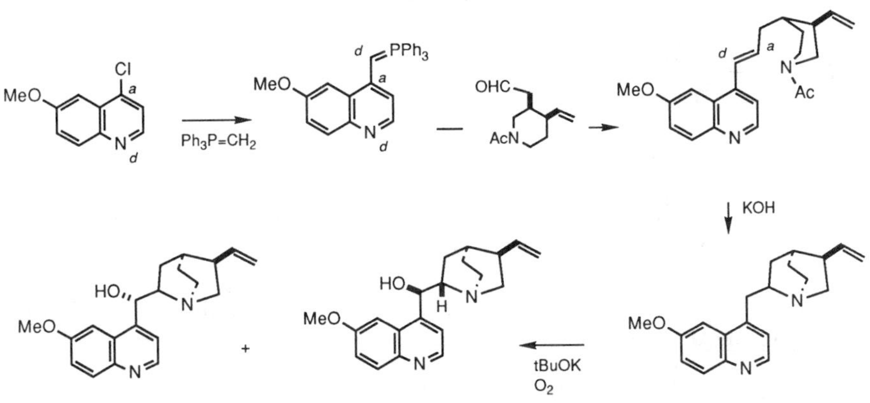

quinidine   quinine

It may seem unconventional to consider the Birch reduction here. It is not unreasonable to consider electrons that add to aromatic systems as nucleophiles.

The cyclohexadiene structures are strongly dependent on nuclear substituents. For example, a donor group favors the generation of the 1-substituted 1,4-cyclohexadiene, while an acceptor group directs the formation of the 3-substituted isomer.

These consequences are easily rationalized by examining the radical anion intermediates [Ho, 1988]. Thus, from a donor-substituted substrate the destabilization of a structure containing a $d-d$ atom pair shifts the equilibrium toward the alternative isomer. On the other hand, such a structure is the favored one when the donor group is replaced by an acceptor, as an $a-d$ union results.

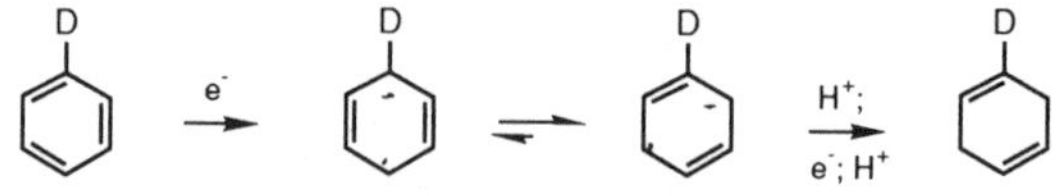

$D = OR, NR_2 \cdots$

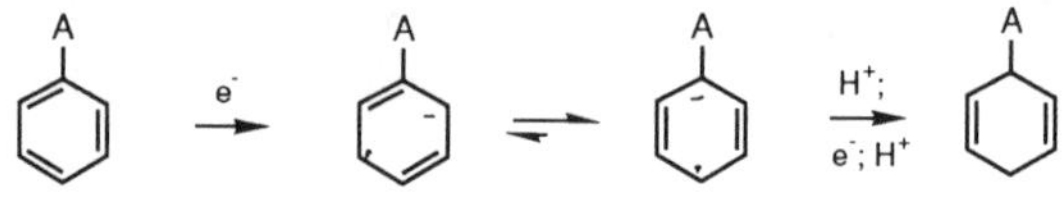

$A = COOH, SiMe_3$

2.3.3. Substitutions via Benzyne Intermediates

Haloarenes may undergo nucleophilic substitution via the benzyne mechanism. For a haloarene containing another substituent, the entering site for the donor is determined by that substituent. The two equations shown below clearly indicate the reactive position of the aromatic ring is an acceptor site [Biehl, 1971; Pansegrau, 1988].

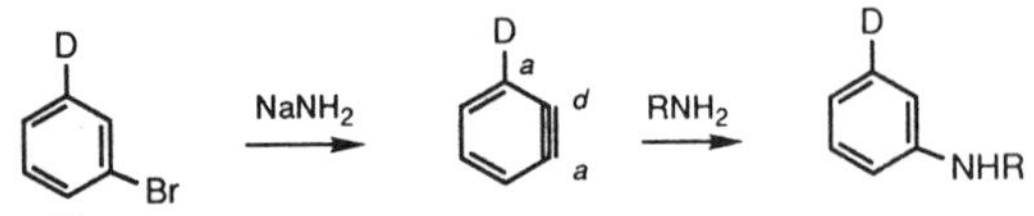

$D = OMe, NMe_2, Cl$

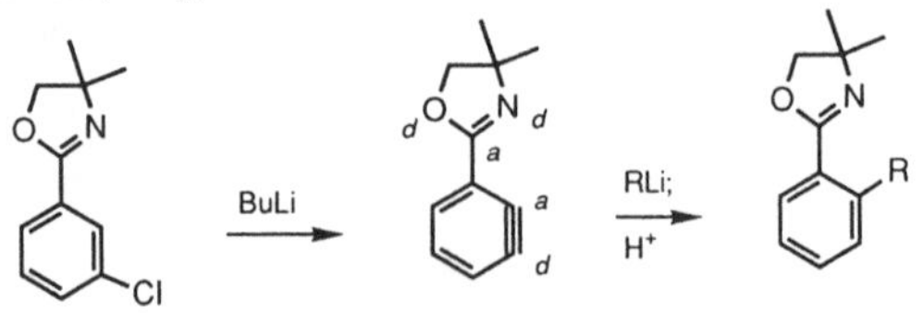

Two substituents may be introduced to the haloarene via this reaction [Birch, 1971; Crenshaw, 1988].

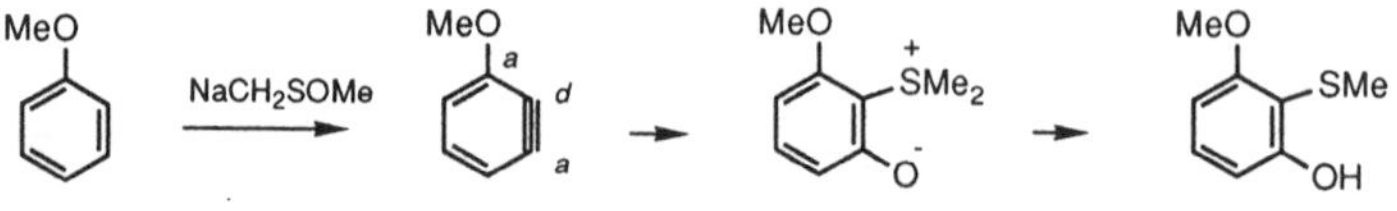

A synthesis of lysergic acid [Julia, 1969] is an example of exploiting the reactivity of benzyne and steric constraint to form a tetracyclic intermediate from relatively readily accessible predecessors. In terms of electronic factors, the cyclization is favored by the conjoint disposition of the acceptor site with the acetamido group, and the donor atom with both the ester function and the amine.

The substrate for this cyclization was obtained from the condensation product of 5-bromoisatin and methyl 6-methylnicotinate. The enhanced acidity of the α-picolinic methyl group by the ester in its *para* position makes the condensation very facile.

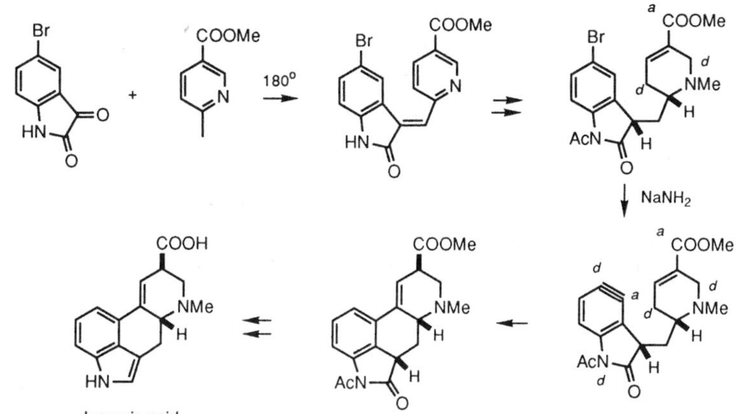

lysergic acid

The benzyne route to corydaline [Saá, 1986] is also highly dependent on the regioselectivity of the initial C—C bond formation. This result could have stemmed from a combination of electronic and steric effects.

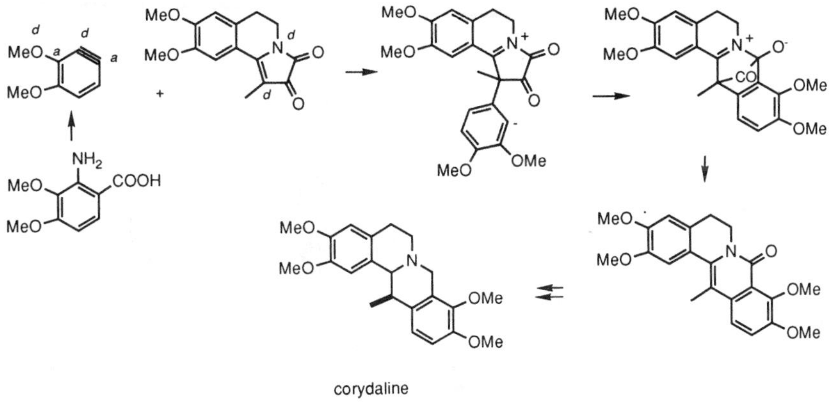

corydaline

2.4. REACTION AT A BENZYLIC POSITION

A nuclear substituent has a profound effect on the reactivity of a coexisting methyl or methylene group. The cationic stabilization of *o*- and *p*-donor-substituted benzylic species, and anionic stabilization of the corresponding acceptor-substituted benzylic sites are well known. There is clearly the existence of a conjoint circuit in each case. The high acidity of α-, γ-picolines and their benzologs is also understandable on the same basis.

Synthetically, these characteristics have been exploited. Thus, the ready ionization of a benzylic alcohol by virtue of its conjoint relationship with a p-hydroxyl was responsible for a facile synthesis of cherylline [Schwartz, 1971].

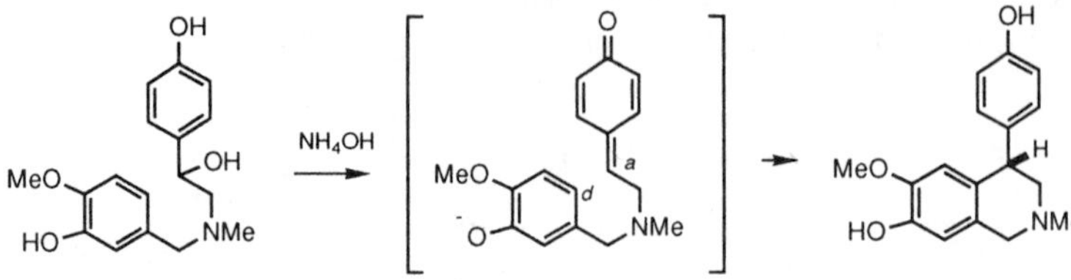

cherylline

The cyclization step of the cherylline synthesis most likely proceeded via a p-quinone methide intermediate which was formed by a facile dehydration. An alternative protocol to access analogous intermediates is by use of p-quinone ketals [Hart, 1978]. Noteworthy is that p-quinone ketals are p-alkoxyaryl cation equivalents.

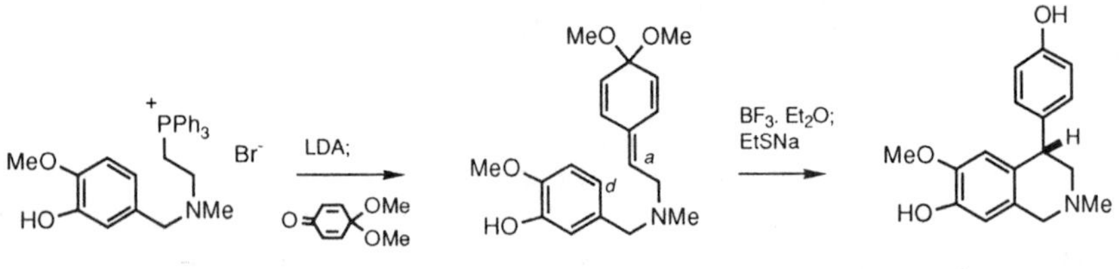

cherylline

An oxidative route to quinone methides has been explored and demonstrated in a synthesis of picropodophyllin [Kende, 1977].

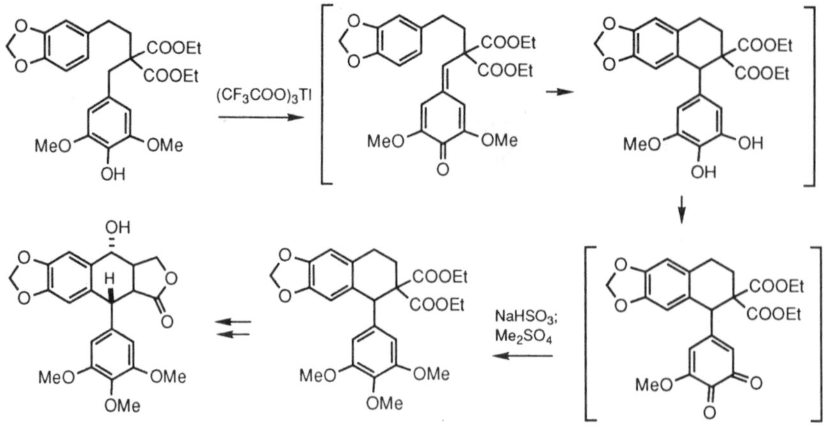

picropodophyllin

Acetophenone N-(o-trifluoromethylanilino)imine forms 4-donor-substituted 2-phenylquinolines on reaction with strong bases via a series of dehydrofluorination, conjugate addition and cyclization [Strekowski, 1990]. Such a reaction pathway is unavailable to the disjoint m-trifluoromethyl anilino analog.

Very subtle yet profound electronic effects of alkoxy substituents on benzylic reactivity have been uncovered in a Polonovski-type reaction of a dibenzo[$c.f$]-azocine. The reaction is a key to another synthesis of cherylline [Nomoto, 1984].

cherylline

The two benzylic methylene units attaching to the nitrogen atom differ only in that one of them is *para* to a methoxy group. Both have a *meta* benzyloxy substituent. The influences of the respective benzyloxy substituents must cancel out, the methylene which is also *para* to the methoxy would acquire some acceptor characters and therefore has a tendency to resist deprotonation that constitutes the initial step of the C—N bond cleavage.

A synthesis of lennoxamine [Moody, 1987] via an intramolecular nitrene addition to an alkene is interesting because the desired benzazepine product is predominant. From a simpler system in which both aryl groups are unsubstituted, an approximately equimolar mixture of the phenylbenzazepine and 1-benzylisoquinoline derivatives was formed.

In the synthesis that requires an ester group in the aromatic ring farther from the nitrene, the acceptor confers some donor character to its proximal benzylic carbon atom, and therefore C—N bond scission from the other direction is more favorable.

lennoxamine

It is interesting to note that 1,3,2-dioxaphospholanes are produced from reactions of trialkyl phosphites with aromatic aldehydes bearing strongly electron-withdrawing groups (e.g., *p*-nitro), as a result of initial attack at the oxygen atom of the carbonyl [Ramirez, 1967]. The zwitterionic adducts would be quickly scavenged by the aldehyde molecules (1,3-dipolar cycloaddition). At least in the case of *p*-nitrobenzaldehyde the oxybenzyl carbanion is conjoint with the nitro group and is stabilized.

A lithiation study on xylene isomers [Klein, 1988] showed a reactivity profile of *m*- ≫ *o*- > *p*-, On the other hand, only the *meta* isomer of methylacetophenone is resistant to dianion formation. All these observations are consistent with the polarity-alternating influence of one donor center to other atoms of the same molecule.

Deprotonation of tricarbonyl(*N*-methyl-1,2,3,4-tetrahydroisoquinoline)-chromium occurs at C-4 and thereby opens a pathway to various 4-substituted tetrahydroisoquinolines [Blagg, 1985]. The site of deprotonation is actually that which would be suggested by the polarity alternation rule. Between the two benzylic centers C-4 is a donor and C-1 is an accept if the nitrogen atom exerts its influences. (Besides offering steric control the tricarbonylchromium only serves to increase the acidity of the benzylic hydrogens, it should have the same effect on the two benzylic methylene groups.)

The following exemplifies the employment of alkylpyridines in synthesis. Thus, the $N_{(a)}$-methyl derivative of sempivirine was assembled from *N*-methylharman and 2-isopropoxylmethylenecyclohexanone [Woodward, 1949].

sempervirine
N-metho salt

The 2,6-lutidine system has been identified as a latent precursor of 3-alkyl-2-cyclohexenone and used in steroid synthesis [Danishefsky, 1975]. Specifically, an approach to estrone successfully exploited this concept.

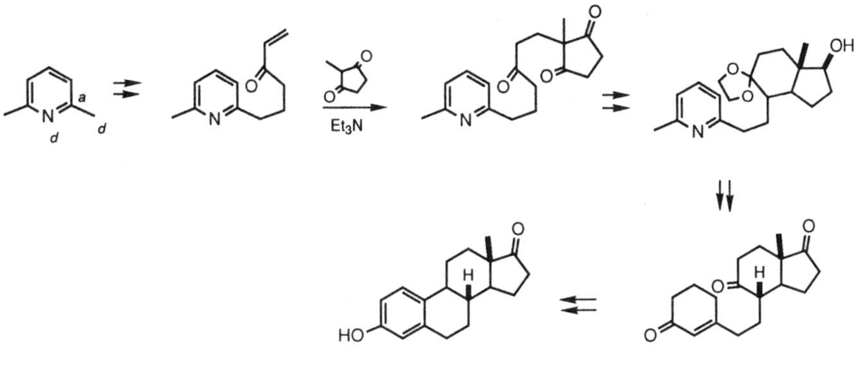

estrone

In the lutidine ring the donor nitrogen is connected to two carbon atoms (necessarily acceptors) which are 1,5-related. 2,6-Lutidine can then provide the structural element of the A-ring and C-6 of estrone. Furthermore, engaging the donor sites (methyl groups) in C—C bond formation is conceivable.

The diketone may be unveiled by Birch reduction and hydrolysis.

The enhanced donor reactivity of the methylene group of methyl pyridine-2-acetate enables a rapid construction of a substituted quinolizidine skeleton, and thence anagyrine [Goldberg, 1972].

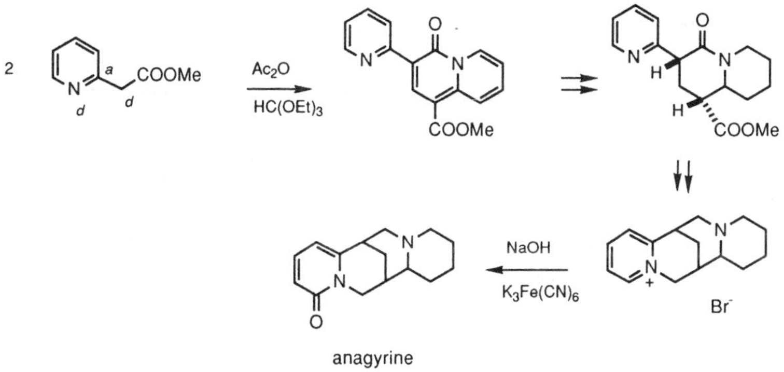

anagyrine

Carbonyl protection as acetal/ketal of 2-(α-pyridyl)-1,3-propanediol [Katritzky, 1987] has the advantage of facile demasking by mild base treatment of the N-metho salts. The methine hydrogen is further acidified in the quaternization step.

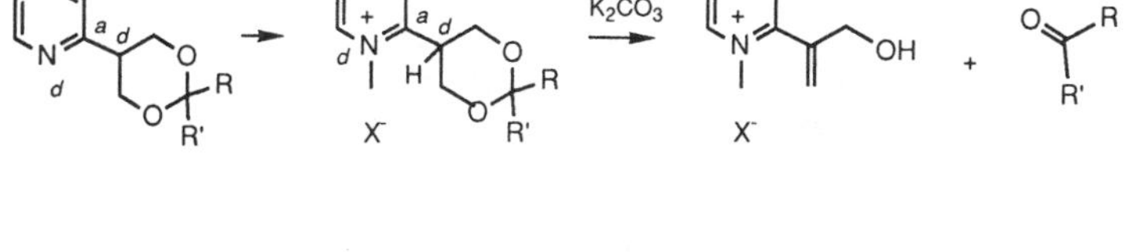

In a synthetic route toward (+)-muscopyridine [Utimoto, 1982] involving asymmetric hydroboration of a vinylpyridine intermediate, the latter compound was formed from a readily enolized bicyclic ketone with methyllithium! It is remarkable that the alkoxylated carbon atom would retain sufficient acceptor character by virtue of its relationship with the pyridine nitrogen atom.

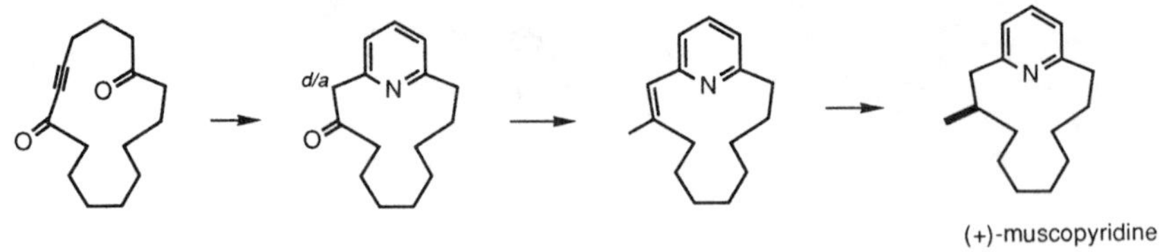

(+)-muscopyridine

In a quinine/quinidine synthesis [Taylor, 1975] an oxygenation at the benzylic position under basic conditions constituted the last step. The success was due to the high acidity of the methylene group bestowed by the nitrogen atom of the quinoline ring.

Methanolysis of isomeric trichloromethylpyridines [Dainter, 1988] shows the dominating effect of the nitrogen atom. Only when the benzylic carbon is an acceptor are its attaching chlorine atoms displaced.

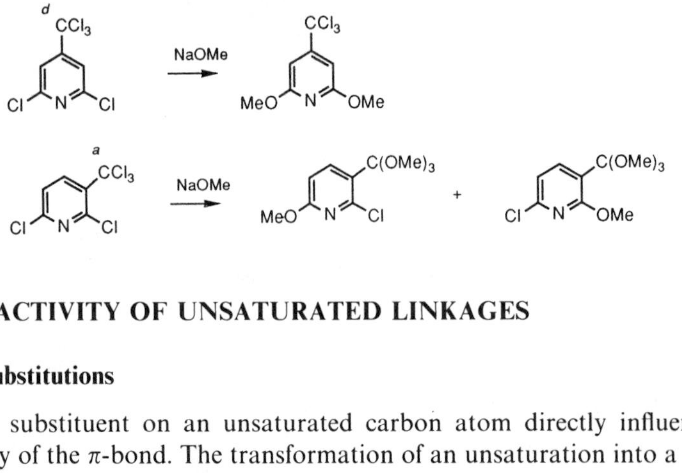

2.5. REACTIVITY OF UNSATURATED LINKAGES

2.5.1. Substitutions

A polar substituent on an unsaturated carbon atom directly influences the reactivity of the π-bond. The transformation of an unsaturation into a Michael acceptor by a substituent capable of electron delocalization is well known. Silyl groups often exhibit the same capability of polarizing the π-bond to which they are attached directly, as shown by the desilylative acylation of vinylsilanes [Fleming, 1981].

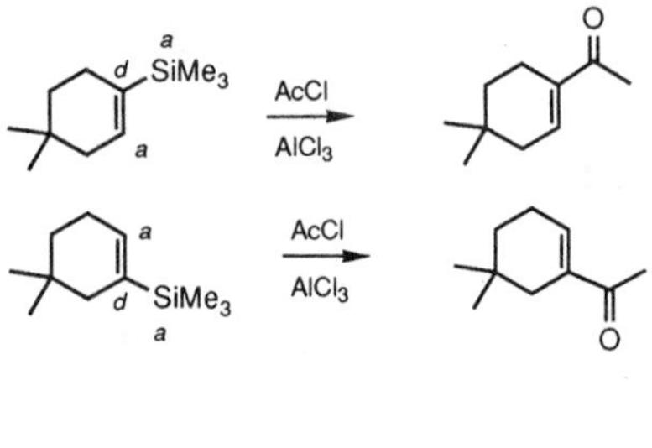

The Nazarov cyclization can be directed by a β-silyl substituent [Denmark, 1982]. In such compounds the silyl group is disjointly connected to the carbonyl, therefore delocalization involving the silylated double bond is highly unfavorable. The synthetic significance is that thermodynamically less stable cyclopentenones may be prepared from such substrates.

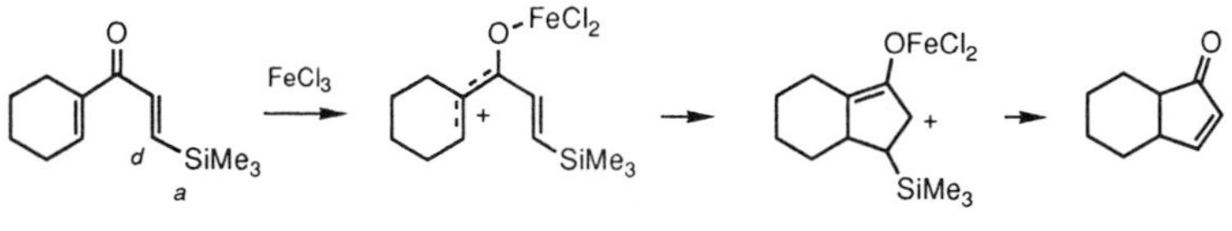

A variety of unsaturated azacycles can be prepared by an intramolecular reaction of iminium ions with vinylsilanes, with the silicon controlling the regiochemistry and stereochemistry of the product double bond. An example is the completion of the indolizidine nucleus of (+)-pumiliotoxin 251D [Overman, 1981]. Very clever manipulations of the heterocyclization initiated by either an iminium ion or an N-acyliminium species deriving from the same precursor led to enantiomeric indolizidines [Heitz, 1989].

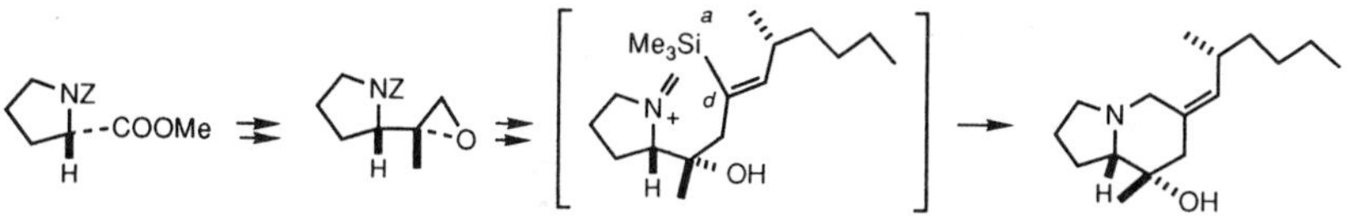

(+)-pumiliotoxin-251D

Ring closure that is exocyclic with respect to the iminium ion initiator is strongly dependent on the vinylsilane stereochemistry. Cyclization of (Z)-vinylsilanes could proceed several thousand times faster than the (E)-isomer, as only the (Z)-alkene substituent can participate initially in $\sigma-\pi$ stabilization of the developing β-silyl cation [Overman, 1984]. This chemistry has found use in synthesis of Amaryllidaceae alkaloids such as epielwesine.

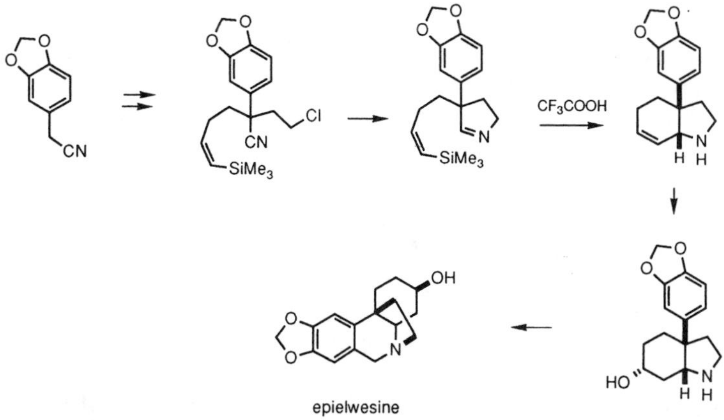

epielwesine

Entry into five-, six-. and seven-membered cyclic ethers has been gained by an analogous cyclization [Overman, 1986].

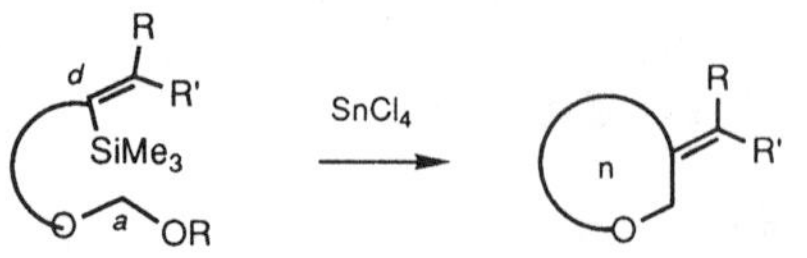

Monoacylation of iron tricarbonyl complexes of 2-trialkyl-1,3-butadienes under the Friedel–Crafts conditions [Franck-Neumann, 1987] occurs predominantly at C-4. The diene terminus nearest to the silyl group has acquired much acceptor characteristics and is less reactive.

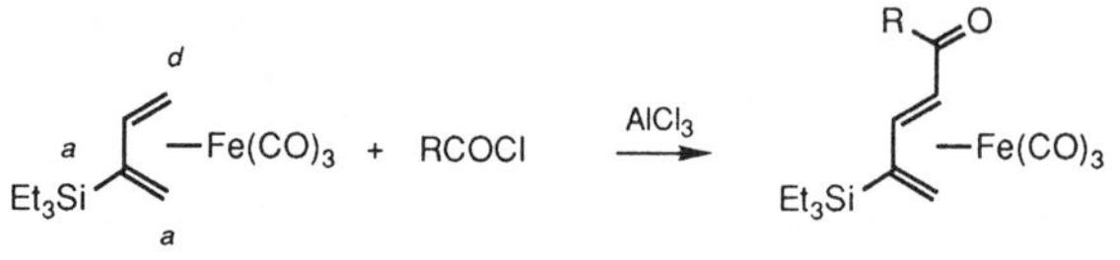

The Heck arylation of alkenes formally displaces a hydride from the trigonal carbon [Heck, 1979]. While these reactions are very sensitive to steric effects, subtle electronic contributions to the regiochemical outcome may be evaluated. It is seen that the aryl group prefers bonding with the acceptor end of the alkene linkage [Trost, 1982b].

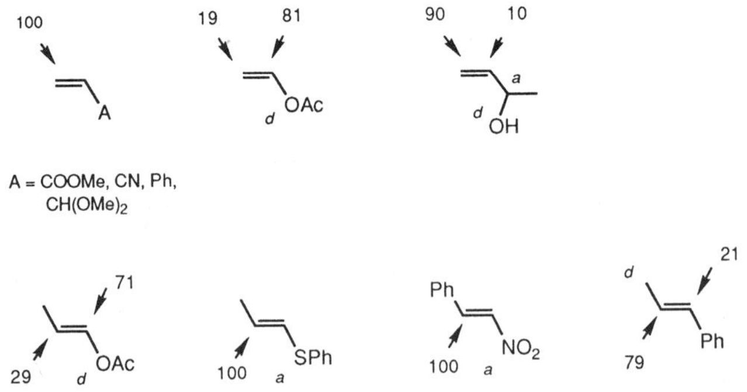

A = COOMe, CN, Ph, CH(OMe)$_2$

Arylation of allylic alcohols leads to 1-aryl-3-alkanones, in accord with the orientation indicated above. The Pd—H elimination involves departure of a hydride from carbon and this process is facilitated by the geminal donor hydroxyl.

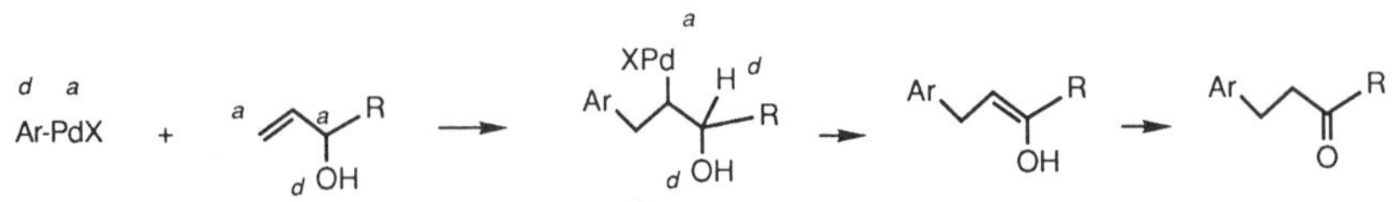

The ene reaction may be considered as a transpositional substitution on the ene part (and addition on the enophile). A synthesis of α-kainic acid [Oppoizer, 1984b,c] showed the regiocontrol of the ene reaction by strategically placing a polar group which in this case most gratifyingly corresponds to a required functionality. Using a bulky chiral auxiliary group, π-face differentiation was maximized, establishing both the relative and absolute stereochemistry.

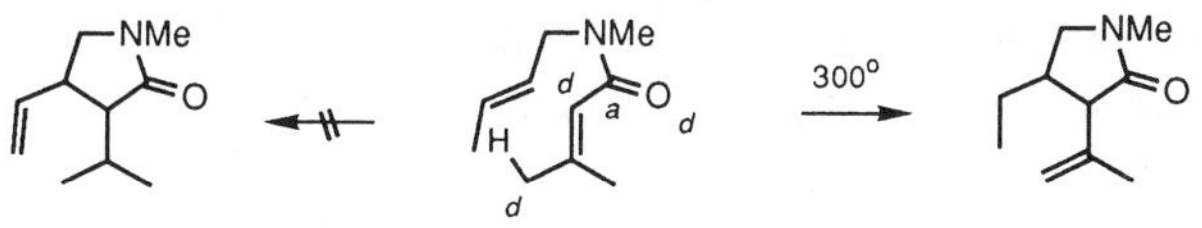

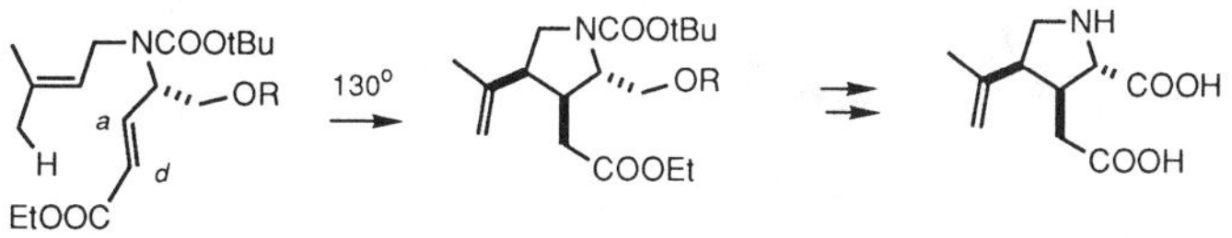

The oxidation of 3-trimethylsilylcycloalkenes (five- to eight-membered) gives 3-trimethylsilyl-2-cycloalkenones [Reuter, 1978]. Perhaps this is an ene-type reaction with its regiochemistry governed by the acceptor substituent.

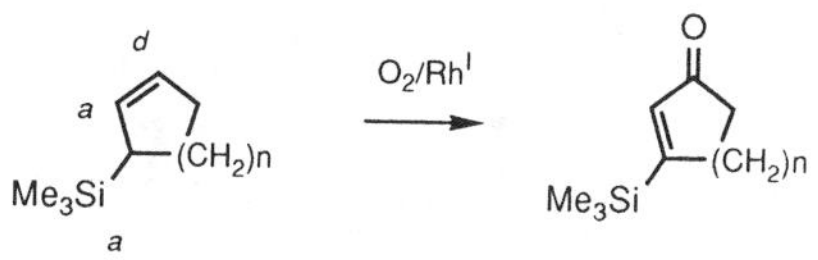

Nucleophilic substitution of arylhaloacetylenes with sodium p-toluenethiolate (α-addition–elimination mechanism) shows a reactivity trend of Cl > Br [Beltrame, 1973]. Apparently a better donor makes its adjoining atom a better acceptor (stronger polarity alternation).

2.5.2. Additions

The empirical Markovnikov rule on the orientation of HX addition to C=C bonds must be evaluated in the light of the substrate structures. The so-called exceptions are those involving alkenes substituted with groups other than simple alkyls. Thus the electronic differences between CH_3 and CF_3 [Henne, 1950] must be recognized in that the methyl carbon belongs to the donor class whereas the trifluoromethyl carbon is acceptor-like (its attaching fluorine atoms being donors).

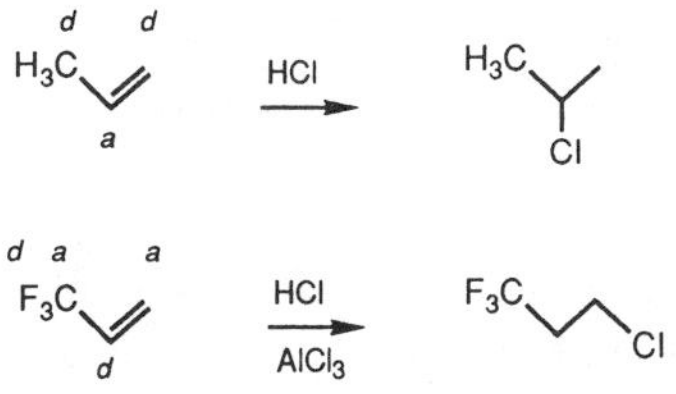

3-Buten-1-yne adds methanol to give 1,3,3-trimethoxybutane. Upon mercury(II) catalysis the reaction first occurs at the triple bond, and the entered methoxy groups then control the regiochemistry of the methanolation at the alkene linkage, in a seemingly anti-Markovnikov manner. Examination of the intermediate structure reveals the normalcy of the reaction. (The trimethoxybutane is a precursor of 2-methoxy-1,3-butadiene which is an excellent diene for Diels–Alder reactions. Its use has now been eclipsed by the introduction of 2-trialkylsiloxy-1,3-butadienes [Brownbridge, 1983].)

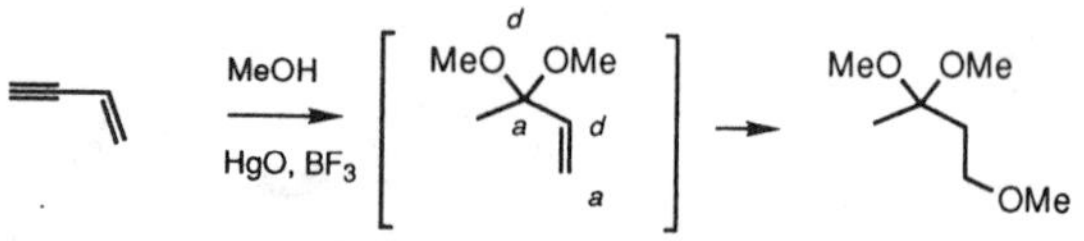

A silicon atom in alkylsilanes is an acceptor and the solvatomercuration of 1-trimethylsilylcyclohexene [Berti, 1984] follows the polarization pattern faithfully.

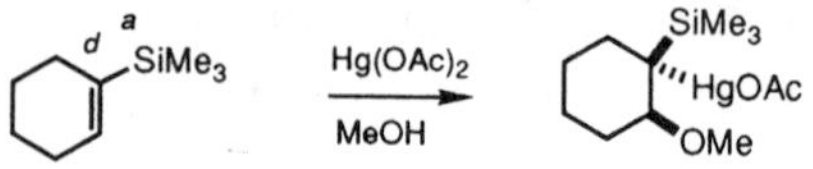

Extraordinary regioselectivity has been observed in the nitromercuration of some simple cyclohexenes [Corey, 1978c]. The results are in full accord with those predicted by the polarity alternation rule.

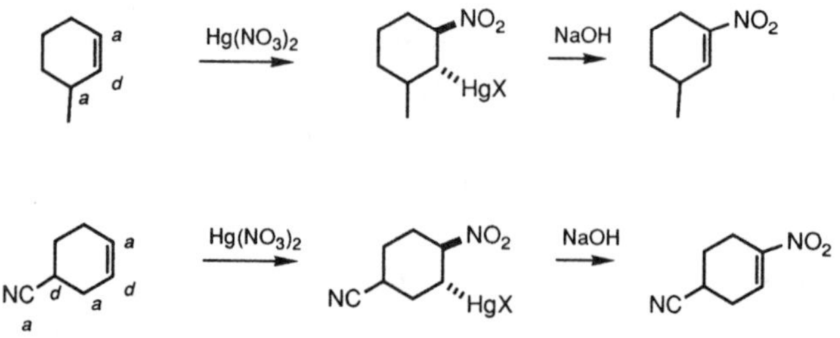

Vinylsilanes and allylsilanes furnish β-aminoalkylsilanes via an aminomercuration–demercuration sequence [Barluenga, 1982]. The striking orientation exhibited by the vinylsilanes can be similarly rationalized.

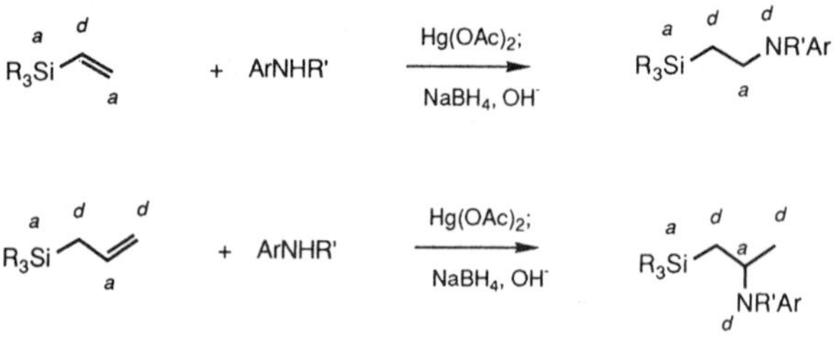

There is an unmistakable directing effect of an allylic silicon in hydroboration of a double bond [Fleming, 1988], it is remarkable that such control is extendable to tetrasubstituted alkenes [Akers, 1989]. On the other hand, an allylic alkoxy group, being a donor, renders an opposite influence. Consequently, hydroboration of such alkenes leads to *vic*-dioxygenated products [S.L. Schreiber, 1989].

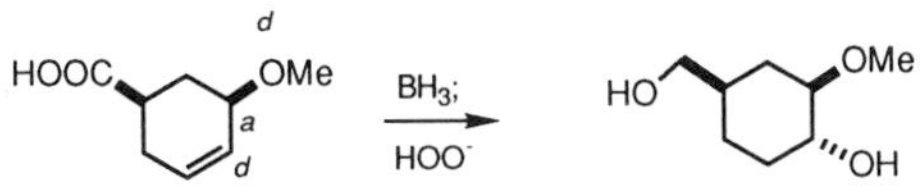

In exactly the same pattern is the allylic substituent-controlled cycloaddition of nitrones to the alkene substrates [Inouye, 1987].

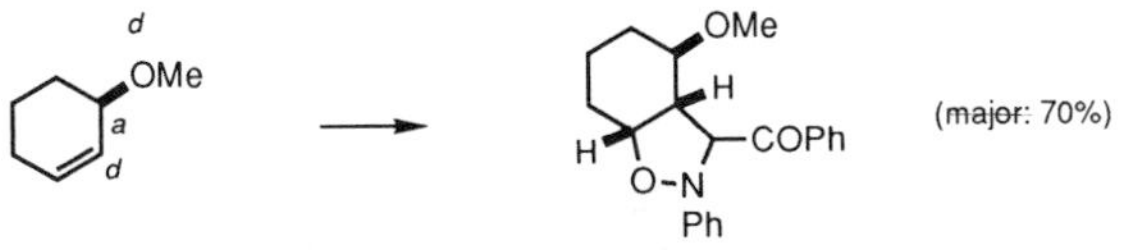

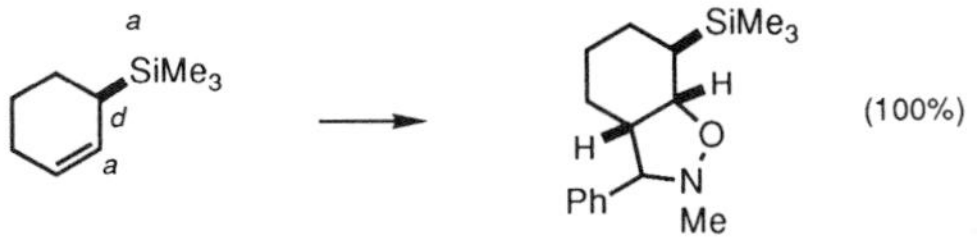

The importance of an allylic silyl group as terminator of intramolecular cation cyclization has been shown by a comparative study of the nonsilylated substrate [Majetich, 1988].

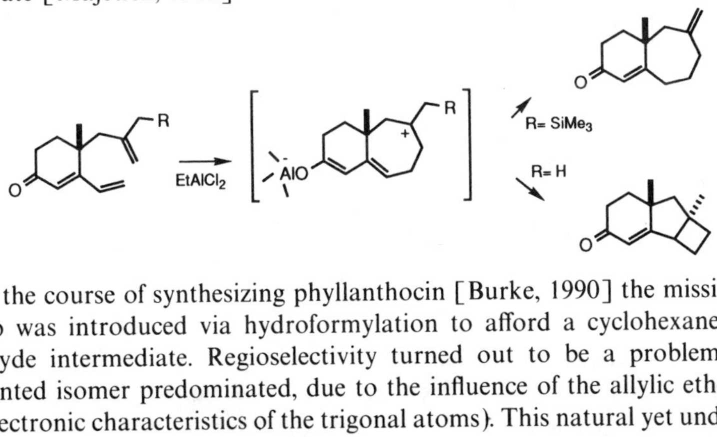

In the course of synthesizing phyllanthocin [Burke, 1990] the missing ester group was introduced via hydroformylation to afford a cyclohexanecarboxaldehyde intermediate. Regioselectivity turned out to be a problem as the unwanted isomer predominated, due to the influence of the allylic ether (note the electronic characteristics of the trigonal atoms). This natural yet undesirable feature was counteracted by intramolecular ligation of the rhodium atom such that steric constraints forced the delivery of the carbonyl group to the *meta* position.

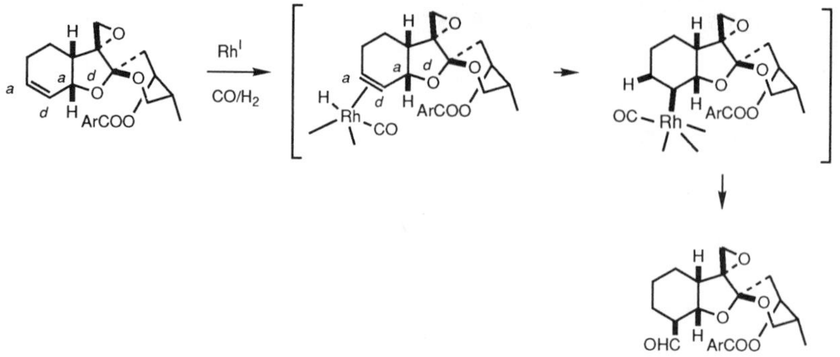

The regiochemistry of the addition of phenylselenenyl chloride to allylic alcohols and their esters [Liotta, 1982] can be explained simply by using the polarity alternation rule except in those cases where steric factors become dominating.

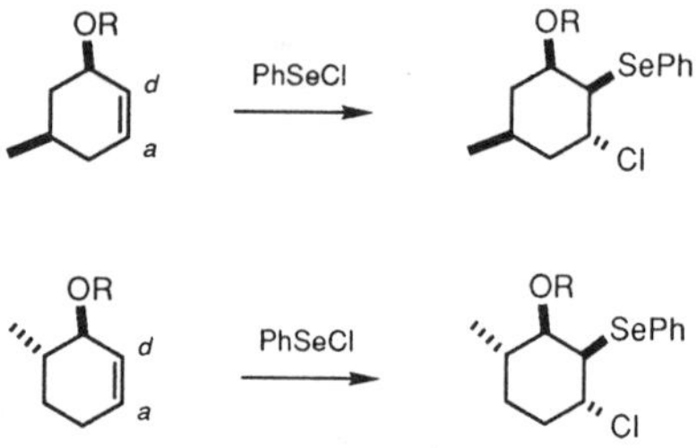

Allylic nitrogen donors have the same effect on the double bond with respect to addition reactions [Krow, 1974, 1988].

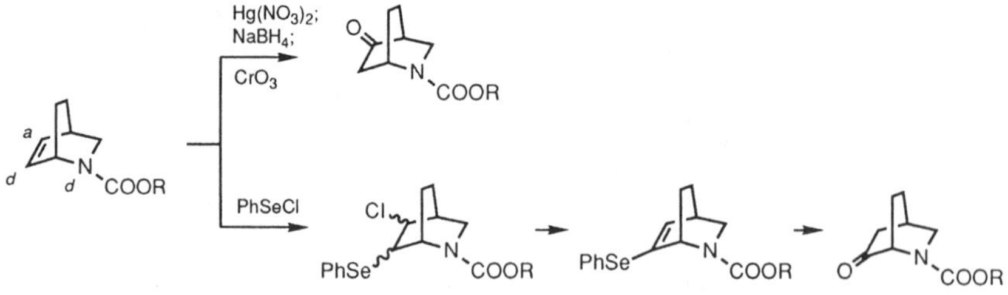

Surprisingly, remote functional control of the regiochemistry for 1,2-addition to bridged bicycloalkenes has been witnessed [Carrupt, 1982; Black, 1986]. Again, the polarity alternation consideration provides a convenient forum for rationalization.

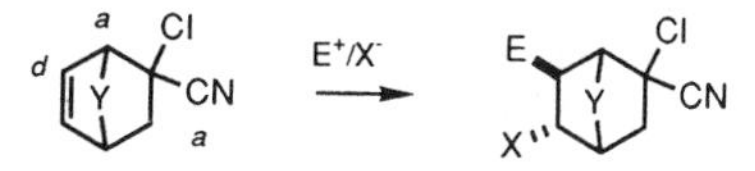

Y = O, CH$_2$, (CH$_2$)$_2$

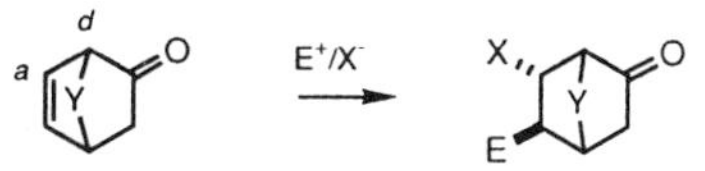

Y = CH$_2$, (CH$_2$)$_2$

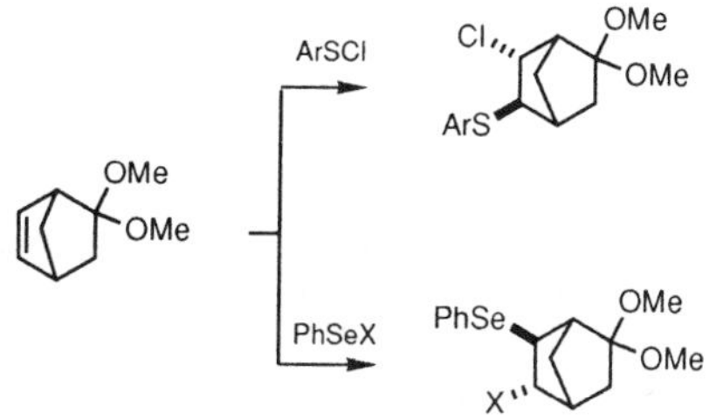

It should be noted that the last reaction is anomalous owing to the soft nature of selenium such that the intermediate is an episelenonium species and the attack of the counterion is subject to steric effects such as the 2,6-*endo* interaction.

Hydration of the cyclohexene during a synthesis of morphine [Gates, 1956] can hardly be predicted with regard to its regioselectivity. However, the donor nitrogen atom and the closest methoxy group, although many bonds away, may have been responsible for the protonation at C-7 instead of at C-6, via a polarity alternation mechanism.

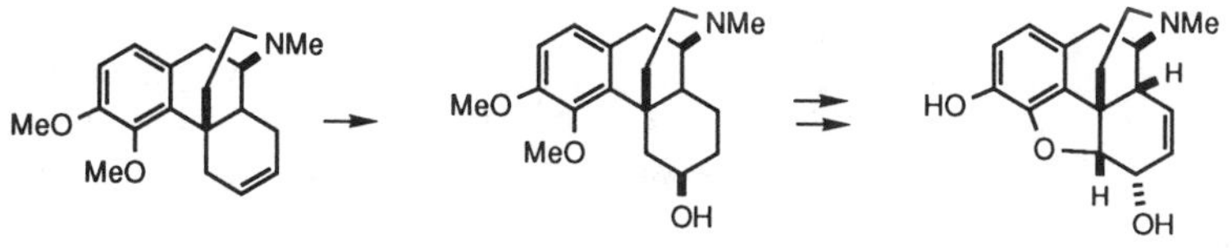

morphine

As expected, addition to alkynes is also subject to control by polar functions in the vicinity. The electronic differences of methyl and trialkylsilyl groups are evident from an examination of the diverse modes of intramolecular capture of an allylic carbenium ion [Kozar, 1977] and of an iminium species [Overman, 1988].

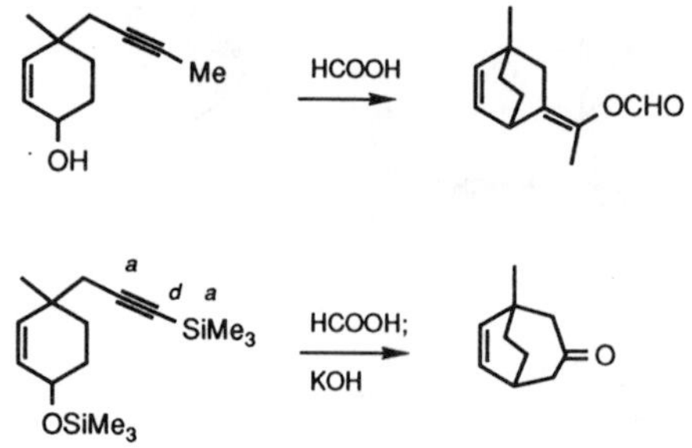

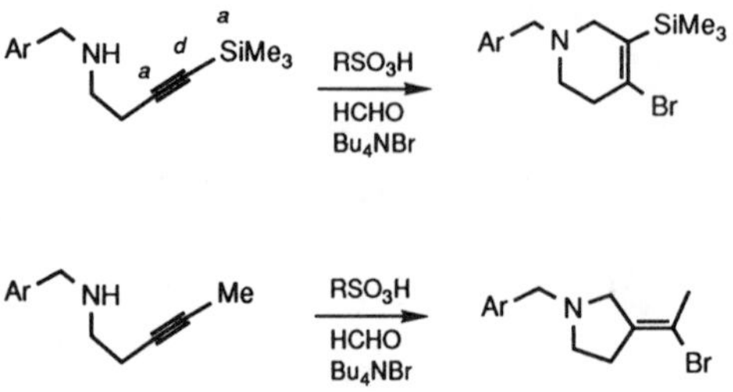

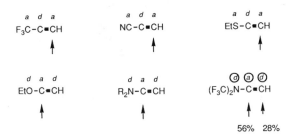

Various reactions of substituted acetylenes with nucleophiles are shown in the following diagram. Note the predominant attack on the acceptor sites [Brandsma, 1968; Freear, 1969; Petrov, 1972; Raunio, 1971; Truce, 1972].

In a series of ynamines the reactivity toward nucleophiles ascends as $(F_3C)_2N—C≡C—R ≪ Ph_2N—C≡C—R ≪ Et_2N—C≡C—R$ [Ficini, 1976]. Hydration of the N,N-bis(trifluoromethyl)acetylenes require mercury(II) ion assistance.

Michael addition to activated CC multiple bonds will be discussed at length later. However, it should be mentioned that a donor reagent always attacks at the β-carbon to the activating acceptor group. In recent years synthetic methodologies have been refined to allow for double alkylation of an α,β-enone system in essentially one operation. The highly convergent three-component coupling synthesis of the prostaglandins [Danishefsky, 1989b; Noyori, 1990] is most enlightening.

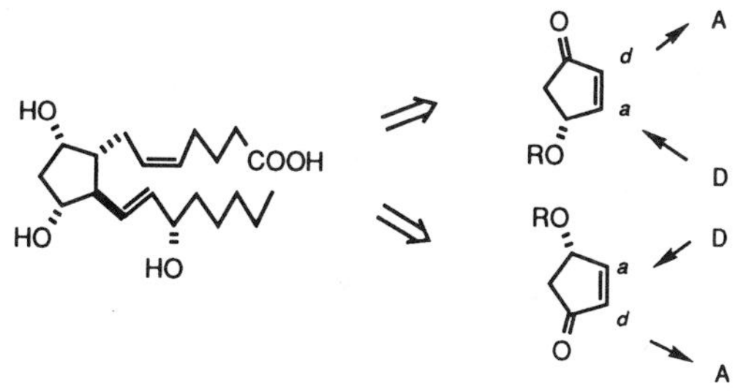

In simple β-alkyl addition to an α,β-enone by the cuprate reagent, coordination of the Michael acceptor with chlorotrimethylsilane is extremely helpful [Corey, 1985a]. The silyl reagent accentuates the acceptor.

On α-substitution of an enone with another acceptor group the acceptor character of the β-carbon is also heightened. For example, a nootkatone synthesis [Pesaro, 1968] from a cross-conjugated cyclohexadienone derivative via β-methyl addition was specifically directed by an additional ester function. The service of this ester was needed once more to help redress the stereochemistry of the secondary methyl group which was delivered axially. Thus reintroduction of the double bond and hydride reduction from the same β-face accomplished the task.

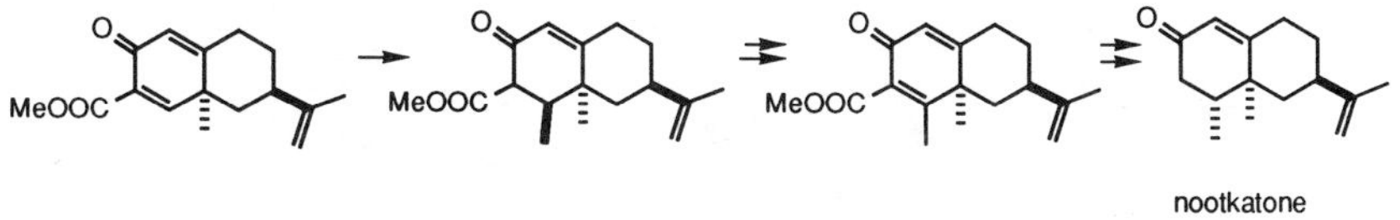

nootkatone

Often an imine is not sufficiently reactive to combine with donors. In such cases a preliminary N-trimethylsilylation is a remedy [Jahangir, 1986]. The silyl group is a readily removable surrogate for an N-alkyl substituent of a reactive iminium species.

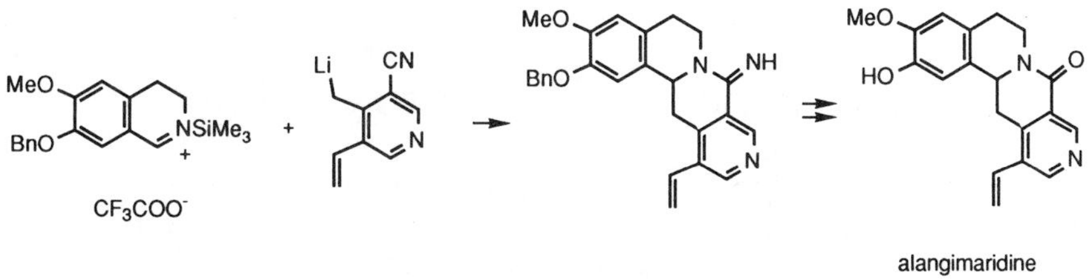

alangimaridine

A wonderful situation conducive to intramolecular cyclization of an extremely reactive iminium intermediate has been created from reaction of a lactone to an ynamine [Ficini, 1981]. Note the trigonal carbon of the iminium function is also the β-carbon of an enone and therefore it is doubly activated.

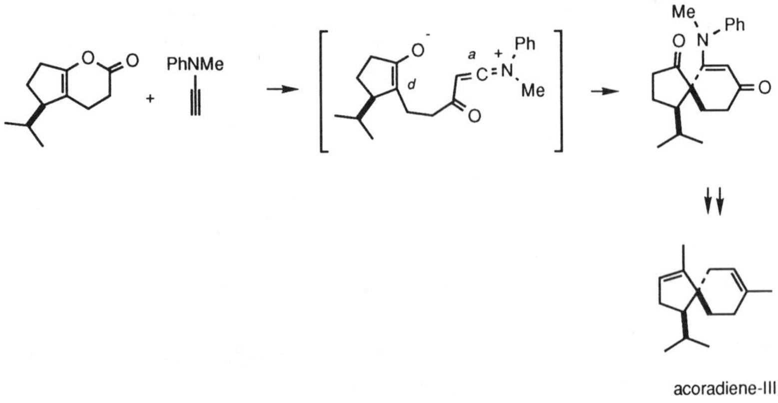

acoradiene-III

The Pauson–Khand cyclopentenone formation exhibits regiocontrol by a remote substituent of the alkene [Krafft, 1988]. As shown in the diagram the better selectivity of reaction of an alkene containing a homoallylic donor with an acceptor-substituted alkyne is in full agreement with electronic matchings. When the donor is one more carbon away, the regioselectivity drops, but the polar effect (in the opposite direction) is unmistakable.

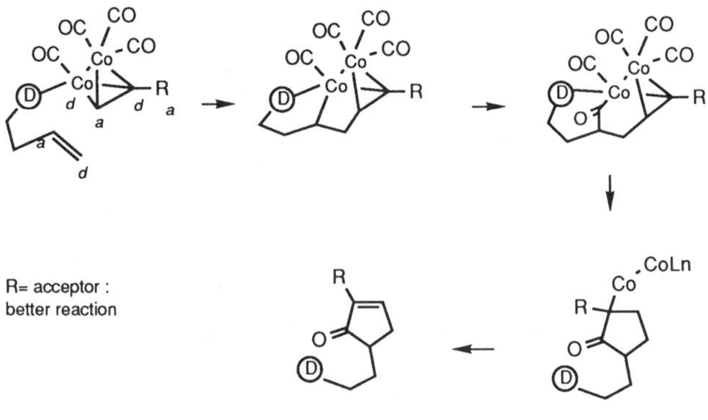

R = acceptor :
better reaction

2.6. ALKENE FORMATION VIA ELIMINATION REACTIONS

Dehydration of an alcohol containing a β-acceptor group proceeds readily. *In situ* elimination of sulfonic acid from the derived sulfonate represents a useful protocol. Interestingly, a gradated facility in the dehydration of certain α-heterosubstituted nitro alcohols has been found [Barrett, 1989]. The more reluctant elimination of methanesulfonic acid from the α-benzyloxy nitro derivative in comparison with the corresponding α-phenylthio analog could be due to a more donor-like character of the α-hydrogen (captodative situation).

Dehydrohalogenation and dehalogenation are the most common reactions to generate alkenes. In the latter process the reagent is often a reducing agent and one of the halogen atoms acts as an acceptor.

Dehydrohalogenation of haloarenes leads to benzynes. The *ortho* hydrogen atom has a higher acceptor character (see polarity alternation circuit originating from the halogen). It should be noted that the facility of benzyne formation from a series of halobenzenes (PhF > PhCl > PhBr ≫ PhI) corresponds to donor strength of the halogen, and the trend is in contrast to leaving ability of the heteroatom.

It is pertinent to address the *syn*-dehydrohalogenation of *trans*-1,2-dihalo-cycloalkanes [J.G. Lee, 1979] in terms of selectivity of the leaving halogen. Again, fluorine is by far the best leaving group.

An explanation may be offered as follows. A halogen atom (donor) would affect its geminal hydrogen by means of a polarity alternation mechanism and the donor character of this hydrogen should be proportional to that of the halogen. In other words, a fluorine atom, compared with a bromine, decrees its geminal hydrogen to be a more pronounced donor. Consequently, abstraction of a hydride-like hydrogen is less favorable.

Exactly the same pattern emerged from dehydrohalogenation of 2,3-dihalo-dihydrobenzofurans [Baciocchi, 1983].

The unusual dehydrofluorination of 1,1,1,3,3,3-hexafluoro-2-propanol with *n*-butyllithium [C.P. Qian, 1988] suggests the overwhelming effects of six fluorine atoms at the β-position on the methine hydrogen, bestowing upon the latter sufficient acceptor character despite its geminal relationship to an (ionized) oxygen atom.

We can attribute the same effect on an acetoxy group in the formation of 5-acetoxy-1-bromocyclopentene from dehydrobromination of 2,3-dibromocyclo-pentyl acetate [Danikiewicz, 1983].

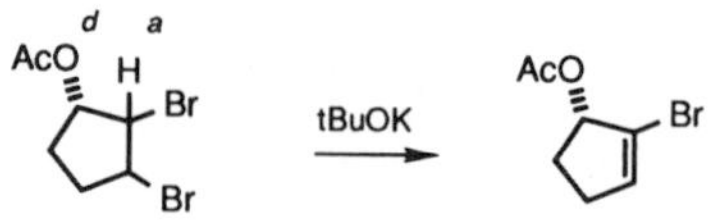

Two photocycloadducts derived from 5-fluorouracil behave dramatically differently [Swenton, 1988] on their exposure to *t*BuOK. Loss of the fluorine was accompanied by opening of the four-membered ring in the *gem*-dimethyl compound, whereas the methoxy-methyl analog (also the *gem*-dimethoxy analog) underwent normal β-elimination.

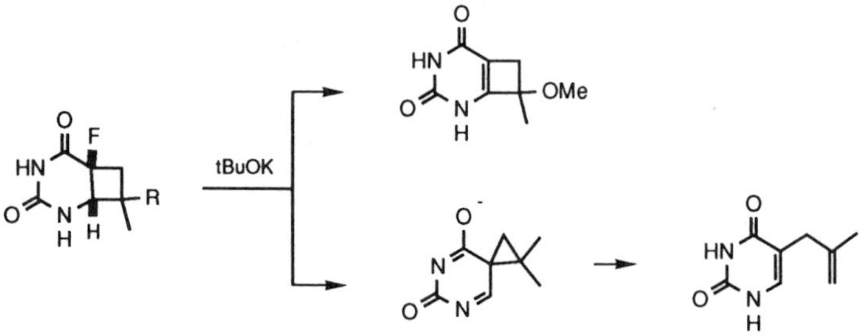

The direct elimination process requires deprotonation of the hydrogen geminal to a donor nitrogen atom. This somewhat unfavorable situation was perhaps ameliorated by a methoxy donor at a β-carbon, i.e., the influence of the nitrogen was reduced to such an extent that deprotonation became possible.

The thermal decomposition of X-oxides has become very important in synthesis because of the mildness of the conditions attainable (e.g., decomposition of selenoxides at room temperature) and the regioselectivity exhibited by such a reaction. Regarding the latter aspect it seems that a vicinal, *trans*-donor substituent favors the formation of an allylic product, whereas an acceptor group strongly biases the generation of a vinylic isomer. The regioselectivity is a result of the same electronic transmission by the polar substituent as that indicated in the polar elimination reactions [Ho, 1989b].

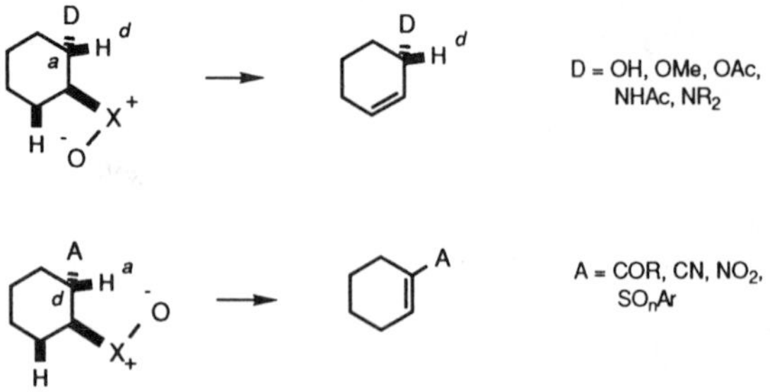

D = OH, OMe, OAc, NHAc, NR$_2$

A = COR, CN, NO$_2$, SO$_n$Ar

A clean conversion of agroclavine to lysergene can be achieved [Timms, 1989] by pyrolysis of an allylic sulfoxide derivative.

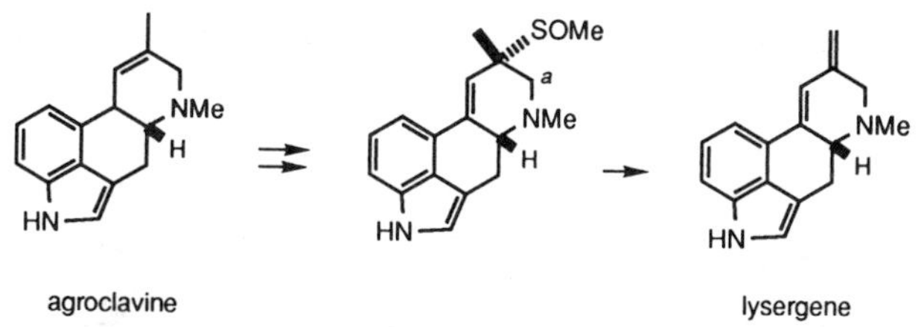

agroclavine   lysergene

2.7. REACTIONS OF EPOXIDES AND AZIRIDINES

Unsymmetrically substituted epoxides react with various donor reagents with definite regioselectivity [R.E. Parker, 1959; Sharpless, 1983] no matter what the reaction mechanisms are (either direct displacement or elimination–Michael addition sequence).

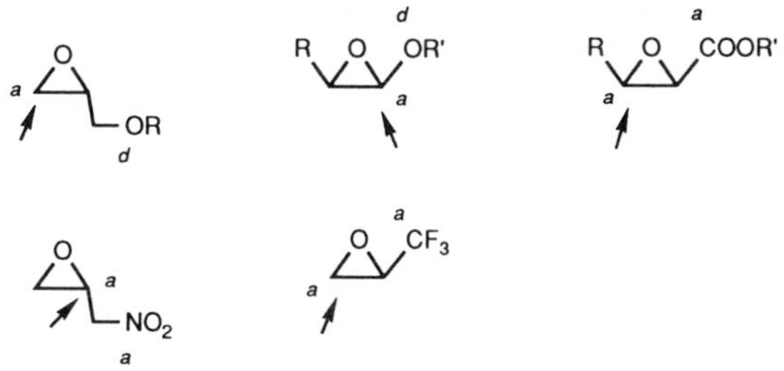

Formolytic cleavage of both a cyclopropane ring and an epoxide constituted a key step toward a synthesis of prostaglandins [White, 1976]. The epoxide ring is vicinally substituted with a cyano group and a cyclopropane and both of these substituents favor heterolysis in the observed direction.

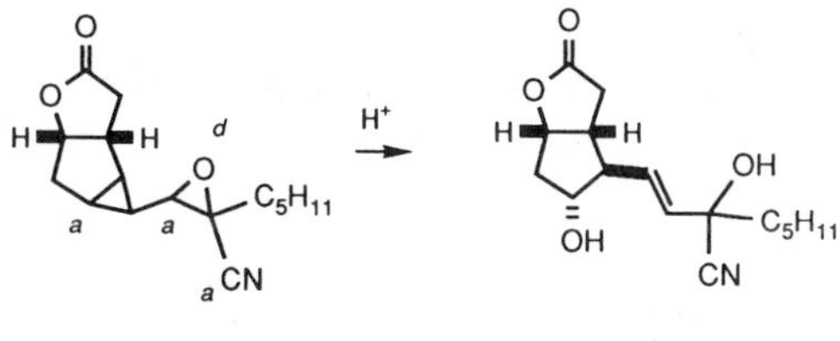

Both 5-exo-tri and 6-endo-tri modes of cyclization are favorable [J.E. Baldwin, 1976]. However, only a δ-lactone was formed by the intramolecular attack of an α-sulfinyl epoxide [T. Satoh, 1987] with an oxide ion. The reason for this predilection seems to be that the alternative opening would involve reaction at a captodative carbon atom with a diminished acceptor character. On the other hand, the benzylic carbon is a good acceptor.

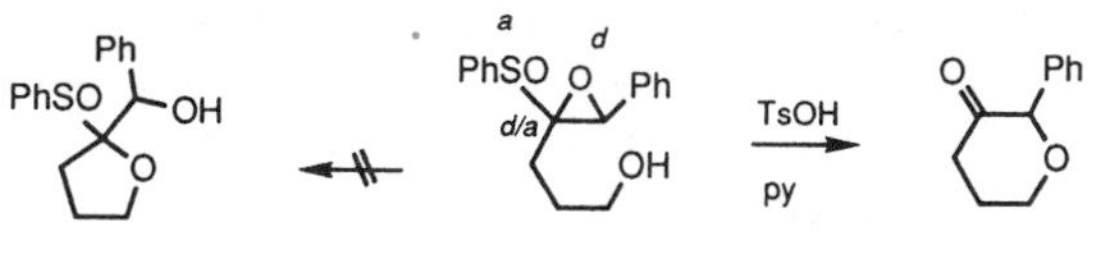

Rather interestingly, the direction of epoxide opening by an internal amide was influenced to a certain degree by the nature of an in-chain nitrogen atom [Evans, 1985].

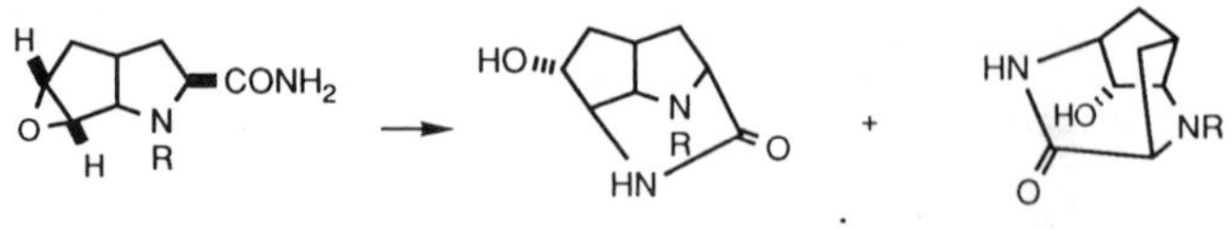

R = Cbz	50	50
R = Me	95	5

During a synthesis of zoapatanol [R. Chen, 1980] in which intramolecular epoxide opening was a decisive step to create the oxepane ring, it was noted that the more convergent approach involving an α-tosyloxymethyl epoxide led to the unwanted tetrahydropyran derivative. The results are consistent with the possible influence of the donor (OTs) group on the acceptor site.

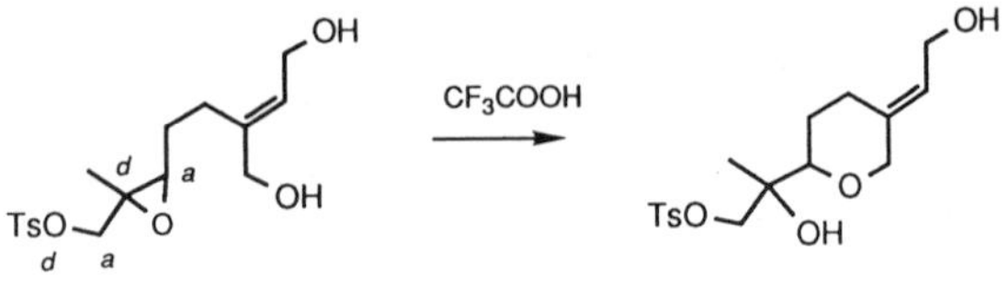

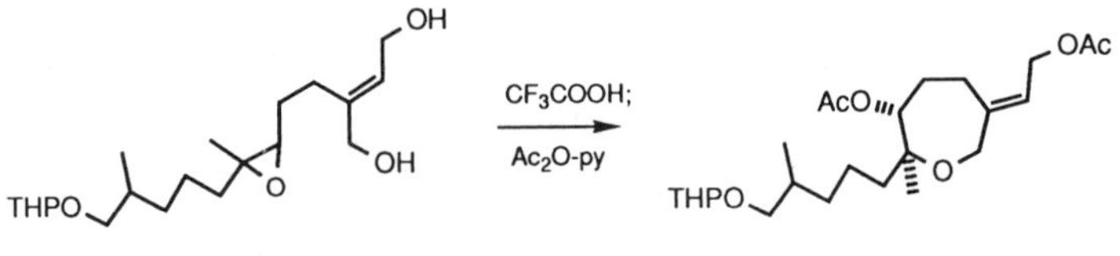

The same rationale can be advanced concerning the regioselective epoxide formation from an anhydroglucose ditosylate [T. Ogawa, 1978].

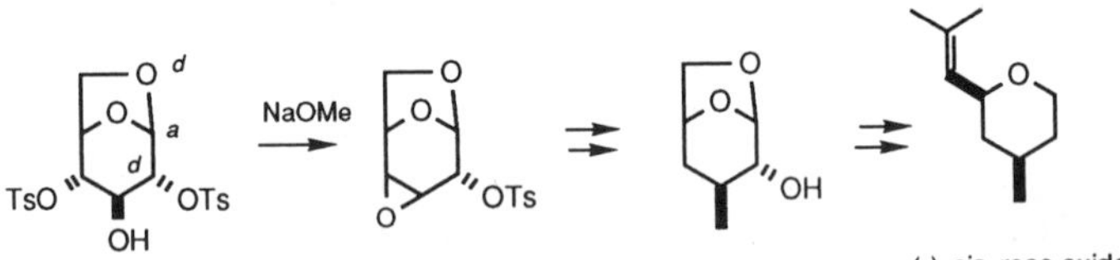

(-)-*cis*-rose oxide

A vinyl epoxide is ionized on contact with palladium(0) reagent to generate a π-allylpalladium complex which may lead to a cyclic product when an appropriate donor is present in the same molecule [Trost, 1989a]. The new bond is formed preferentially at the distal end of the allylic system, apparently due to the influence of the hydroxy group which dictates such a position to be the donor site. The distal carbon is two more atoms apart, and it would experience a lesser polar effect.

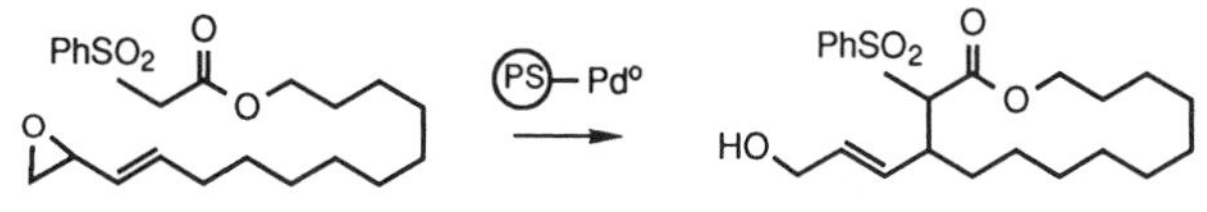

The major attacking site of diethylamine on glycidic acids can be diverted by the addition of tetraisopropyl titanate [Chong, 1985]. Actually, the catalyst ensures the normal reactivity pattern by forming a five-membered chelate, while in its absence an ion pair between the two reactants brings the donor closer to the α-carbon of the epoxycarboxylic acid.

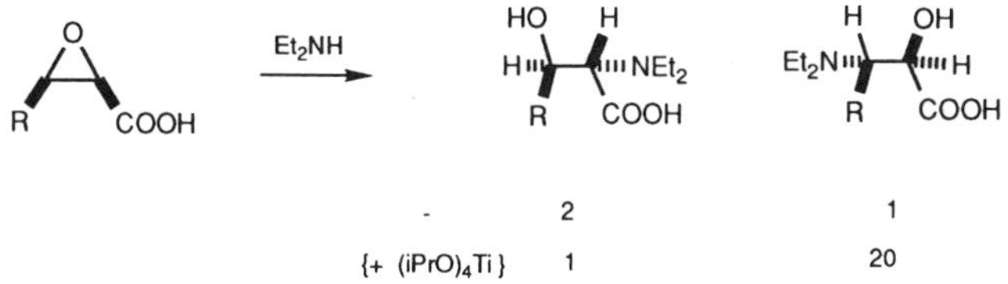

α-Lactones may be regarded as electronically extremely biased epoxides in their reactions with nucleophiles. The *in situ* alcoholysis of α,α-dimethyl-α-lactone [W. Adam, 1972] and α,α-bis(trifluoromethyl)-α-lactone [W. Adam, 1973] led to an α-alkoxy carboxylic acid and an α-hydroxy carboxylic ester, respectively. These findings point to the effect of the fluorine atoms that accentuate the donor properties of the α-carbon.

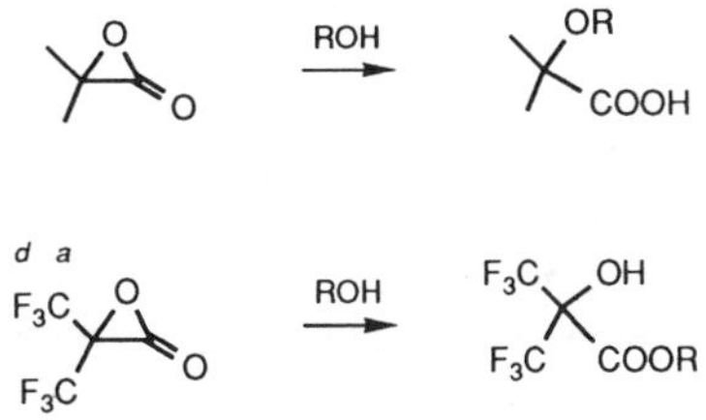

Aziridines behave in an analogous manner to epoxides with respect to their reactions with donors. One example [Furuya, 1988] should be sufficient to illustrate the regioselectivity.

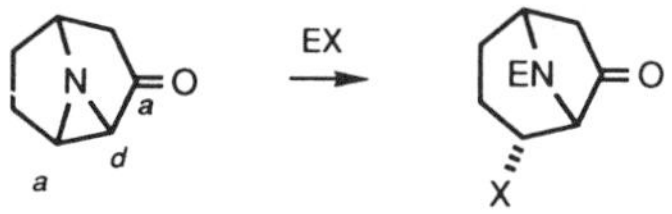

2.8. DEVIATIONS FROM THE POLARITY ALTERNATION RULE

This is a good place to indicate certain limitations of the empirical polarity alternation rule to rationalize or predict the outcome of a reaction. In general, breakdown of the polarity alternation assessment stems from neighboring group participations, chelation effects, steric constraints, and strong through-space interactions.

Neighboring group participation is evident in the hydration of 2-cyclohexenol via an oxymercuration–demercuration protocol in the presence of chloral [Overman, 1974]. The donor that attacks the mercuronium ion intermediate is the alcoholate of chloral.

It has been discovered that hydrogen bonding in the transition state of 1,3-dipolar cycloaddition involving nitrile oxides determines the stereo- and regioselectivities [Curran, 1990]. Reversal of the usual reaction pattern can be achieved.

R = H, R' = Ph	90	10
R = tBu, R' = Me	19	76

The very active research of chelation effects in carbanion chemistry [Beak, 1982] has produced many useful synthetic methods for regioselective alkylations and acylations. Substrates that provide good ligand(s) for metal direct the deprotonation at a site which is maximally benefited by chelation. The facility in the formation of o-lithiobenzamide derivatives has been noted.

Situations are sometimes confusing. For example, lithiation of N-phenylsulfonylindole with t-butyllithium at −100°C occurred at C-3 [Saulnier, 1982]. However, the 2-lithio isomer which may form an intramolecular chelate with the sulfonyl oxygen is favored at higher temperatures.

A contrary case is the isomerization of (E)-β-lithio-β-(1-pyrrolidinyl)acrylonitrile to give the α-lithio derivative [Schmidt, 1971] with restoration of a polarity alternation sequence.

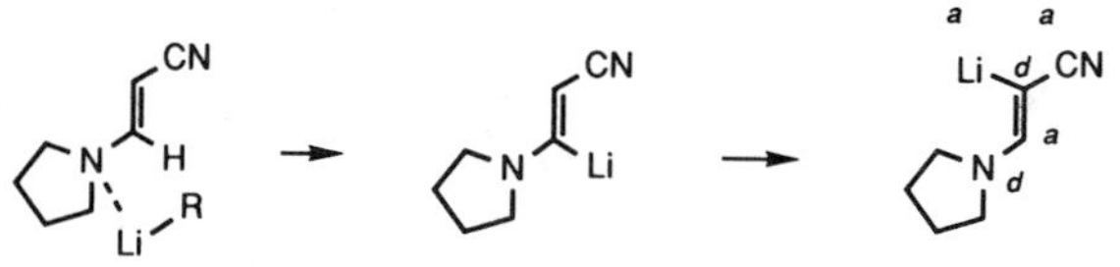

The kinetic lithiation product of (E)-4-tosylbut-3-en-2-one dimethylacetal is stable at $-20°C$ [Najera, 1987]. Here chelation works cooperatively with polarity alternation.

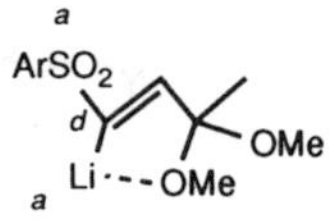

A form of steric constraint has been observed in acetal-initiated cyclization of vinylsilanes [Overman, 1986]. An (E)-isomer giving an alkylidenetetrahydrofuran is both electronically and sterically favored. However, a tetrahydropyran product has been obtained from the (Z)-vinylsilane. Apparently the chair-like transition state for the (Z)-vinylsilane that disposes an equatorial alkyl group is much preferred.

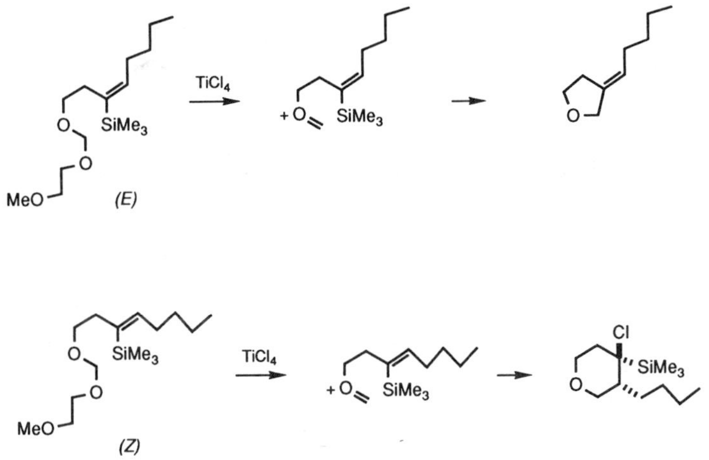

Steric constraints which subvert the matching of polar substituents in intramolecular Diels–Alder reactions are known. Thus syntheses of chelidonine [Oppolzer, 1983b], eupolauramine [Levine, 1983], and oxogambirtannine [Martin, 1989a] involved electronically disfavored cycloadditions. At least in the chelidonine synthesis it has been shown that an intramolecular version of the Diels–Alder cycloaddition afforded predominantly the undesired but predicted regioisomer(s).

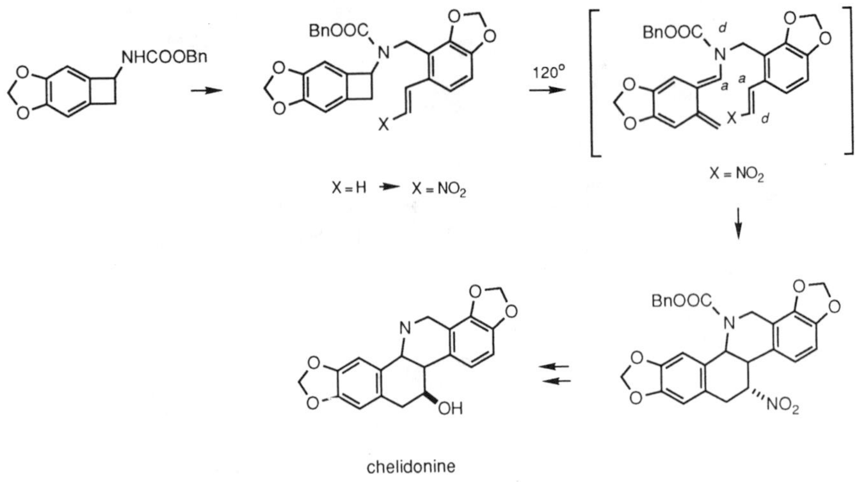

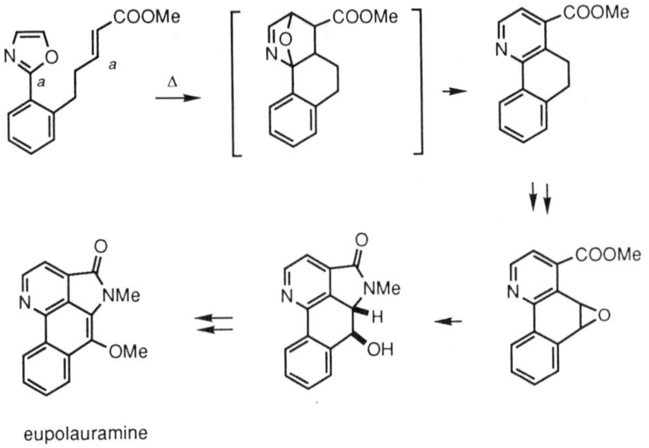

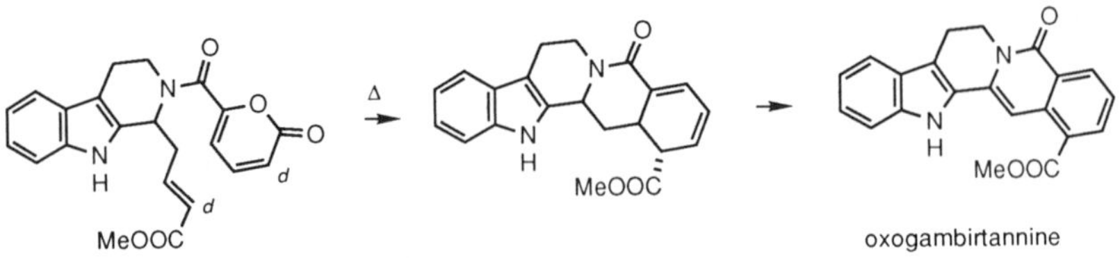

chelidonine

eupolauramine

oxogambirtannine

For intermolecular Diels–Alder reactions unfavorable spatial interactions of substituents can also lead to anomalous results (see Chap. 10).

3

SYNTHESIS OF 1,2-DIFUNCTIONAL COMPOUNDS

3.1. WITTIG, PETERSON, JULIA AND RELATED REACTIONS

Alkenes and alkynes may be considered as 1,2-difunctional compounds, in view of their π-bond reactivity. Methods for their formation include isomerization, elimination, alkylidenation of carbonyl substances, Wittig and related reactions, and metal-promoted coupling.

Generally speaking, alkylidenation is aldol condensation (see Chap. 4) with a dehydrative aftermath, and retention of the carbonyl group. The Wittig reaction [Maercker, 1965] is a truly versatile olefination method that regio-selectively extends a carbonyl group into a $C=C$ bond by coupling with a phosphonium ylide (alkylidenephosphorane) under mild conditions. The α-phosphoryl-stabilized carbanion version [Wadsworth, 1977] is complementary.

A wide range of phosphonium ylides and carbonyl substrates have been employed in the Wittig reaction. Reaction conditions vary according to the nature of the reactants. For the preparation of typical alkylidenephosphoranes, the precursors are derived from triphenylphosphine and alkyl halides. Organo-lithiums, sodium methylsulfinylmethide (NaH/Me_2SO), and alkali metal alkoxides may then be used to deprotonate the phosphonium salts.* Much milder bases are sufficient in the cases that the carbanionic centers are stabilized by an acceptor substituent (polarity alternation stabilization). Such ylides (e.g., $Ph_3P=CHCOOR$) are quite stable, and their nucleophilicity is diminished

---

* A few other ways of generating the phosphoranes include alkyl transfer (Michael reaction) from a cuprate reagent to vinyltriphenylphosphonium bromide, ring opening of a 1-carboalkoxycyclo-propyltriphenylphosphonium salt with a C-nucleophile.

such that they are reactive toward aldehydes but not to ketones. A remedy for these situations is to employ the phosphoryl-stabilized carbanions. It should be noted that ylides do not have net charges.

Because of thermodynamic control, (E)-alkenes are formed predominantly from reactions involving stabilized ylides and the phosphoryl-stabilized carbanions, as a result of syn-elimination after betaine equilibration. On the other hand, reactions of nonstabilized ylides can be highly (Z)-stereoselective under salt-free conditions and in polar aprotic solvents at low temperatures.

Concerning the mechanism of the Wittig reaction there has been a controversy as to whether a betaine is formed prior to the appearance of the oxaphosphetane intermediate. Some evidence seems to indicate operation of a [2 + 2]cyclo-addition pathway [Vedejs, 1990]. (Note the difference between the Wittig reaction proper and the stabilized ylide version and modifications.)

The value of the Wittig–Horner reaction with diphenylphosphinoyl-stabilized carbanions lies in the stability of the adducts. This properly implies that their isolation and controlled decomposition (with NaH in DMF) would ensure configurational purity of the alkenes. Since this reaction is erythro-selective, (Z)-alkenes are available. The preparation of the (E)-isomer precursors can be achieved via acylation followed by the threo-selective reduction with $NaBH_4$.

A most useful modification of the α,β-unsaturated ester synthesis based on stabilized carbanions is by changing the alkoxy groups of the phosphonate to trifluoroethyl [Still, 1983]. Reversal of the stereoselectivity from (E) to (Z) may reflect an enhanced stability of the adducts and/or facilitation of the elimination step in the presence of minimally complexing counterions.

The synthetic applications of the Wittig reaction are innumerable. The intramolecular version of this reaction has gained much popularity because of its effectiveness in closure of very strained ring systems (e.g., bicyclo[3.2.1]oct-1-en-3-one as a transient species [Bestmann, 1982]) as well as macrocycles (e.g., in synthesis of amphoteronolide B [Nicolaou, 1988]).

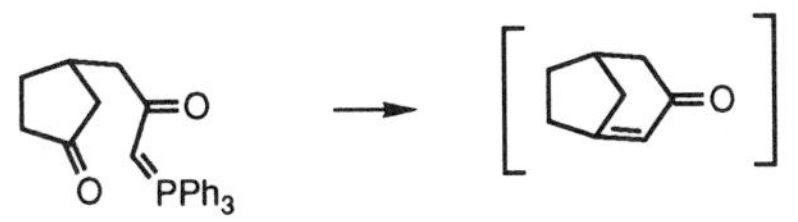

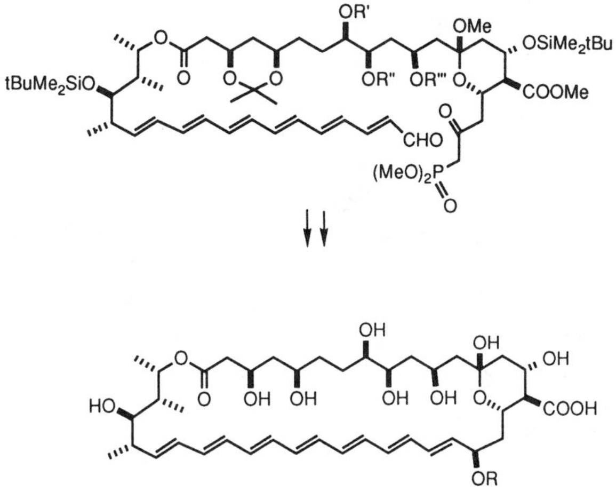

(R = H)   amphoteronolide-B

The Wittig reaction even finds occasional use in the access of the corresponding saturated compounds. In compounds such as ramulosin the carbocycle may be constructed via a tandem Michael–Wittig reaction sequence between the ylide anion of an ethyl acetoacetate γ-phosphonium salt and an unsaturated aldehyde, followed by reduction [Pietrusiewicz, 1988]. The critical requirement is that the Michael addition step must precede the Wittig reaction.

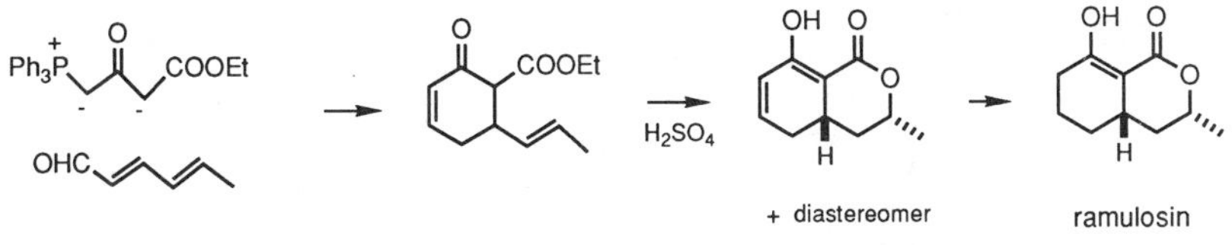

+ diastereomer   ramulosin

Certain esters also undergo Wittig reaction to afford enol ethers. An example is found in a route to trichodiene [Suda, 1982]. However, a popular reagent for methylene transfer to a lactone carbonyl is the Tebbe reagent, $Cp_2TiCl(CH_2)AlMe_2$ [Tebbe, 1978].

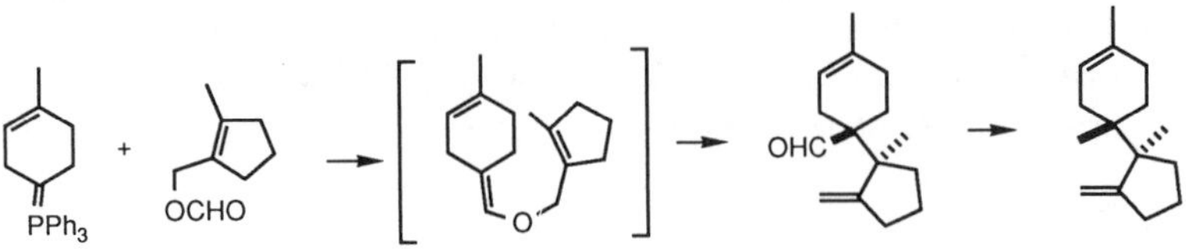

trichodiene

The silicon-based analog of the Wittig reaction is the Peterson reaction [Ager, 1984]. The simplest cases are those involving (trimethylsilyl)methyl-magnesium chloride as the reagent. The β-hydroxy silane intermediates decompose via *syn*-elimination in the presence of a base.

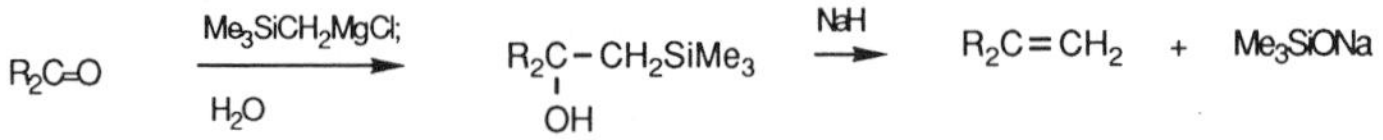

The olefin-forming step is stereoselective. Thus, the *erthro*-hydroxysilane gives a (Z)-alkene whereas the *threo*-hydroxysilane gives the (E)-alkene predominantly on treatment with a base such as NaH or KH. A reverse stereoselectivity is generally observed upon decomposing the hydroxysilane under acidic conditions (H_2SO_4, $BF_3.Et_2O$) [Hudrlik, 1975]. However, as the first step of the Peterson reaction is under kinetic control and irreversible, the stereoselective advantage of the elimination step cannot be taken directly in the synthesis of alkenes. The key intermediates can be obtained from the reaction of α,β-epoxysilanes with organocuprate reagents.

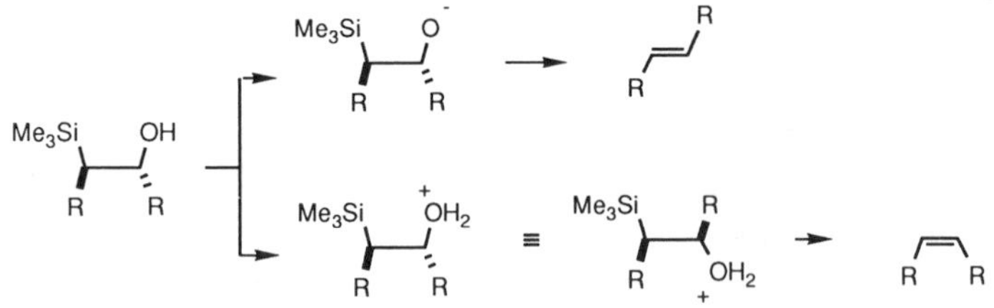

An application of the Peterson reaction is the ethylidenation of a sidechain during a synthesis of vallesiachotamine [Spitzner, 1984]. The corresponding Wittig reaction would not work.

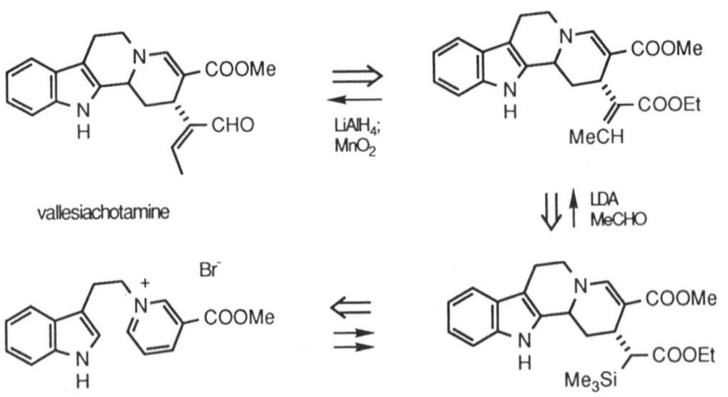

The Julia olefination involves alkylation of a sulfone ketone with subsequent elimination of the sulfinic acid. Usually an (*E*)-alkene is produced. A variant is the condensation of the sulfone with a carbonyl compound and reductive removal of the element of a sulfonic acid (e.g., via the siloxy derivative) from the adduct.

The Julia transform on vitamin D defines (+)-8α-phenylsulfonyl-des-AB-cholestane as the donor moiety. A retrosynthetic pathway evolved from this consideration is totally different from a more conventional Wittig-type approach, and it has inspired a novel access to the sulfone [Nemoto, 1985a].

Thus, logical disconnection of the bicyclic sulfone led to a well-known hydrindenedione. The synthetic maneuver then entails cleavage of the six-membered ring and reconstruction of another one with one of the liberated carbon chains. This has the main purpose of creating stereoselectively the aliphatic portion of the target molecule.

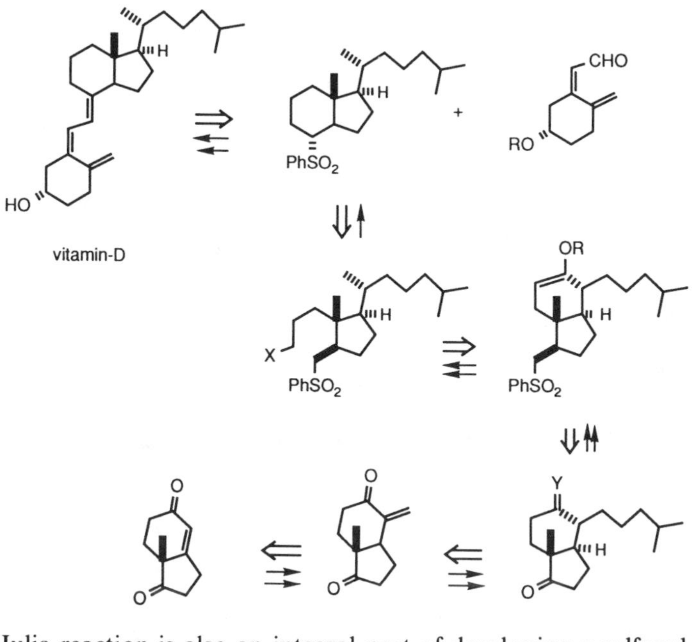

The Julia reaction is also an integral part of developing α-sulfonyl ketones into a RCH=CH⁻ synthon. The *gem*-acceptor substituents facilitate the alkylation step and they also enable fixation of the double bond. A synthesis of pleraplysillin I [Masaki, 1984] succinctly illustrates this strategy of alkene formation.

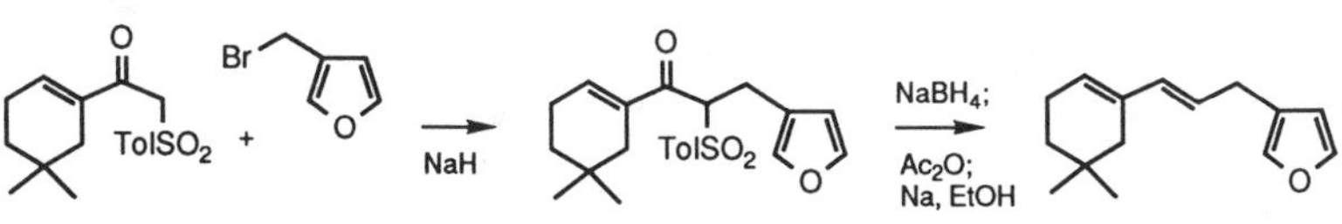

3.2. McMURRY REACTION

One of the most important C—C bond-forming reactions that appeared in recent years is the reductive coupling of two carbonyl groups by low-valent titanium species to afford either alkenes or *vic*-diols [Lenoir, 1989]. Intramolecular coupling of an ω-keto ester leads to a cyclic enol ether and thence a cycloalkanone. Because of the apparent insensitivity of this coupling to ring size, it constitutes one of the best methods for the synthesis of mesocycles and macrocycles. There is ample evidence for a free-radical mechanism, and the reaction likely occurs on the metal surface. Consequently the similarity of this coupling to the acyloin condensation is apparent.

Many different methods for producing the low-valent titanium reagents have been developed. These include $TiCl_3 + Zn/Cu$, $TiCl_3 + Bu_2AlH$, $TiCl_3 + Li$ or K or Mg, $TiCl_3$ or $TiCl_4 + LiAlH_4$, $TiCl_4 + Zn$. Some of these reactive titanium species may be titanium (0), others may be titanium (I) or titanium (II).

Among the numerous applications of this coupling in synthesis, it is sufficient to single out the synthesis of humulene [McMurry, 1982], isocaryophyllene [McMurry, 1983], and gibberellic acid [Corey, 1978a,b].

humulene

isocaryophyllene

gibberellic acid

Let us examine a specific disconnection of a diene system leading to the identification of a McMurry transform for the hexalin portion of compactin [Anderson, 1983]. The disconnection at the B-ring is by far the most advantageous since it reveals a β-hydroxy ketone which can be readily assembled by a kinetic aldol condensation. The substituents of the cyclohexenone would determine the relative configuration of the new stereocenter at the α'-position.

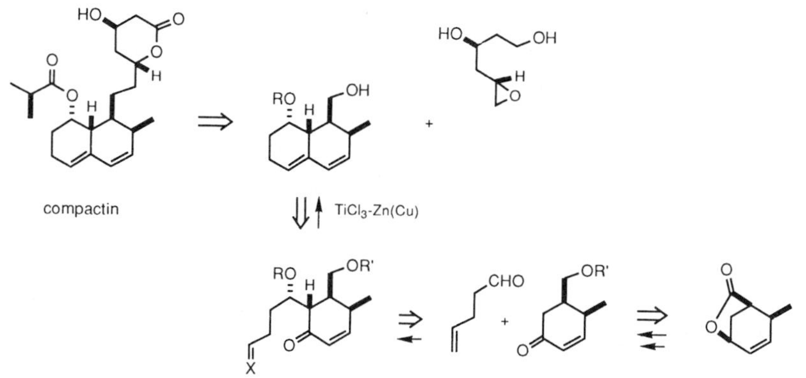

compactin

TiCl$_3$-Zn(Cu)

The aldehyde acceptor for the aldolization must be protected at one end, and 4-pentenal proved to be an adequate surrogate because the vinyl group could be selectively ozonized in the presence of the conjugated double bond.

The development of a simplified synthetic scheme for khusimone [Wu, 1988] depended on the previous work concerning the conversion of an endocyclic double-bond isomer, isokhusimone, back to the norsesquiterpene. A McMurry transform of isokhusimone gives rise to a spirocyclic triketone. This is a valid precursor because the reductive coupling is expected to be uncomplicated due to its intramolecular nature to form a cyclohexene (entropic facilitation) and the local symmetry of the cyclopentanedione segment which always provides a sterically accessible carbonyl (*syn*) for the reaction.

Spiroannulation of a cyclopentanone ketal with 1,2-bit(trimethylsiloxy)cylobutene in the presence of BF$_3$ etherate is an excellent method for securing the key intermediate. The cyclopentanone ancestry is itself easily traced to nor-camphor.

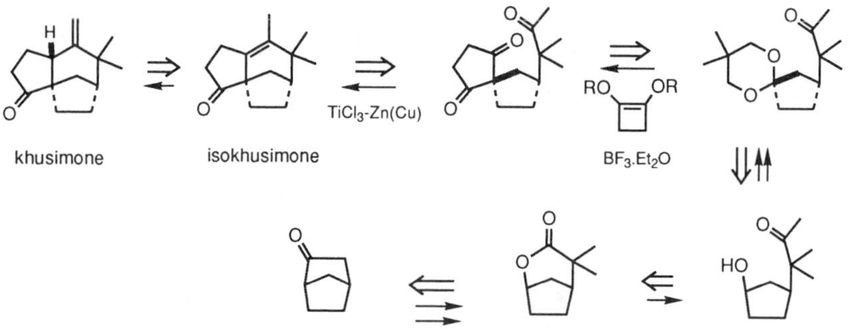

khusimone   isokhusimone

TiCl$_3$-Zn(Cu)

BF$_3$.Et$_2$O

3.3. ACYLOIN CONDENSATION AND OTHER REDUCTIVE COUPLINGS

All the 1,2-donor disubstituted compounds are disjoint and the most general methods for their synthesis involve redox reactions.

Until recently this type of C—C bond-forming reaction was difficult to control, and accordingly, its use in the synthesis of complex molecules was not extensive. The exception may be the acyloin condensation [Bloomfield, 1976].

The acyloin condensation couples two ester groups to form an α-ketol via the radical anions. The most attractive feature of this process is that, for the intramolecular version, there is hardly any ring-size effect. In other words, cyclic structures from four-membered and above may be conveniently prepared.

The best reaction conditions consist of a fine dispersion of metallic sodium and chlorotrimethylsilane in an aromatic hydrocarbon (toluene, xylene). The silane is a trapping agent for the acyloin. Not only does the disilylation stabilize the product, the chlorosilane also suppresses unwanted reactions, such as Claisen and Dieckmann condensations, by scavenging all the alkoxide species. This is an example of hard–soft reactivity discrimination [Ho, 1977]; the soft reagent (Na) is not appreciably attacked by the hard scavenger (Me$_3$SiCl).

2-Hydroxy-3-methyl-2-cyclopentenone is a valuable flavoring material, therefore its synthesis has attracteed considerable effort [Strunz, 1983]. The α-diketone system is a partial acyloin retron, and an α-methyleneglutaric diester is easily identified as a precursor.

According to this general scheme, an oxidation is required after the acyloin condensation to reach the target molecule. It is also noted that the glutaric diester is a *reduced* dimer of an acrylic ester. It seems that two unproductive steps may be saved if an acrylic ester could afford the α-methyleneglutaric diester and the latter cyclize without affecting the conjugated double bond.

A solution to this problem [Cookson, 1979] consists of conducting dimerization of methyl acrylate in the presence of a trialkylphosphine [Nagato, 1973]. The phosphonium zwitterion derived from addition of the phosphine to the acrylic ester starts a Michael reaction with a second molecule of the acrylic ester, and the resulting diester eliminates the phosphine (retro-Michael fission) to renew the reaction cycle. Protection of the unsaturation with a secondary amine against reduction is effective, and the amine is thereafter eliminated (and possibly recovered). Thus, the only chemicals consumed are the acrylic ester, sodium, and chlorotrimethylsilane.

Creation of the tropolone ring precursor in a synthesis of colchicine [van Tamelen, 1959, 1961] was accomplished by an intramolecular acyloin condensation of a δ-lactone δ-acetic ester in liquid ammonia. Such a substrate and therefore the method of ring closure was predicated on the chain attachment method of cyanoethylation at the benzylic position of a benzosuberone and the expediency of adding an acetic acid chain to the activating ketone by the Reformatskii reaction.

It is interesting to compare this approach with a synthesis [Woodward, 1963–4] which involved a Dieckmann cyclization. In this latter approach a four-carbon chain was established upon cyclization of a dienoic ester by an intramolecular Friedel–Crafts alkylation. The pro-benzylic carbon is an acceptor site, whereas the benzylic carbon of the benzosuberone in the former synthesis is a donor.

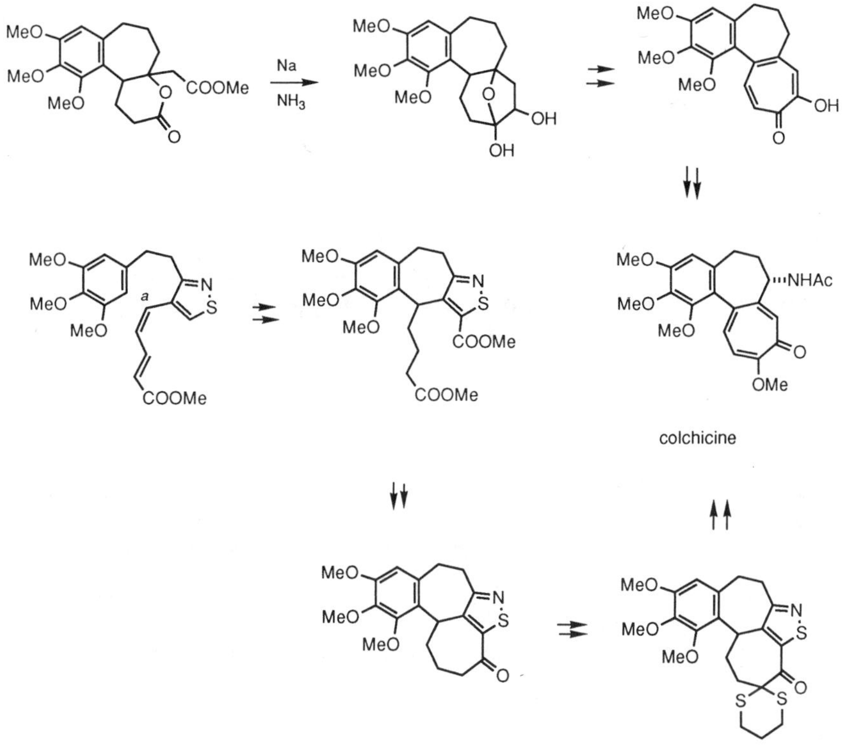

colchicine

Cephalotaxinone, a minor alkaloid co-ocurring with cephalotaxine as well as being a synthetic precursor of the latter, is an α-diketone monomethyl ether. Its synthetic scheme can be simplified to a union of an arylethyl group and an azaspiro[4.4]nonene derivative with the crucial cyclization step toward the aromatic ring [Semmelhack, 1972]. It was deemed feasible to perform a nucleophilic displacement of a suitable leaving group by the ketone enolate. (The most efficient method for this transformation turned out to be a photoinitiated reaction.)

The azaspirocyclic diketone intermediate possesses a plane of symmetry and it was readily prepared by an acyloin condensation and mild oxidation.

cephalotaxinone

Samarium(II) iodide, generated from the metal and 1,2-diiodoethane in an inert atmosphere, can be used to achieve reductive coupling of acid chlorides with aldehydes and ketones to give α-hydroxy ketones [Souppe, 1984]. It is believed that reduction of the acid chloride (in two stages) to an acyl anion, which reacts with the carbonyl compound immediately, is involved.

An intramolecular version of this reaction succeeded in forming an α-hydroxy cyclopentanone from a δ-keto nitrile [Kraus, 1989], when reagents such as Zn/Me_3SiCl [Corey, 1983] and Mg/Me_3SiCl [Hutchinson, 1985] failed.

3.4. OXIDATION OF ALKENES AND ALKYNES

1,2-Dioxy compounds are readily obtained by oxidation of alkenes and alkynes. One of the oldest reagents is potassium permanganate, and its oxidizing power is strongly dependent on reaction conditions such as pH and solvent. For the conversion of alkenes to α-diketones, acetic anhydride is a useful medium, and there is a synthesis of (−)-gloeosporone [S.L. Schreiber, 1988a] relying on this reaction to introduce the diketone function. (This functionalization step was achieved on the corresponding acyclic alkyne by $RuO_2/NaIO_4$ in a synthesis of the enantiomeric (+)-gloeosporone [G. Adam, 1987]).

R = SiMe₂tBu

(-)-gloeosporone

Palladium(II)-mediated 1,2-difunctionalization of alkenes may lead to *vic*-amino alcohols and diamines [Bäckvall, 1978]. The step of C—Pd bond oxidation is contrapolarizing and thereby creates a carbon acceptor.

Several methods are available for indirect oxygenation of an α-carbon of carbonyl compounds. Bisalkylsulfenylation followed by cleavage of the dithio-ketal group is now a routine protocol. Generally, the carbonyl compound is activated in the form of an α-hydroxymethylene derivative, and reacted with propylenedithiol ditosylate, as shown in a synthesis of colchicine [Woodward, 1963–4].

It should be noted that the sulfenylation, like an ordinary alkylation step, exploits the α-carbon as a donor (and the sulfur atom as acceptor), but in the hydrolysis its role is changed transiently into an acceptor. This is consistent with all other redox transformations in which polarity inversion is involved.

3.5. OXIDATIVE MANEUVER OF CARBONYL SUBSTRATES

Enolizable carbonyl compounds may be hydroxylated with $MoO_5.HMPT.py$ [Vedejs, 1974]. Alternatively, treatment of their enol silyl ethers with a percarboxylic acid [Rubottom, 1974] or lead(IV) benzoate [Rubottom, 1976] leads to α-hydroxy ketones and α-benzyloxy ketones, respectively. The possibility of generating both a kinetic and a thermodynamic enol silyl ether from an unsymmetrical ketone makes the latter oxidation pathway particularly useful.

α-Halogenation of carbonyl compounds involves contrapolarization, there-fore it can be considered as an oxidation reaction. Displacement of the halogen by a donor furnishes a disjoint product.

This well-known manipulation has been included in many synthetic opera-tions. For example, it is the key process in a construction of the bicyclic core of anatoxin-a [Wiseman, 1986]. The notable change is the skeletal desym-metrization accompanying the disruption of functional group conjointment.

anatoxin-a

Promising intermediates for the synthesis of securinine and allosecurinine [Coté, 1989] have also been obtained by the intramolecular *N*-alkylation protocol.

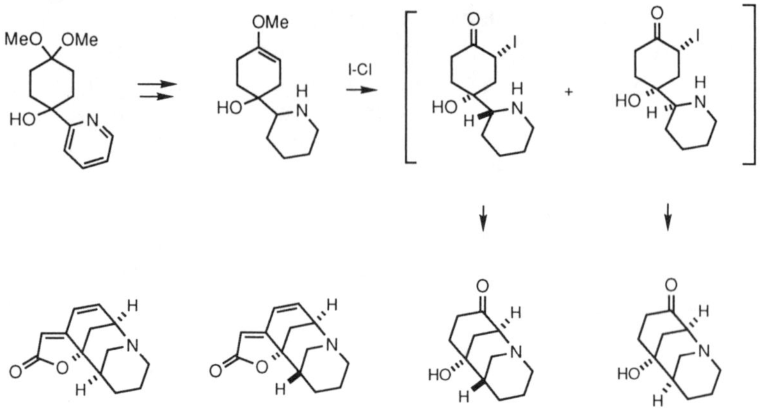

allosecurinine   securinine

In the late stages of a total synthesis of quinine [Woodward, 1945], quinotoxine was assembled via a Claisen condensation of ethyl 6-methoxy-quinoline-4-carboxylate and N-benzoyl-homomeroquinene, and resolved. Ring closure was brought about with alkaline sodium hypobromite.

This last reaction may have involved N-bromination and 1,6-dehydro-bromination. Umpolung took place at the nitrogen atom.

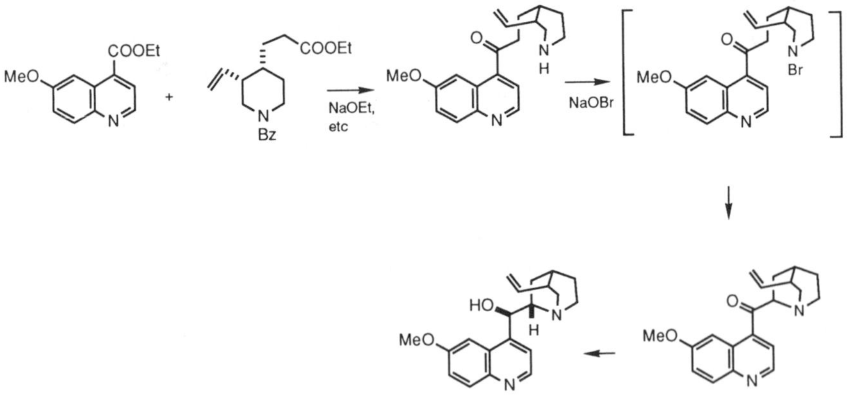

quinine

With the development of a versatile method for alkaloid synthesis via the tandem aza-Cope rearrangement–Mannich cyclization [Overman, 1991] the access of (protected) α-amino ketones is of paramount importance. Generally, these substrates may be obtained from N-acyl enamines by the hydroboration–oxidation sequence, as described in the context of a 16-methoxytabersonine synthesis [Overman, 1983].

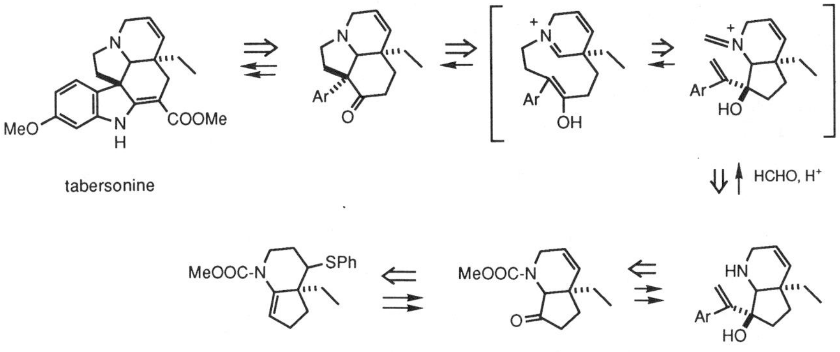

tabersonine

As noted, the disjoint nature of the α-amino ketones requires umpolung manipulations in their preparation. Accordingly, an oxidation step is needed in the methods described in the following equations [Hiemstra, 1983; M.J. Fischer, 1990].

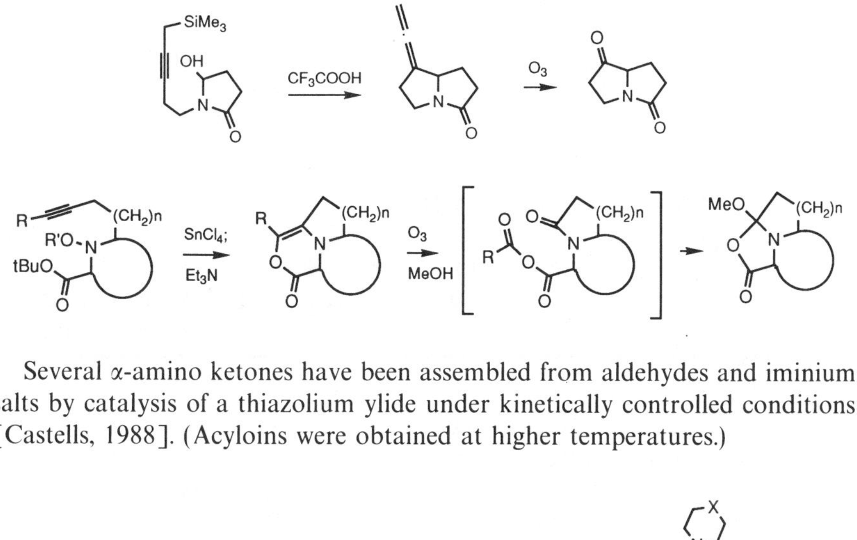

Several α-amino ketones have been assembled from aldehydes and iminium salts by catalysis of a thiazolium ylide under kinetically controlled conditions [Castells, 1988]. (Acyloins were obtained at higher temperatures.)

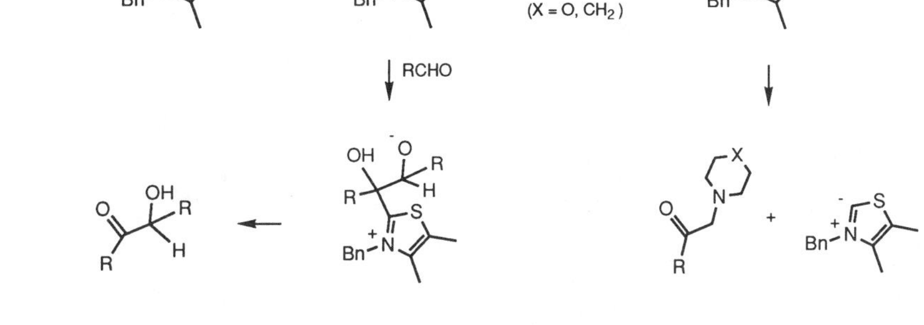

The donor of this Mannich-type condensation is a stabilized α-hydroxy carbanion, consequently, umpolung is involved.

1,2-Diamines are difficult to obtain. Thus, in contrast to the facile synthesis of *Strychnos*-type indolines by an intramolecular Mannich reaction, synthesis of 2,2,3-trialkylindolines such as vallesamidine poses special problems because the two nitrogen atoms are disjoint. In other words, an internal bond formation must feature a contrapolarization–umpolung strategy.

A solution to this problem is by an oxidative ring closure [Dickman, 1989].

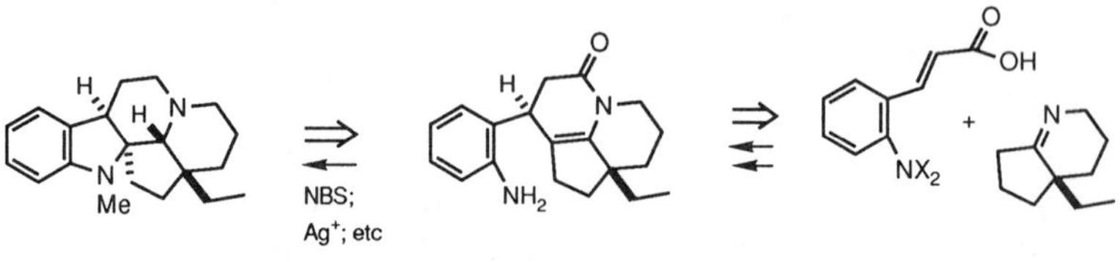

vallesamidine

Reductive coupling of imines with carbonyl compounds is promoted by niobium(III) chloride [Roskamp, 1987]. This method should find uses in synthesis.

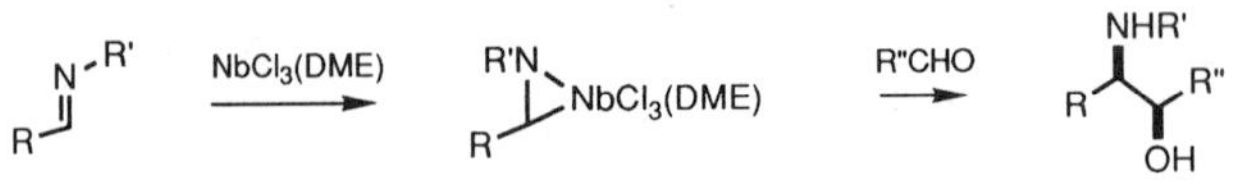

3.6. REARRANGEMENT METHODS

The rearrangement route to 1,2-diamines takes advantage of the facile preparation of β-amino carbonyl compounds by various methods. The easily controlled rearrangement step (e.g., Beckmann and Schmidt rearrangements) converts a conjoint molecule into a disjoint species.

An inspiring approach to biotin [Confalone, 1980] employed an intramolecular 1,3-dipolar cycloaddition to set up a 1,3-N,O segment which on redox and protection maneuver led to a bicyclic keto carbamate. Beckmann rearrangement in this case was due to a competing fragmentation process originating from sulfur atom participation.

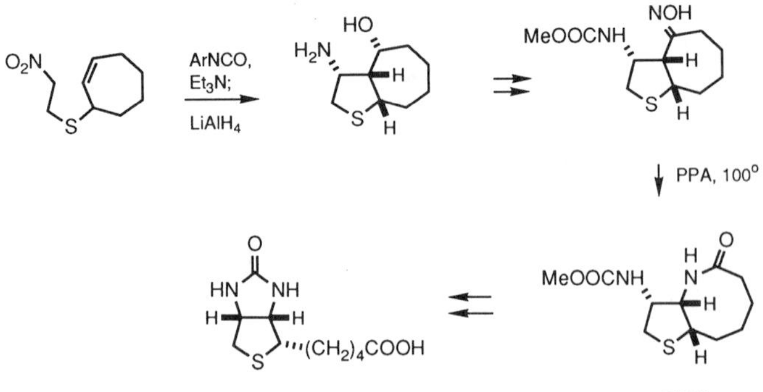

biotin

The monumental vitamin B_{12} synthesis [Woodward, 1968] consisted of elaborating the AD-ring portion from the corrnorsterones. (The name is a pun suggesting a skeleton to a B-norsteroid as a cornerstone for a corrin.) This is a supreme example that demonstrates stereocontrol of peripheral substituents as they are being unfolded from a polycyclic predecessor.

A highly abbreviated retrosynthetic logic is delineated here. The prominent feature of the synthetic pathway is the reconnection of donor and acceptor sites that are 1,5- or 1,6-related.

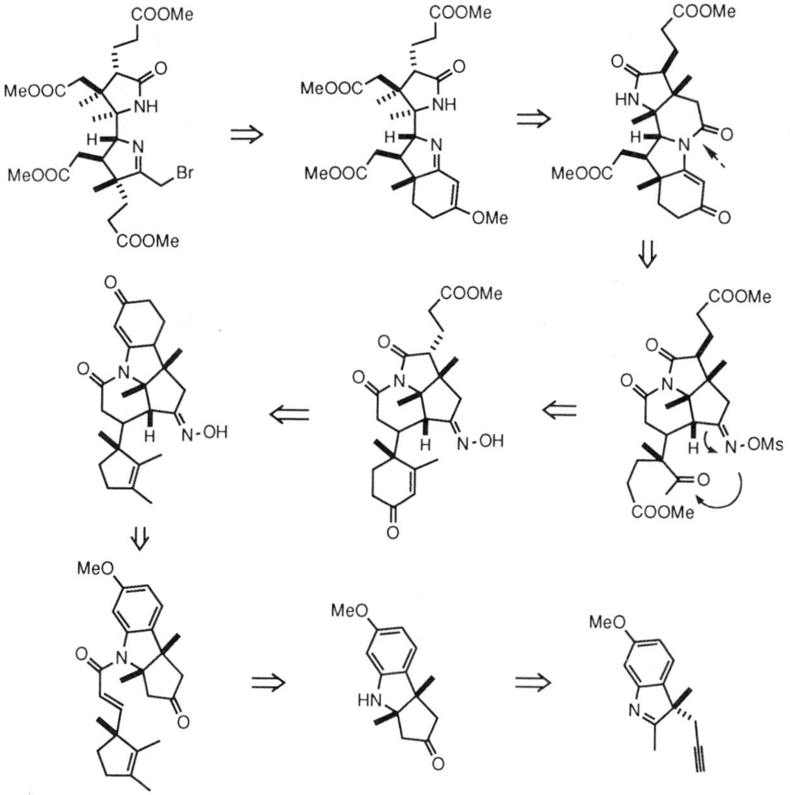

For the top moiety of the A-ring, the five-carbon subunit ranging from the lactam carbonyl to the propionyl chain, was fashioned from an aromatic ring which contains a methoxy group. Oxidative cleavage of the enone derived from a Birch reduction delineated the structural array.

The vicinal *cis*-dimethyl group in the A-ring and the terminus of the acetic sidechain which is four carbons away from the B-ring nitrogen atom indicates an ideal starting point at a symmetrical 2,3-dimethyl-2,3-(oxocyclopentano)-indoline. The 5:5 ring fusion ensures a *cis* stereochemistry, and a Beckmann rearrangement would install both the nitrogen atom and the carbon chain in one operation.

Since the acetic chain terminus of the B-ring is conveniently separated from the A-ring nitrogen by four atoms, C—C bond formation between the inner set of the α and β atoms of the pyrroline system (B-ring) in a precursor may be advantageously achieved through amidation of the indoline and an internal Michael addition from the proximal cyclopentanone. The desymmetrization of the ketone group ensured oximation and hence the Beckmann rearrangement in the desired regiochemical sense, when the B-ring nitrogen atom was to be inserted later.

The Michael addition created a new stereogenic center. The all-*cis* stereochemistry, demanded by strain minimization, also had the effect of placing the large group (which contained elements for the rest of the B-ring) in an *exo* orientation. This orientation was desired.

vic-Aminohydroxylation of a 1,3-diene system consisting of an intramolecular hetero-Diels–Alder reaction with *N*-sulfinylcarbamate as dienophile has been reported [Remiszewski, 1985]. Cleavage of the product with a Grignard reagent followed by thermal rearrangement of the resulting allylic sulfoxide accomplished the process.

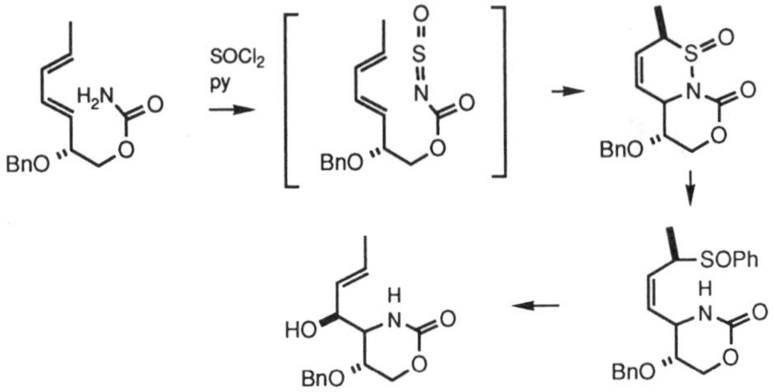

3.7. MISCELLANEOUS UMPOLUNG TECHNIQUES

Inherited umpolung reactivity or disjoint feature from a suitable substrate facilitates the assembly of compounds such as *vic*-diamines. Thus, in a synthesis of vinoxine [Bosch, 1984] the nucleophilicity of the indole α-position works hand in hand with the steric constraint of an intramolecular cyclization.

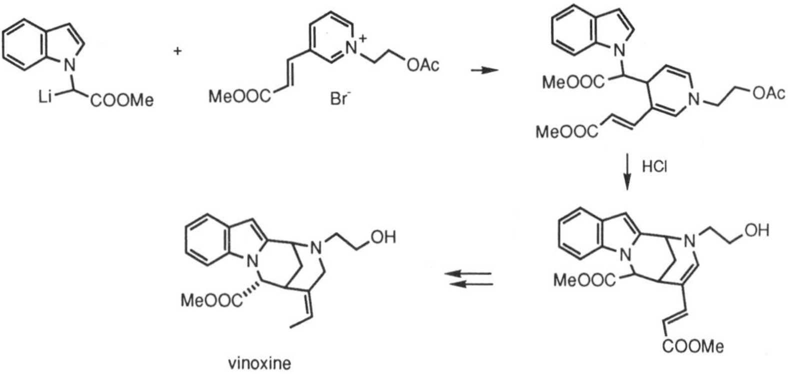

vinoxine

β-Nitro alcohols are accessible by combining the highly acidic nitroalkanes with carbonyl compounds under mild conditions. Alkoxide bases and fluoride ion are excellent catalysts for the condensation. Thus, the two-step sequence involving the condensation and reduction represents a general method for the synthesis of 1,2-amino alcohols. In a synthesis of pseudoconhydrine [E. Brown, 1973] the reduction also accomplished the ring closure.

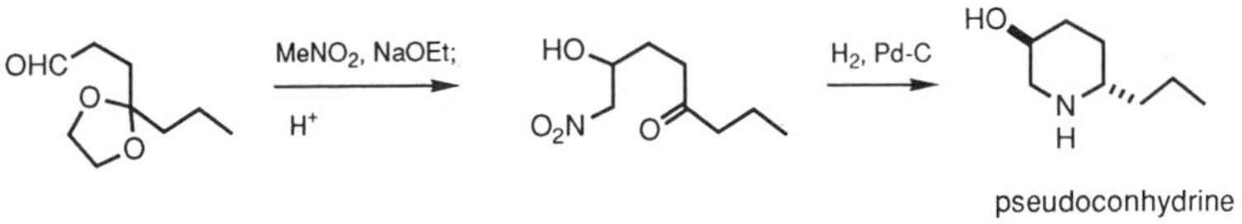

pseudoconhydrine

Organyl azides are precursors of nitrenium ions and their reactions with appropriate donors lead to amine derivatives, as exemplified by a preparation of proline derivatives [Moss, 1990]. There is contrapolarization at the non-leaving nitrogen atom during the reaction.

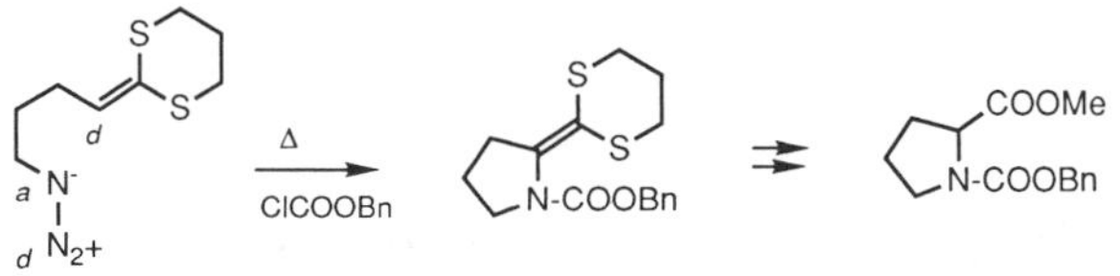

An expedient approach to arcyriaflavin A [Bergman, 1989] which contains an indolocarbazole skeleton consists of double Fischer indolization. The necessary contrapolarization was prearranged in the acyloin condensation of dialkyl succinate which provided the 2,3-bis(trimethylsiloxy)-1,3-butadiene.

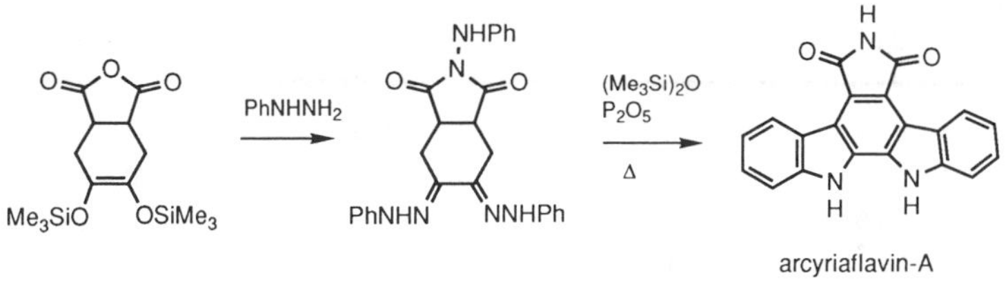

arcyriaflavin-A

4

CARBONYL CONDENSATIONS

4.1. ALDOL CONDENSATION AND VARIANTS (REFORMATSKII, KNOEVENAGEL, AND RELATED REACTIONS)

The aldol condensation [Nielsen, 1968] is the time-honored favorite among the many methods available for the preparation of 1,3-dioxygenated compounds. Because the products of this reaction possess two different functionalities, they can serve as pivotal intermediates for the access of other important classes of substances including 1,3-diols, allylic alcohols, dienes, and enones, and the various nitrogenous derivatives.

The condensation unites two carbonyl molecules in which at least one of them contains an α-hydrogen atom. Variations exist that involve two aldehyde molecules, an aldehyde and a ketone, and two ketones. Molecules in cases 1 and 3 may be the same or they may be different, although the chemoselective issue arises when different molecules are allowed to react. The cross-aldol reaction is known as the Claisen–Schmidt reaction.

Most frequently an aldol condensation is conducted with a base (e.g., alkali hydroxide, alkoxide) as catalyst. Depending on the structures (acidities) of the substrates, various bases are required, and other reaction parameters such as solvent and temperature also play important roles. The cross-condensation of two different ketones is more tedious and it often necessitates preforming enolate from one of them.

In this context, zinc enolates generated from lithium enolates and zinc chloride are particularly valuable [House, 1973]. In recent years it has become quite popular to employ an enol silyl ether as the donor to condense with another carbonyl compound or its acetal to achieve regiospecificity [Mukaiyama, 1977].

Lewis acids such as TiCl$_4$ are required as catalyst. Enol borinates have also been developed as the donors [Inoue, 1980b, Kuwajima, 1980].

The added advantage of preformed enol derivatives is stereocontrol in producing one pair of diastereoisomers predominantly [Heathcock, 1984]. Generally, the (E)-enol derivatives afford the *threo* (*anti*) pair and the (Z)-enol counterparts provide the *erythro* (*syn*) pair of aldols. Enantioselectivity may be achieved starting from chiral enol derivatives, chiral acceptors, or both [Heathcock, 1981]. Research in enantioselective aldol methodology has been prodigious [Corey, 1989c; Evans, 1982; Patterson, 1988], and this activity is surely to continue for some time.

erythro
syn

threo
anti

An early effort to control chemoselectivity is Wittig's directed aldol reaction [Wittig, 1963] in which an α-lithiated aldimine is reacted with a ketone. Mild hydrolysis of the condensed product furnishes the aldol. Nowadays NN-dimethylhydrazones of aldehydes and ketones are the preferred donors [Corey, 1976b].

The chemoselective intramolecular aldolizations of the ketoaldehyde derived from limonene are of great interest in synthesis. These examples show a divergence of reaction pathways under different reaction conditions. Thus, the exposure of the ketoaldehyde to KOH results in a cyclopentenyl methyl ketone [Wolinsky, 1960], whereas the reaction catalyzed by piperidine/acetic acid produces the conjugated aldehyde [Wolinsky, 1964]. In the latter circumstance the aldehyde is selectively converted into the enamine donor to initiate C—C bond formation.

Acid-catalyzed aldol condensation is also known. Sometimes different products from the base-catalyzed reaction arise as the result of preferential formation of the alternative enols. For example, the Robinson annulation leading to a fused ring system may be diverted to a bridged ring. It is also frequently observed that dehydration occurs under the acidic conditions.

Enolate generation under neutral conditions constitutes a major advantage in synthesis. Palladium-promoted cleavage of allyl β-ketocarboxylates [Nokami,

1989] is one of such new developments. Thus, an intramolecular aldol condensation leading to five- and six-membered fused or spirocyclic hydroxy ketones has been reported.

erythro:threo
=1:5

The aldol reaction is the middle of the tripartite steps of the Robinson annulation* [Jung, 1976; Gawley, 1976]. Nowadays, many experimental modifications on the reaction sequence have been devised, and occasionally it is advantageous to perform the steps separately.

Although alkyl vinyl ketones are commonly used in the Robinson annulation, the tendency of these compounds toward polymerization in the presence of strong bases in nonhydroxylic solvents restricts their applications. In earlier work the Michael acceptors were generated *in situ* from the masked species such as the Michael adducts, $(Et_2MeNCH_2CH_2COR)^+I^-$.

When the Robinson annulation is conducted at low temperatures, and by adding methyl vinyl ketone slowly to the ketone and the catalyst, the polymerization of methyl vinyl ketone is slowed down and the Michael addition becomes favored [Marshall, 1964]. However, the dehydration step also stops.

Let us examine a synthesis of trisporic acid [Prisbylla, 1979] which was keyed by the Robinson annulation (in a stepwise fashion). It appears possible to generate many retrosynthetic pathways for this compound in view of its multiple functionality. However, an appealing disconnection (Wittig transform) is at the disubstituted double bond, yielding an aliphatic Wittig reagent, and an oxocyclohexenecarboxaldehyde. A Robinson annulation transform reduces the latter into 1-penten-3-one and β,γ-dioxo-α-methylbutyric acid in which the aldehyde group must be masked. It is not difficult to identify α-methyltetronic acid as a synthetic equivalent for the latter compound for obvious reasons.

trisporic acid

* The Robinson annulation was initially observed by Knoevenagel in 1894. Robinson investigated this reaction extensively in his effort in steroid synthesis.

An acid version of the Robinson annulation of cyclohexanones has also been developed [Heathcock, 1971].

Except when severe steric interactions prevail, the donor site for the Michael addition step of an α-monosubstituted cycloalkanone occurs at the more highly substituted carbon atom, and it is almost invariably the case when the substituent is an acceptor group (COOR, etc.) This regioselectivity can often be changed by employing kinetic enolates or the enamines.

The aprotic conditions for the generation of kinetic enolates are usually incompatible with alkyl vinyl ketones, but a more stable α-silylvinyl derivative may be used [Boeckman, 1973; Stork, 1973]. The increased stability of the silylated vinyl ketones may be due to both steric and electronic effects. The steric bulk of the silyl substituent would discourage bimolecular reaction of the Michael adduct, and the enolate is also stabilized against oxidation to a free radical with the additional acceptor.

Synthesis of adrenosterone requires an efficient access of a *trans*-hydrindene-dione. An excellent intermediate is related to an enone aldehyde by a special Robinson annulation [Stork, 1982b].

adrenosterone

Dehydrofuropelargones can be disconnected to yield a 1,6-diketone. The aldol transform is unambiguous as the alternative aldol condensation would give an aldol which is not as readily dehydrated and the retro-aldol fission of which would prevail. In fact this alternative aldol represents a useful predecessor [Büchi, 1968] which in turn may be prepared from the corresponding β-diketone by Grignard reaction at the more reactive, nonconjugated carbonyl.

3-Isopropylfuroic acid has been obtained by a Darzens glycidic ester condensation.

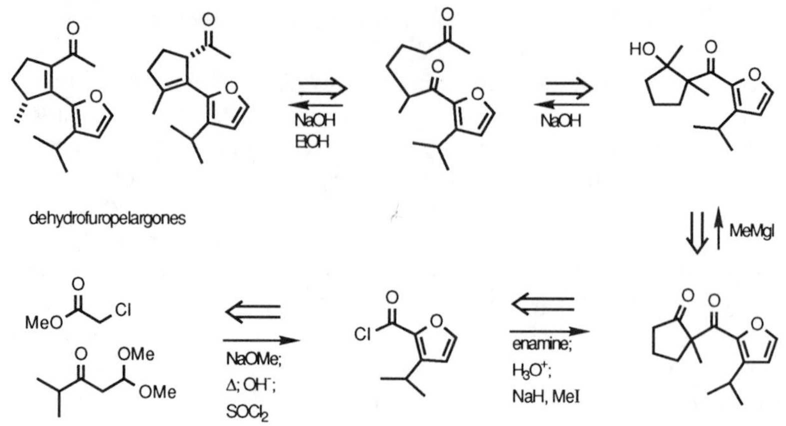

dehydrofuropelargones

The molecular framework of illudin S suggests an aldol step to close the six-membered ring for its synthesis [Matsumoto, 1971]. Furthermore, there is a slightly veiled symmetry in the noncyclopentane portion of the sesquiterpene. Thus, removing the carbonyl oxygen and the tertiary methyl group transforms it into a *gem*-diacetylcyclopropane.

With proper modification of this building block, its union with a cyclopentenone by a sequence (not necessarily consecutive) of Michael and aldol reactions would serve to assemble the carbon skeleton. Ample functionalities enable the access of the target molecule.

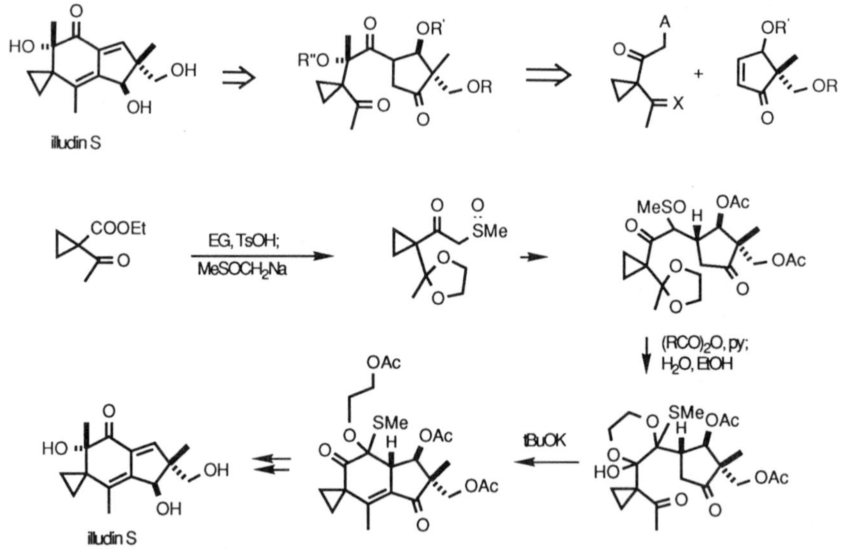

illudin S

illudin S

A transketalization occurred after a Pummerer rearrangement which was intended to generate an α-diketone subunit. This unexpected event only forced a change of the reaction sequence order in the rest of the synthetic steps.

The homoallylic system present in silphiperfol-5-en-3-ol indicates the possibility of an aldol route to construct the diquinane portion [Brendel, 1989].

Since the angular positions are fully substituted, no regiochemical ambiguity would arise to complicate the synthesis.

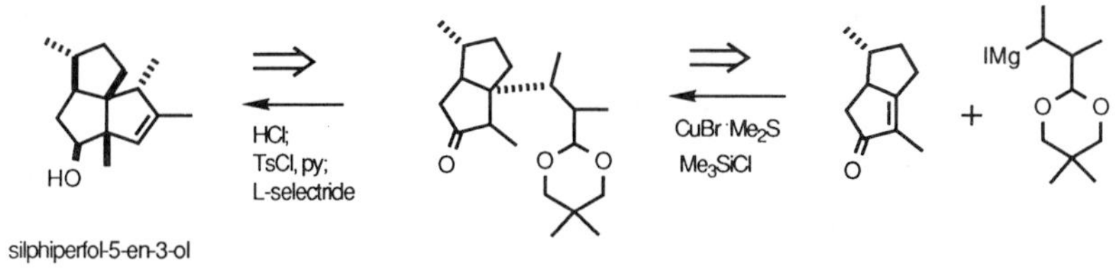

silphiperfol-5-en-3-ol

It is significant that the secoilludane sesquiterpene hypacrone has been assembled [Hayashi, 1975] via an aldol (Mukaiyama) route using 1,1-diacetyl-cyclopropane as the acceptor. The approach was made possible by the conjoint 1,7-diketone array.

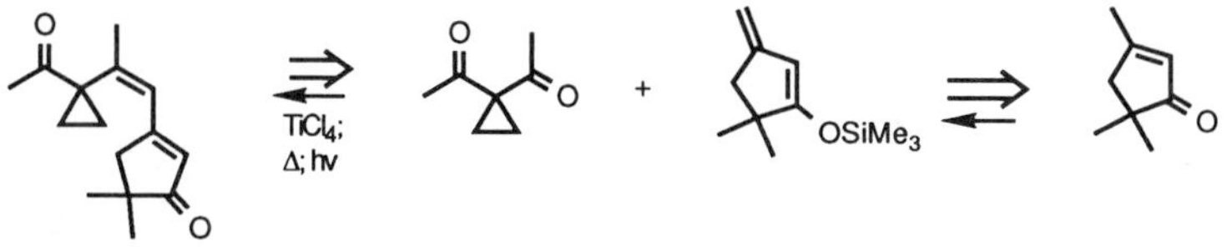

hypacrone

The molecular network of forskolin can be disconnected into an octalone in which the double bond is provided for bishydroxylation and also a carbonyl (of a lactone) for elaboration of the pyrone nucleus [Somoza, 1989]. The enone moiety of the octalone is related to a trimethyl hydroxycyclohexenecarboxalde-hyde through a tandem aldol–Michael transform. It is important to note that three new asymmetric carbons of the octalone could be generated in the correct sense by anchoring the Michael donor to the C-1 hydroxyl, as five-membered ring closure must give a *cis*-fused lactone, and the final product from the aldol condensation is subject to equilibration, even if the aldehyde group should assume a less likely axial orientation.

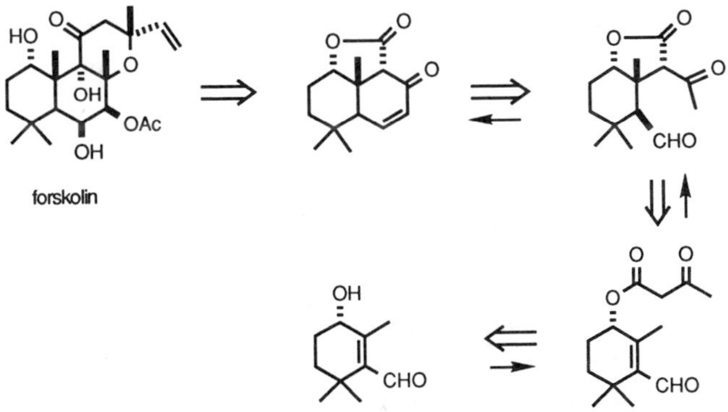

forskolin

The stereocontrolled introduction of the carboxyl residue for subsequent lactonization in synthetic approaches to quadrone is not as easy as one might expect. One solution, evolved from one of the many endeavors, took advantage of a cyclohexenone intermediate [Burke, 1984]. Thus, a Wharton rearrangement (epoxide + hydrazine) reductively transposed the oxygen function of the enone, and a Claisen rearrangement furnished an acetaldehyde chain with the desired configuration. Instead of degradation, the acetaldehyde was put to full use in an intramolecular aldolization. The acetate of the aldol was pyrolyzed to give predominantly an unconjugated ketone. Redox maneuvering of the latter substance completed the synthesis.

One drawback of this route is the formation of an isomeric lactone during oxidation of the 1,5-diol intermediate.

quadrone

Concerning the access of the tricyclic diketone [Burke, 1982] it is noted that the two carbonyl groups bear a disjoint relationship along the short span. However, they are conjoint when the long span is followed, and therefore some facile condensation reactions may be employed for the assembly of the diketone from seco predecessors that can be generated by disconnection at the latter path. Accordingly, disconnection of the cyclohexanone results in a bicyclic system containing a 1,5-ketoaldehyde segment. Performing a Michael transform and carbonyl reconnection reveals an initial target as a spirocyclic dienone.

quadrone

Development of an excellent method for acquisition of γ-acyl-δ-valerolactones by conjugate addition of 2-(alkylthio)carboxylic esters to enones with *in situ* trapping of the keto enolate by formaldehyde (aldolization) has had implication to another approach to quadrone [Bornack, 1981]. Although, for steric reasons, the reaction did not afford the lactone directly, the introduction of the one-carbon fragment in this manner deflected any regiochemical issue relating to the requisite operation.

quadrone

III

PhSCH(Li)COOR;
HCHO;
NaBH₄;
Me₂C(OMe)₂, H⁺

R'+R' = CMe₂

The formation of two rings by a "zipper" cyclization is the most outstanding feature of an approach to aklavinone [Uno, 1984]. The experimental conditions are very critical for the optimal production of the stereoisomer with the correct relative configuration of the β-hydroxy ester subunit of the A-ring.

The Michael addition occurred even at −78°C. The reason for this reactivity is the release of charge repulsion of the dianion. In the presence of a cryptand the weak chelation of the cation to the enolate and the acceptor carbonyl caused retardation of the following aldolization. The retardation permitted conformer equilibration, and the weak chelation rendered both dipole and steric interactions less tolerable. The combined effect was that an isomer with axial hydroxyl and ester groups became the major product.

aklavinone

CAN;
Br₂

KH, K222
HMPA, THF

Huperzine A (selagine) shows potent anticholinesterase activities and has therefore attracted many synthetic efforts. Retrosynthetically, a tricyclic keto ester is easily conceived to be the key intermediate [L. Qian, 1989]. This skeleton has the attribute of being readily assembled by a Michael–aldol condensation sequence (note the cyclic double bond of the target correlates directly with the hydroxyl group of the ketol intermediate). A Wittig reaction is expected to afford the exocyclic (E)-alkene due to a steric effect of the bridgehead ester.

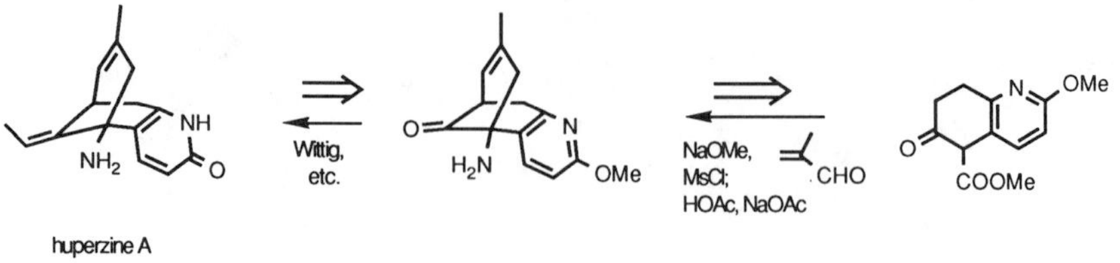

huperzine A

A synthesis of erythromycin [Woodward, 1981] started from a preparation of erythronolide A seco acid which contains 10 asymmetric centers. The task was somewhat simplified by recognizing the C-8/C-9 bond as strategic and the asymmetric segments of C-4, 5, 6 and C-10, 11, 12 are the same. A *cis*-fused dithiadecalin was chosen as a common intermediate.

A most gratifying finding was the the D-proline promoted intramolecular aldolization leading to an enantiomerically enriched compound which was readily purified to a homochiral state. The aldol was dehydrated, reduced, and dihydroxylated.

Union of the two moieties to create the major portion of the seco acid's backbone also enlisted the service of an aldol reaction. The donor segment retained the bicyclic structure whereas in the aldehyde acceptor the sulfur atoms had been deleted. Preservation of the bicyclic structure of the donor is of tactical importance in establishing the C-8 stereochemistry from a Δ^7-en-9-one. The latter compound was elaborated from the aldol via a series of oxidoreduction processes.

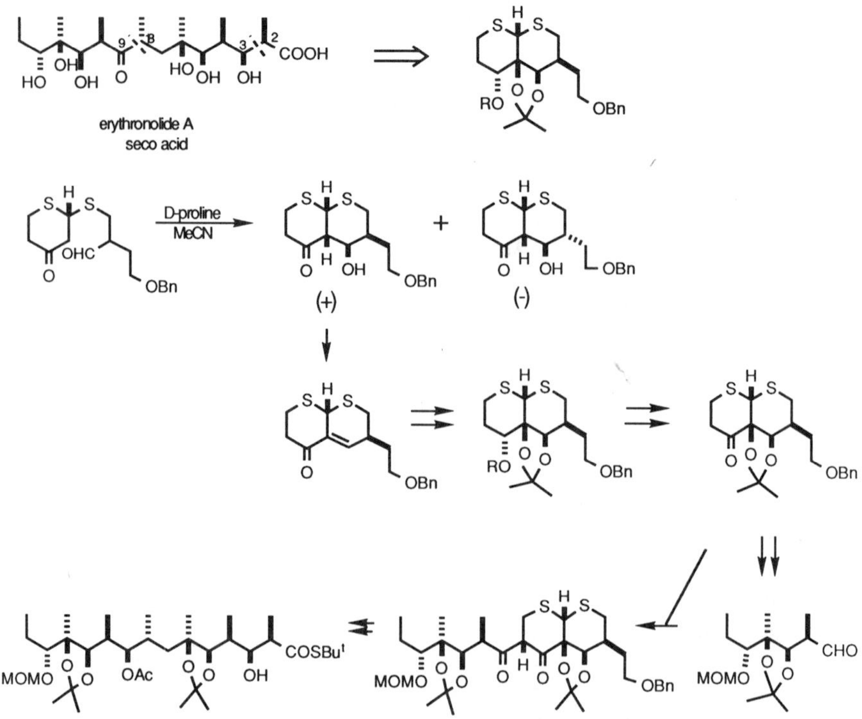

erythronolide A
seco acid

α,β-Enones are practically equivalent to aldols in terms of their synthetic accessibility. Consequently the synthesis of eremophilone [Ziegler, 1974; Ficini, 1977] based on an aldol transform is viable.

Since 1,3-transposition of β-monosubstituted enones can be effected via the epoxy ketones by Wharton rearrangement and oxidation, a transposed enone/ aldol of a synthetic target may be set up instead. This indirect approach [Ziegler, 1974] was predicated on a stereoselective establishment of the *vic*-dimethyl sector from 3,4-dimethyl-2-cyclohexenone using organocopper chemistry.

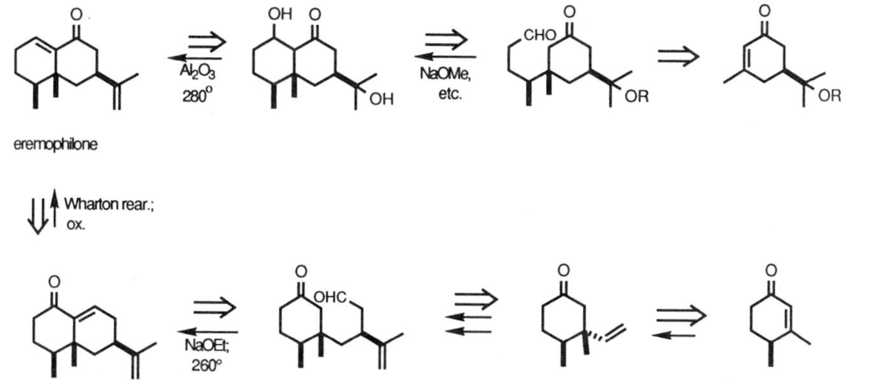

It is also easily conceivable that the aldol route may supplant an alkylation route to obtain an α-alkyl ketone. This alternative has the merit of avoiding disubstitution and/or achieving regioselection. The most important consideration may be the availability of the starting material. In this context, a synthesis of cryptomeridiol [Kawamata, 1988] is worthy of discussion.

This sesquiterpene has a conjoint circuit between the two hydroxyl groups. This favorable electronic arrangement may be exploited in an intracircuit C—C bond formation. The synthesis under discussion involves two Michael additions, an aldol condensation, and conventional redox and other reactions. The possibility of employing the Michael processes are due to the conjoint relationship.

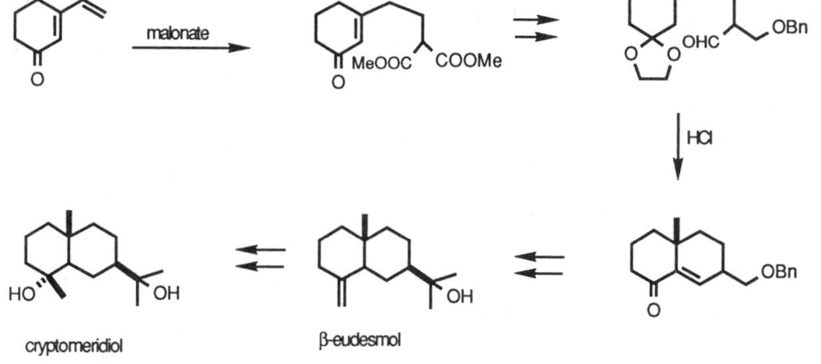

Trihydroxydecipiadiene is a challenging synthetic target. it is obvious that the major task of such an endeavor is the construction of the tricyclo[5.3.1.0^{3,11}] undecane system. Except for the secondary methyl group, functional groups must be supplied to the proper core atoms so that the three other sidechains may be attached. This analysis would lead to a 1,3-dioxygenated compound and thence a bicyclo[4.2.0]octanone to which a propanal chain is appended. In other words, an aldol transform is the key disconnection [M.L. Greenlee, 1981].

trihydroxydecipiadiene

Cycloaddition involving dichloroketene is the most expedient method for securing cyclobutanone of this type. The choice of a 1,3-cyclohexadiene instead of the monoolefin provided a solution to the stereochemical problem in that a subsequent hydrogenation at the tricyclic level gave rise to an all-*cis* orientation of the methine hydrogens.

It is instructive to trace and compare the involvement of the aldol condensation in the synthesis of two sesquiterpenes with a hydrazulene skeleton. The outline of a route to parthenin [Heathcock, 1982c] is clearly identifiable with an aldol methodology on the basis of the ketone/lactone relationship, although targeting a transposed bicyclic aldol needed some thoughts on the facility in acquiring the precursor and the introduction of the remaining functionality, notably the lactone moiety. On the other hand, discerning an aldol transform for synthesis of aphanamol I [Mehta, 1989] must come from a stepwise analysis.

It is possible that the synthetic approach to the latter sesquiterpene was perceived bidirectionally, i.e., the desire to utilize a previously obtained intermediate might have promoted a forward-looking analysis. Nevertheless, the transposed norketone of aphanamol I, easily arrived by retrosynthetic analysis, presents a stereochemical problem in its construction which may be resolved at the stage of a cyclopentane precursor. A better solution is to establish such stereochemistry from the corresponding dehydro compound. An aldol transform is now applicable. Since the resulting triketone harbors both a 1,4- and a 1,5-dicarbonyl subunit, reconnection of the latter leads to another hydrazulene which is derivable from limonene.

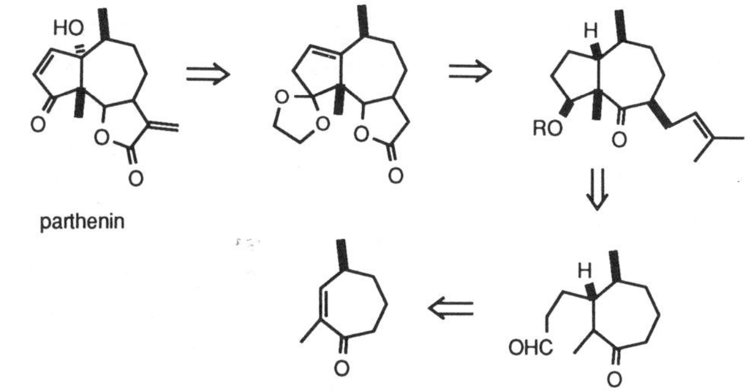
HO
parthenin
RO
H
OHC

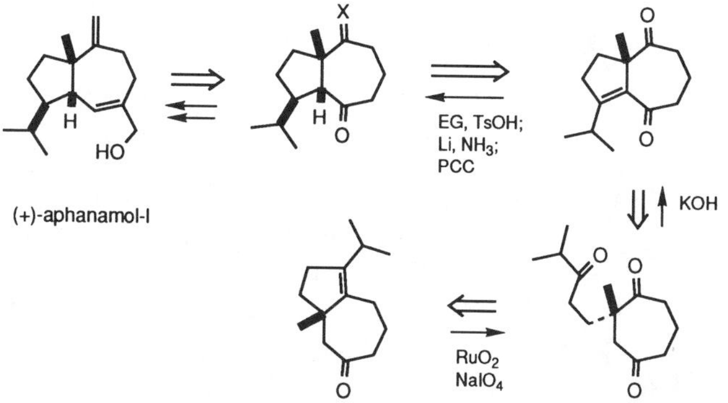
(+)-aphanamol-I
X
EG, TsOH;
Li, NH3;
PCC
KOH
RuO2
NaIO4

Pumiliotoxin C is a perhydroquinoline derivative whose topography permits an assembly from a 4-methoxypyridinium salt [Comins, 1990]. The α-position of the heterocycle is an acceptor site which reacts readily with a Grignard reagent. Mild hydrolysis of the dienol ether so obtained resulted in an enone system to which the propyl chain was introduced. The conjugate alkylation was also used in the stereoselective attachment of the methyl group after an intramolecular aldol condensation. Thus, the conjoint 4-methoxypyridine is a versatile building block in which C—C bond formation can be directed to C-2, 3, and 6, and possibly other positions also.

The importance of functional group addition (FGA) operation which delineated the aldol transform is clear.

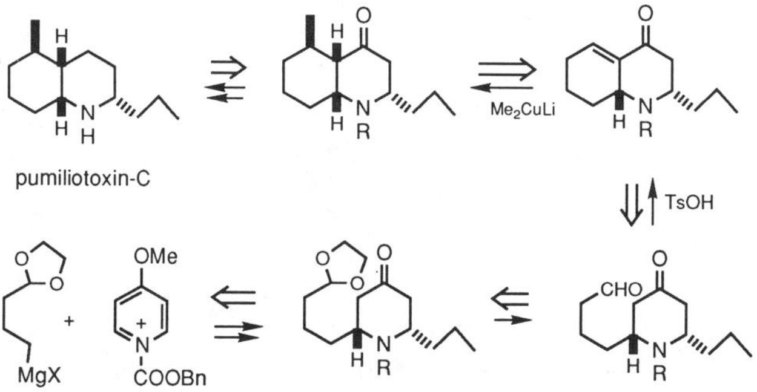
pumiliotoxin-C
Me2CuLi
TsOH
MgX
OMe
COOBn
CHO

A route to ferruginol [Ohashi, 1968] illustrates the power of careful analysis of hidden symmetry elements in a known precursor. Thus, correlation of the aromatic portion with a 1,5-diketone in which the sidechain is also substituted with an acetyl subunit is revealed. An excellent latent group for this sidechain is 1,3-dimethylisoxazol-4-ylmethyl. Conditions for unveiling the β-diketone are also conducive to the aldol cyclization.

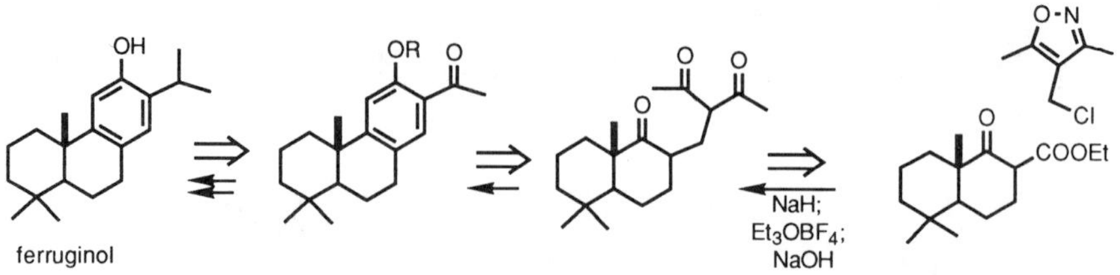

The obvious retrosynthetic disconnections of cinchonamine are the hydroxy-ethyl chain and the C—N bond between the basic nitrogen atom and the carbon branch attaching to the indole nucleus. To effect the C—N bond formation there are several options available to an amino ketone, and in the present case a 2-acylindole would be an excellent subgoal for synthetic operations [Grethe, 1976].

The Fischer indole synthesis may not be the ideal method to create the required intermediate which contains sensitive substituents such as the vinyl group. Furthermore, regioselectivity may also be a problem. Based on these considerations, an aldol method to form the indole ring system appears to be a superior choice.

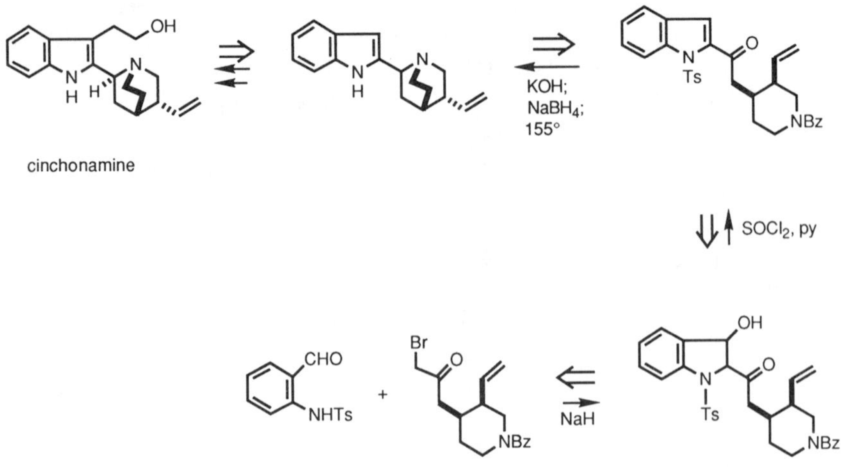

In fredericamycin A, a spirocyclic 1,3-diketone is present. It has been found advantageous to construct this structural entity by an intramolecular aldol condensation followed by oxidation [T.R. Kelly, 1988].

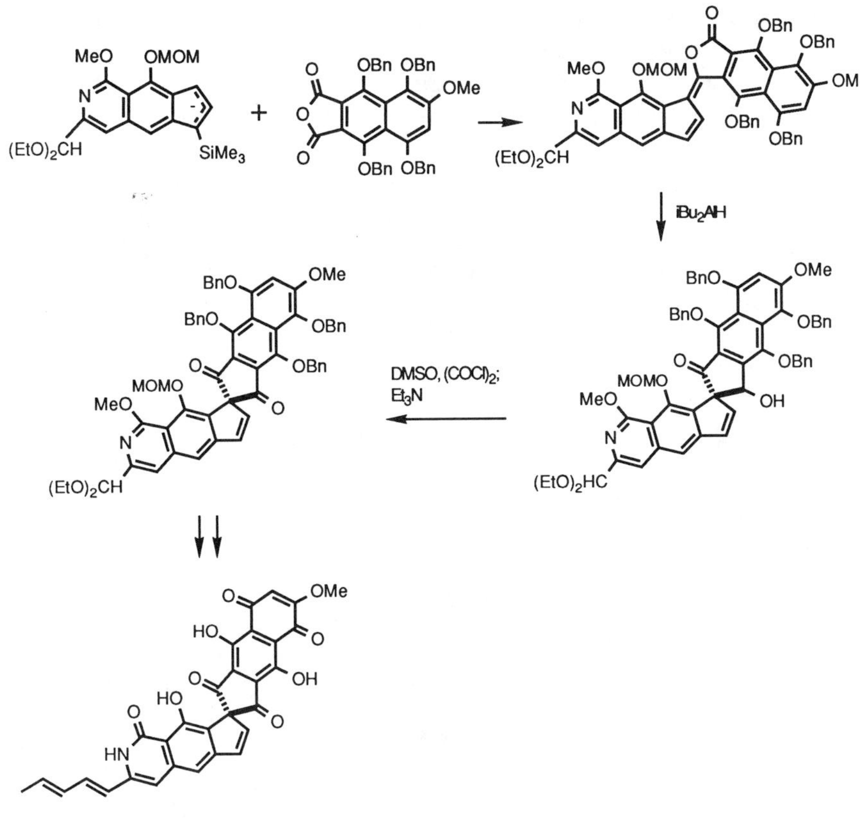

fredericamycin A

The bridged ring segment of grayanotoxin II contains a 1,3-diol subunit, therefore it is natural to assemble it by the aldolization technique [Hamanaka, 1972, Gasa, 1976].

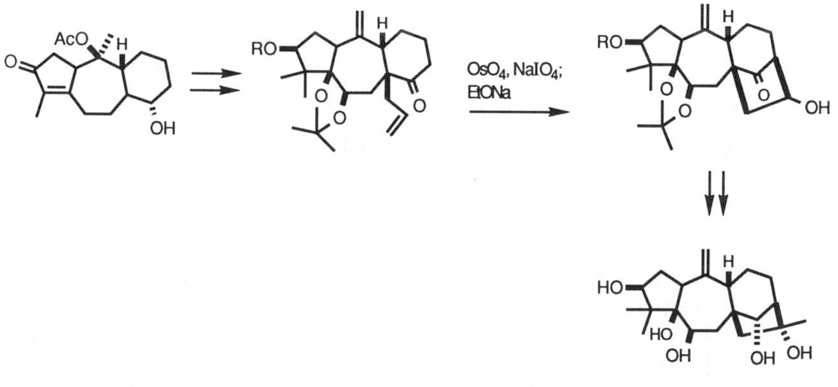

grayanotoxin II.

More intriguing applications of the intramolecular aldol reaction which take advantage of the differentiated functionalities pertain to synthesis of stemarin [R.B. Kelly, 1980] and aphidicolin [Marini-Bettolo, 1983]. These functionalities can be fashioned into appropriate bicyclo[2.2.2]octa(e)ne derivatives to allow specific rearrangements.

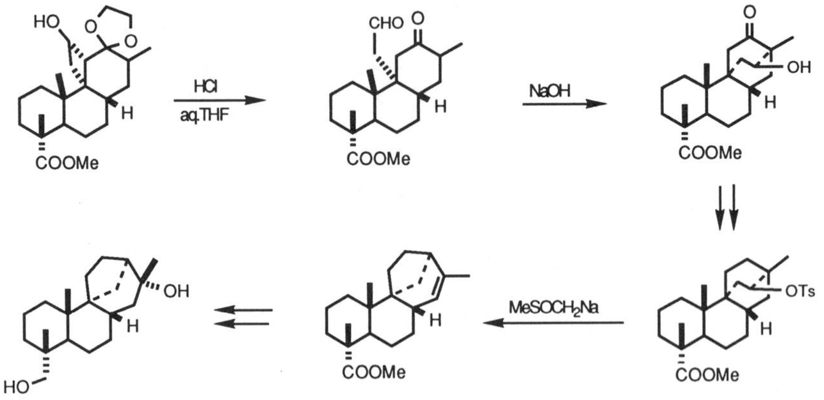

stemarin

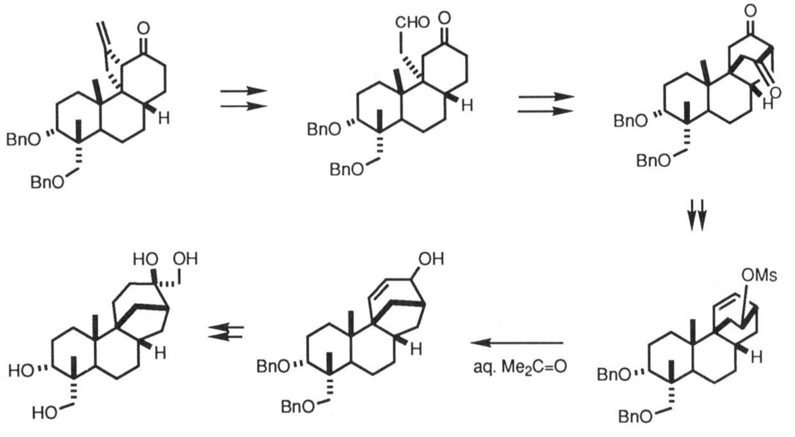

aphidicolin

Iboga-type alkaloids are most expeditiously constructed from a tryptophyl residue and an azabicyclo[2.2.2]octane subunit as an intramolecular electrophilic alkylation can effectively create the C—C bond at the α-carbon of the indole nucleus. The identification of the acceptor site directs the mode of assembly of the bicyclic system. An aldol transform not only becomes evident, it is superior in the sense that an oxygen function is left at the site to which an ethyl group must be attached. This arrangement means that a simple substrate may be used, yet it would still allow the introduction of the ethyl group at a later stage.

A synthesis of catharanthine [Imanishi, 1980] has been carried out on the basis of this rationale.

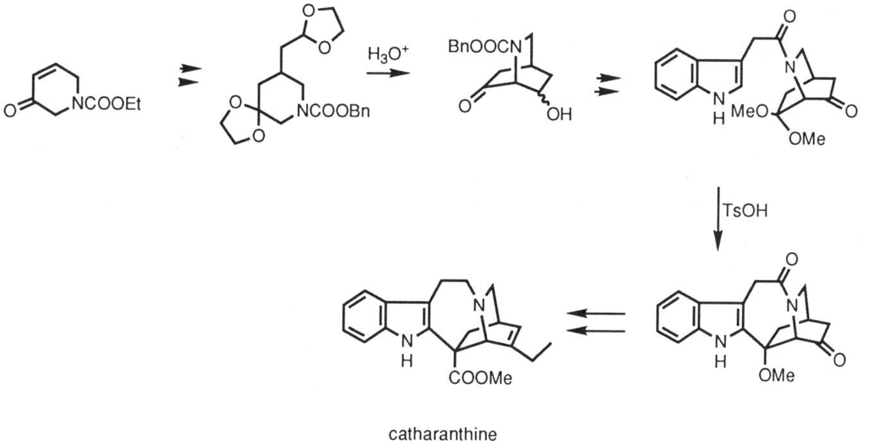

catharanthine

Disconnection of 2- and 9-isocyanopupukeanane by functional-group-keyed analysis identifies C-2 and C-9 as acceptor atoms. By specifying C-9 as a carbonyl site the synthesis of both terpene molecules may be accomplished using a common intermediate [Corey, 1979a,b].

Because C-2 and C-9 are 1,3-related, aldolization of a bicyclic precursos is convenient. (The synthesis of 9-isocyanopupukeanane was carried out via an irreversible intramolecular alkylation step.)

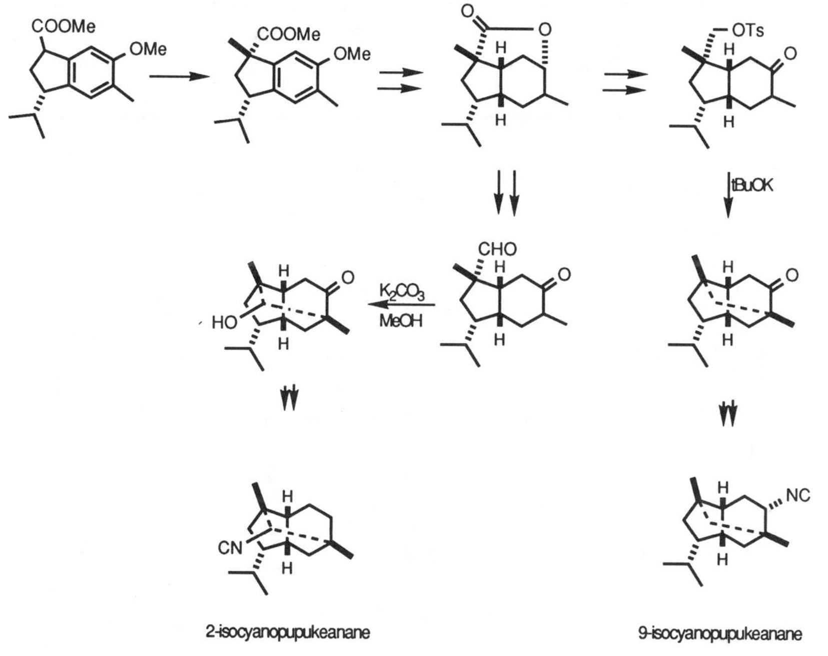

2-isocyanopupukeanane   9-isocyanopupukeanane

Retrosynthetic analysis of helminthosporal suggests a C_{14} ketone as a subtarget molecule [Corey, 1963]. Because carbonyl homologation is readily achievable and the secondary formyl group to be established is equatorial and projecting into a less hindered space, no stereochemical problem is associated with its introduction. Of the various disconnection pathways the one leading to a bicyclo[3.3.1]nonenone is most favorable in terms of its accessibility. The identification of the fully substituted alkenic carbon as an acceptor site equates it with a carbonyl center, and a Michael–aldol (the latter with acid catalysis) reaction sequence becomes a reasonable lead to initiate the construction of the bridged ring system.

helminthosporal

carvomenthone

NaOEt, HCOOEt;
MVK, Et₃N;
K₂CO₃, MeOH

Trichodermin and calonectrin differ mainly in the position of the secondary acetoxy substituent in the trichthecane framework. Retrosynthetic analysis of the two molecules points to an aldol transform for both compounds, with their respective predecessors having an acetaldehyde chain at either the α- and γ-position of the 3-pyranone ring.

It is interesting to note that the aldol reaction leading to the calonectrin skeleton [Kraus, 1982] proceeded smoothly, whereas the corresponding reaction used in the trichodermin synthesis [Colvin, 1973] was poor.

The reasons for the diverse behavior may be twofold. Ring closure to give the trichodermin intermediate requires an axial acetaldehyde chain, and in one of the conformations an epimerization of this chain must precede the aldol condensation. Enolization–epimerization of α-alkoxy ketones is less favorable than the all-carba analogues because the α-carbon is captodative. Enolization toward the α'-position is preferred. In the calonectrin case the molecule probably exists mainly in the active conformation (axial aldehyde chain) also unproductive competitive enolization is not possible.

calonectrin

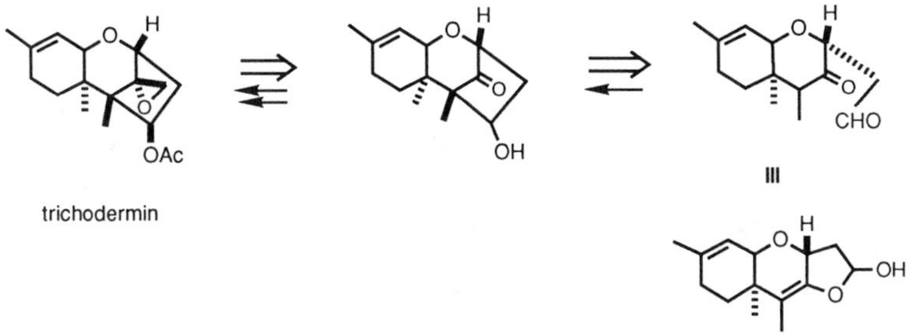

trichodermin

III

There is a 1,3-dioxygenated segment in damsin and it is logical to expect the utility of an aldol or a Claisen condensation in the pursuit of its synthesis. A rather unusual route [Kretchmer, 1976] started from a tetralin. After redox reactions of the tetralin to give a cyclodecanetrione, a combined transannular aldolization–methylation process was effected to generate the functionalized hydrazulene.

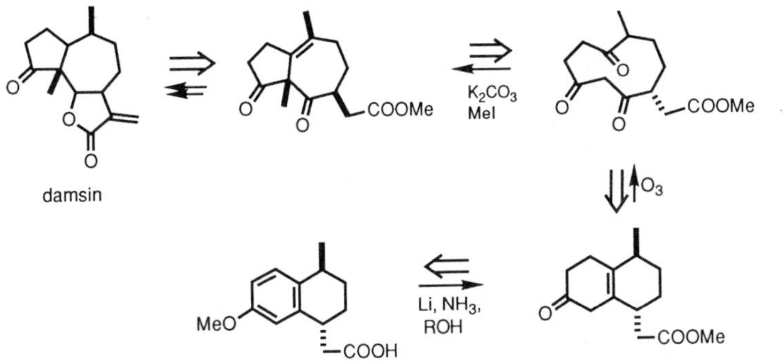

damsin

Aldol condensation between an aldehyde and a ketone usually takes the course of producing an enone. A reversal of the donor–acceptor roles is critical to a ring closure directed toward construction of the hexahydrobenzofuran subunit of the avermectin/milbemycin antibiotics [Hirama, 1988].

The reason for this abnormal pattern of reactivity is steric constraint. The other modes of reaction would result in the more strained bridged ring compounds.

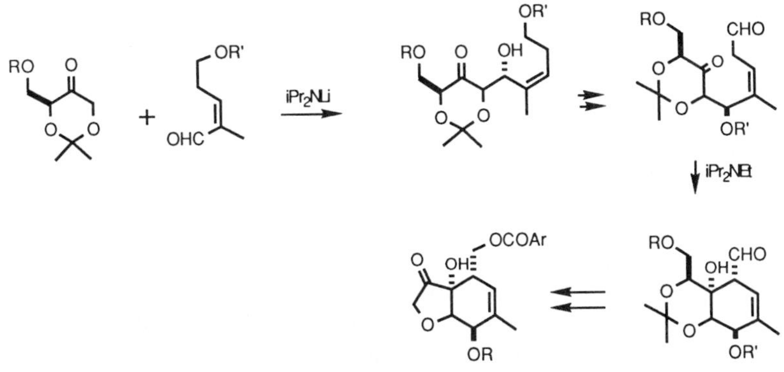

The hydrobenzofuran moiety present in the avermectins/milbemycins actually contains an α,β-ene- α'-hydroxyalkyl carbonyl function in which the carbonyl group cannot facilitate dehydration. The same structural features also make it more difficult to synthesize these substances by a conventional aldol process involving enolization, despite the conjointment of the two oxygen functions. An alternative pathway is via trapping of an appropriate Michael adduct and eliminating the addend *in situ*. The Nozaki method involving (phenylthio)-dimethylaluminum appears to be an effective process to achieve the C—C bond formation [Danishefsky, 1989c].

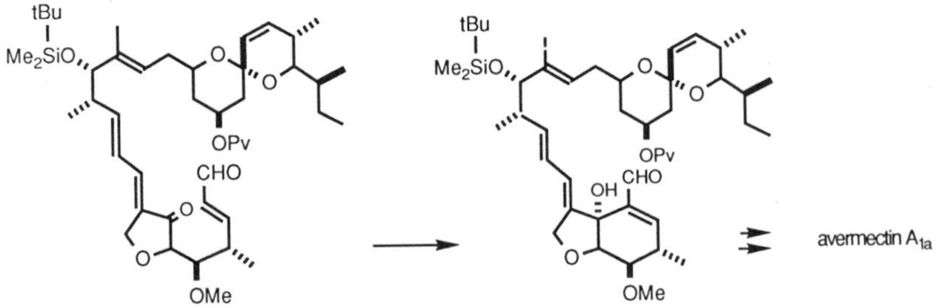

Although less efficiently, PhSLi has been used to cyclize 3-(3-oxobutyl)-2-cyclohexenone [Myers, 1989].

A fragmentation route to norpatchoulenol [Niwa, 1984a] predicated upon the convenient stereocontrol as conferred by a bicyclo[2.2.2]octane skeleton is very efficient because a tandem Michael–aldol process forms two rings from the cyclohexenone in one step.

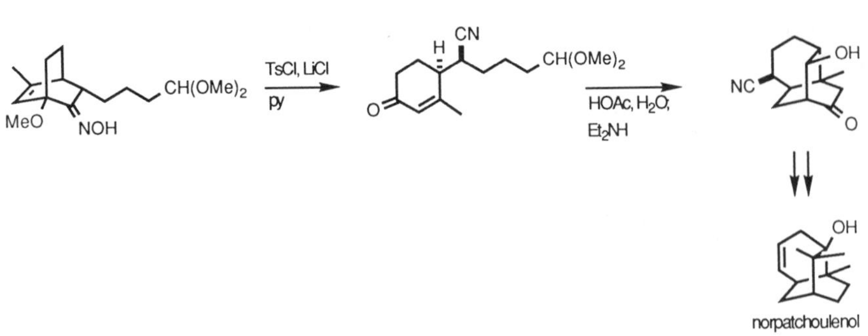

norpatchoulenol

The Friedländer condensation involving *o*-aminobenzaldehydes (and ketones) with carbonyl compounds is a directed aldol reaction. It may proceed via enamines. Notable application of this reaction in natural products synthesis is in the area of camptothecin, as many reports described the formation of the pyrrolidinoquinoline ring system [Stork, 1971; and others].

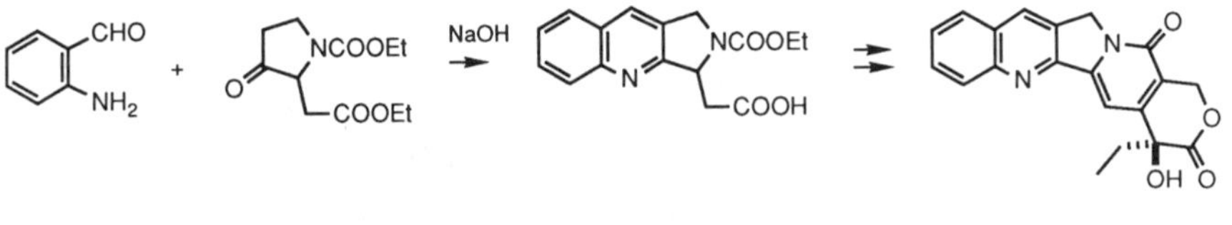

camptothecin

Most of the aldol reactions mentioned above employ basic catalysts. The complementary Mukaiyama version is becoming more important. For example, an asymmetric synthesis of aklavinone [McNamara, 1984] was readily accomplished using a homochiral acetal to condense with 1-trimethylsilyl-2-butanone and closing the hydroaromatic ring by base treatment. The predominant formation of the desired stereoisomer in the aldolization step may be explained on the basis of the Johnson–Bartlett model [Bartlett, 1983].

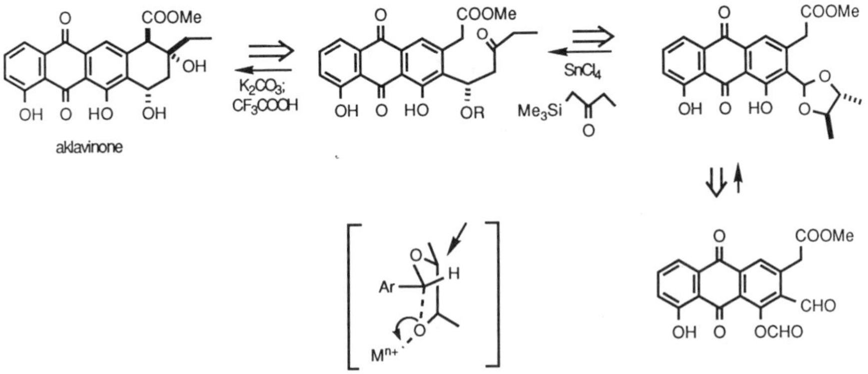

aklavinone

An acetone α,α'-dianion synthon for aldolization can be generated from 2-trimethylsiloxy-3-trimethylsilyl-1-propene in the presence of a Lewis acid [Hosomi, 1989].

In coriamyrtin there is an α,β-hydroxy lactone subunit. It has been found that the aldol-type condensation can be effected during the introduction of the isopropenyl pendant by conjugate addition to an acrylic ester [K. Tanaka, 1982]. The acrylic ester chain is in a perfect disposition to form the C—C bond with the ketone group.

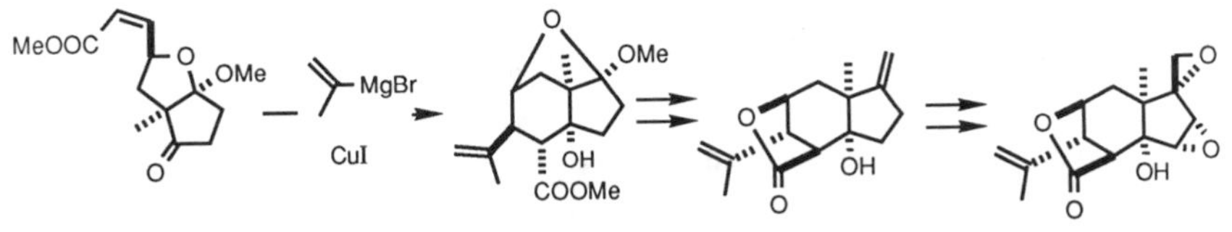

coriamyrtin

The focal point of a lysergic acid synthesis has always been the construction of the tetrahydronicotinic acid portion. Two-group disconnection of lysergic acid reveals a donor species derived from β-alanine and a tricyclic indole nucleus in which two acceptor sites are allylically linked. By FGI the latter synthon is represented by an α,β-unsaturated carbonyl system.

Two options are available to the union of the two synthons. Carbon–carbon bond formation preceding N—C bond formation which prevents the undesirable Schiff base coupling is preferred. Necessarily the secondary amine must be protected prior to the aldolization [Kurihara, 1986].

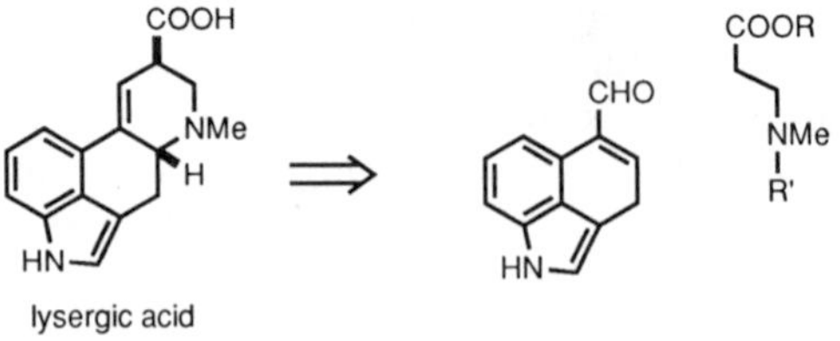

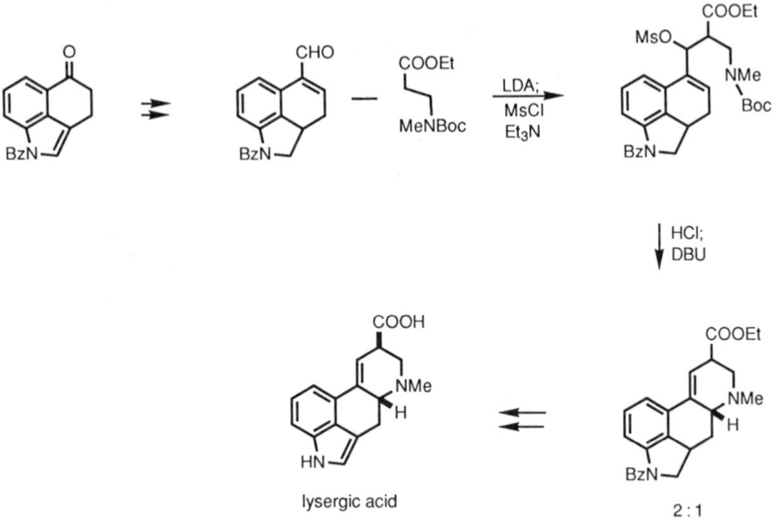

The clavicipitic acids are very unusual ergot alkaloids. Their synthesis from a tryptophan derivative by pericyclization toward the benzene ring is more difficult because the donor atom (C-4) is at best nonactivated (it is *meta* to the indolic nitrogen atom and has acceptor character). A more practical approach is to anchor a functionalized chain to C-4 first [Muratake, 1983].

With this consideration the formation of the seven-membered heterocycle can be easily visualized to involve an intramolecular aldol (Knoevenagel) reaction. 3-Formyl indoles are readily accessible by the Vilsmeier–Haack reaction.

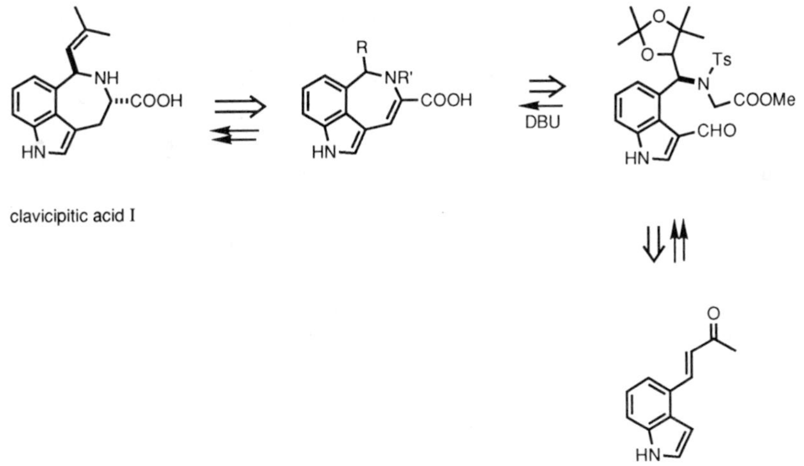

clavicipitic acid I

The previous examples involve reactions with ester α-carbanions. Before the development of dialkylamide bases such reactions were limited to donors with highly acidic hydrogens (Knoevenagel reaction, etc). The alternative is the Reformatskii reaction [Rathke, 1975; Fürstner, 1989]. An improved procedure prescribes the addition of a dialkylaluminum chloride as a co-reagent [Maruoka, 1977]. Its purpose is to coordinate and hence activate the acceptor carbonyl, and also to stabilize the product by chelation.

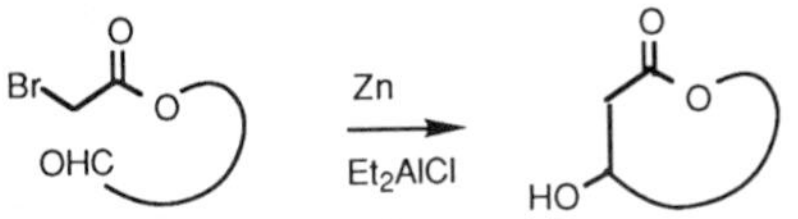

ester α-carbanions have also been generated under nonbasic conditions from the α-silyl esters with fluoride ion. A synthesis of 17-O-methyllythridine [Seitz, 1982] relied on this chemistry to form the macrocyclic lactone.

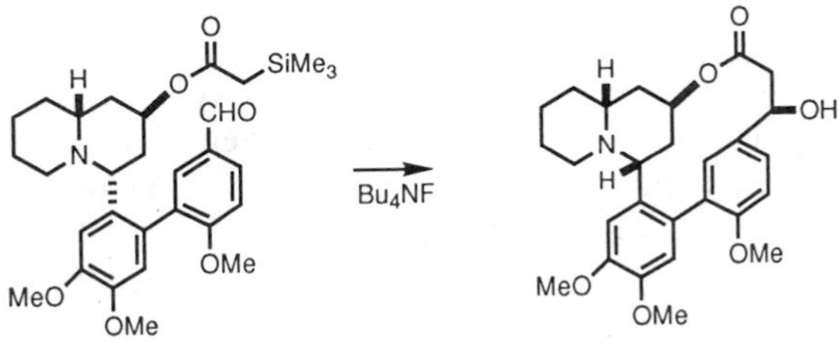

17-O-methyllythridine

Aldol condensation could initiate a reaction sequence leading to conjugate dienes. The position of the diene may be regulated by varying the set of transformations subsequent to the aldolization. For effecting α-alkylidenation and deoxygenative elimination toward the α′-carbon of a ketone, the dehydrated aldol may be subjected to decomposition of the corresponding arenesulfonhydrazone with a potent base such as an alkyllithium [Shapiro, 1976].

Applying this series of reactions to a synthesis of the hexalin portion of compactin [Heathcock, 1982b] is advantageous in view of what structural features the keto aldehyde precursor offers. Thus, the protruding point of the ester group must become attached to a β-carbon of the ketone and conjugate addition of a 2-lithio-1,3-dithiane to an enone is suggested. The latter reaction should be under steric control by the substituent at C-4 of the cyclohexenone. The required enone in turn is apparently obtainable by a Diels–Alder reaction involving an alkyl (Z)-crotonate.

4.2. CLAISEN CONDENSATION

Five- and six-membered 1,3-cycloalkanediones are obtained readily by an intramolecular Claisen condensation of the proper keto esters using bases such as sodium hydride and sodium alkoxides. Since these diketones are normally extensively enolized, they are adequate precursors of cycloalkenones (via O-alkylation, reduction, and hydrolysis). Occasionally this reaction sequence involving the Claisen condensation is preferable to the aldol route because it is very efficient and amenable to large-scale preparation.

The Claisen route has been used in a synthesis of bilobanone [Escalone, 1980]. Advantage was taken of the symmetrical 1,3-cyclohexanedione intermediate so that worrisome isomeric mixtures did not result.

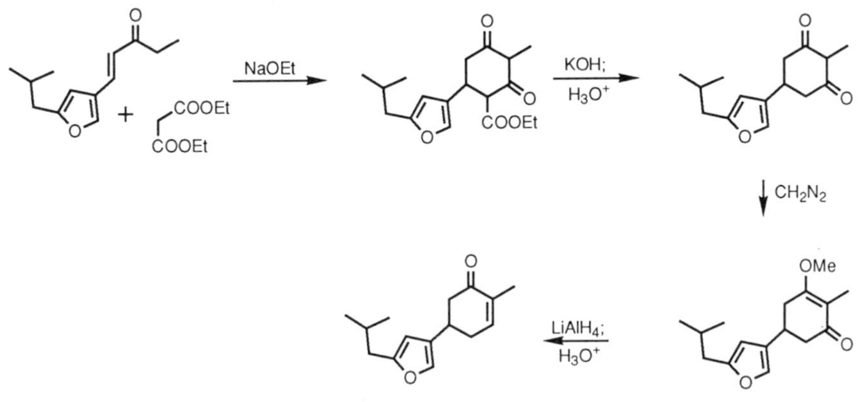

bilobanone

A synthesis of lycoramine [Misaka, 1968] demonstrates the classical guiding principle for the construction of a moderately complex molecule. The presence of a 1,3-dioxy segment in a six-membered ring suggests a Claisen condensation approach. With a properly functionalized carbon chain anchoring at the quaternary benzylic position, the condensation product is well disposed toward a smooth transformation to the target compound. The two carbonyl groups of the 1,3-diketone system are easily differentiable as one of them is at a neopentyl location.

In the actual synthesis the benzylic chain was eventually involved in an intramolecular acylation (Friedel–Crafts) and submitted to a Schmidt reaction to introduce the nitrogen atom. N-Methylation and metal hydride reduction completed the effort.

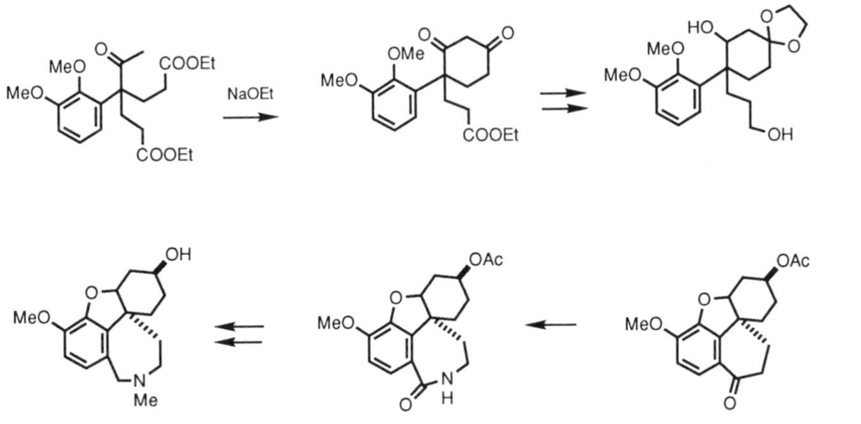

lycoramine   + isomer

Vernolepin and curzerenone are secoeudesmane sesquiterpenes. As both compounds contain a 1,3-dioxygenated cyclohexane nucleus, the evocation of synthetic plans [Kieczykowski, 1978; Miyashita, 1981] based on Claisen condensation is not surprising.

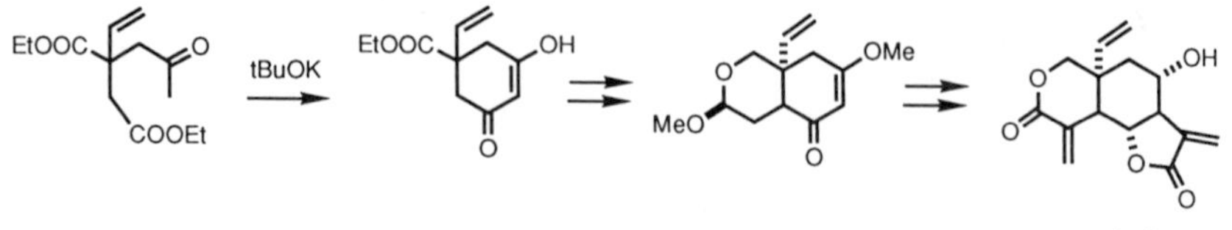

vernolepin

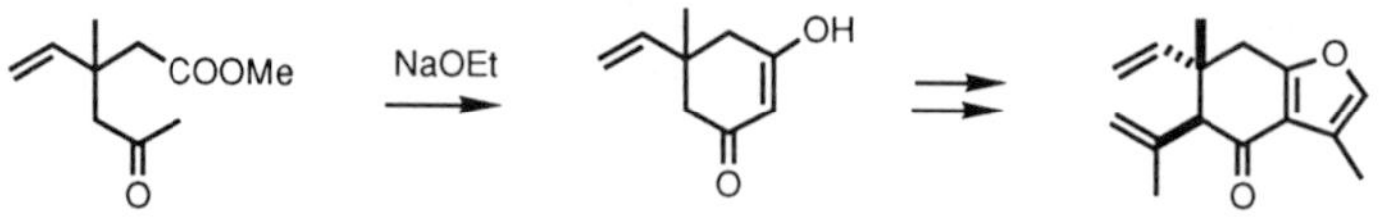

curzerenone

A new route to (+)-yohimbine [Hirai, 1990] also employs a Claisen condensation to construct the E-ring with the benefit that the enone derived from the cyclohexanedione can be regioselectively and stereoselectively carbomethoxylated (under thermodynamic control).

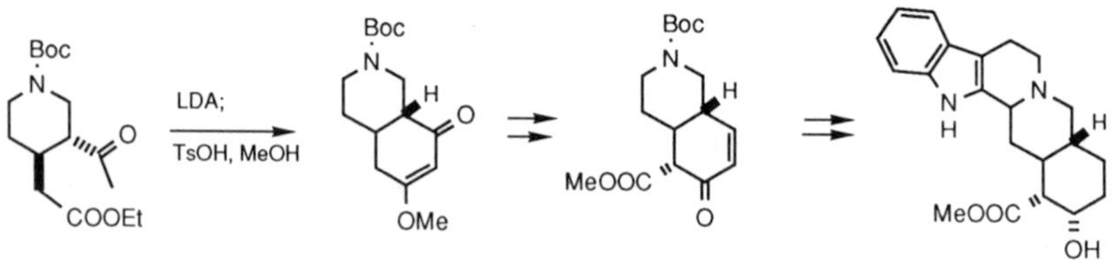

(+)-yohimbine

The presence of a triacylmethane function in tenellin facilitates its retrosynthetic analysis. In other words, a Claisen condensation would be extremely useful for the assembly of the molecular skeleton.

Two syntheses of tenellin have been reported. One of these involves an intramolecular Claisen condensation and extension of the sidechain by an aldol reaction [D.R. Williams, 1982]. The other synthesis [Rigby, 1989] exploits the C-carbamation of β-keto esters with isocyanates which may be considered as a variant of the Claisen condensation. Since the isocyanate used was unsaturated, the resulting acylenamine was well disposed to an internal ring closure.

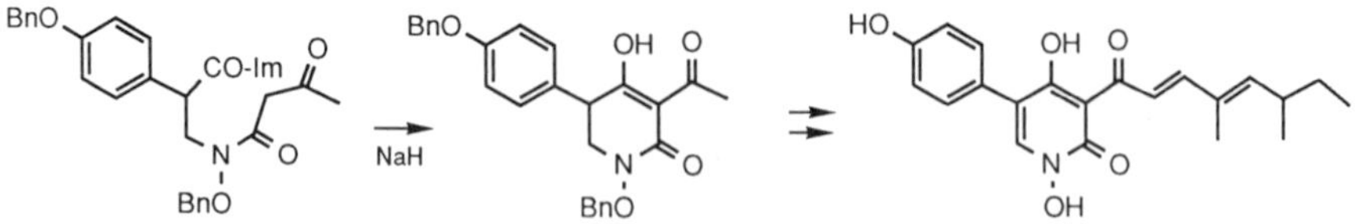

tenellin

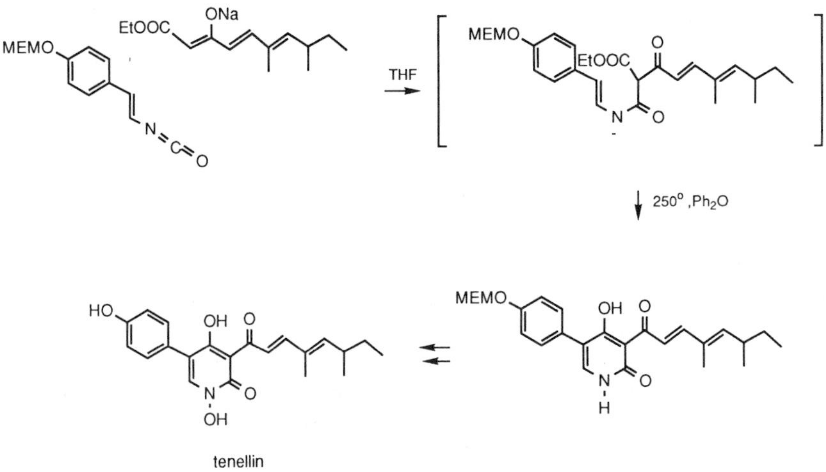

tenellin

1,3-Dioxygenated benzoic acid derivatives may sometimes be more readily prepared by a Claisen route. This method simplifies a synthesis of lasiodiplodin [Gerlach, 1977].

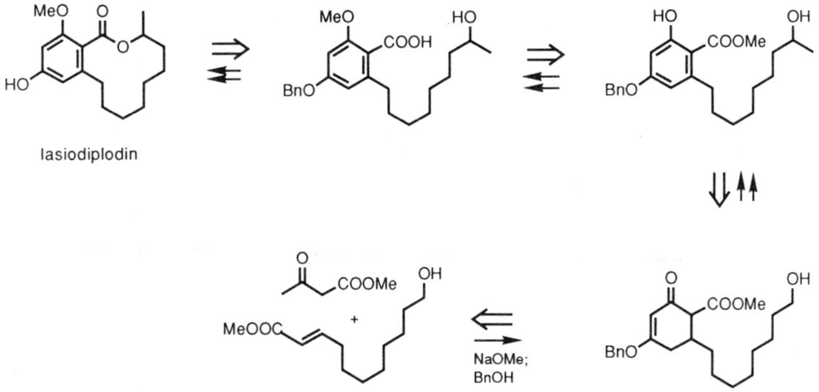

lasiodiplodin

In contemplating cedrol as a synthetic target, one's attention would immediately be drawn to the tertiary alcohol. Naturally, the norketone is a most suitable subtarget.

Pursuit of a cedrol synthesis [Stork, 1961] based on a Claisen transform of the norketone was perhaps greatly influenced by the fact that norcedranedicarboxylic acid was a known degradation product and could therefore serve as a checkpoint and a relay compound to alleviate supply of a late intermediate in a synthesis.

It turned out that steps after the Claisen condensation were sterically facilitated, as one of the carbonyl groups of the 1,3-diketone system is neopentyl which caused unidirectional enolization.

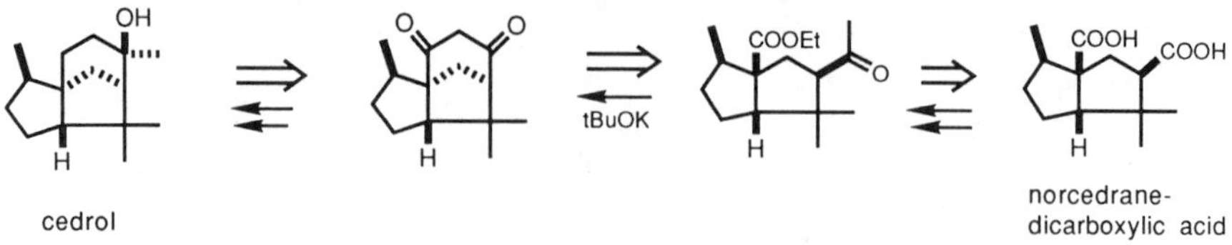

cedrol

norcedrane-
dicarboxylic acid

Claisen condensation was found to be superior to the aldol approach for closing the bridged ring of gymnomitrol [S.C. Welch, 1979]. These observations can be explained on steric grounds too. The cyclization is from the *endo* face, and nonbonding compressions of the propionyl group with the existing cyclopentane are unavoidable. The steric interactions could be largely eliminated after the Claisen condensation by enolization of the newly created ketone, away from the bridgehead. This enolization is not contrathemodynamic, since the bridged ring system does not permit the formation of a resonance-stabilized enol. The aldol is not afforded the same opportunity to lessen its steric strain, and the equilibrium lies toward the side of the open-chain keto aldehyde.

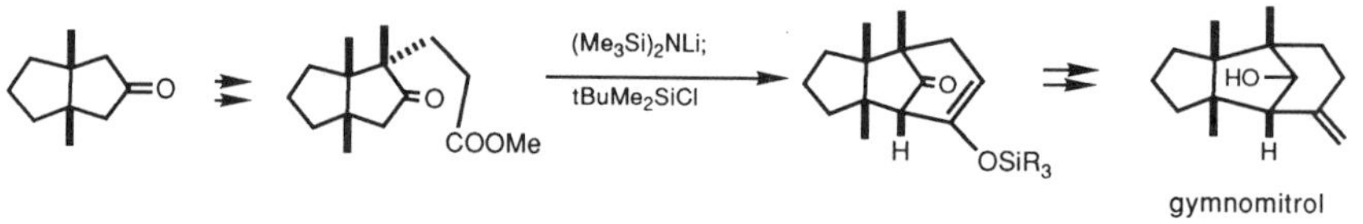

gymnomitrol

Tetracycline antibiotics retain in their structures most of the polyketide functionality (with some modifications). In a sense the Claisen condensation mimics the biosynthetic process and it should be of great value in the synthesis of such compounds. Indeed, an approach to 6-demethyl-6-deoxytetracycline [Woodward, 1963a] utilized the Claisen condensation for the formation of both the B- and the A-rings.

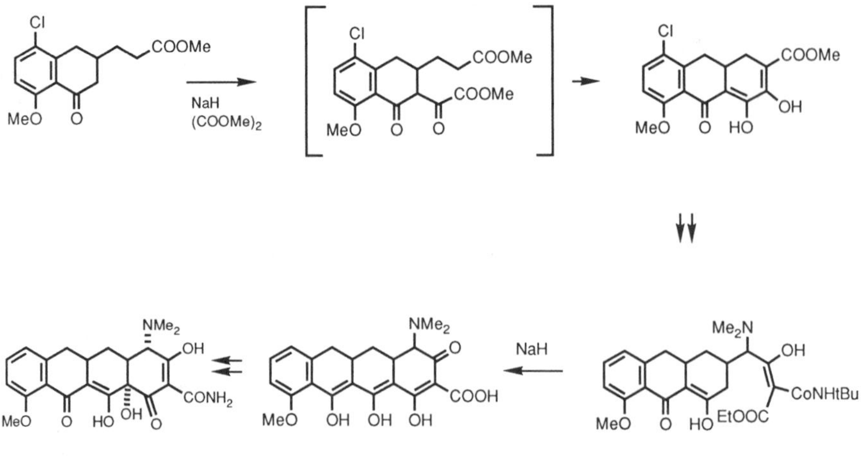

Another route [Stork, 1978] consists of A-ring closure by the same reaction on an isoxazole ester. The heterocycle served as a masked β-ketoamide.

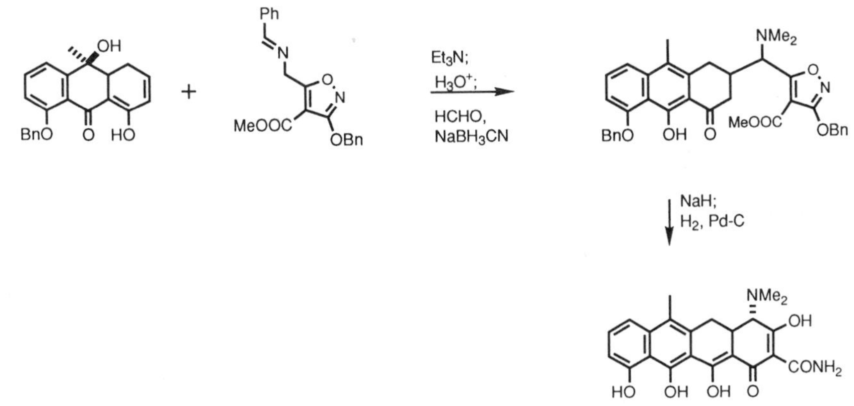

The key step of a terramycin synthesis [Muxfeldt, 1968] is the bisannulation via condensation of a thiazolinone with an acetonedicarboxylic acid derivative. The A-ring was formed by a Michael–Claisen tandem reaction, it was then followed by another Claisen condensation to complete the linear tetracycle.

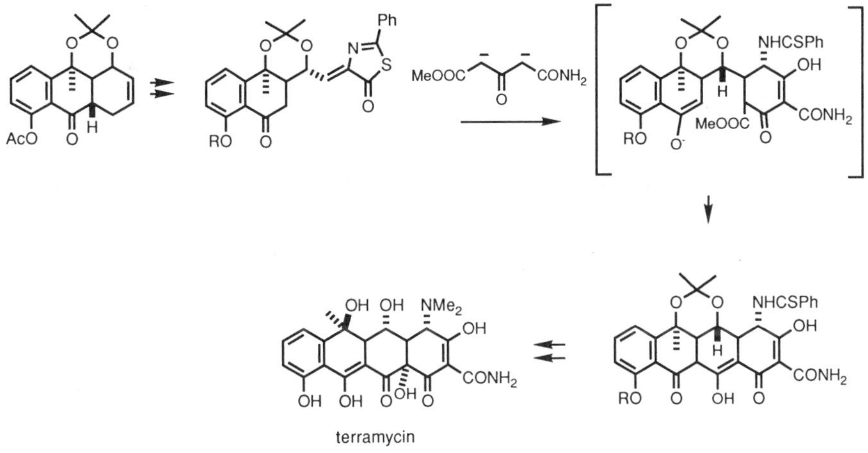

terramycin

Naphthalenones have found many applications in the synthesis of the anthracycline nucleus. Thus, an approach to aklavinone [Hauser, 1989] a Michael–Claisen tandem was envisaged. The requisite naphthalenone can be disconnected to yield a keto aldehyde and ultimately methyl vinyl ketone and a C_8-diene.

The central concept of this synthesis is very similar to the one reported eight years previously [Li, 1981].

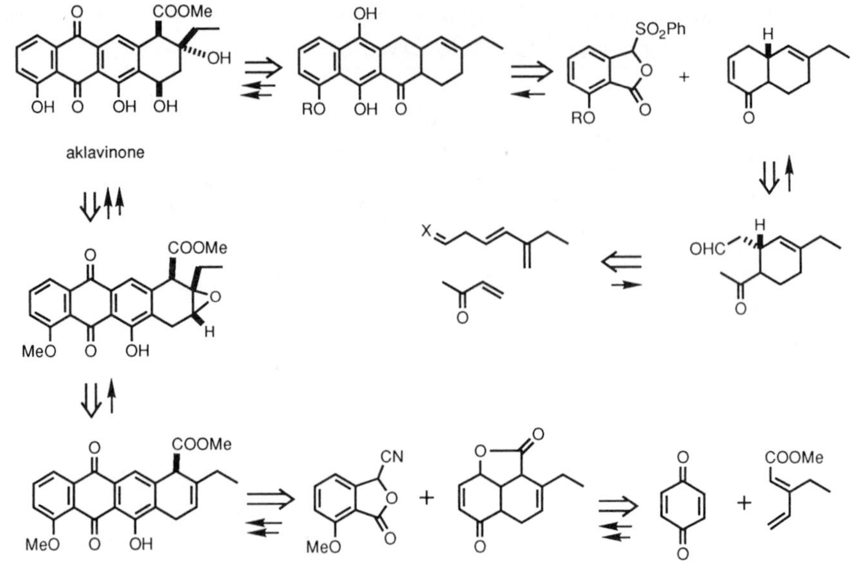

aklavinone

For synthesis of crytosporin [Brade, 1989] a very useful Michael acceptor is nitrofucal. The nitro group served a dual role of activation and a leaving group in the elimination.

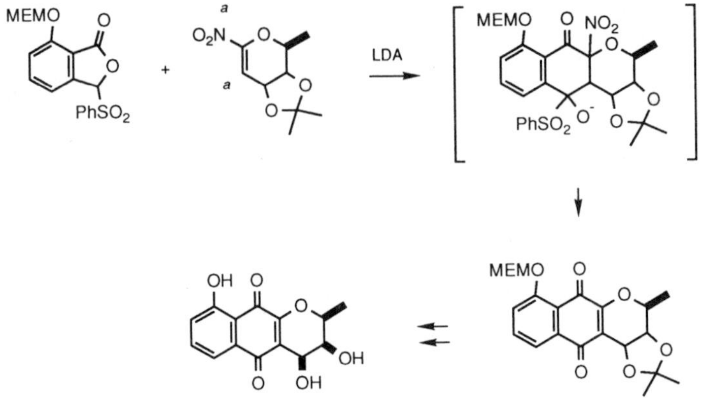

ent-cryptosporin

A Michael–Michael–Claisen sequence was designed to effect the assembly of the nor-indolic portion of the aflavinine ring system [Danishefsky, 1983].

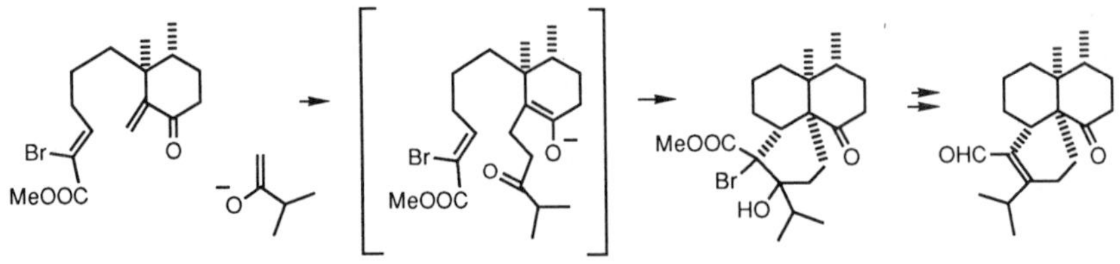

Judging from the numerous examples listed above, intramolecularity certainly contributes to the efficiency of the Claisen condensation. However, the inter-molecular version has not been totally eclipsed.

A chain elongation method specifically developed for polyketide synthesis involves C-acylation of acetone α,α'-dianion [Bringmann, 1985] or dianions of β-diketones or β-keto esters [Harris, 1986].

chrysophanol

eleutherin

A special case of the Claisen condensation is carboalkoxylation of enolates. Often in synthesis this reaction is employed in activation and/or regiocontrol. The need for homologation with branching is also satisfied by this method. Generally, excess dialkyl carbonate is mixed with a ketone in the presence of sodium hydride or a sodium alkoxide. Glyme and dioxane are good solvents for this reaction. Methyl cyanoformate appears to be gaining popularity as a carbomethoxylating agent [Mander, 1983].

Enamines are also excellent donors toward alkyl chloroformates [Stork, 1963a].

Two examples are mentioned here to show the utility of carboalkoxylation. A synthesis of occidol [Ho, 1971] from 5,8-dimethyl-α-tetralone relied on it to introduce the isopropanol pendant. In the brilliant synthesis of caryophyllene/isocaryophyllene [Corey, 1964b] the carbomethoxylation transcended the purpose of activation (in the subsequent methylation). The ester group played a critical role in a Dieckmann condensation and the ketone into which it transformed was also indispensable for unraveling the nine-membered ring in a fragmentation step.

occidol

caryophyllene

Condensation of ketone with a nitrile acceptor is also known [Yeh, 1988]. A pleasing approach to daunomycinone [Sih, 1979] capitalized on the dynamic equilibrium of the dihydroxyquinone moiety and the oxidizability of the reduced tetraol. When the latter compound is written in the dihydroquinone form its disconnection reveals a 2-acylnaphthalene-1,4-diol, with a 1,2,4-trisubstituted cyclohexane constituting the acyl portion. Recombination of the C-2 and C-4 substituents gives rise to a bicyclo[2.2.2]octenecarboxylic acid derivative. A Claisen condensation of 5-methoxy-α-tetralone successfully initiated a synthesis of the anthracycline network.

daunomycinone

Lactams may serve as donors in a Claisen condensation. Proxiphomin features an α-acyl-γ-butyrolactam moiety, and the Claisen condensation paved the way to the elaboration of the complete skeleton of this substance by an intramolecular Diels–Alder cycloaddition [Taploczay, 1985].

proxiphomin

The pathway involving α-acylation and retro-Claisen ring opening (and decarboxylation) of the product has been recognized as a useful synthetic approach to the nicotine alkaloids. A more recent exploitation of this reaction sequence is shown in a synthesis of myosmine [Brandange, 1976] in which the lactam nitrogen was protected by a vinyl group.

myosmine

A further application of the Claisen condensation is seen in a synthesis of *ent*-thiolactomycin [Chambers, 1989] from (2S)-ethyl lactate.

ent-thiolactomycin

Furanoquinoline alkaloids such as maculine contain a 1,3-diketone which is masked as enol ethers. The imino ether subunit may be derived from an amino lactone or a hydroxy lactam via cyclodehydration. With respect to the former option, the precursor would be an α-aroyl-γ-butyrolactone. Although by this synthetic method a dehydrogenation step at a later stage would be required to establish the furan ring, the availability of starting material augurs a Claisen condensation with butyrolactone as the donor moiety [Zimmer, 1963]. (Conceptually the same as the conventional Claisen condensation but higher yielding is the method consisting of acylation of α-acetyl-γ-butyrolactone with the aroyl chloride, followed by deacetylation with ammonia.)

maculine

4.3. DIECKMANN CYCLIZATION

Dieckmann cyclization [Schaefer, 1967] is referred to the base-promoted intramolecular condensation of a diester to form an α-carboalkoxycyclo-alkanone. Somewhat surprisingly, phenyl esters of dicarboxylic acids are generally unreactive. The reaction is one of the most popular and efficient methods for the construction of five- and six-membered ring compounds, especially ketones.

Closely related to, but much less popular than the Dieckmann cyclization is the Thorpe cyclization of dinitriles. The rare application of the latter reaction in synthesis is due to lack of available substrates, requirement of stronger bases (e.g., amide anions), and the lesser versatility of the imino nitrile (or keto nitrile) products.

Like other Claisen-type condensations, the Dieckmann cyclization is driven by enolate formation of the products. This stabilization suppresses the reversible decomposition, and it explains the failure of diesters to cyclize when the cyclic β-keto esters are inhibited to enolize.

The dimerization of diethyl succinate under the influence of sodium ethhoxide involves the Dieckmann cyclization in the second stage.

In the area of natural product synthesis one notes an efficient method for elaboration of camphor [Ruzicka, 1920] using the Dieckmann condensation.

camphor

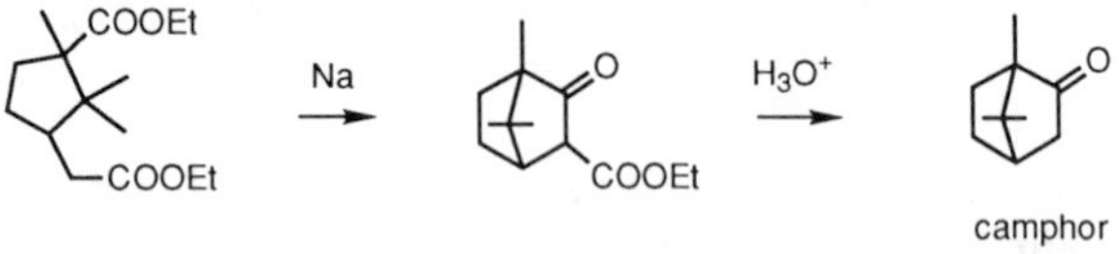

An early elaboration of equilenin [Bachmann, 1940] testifies to the importance of the Dieckmann cyclization in relatively complex synthetic operations before the advent of modern methodology. In fact, for the construction of the steroid D-ring, few methods are superior.

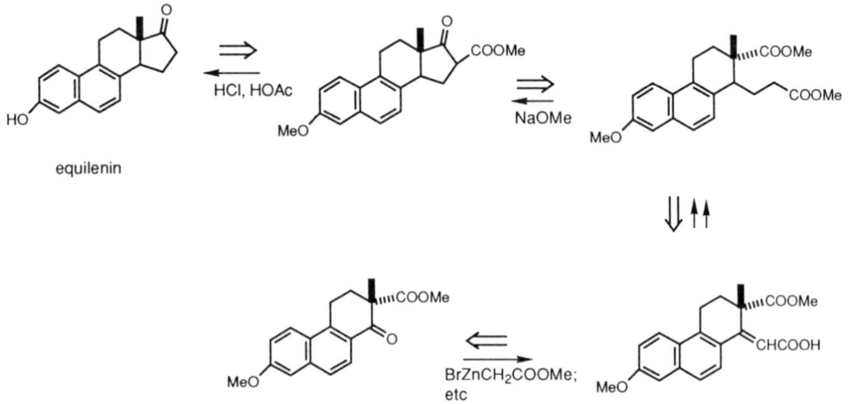

equilenin

It is interesting to note that in a synthesis of the culmorin diketone [B.W. Roberts, 1969] from tetrahydroeucarvone the two functionalized five-membered rings were established by a Claisen and a Dieckmann condensation, respectively.

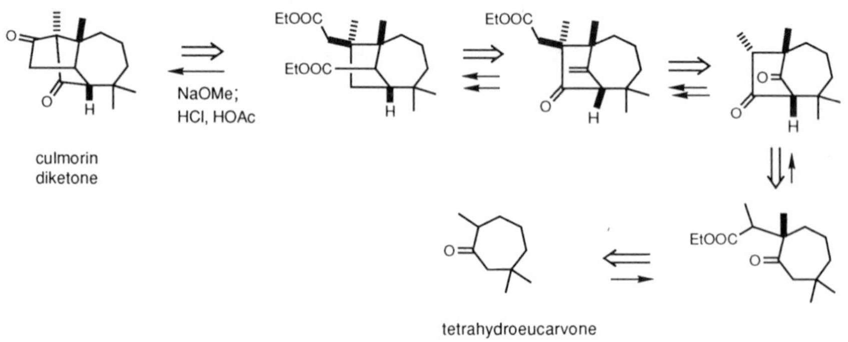

culmorin
diketone

tetrahydroeucarvone

The presence of the β-keto ester subunit in a Dieckmann cyclization product enables an efficient introduction of a carbon chain to the α-position of a cyclic ketone. An example is described in an approach to nootkatone [Marshall, 1970]. In fact, the Dieckmann condensation was achieved in conjunction with a Wittig olefination step. Note the chemoselectivity resulting from the enolization which protected the β-keto ester moiety from attack by the Wittig reagent.

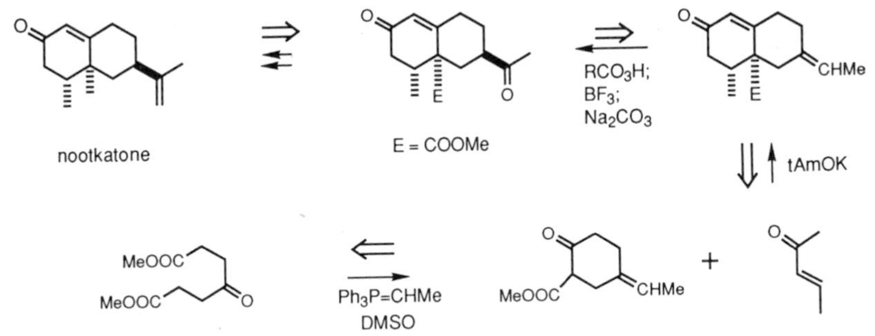
nootkatone
E = COOMe
RCO3H;
BF3;
Na2CO3
CHMe
tAmOK
MeOOC
MeOOC
Ph3P=CHMe
DMSO
MeOOC
CHMe

The structure of gascardic acid features a cycloheptenecarboxylic acid which is fused to a cyclohexane and spiroannulated to a five-membered ring simultaneously. The unsaturated acid system reminds one of its accessibility from a β-keto ester, and hence a Dieckmann transform ranks high in the synthetic design of the sesterterpene [Boeckman, 1979]. The selectivity of this cyclization was caused by steric hindrance to deprotonation at the neopentyl methylene group.

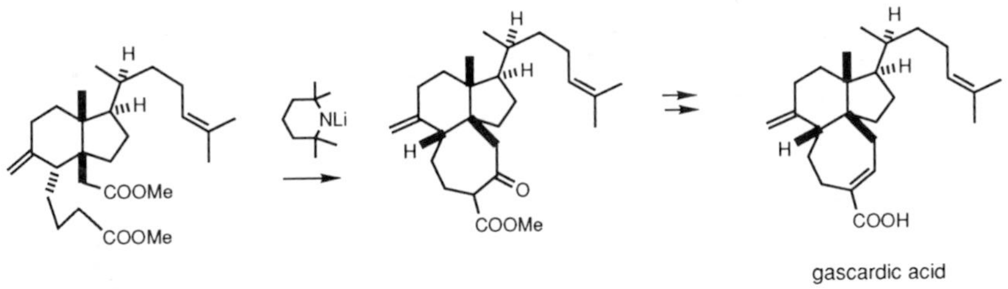
COOMe
COOMe
NLi
COOMe
COOH
gascardic acid

Gnididione is an interesting synthetic target in view of its multifunctionality, the presence of a hydrazulene framework, and a furan ring. A disconnection that has been successfully followed through is a sequence of aldol, alkylation, and Dieckmann transforms [Buttery, 1990], the last one with the prior FGA operation.

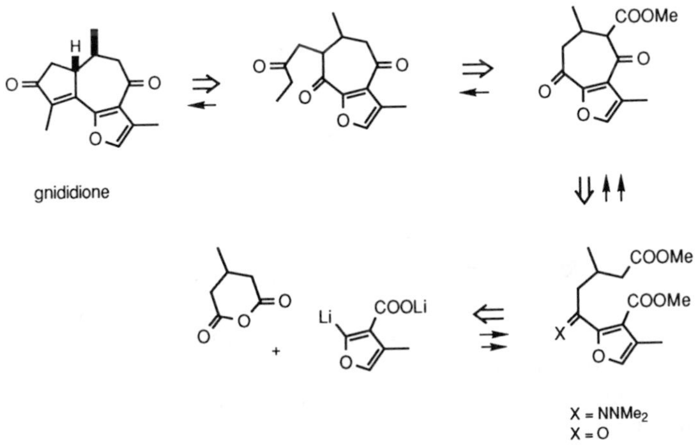
gnididione
COOMe
COOLi
COOMe
COOMe
X = NNMe2
X = O

The major advantage of this route pertains to reactivity differentiation of the two ketone groups in the alkylation step necessary for cyclopentenone formation. The diketo ester dianion would undergo monoalkylation at the nonstabilized enolate selectively. The precursor of the Dieckmann cyclization substrate is readily conceivable, and most gratifyingly it can be derived, in a few steps, from coupling of the symmetrical methylglutaric anhydride with a furan derivative.

The synthesis of olivin [Roush, 1987], the aglycon of olivomycin A, was keyed on allylboronate chemistry to elaborate the polyoxygenated sidechain, and two Dieckmann cyclizations to construct the α-tetralone moiety of the hydroanthracene nucleus.

The ester pendant left from the first Dieckmann condensation became the ketone group of the target molecule. This is a pleasing aspect as most often the ester pendant would be removed via hydrolysis (this is also the case after the second Dieckmann step is realized in the present synthesis).

(+)-olivin

Yohimbe alkaloids are amenable to synthesis which is keyed on E-ring formation from tetracyclic intermediates, because the terminal ring is substituted with a β-hydroxy ester subunit. This structural requirement suggests a Dieckmann transform as a viable disconnection.

The generalized scheme for alkaloid synthesis has been extended to an access of pseudoyohimbine and a predecessor of deserpidine [Wenkert, 1984]. Thus, the N-tryptophylpyridinium bromide derived from methyl α-methoxy-β-(3-pyridyl)acrylate was converted into two diesters which differed in the configuration of C-20 (yohimbe alkaloid numbering). Dieckmann cyclization of each isomer culminated in the pentacyclic products.

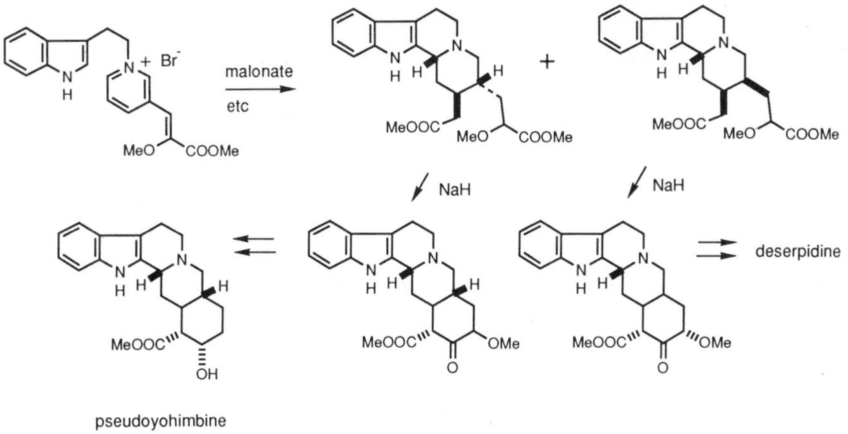

pseudoyohimbine

The Dieckmann cyclizations are under both kinetic and thermodynamic control. Enolization at the alternative site is disfavored by the donor α-methoxy group (captodative effect). The β-keto esters from those ester enolates would be deprived of stabilization through metal chelation.

In gelsedine the C—C bond joining the spirocyclic center to the ethereal carbon is strategic for disconnection and an aldol condensation should be useful for its reconstitution. Further disconnection of the simplified target gives rise to a pyrrolidine which is annulated to a lactol ring, and ultimately a Michael–Dieckmann transform can be identified.

With respect to the actual synthesis work [Kende, 1990] it is unfortunate that the final step of aldol condensation yielded only an epimer. The reason is that the formation of the gelsedine skeleton is electrostatically disfavored.

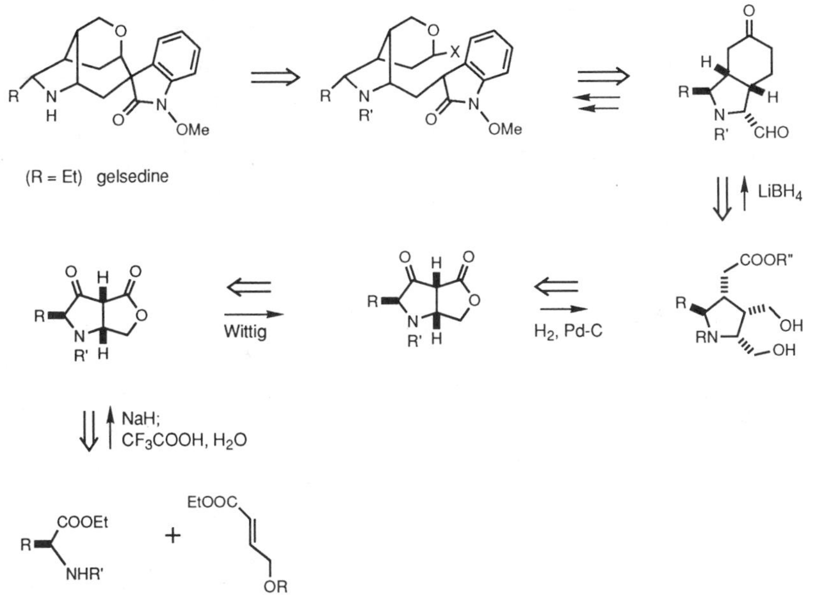

The Dieckmann cyclization is not limited to the formation of carbocycles. Many alkaloid syntheses have employed this reaction to obtain the desired intermediates. For example, the intriguing transannular closure approach to the necine bases [Leonard, 1969] relied on the Dieckmann cyclization to form an 8-membered azacycle. It should be noted that despite the lower yields such products are obtainable from symmetrical acyclic diesters. This feature simplifies the preparation of starting materials. (When dialdehydes are used instead of the diesters to construct the necine bases such as trachelanthamidine [Leonard, 1960] the process is a Mannich reaction [Takano, 1981].)

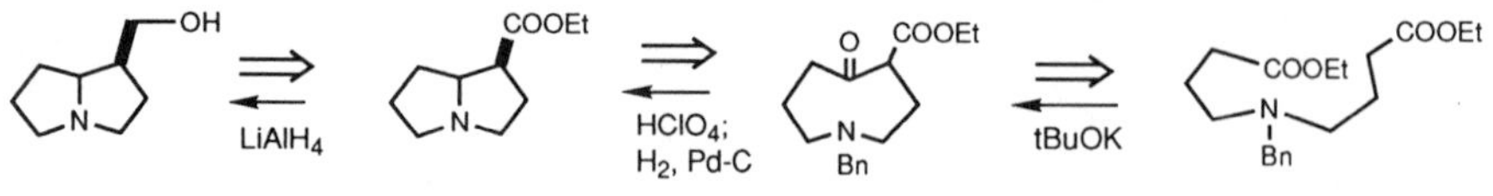

isoretronecanol

Tropinone has also been synthesized by a Dieckmann cyclization of dimethyl pyrrolidine-2,5-diacetate, followed by removal of the ester group [Willstätter, 1921a; W. Parker, 1959]. Again, the value of molecular symmetry in facilitating synthesis is evident.

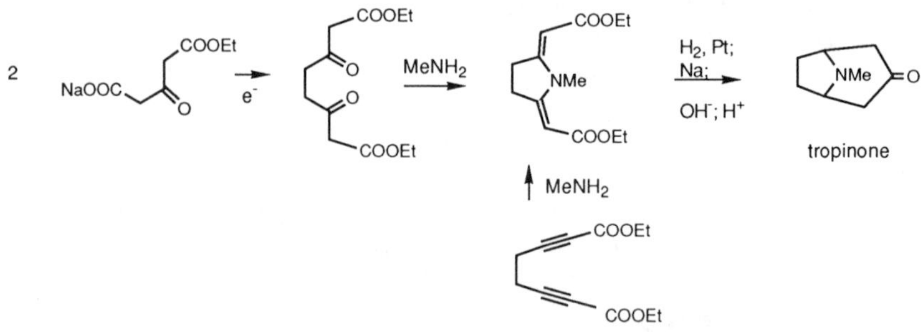

tropinone

There are many ways to disconnect ipalbidine. Unraveling the nature of the nonaromatic trigonal carbon atoms as *preferred* by the nitrogen atom is helpful in formulating efficient synthetic pathways. As the anchoring point of the methyl group turns out to be an acceptor, an FGI operation with a carbonyl, then FGA of an ester permitted application of a Dieckmann transform [Govindachari, 1970]. (A Mannich transform is also conceivable.)

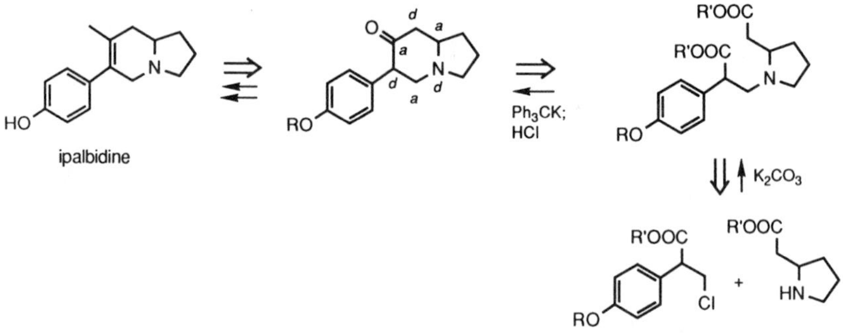

ipalbidine

R= Me, R'= Et

The first successful synthesis of the antitumor and antileukemic alkaloid camptothecin [Stork, 1971] exploited the 1,5-relationship of the two carbonyl functions. Thus, a Michael-type reaction served to create this important structural segment. Using a mixed carbonate to protect the α-hydroxy butanoic ester made it possible to form a γ-lactone by an intramolecular acylation. The lactone carbonyl group was to be reduced at a later stage to provide the methylene group of the δ-lactone. The tetracyclic Michael acceptor was formed by a Dieckmann cyclization followed by reduction and decarboalkoxylation.

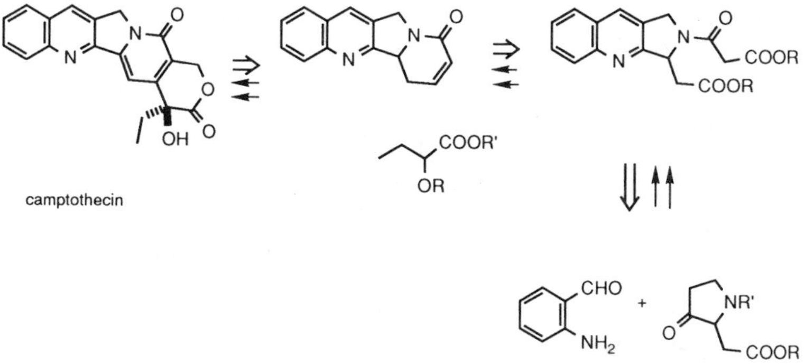

A synthesis of yohimbine [Stork, 1972a] started from a 4-piperidinone which is easily obtainable by a Dieckmann cyclization route. In this synthesis four contiguous asymmetric centers were created during annulation of this ketone and the subsequent reduction steps. Thus, thermodynamically controlled reduction of the enone led to a product with a *trans* ring juncture and an equatorial ester pendant. The secondary hydroxyl group was established by catalytic hydrogenation of the ketone.

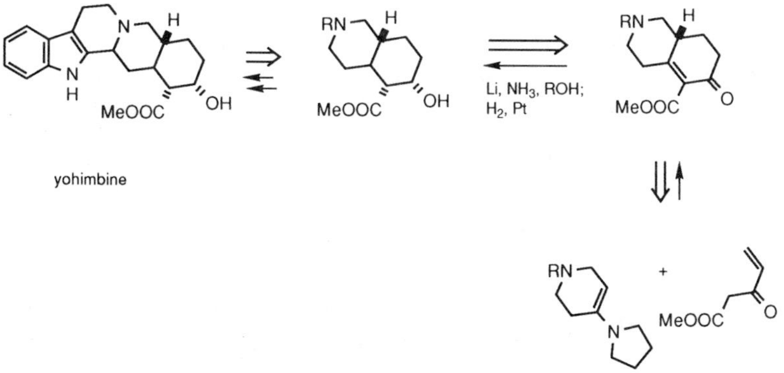

The Dieckmann cyclization has found excellent use in the convergent and enantioselective approach to (+)-ikarugamycin [Boeckman, 1989b; Paquette, 1989]. The formation of a keto lactam within a macrocyclic framework actually resulted in a bridged ring system.

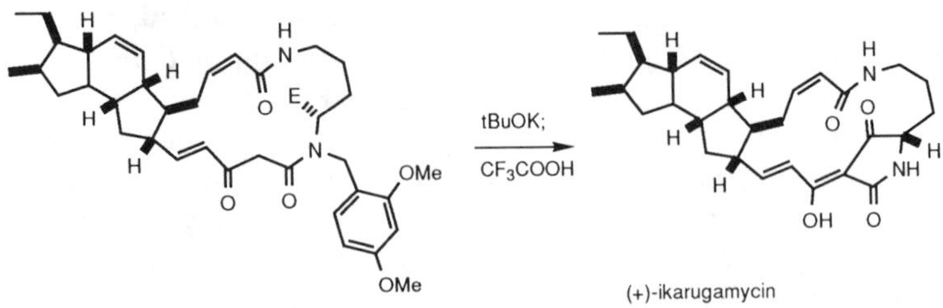

(+)-ikarugamycin

α-Cyclopiazonic acid is a mycotoxin which contains a tetramic acid moiety. In a synthesis [Kozikowski, 1984b] the assembly of this structural subunit also relied on the Dieckmann cyclization.

Luciduline contains a β-amino ketone subunit. As an alternative to a more obvious Mannich transform, an intramolecular acylation route may be considered [Szychowski, 1979; Schumann, 1984]. The proper intermediate must possess an additional carbonyl group, i.e., a lactam function.

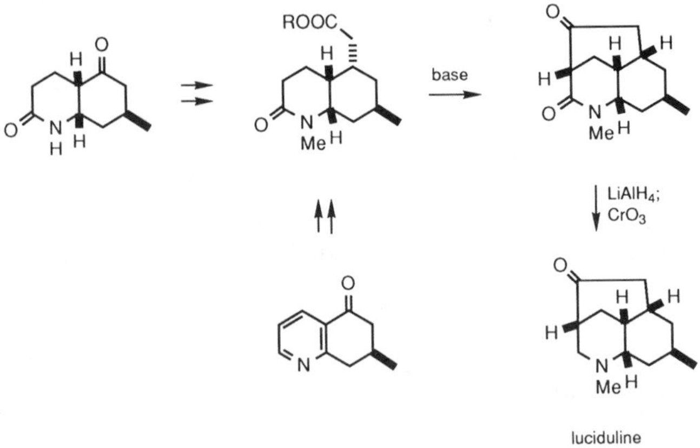

luciduline

A synthesis of 8-methoxy-α-tetralone by a tandem Michael–Dieckmann condensation has been reported [Tarnchompoo, 1986]. The Michael donor is a slightly stabilized (by an ester group at an *ortho* position) benzylic anion, and the accept is an acrylic ester.

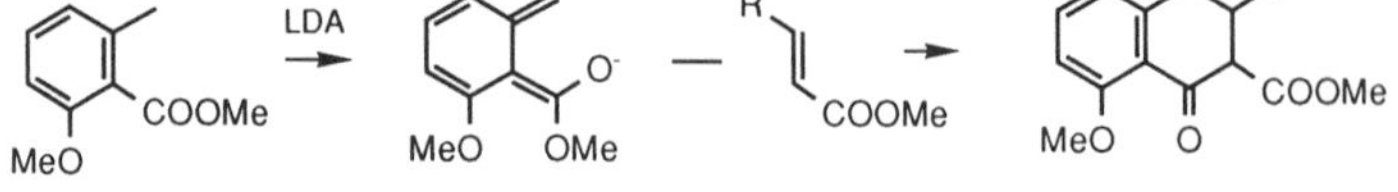

In a synthesis of equilenin [Johnson, 1947] a Stobbe condensation [Johnson, 1951] was used to introduce the D-ring. The ring closure stage is a Dieckmann-type cyclization in which the acceptor is a cyano group. Noteworthy is the fact that the tricyclic β-keto ester failed to react in a similar fashion, presumably due to the much more severe steric hindrance surrounding the ketone. It should also be mentioned that hydrogenation of the C-15/C-16 double bond gave a 2:1 C/D *trans:cis* mixture.

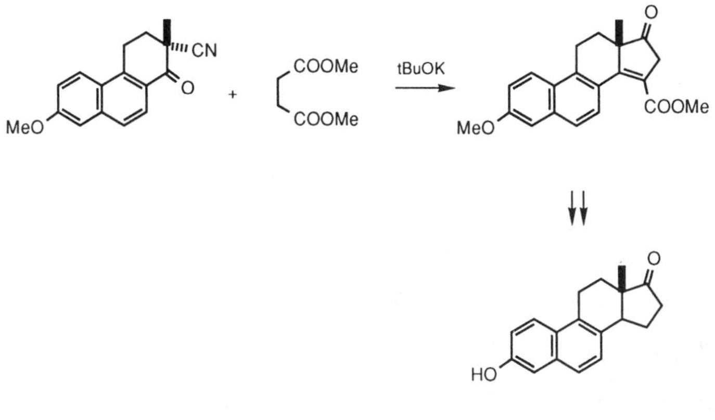

equilenin

A formal [3 + 2] cycloadditive assembly of 2-carboalkoxy-2-cyclopentenones can be achieved by reaction of carboalkoxyethylzinc derivatives with acetylenic esters [Crimmins, 1990]. The reaction undoubtedly proceeds by a Michael–Dieckmann reaction tandem. Interestingly, with 2-cyclohexenone as Michael acceptor the intramolecular Claisen condensation did not occur.

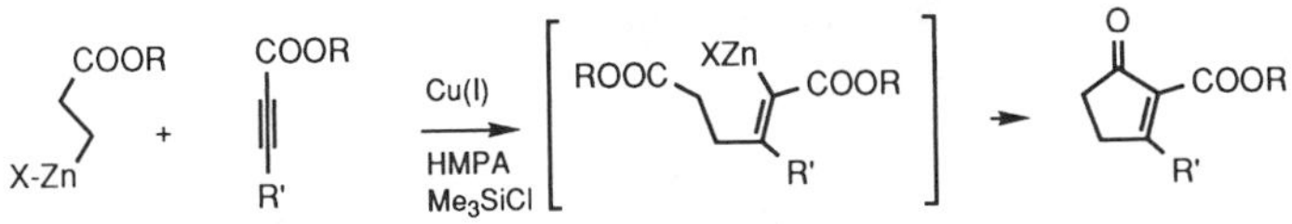

The acid version of the Claisen–Dieckmann cyclization is the intramolecular Friedel–Crafts acylation which is exemplified by the closure of the cyclopentenone ring of aflatoxin B_1 [Büchi, 1967; J.C. Roberts, 1968].

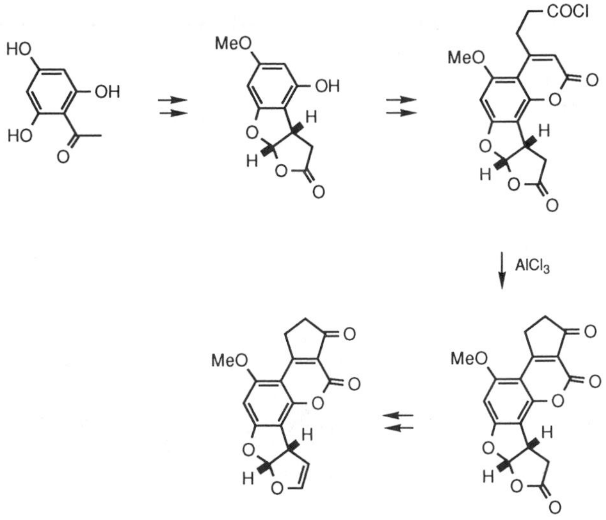

aflatoxin B_1

5

OTHER METHODS FOR THE PREPARATION OF 1,3-DIFUNCTIONAL COMPOUNDS

5.1. 1,3-DIOXY COMPOUNDS FROM α,β-UNSATURATED CARBONYL SUBSTANCES

Various factors may dictate the ways that 1,3-dioxy compounds are synthesized. It is possible that the functional array be created from a monofunctional predecessor such as a carbonyl compound. A case in point is ptaquilosin which possesses an aldol moiety. A synthesis [Kigoshi, 1989] actually was conducted by an elaboration of a hydrindenone in which the ketone group is located in the six-membered ring. In this design the introduction of the cyclopropane and the placement of the unsaturation was readily accomplished. The peri carbon was oxygenated via epoxidation of the corresponding conjugate enone. The reported synthesis resulted in the enantiomer of the natural product because of the (+)-menthyl cyclopentanecarboxylate used is in the opposite optical series.

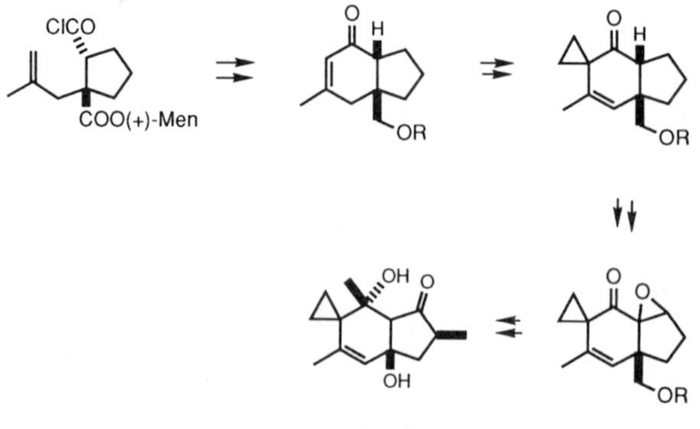

ent-ptaquilosin

Hydration of conjugated ynones leads to 1,3-diketones. In view of the proclivity of 1,3-diketones toward fragmentation, it might be advantageous to add an alcohol to the triple bond to afford enol ether intermediates. Twofold addition of ethylene glycol leads to the monoketals of 1,3-diketones.

A reaction sequence encompassing methanolation of an ynone and mild hydrolysis concluded the penultimate stages of a synthesis of (−)-talaromycin A [Crimmins, 1989]. Although the final elaboration of the target molecule is unrelated to the chemistry under discussion, it is notable that an intramolecular free-radical cyclization into the dioxaspiro[5.5]undecane template solved a very difficult problem of introducing an axial hydroxylmethyl group vicinal to the equatorial alcohol.

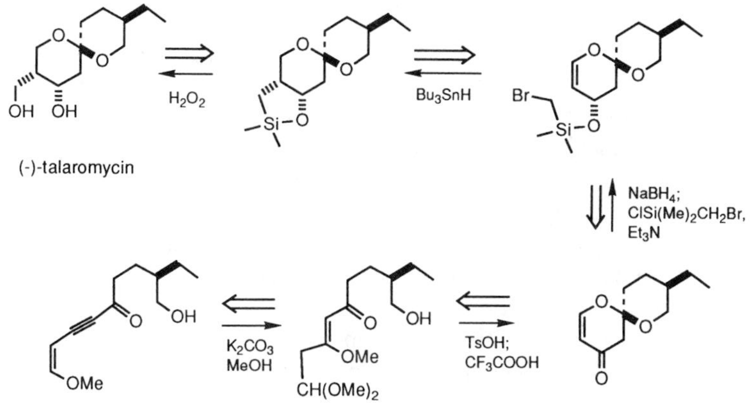

A forskolin synthesis [Corey, 1988a] also proceeded from an ynone to create the heterocyclic portion of the molecule.

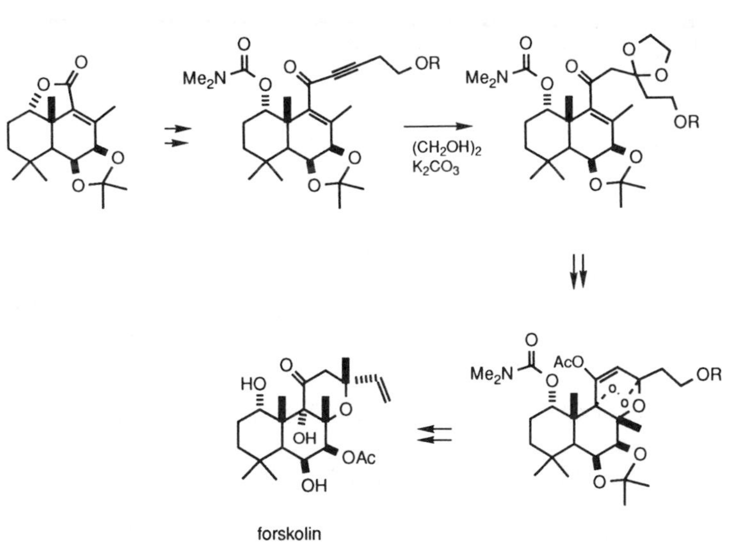

forskolin

Lewis acids promote rearrangement of α,β-epoxy ketones to 1,3-diketones. Cyperolone has been obtained from 4β,5β-epoxy-α-cyperone on exposing the latter compound to BF₃ [Hikino, 1966].

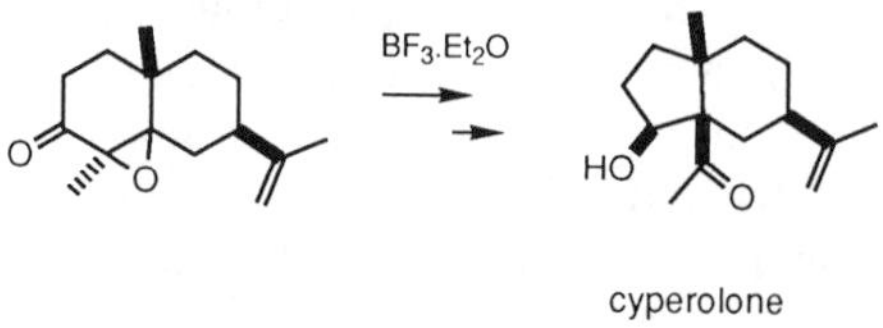

cyperolone

5.2. 1,3-DIOXY COMPOUNDS FROM HOMOALLYLIC ALCOHOLS

Homoallylic alcohols are latent aldols that could be generated by cleavage of the double bond. A method for aldol synthesis is by submitting carbonyl compounds to allylorganometallic reactions followed by ozonolysis or other oxidative manipulations.

Vermiculine has two acetone subunits attaching to the ethereal carbons of the 16-membered diolide nucleus. These β-oxy ketone subunits probably would not survive the many reaction conditions along a synthetic pathway. Consequently, it is essential to introduce them in a masked form or protect them immediately after their appearance. A homomethallyl alcohol derivative proved adequate for the synthetic purpose [Corey, 1975c].

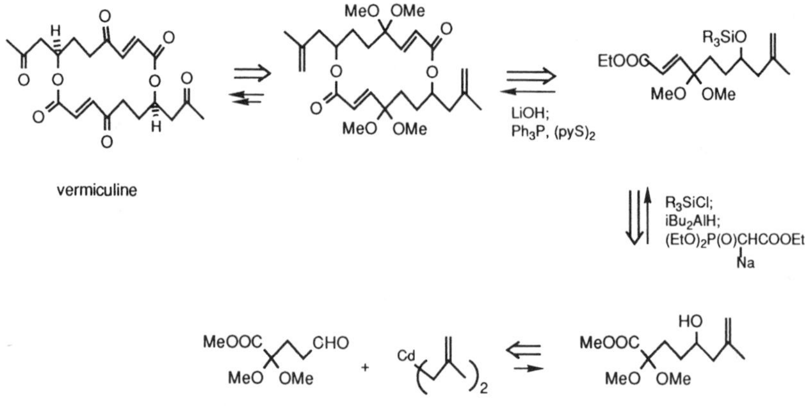

vermiculine

In one approach to monensin [Collum, 1980] the left region of the molecule was constructed by an aldol reaction and a condensation with a (Z)-crotyl-aluminum reagent. The homoallylic alcohol obtained from the latter reaction was protected and ozonized to reveal the aldehyde group which participated in another aldol process to complete the entire skeleton of monensin.

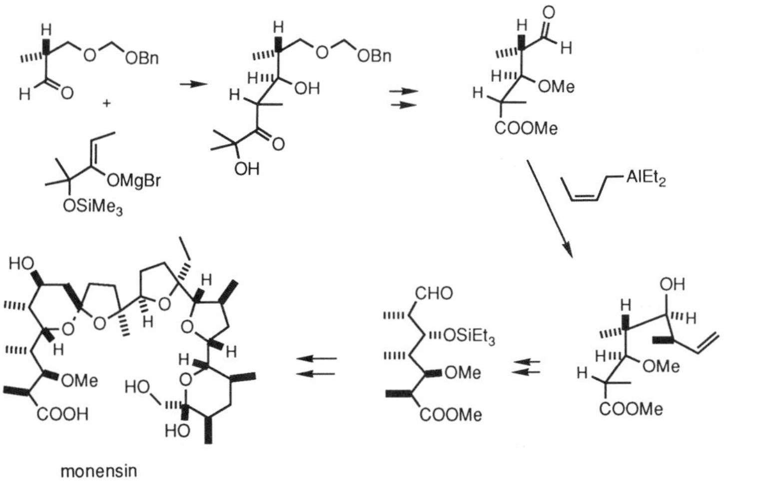

monensin

5.3. 1,3-DIOXY COMPOUNDS BY HOMOLOGATION OF THE CARBONYL GROUP

β-Keto esters are formed by reaction of ketones with a diazoacetic ester under the influence of a Lewis acid. This C—C bond insertion has become a method of choice for ring expansion of cyclic ketones. Their stability, in comparison with diazomethane, their commercial availability (far more economical than trimethylsilyldiazomethane), the regiocontrol they render [Liu, 1973, 1975], and the formation of difunctional products in their reactions contribute to the superiority of the diazoacetic esters as reagents.

The regioselectivity of the ring expansion may stem from steric interactions. Apparently one diastereomeric intermediate is formed preferentially in the reversible step. This diastereomer rearranges with migration of the less highly substituted carbon. The regioselectivity is crucial to syntheses such as that of (−)-khusimone [Liu, 1979] which was planned on the premise of a photocyclo-addition–ring expansion–intramolecular alkylation sequence.

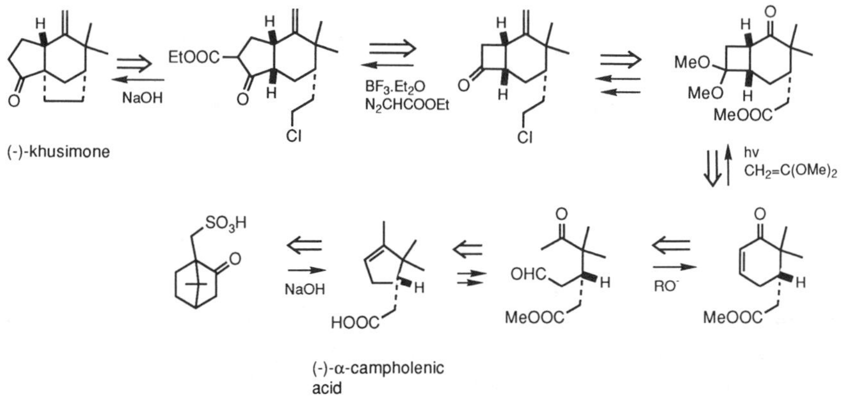

(-)-khusimone

(-)-α-campholenic
acid

The analogous reaction with aldehydes leads to α-unsubstituted β-keto esters [Holmquist, 1989].

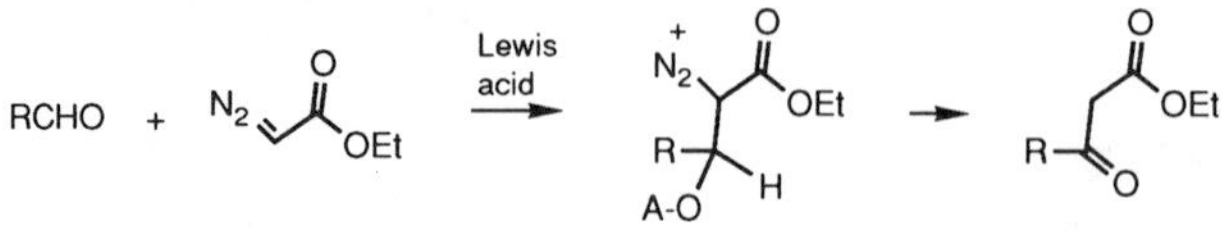

5.4. 1,3-DIFUNCTIONAL COMPOUNDS VIA REDUCTION

Reaction of α,β-epoxy ketones with zinc, chromium(II), and other mild reducing agents can lead to aldols. As expected, this umpoled reaction (all redox reactions involve contrapolarization) on a disjoint molecule (-a–a-) furnishes a conjoint product.

B-Selenyl borates exhibit an excellent potential for effecting the same transformation [Miyashita, 1988]. The mechanism may be different.

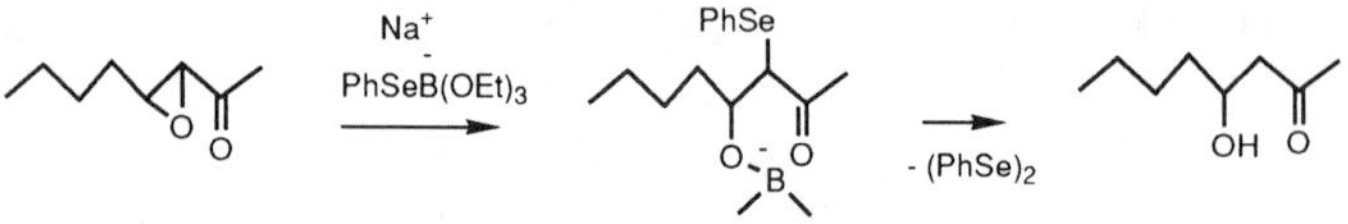

1,3-Diols are generated by reduction by glycidols with complex metal hydrides. The hydride delivery likely proceeds intramolecularly via a five-membered transition state. A closed segment of five atoms is necessarily disjoint, and it is interesting to note that this umpoled reaction on a disjoint substrate (epoxide) leads to a conjoint 1,3-diol. The double inversion of polarity is general, another example being the assembly of β-hydroxy carbonyl compounds from 1,3-dithianes and epoxides.

A 1,3-diol is the key intermediate in a synthesis of (+)-α-lipoic acid [Page, 1986].

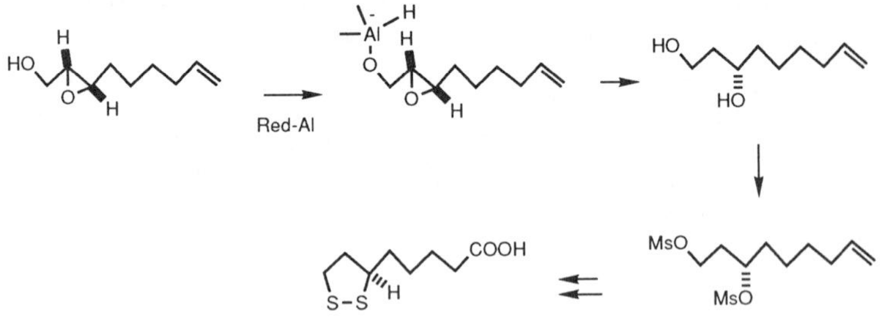

α-lipoic acid

A convenient preparation of 1,2-dioxolane is by O_2 insertion of *trans*-1,2-disubstituted cyclopropanes [K. Feldman, 1989]. Reductive O—O bond cleavage with samarium(II) iodide results in syn-1,3-diols. Yashabushitriol has been acquired from an intermediate thus synthesized.

yashabushitriol

Perhydrohistrionicotoxin is a 1,3-amino alcohol amenable to assembly not only by straightforward manipulation in terms of laying the two hetero atoms in a conjoint relationship. A scheme consisting of double polarity inversion has been devised [Godleski, 1983]. Thus, palladium(0)-catalyzed intramolecular displacement of an allylic acetate by an amine was followed by hydroboration which introduced, after oxidation of the organoborane intermediate, the requisite oxygen function. Although the hydroboration step was essentially controlled by the substitution pattern of the double bond, the unmistakable fact is that such an anti-Markovnikov addition entails contrapolarization of the carbon atoms of the transposed double bond.

perhydro-
histrionicotoxin

5.5. 1,3-DIPOLAR CYCLOADDITIONS

Certain types of 1,3-dipolar cycloadditions [Padwa, 1984] are very useful for assembling 1,3-difunctional molecules. There has been a phenomenal growth in the application of these reactions to complex synthesis.

A 1,3-dipolar cycloaddition involves 6π electrons (cf. Diels–Alder reaction), four from the 1,3-dipole and two from the dipolarophile. There are three types of 1,3-dipoles [Sustmann, 1971]: type I dipoles have relatively high-lying HOMOs and LUMOs, and are generally referred to as HOMO-controlled, or nucleophilic 1,3-dipoles. Type III dipoles possess lowest-lying frontier orbitals, they are electrophilic, undergoing LUMO-controlled reactions. It is easy to imagine that type III 1,3-dipoles prefer to cycloadd to donor-substituted dipolarophiles.

Interestingly, reactions of the ambiphilic type II, 1,3-dipoles are accelerated by either a donor or acceptor-substituent on the dipolarophile.

5.5.1. Cycloadditions of Nitrile Oxides and Nitrones

Nitrile oxides and nitrones are particularly versatile 1,3-dipoles because they are readily prepared and the heteroatoms of the cycloadducts may be transformed into various functional groups [Torssell, 1988]. For example, a synthesis of hernandulcin [Zheng, 1989] shows a direct translation of a double-bond configuration into two asymmetric centers with simultaneous introduction of a β-hydroxy ketone subunit in a masked form. Although the unmasking of the ketone also destroyed the conjugate double bond, its reintroduction was not problematical.

hernandulcin

While the hernandulcin synthesis is concerned mainly with the establishment of a β-hydroxy ketone subunit, it must be emphasized that the 1,3-dipolar cycloaddition route differs from, but complements the aldol condensation. Note the C—C bonds formed in the two methods are C-*ipso*/C-α, C-α/C-β, respectively.

The major problems concerning synthesis of phyllanthocin are regio- and stereocontrol in the attachment of the ester group and the tetrahydrofuran ring to the cyclohexane nucleus. Nitrile oxide cycloaddition to a suitable cyclohexene, if one could be found, would be very effective for the establishment of the tetrahydrofuran moiety that contains an oxygenated site at the β-carbon, and also unambiguously the *cis* ring juncture. The challenge is to find a cyclohexane derivative equipped with a latent carboxyl function and capable of directing the cycloaddition in a sterically and orientationally correct sense.

It is not expected that any arbitrary 4-substituted cyclohexene could fulfill these two criteria. Very strong steric discrimination toward one of its faces during the cycloaddition would be questionable. Additionally, a 4-carbonyl derivative would also favor the undesirable isomer (polarity alternation). On the other hand, a bicyclic lactone has proved to be an ideal cycloaddend [Martin, 1987], as the lactone bridge shields the approach of the nitrile oxide from the *syn* face, and electronically or electrostatically, the allylic oxygen atom exerts a major influence on the regioselectivity of the cycloaddition. The carbon group is farther away, and accordingly less effective in comparison. Needless to say, the ethereal oxygen atom of the lactone is extraneous, it was to be removed subsequently in this synthetic pathway.

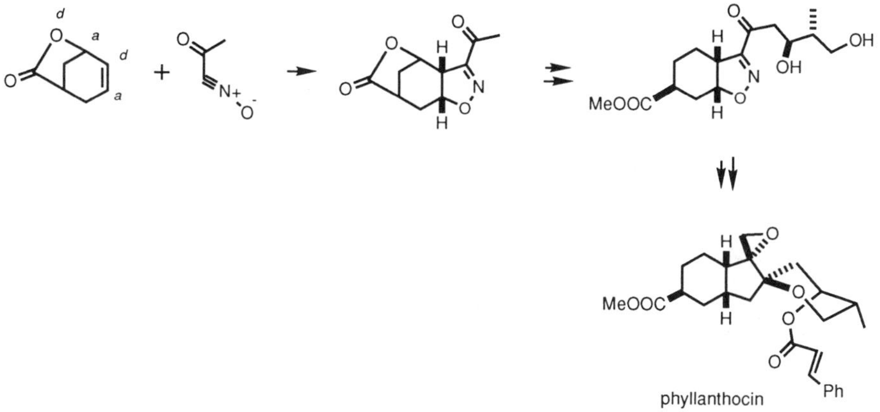

phyllanthocin

More straightforward is a synthesis of bisabolangelone [Riss, 1986] based on the same reaction which took full advantage of the pronounced polarization of the cyclohexenone double bond.

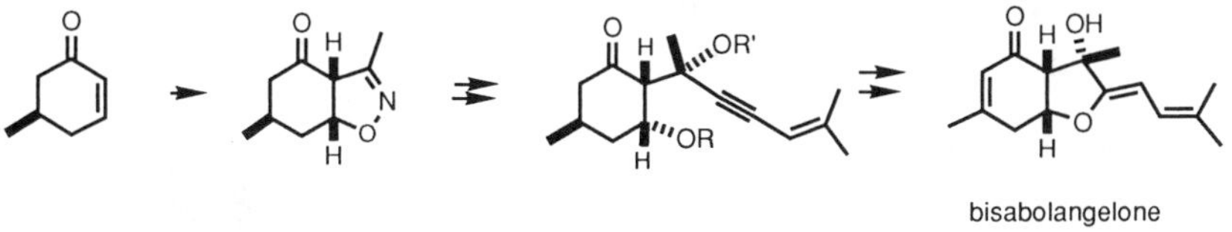

bisabolangelone

A synthesis of (+)-[6]-gingerol [Le Gall, 1989] demonstrates an asymmetric induction by a diene-iron tricarbonyl residue directly bonded to the dipolarophilic alkene. In terms of synthetic equivalency the nitrile oxide corresponds to an acyl anion and the dipolarophile to a chiral epoxide.

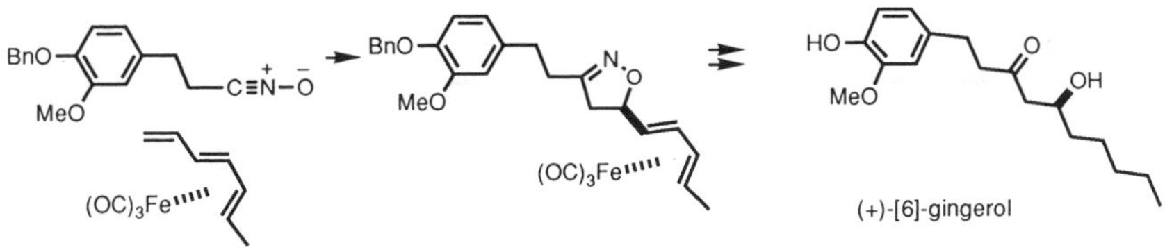

(+)-[6]-gingerol

It is expected that many enones can be prepared from the nitrile oxide/alkene adducts. There are reports on the synthesis of sarkomycin [Kozikowski, 1982] and prostaglandin $F_{2\alpha}$ [Kozikowski, 1984b]. A more interesting application concerns dimeric condensation leading to a masked pyrenophorin [Asaoka, 1981] because the formation of an eight-membered ring is disfavored. It is of interest also that the adduct is a distinctly disjoint system.

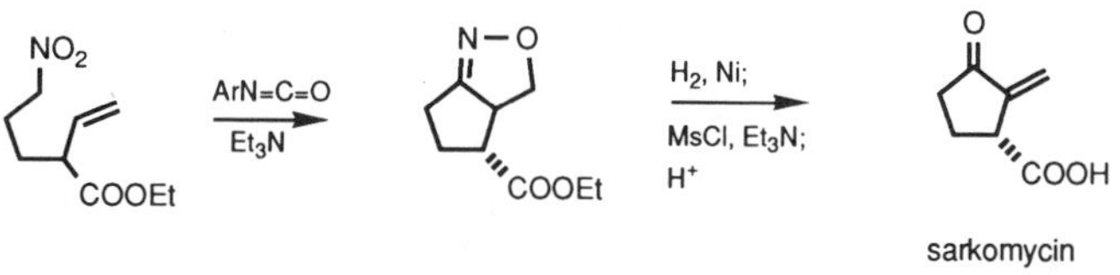

sarkomycin

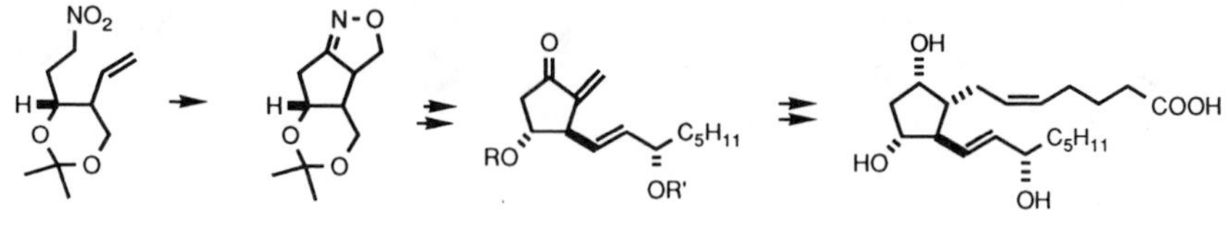

prostaglandin F$_{2\alpha}$

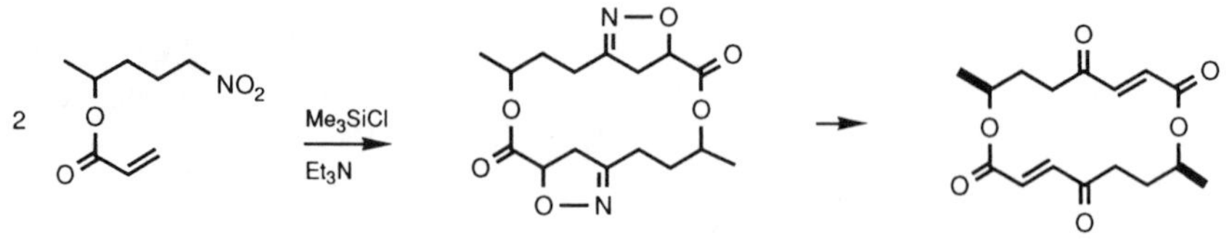

pyrenophorin

The 1,3-dicarbonyl substances are at a higher oxidation state. As they can be obtained by reductive ring cleavage of isoxazoles (with imine/enamine hydrolysis), the nitrile oxide cycloaddition to alkynes represents an alternative route to these compounds. A facile assembly of flavones and isoflavones [Thomsen, 1988], which are dehydro-1,3-diketones, has been accomplished.

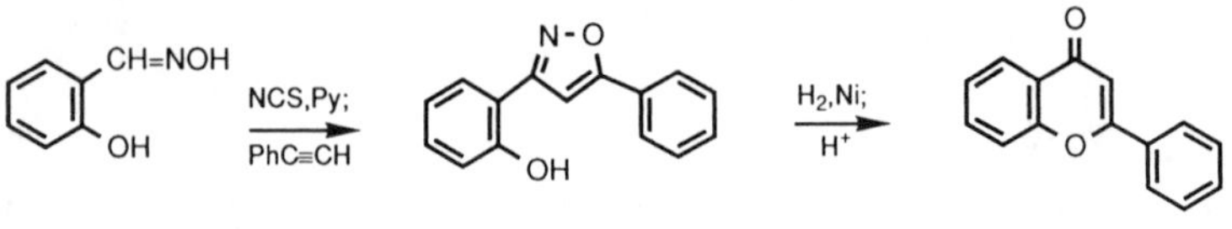

flavone

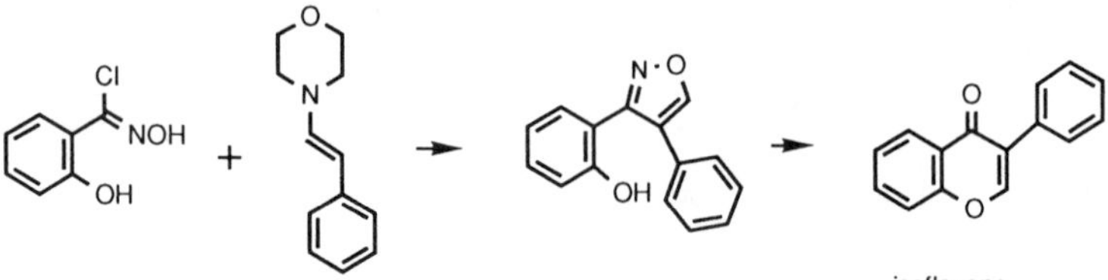

isoflavone

The latent 1,3-diketone in the form of a 3(2H)-furanone, as existing in geiparvarin, has also been recognized. Accordingly, a synthesis of geiparvarin via an isoxazole derivtive has been carried out [Baraldi, 1983].

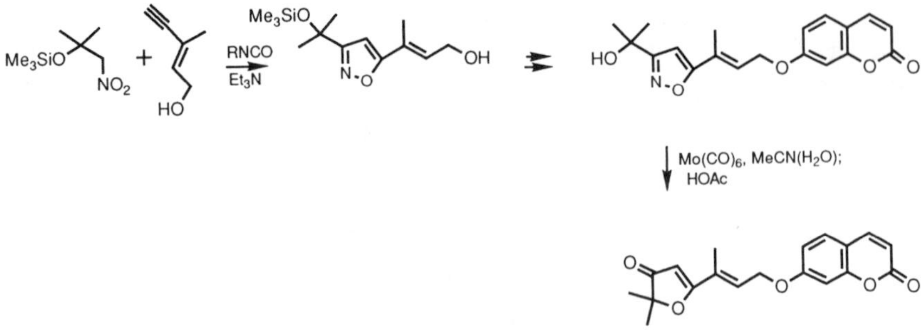

Mo(CO)$_6$, MeCN(H$_2$O);
HOAc

geiparvarin

The use of a preformed isoxazole ring in synthesis has been popularized [Stork, 1967]. A more recent example is the synthesis of curcumin [Kashima, 1977] from a 2,3,5-trimethylisoxazolium salt and vanillin.

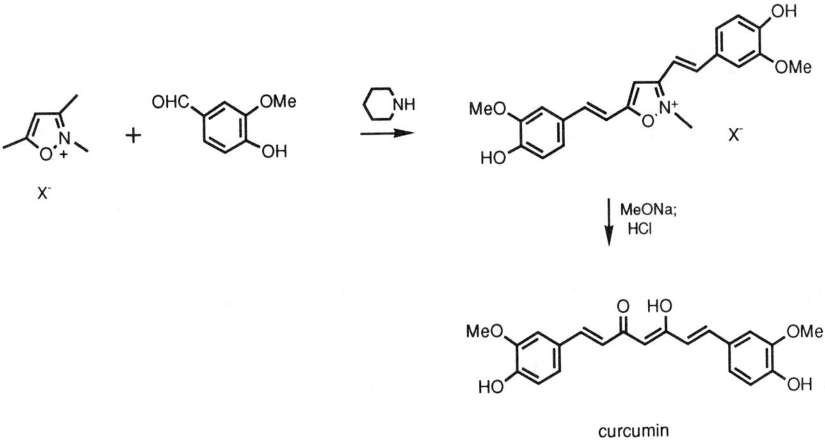

curcumin

Admittedly, the dehydrogenation of isoxazoline to the aromatic isoxazole is a facile process. This indirect approach may be advantageous to the preparation of β-diketones. An eminent example is a synthesis of 4-acylresorcinols [Auricchio, 1974].

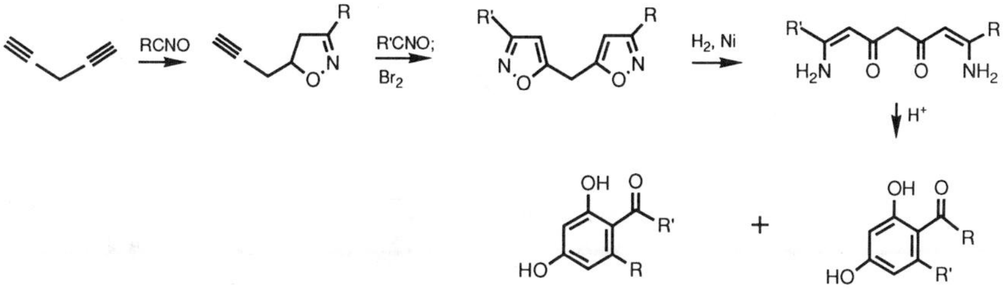

1,3-Diamino groups are also accessible from the isoxazolines procured from cycloaddition of nitrile oxides. Two syntheses of ptilocaulin [Walts, 1985; Hassner, 1986] were accomplished via intramolecular annulation onto either a cyclopentene or a cyclohexene.

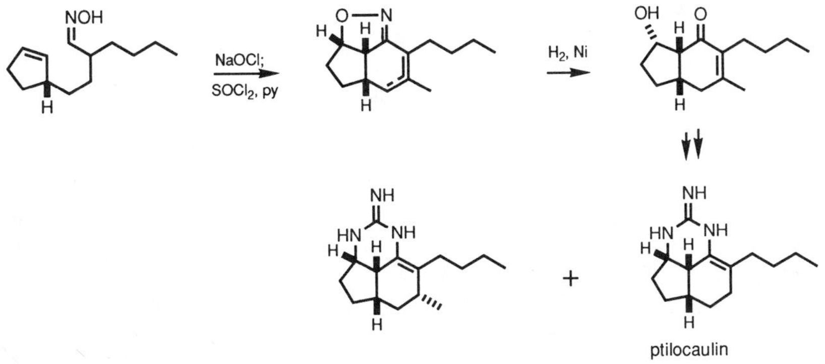

ptilocaulin

1,3-Amino alcohols are even more directly available from the isoxazolines. In turn, these compounds may be employed in other synthetic operations. Thus, one method for the construction of the northern portion of milbemycin β_3 was based on the nitrile oxide cycloaddition [Schow, 1986]. The strategy was predicated on the homoallylic relationship of the sidechain double bond with the nearest oxygen atom of the spiroketal subunit, and also the three-carbon separation of the same oxygen atom with that involved in lactonization. Accordingly, reduction of the isoxazoline adduct in which a protected acetaldehyde chain is attached to C-3, to the amino alcohol reveals the lactonic oxygen and an easily eliminatable amino function (β to the masked aldehyde). Eventually an intramolecular Michael addition of the lactol to the liberated enal completed this important phase of the synthetic effort.

An access of sedridine [Tufariello, 1978a] demonstrates the possibility of achieving regio- and stereospecificity in a remarkably simple manner. The cycloaddition apparently passed through an *exo* transition state.

In a route to supinidine [Tufariello, 1971] the oxygen atom of the nitrone was transferred to the dipolarophile moiety and eventually it was eliminated. (An alternative entry to the pyrrolizidine skeleton involves 2,3-dihydrofuran in an *exo* cycloaddition [Iwashita, 1979]. Further extension of this latter approach culminated in a short synthesis of α-isosparteine [Oinuma, 1983].)

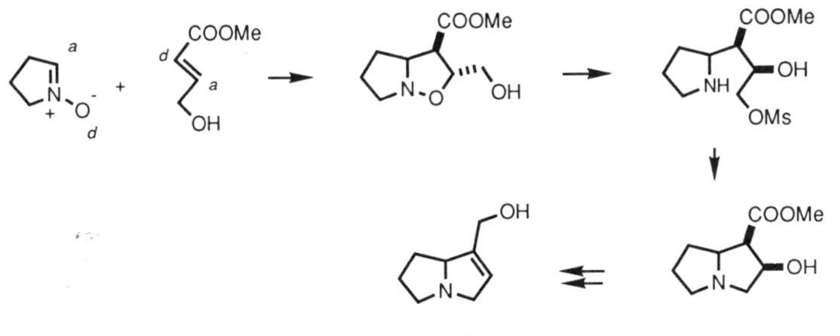

supinidine

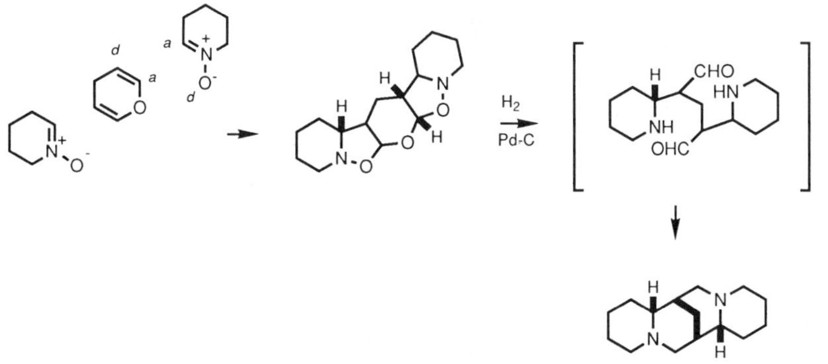

α-isosparteine

In other circumstances the transferred oxygen atom may be retained and modified. For example, an oxidation step is required to complete the synthesis of luciduline [Oppolzer, 1976]. An oxidation of the amino alcohol induced an intramolecular Michael cyclization which was planned as the final step of a myritine synthesis [Tufariello, 1978b].

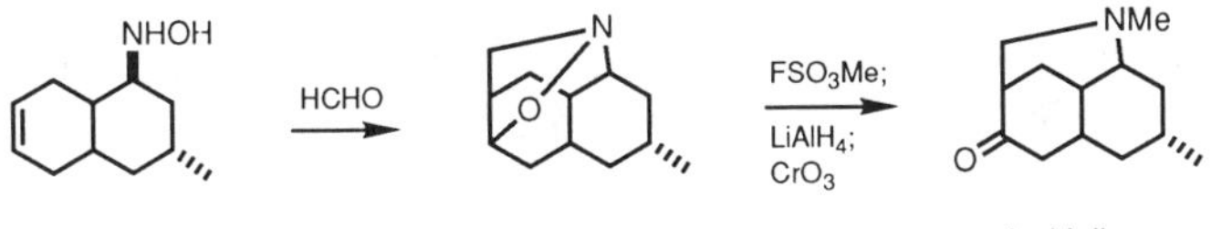

luciduline

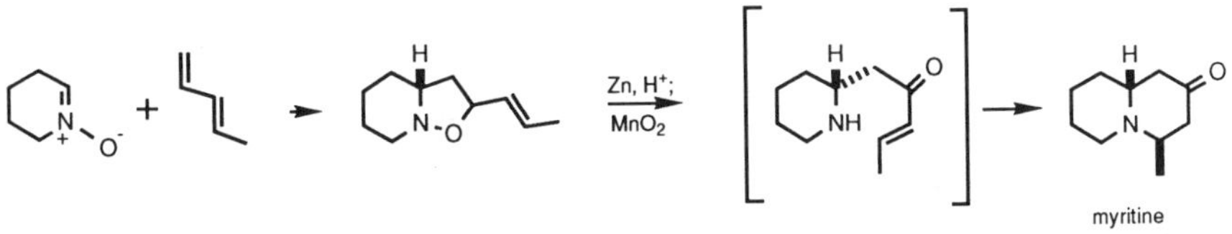

myritine

Elaeokanine A, when retrosynthetically analyzed, yields successive aldol and Michael transforms. The resulting 2-(2-oxopentyl)pyrrolidine is a β-amino ketone and therefore it may be assembled by the nitrone cycloaddition method [Tufariello, 1979a].

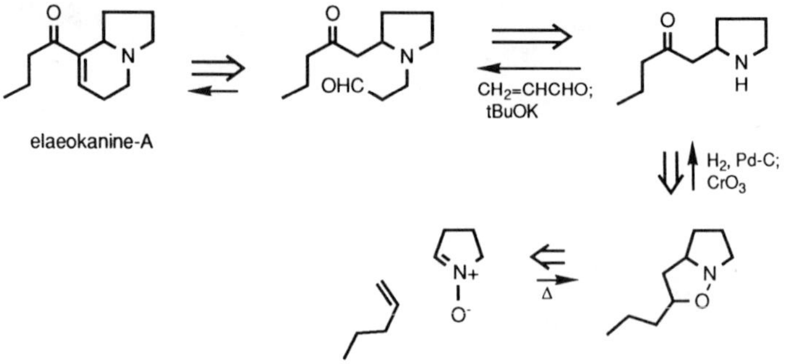

1-Pyrroline-*N*-oxide and methyl 3-butenoate supply all the elements of cocaine except the *N*-methyl group and the *O*-benzoyl residue. A synthesis [Tufariello, 1979b] starting from the combination of the two addends actually involved two cycloadditions. The second reaction was intramolecular.

The synthesis is only slightly complicated due to the need to protect the nitrone upon its regeneration. However, the deprotection conditions were the same as the thermal cycloaddition.

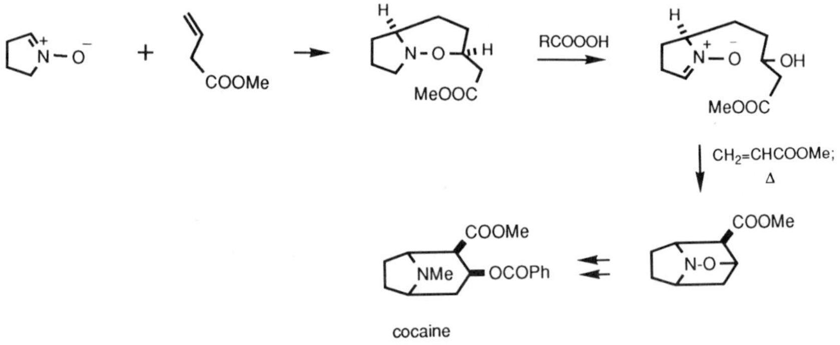

A facile synthesis of nitramine [Snider, 1984] which is a spiroannulated piperidine, exploited an intramolecular version of the nitrone/alkene cyclo-addition.

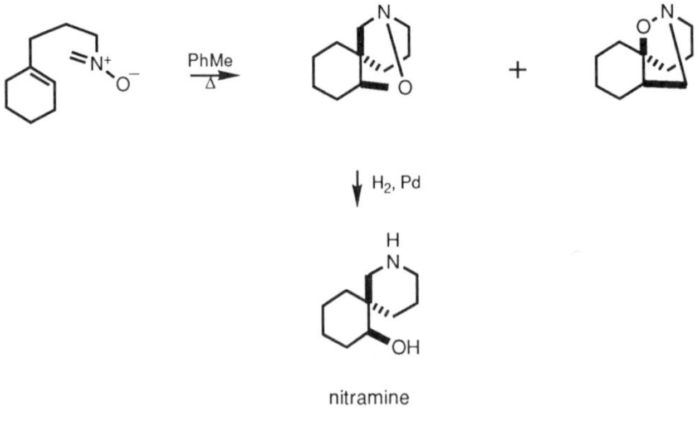

There is an obvious 1,3-amino alcohol moiety in paliclavine and as such it is amenable to assembly by nitrone or nitrile oxide cycloadditions [Kozikowski, 1981].

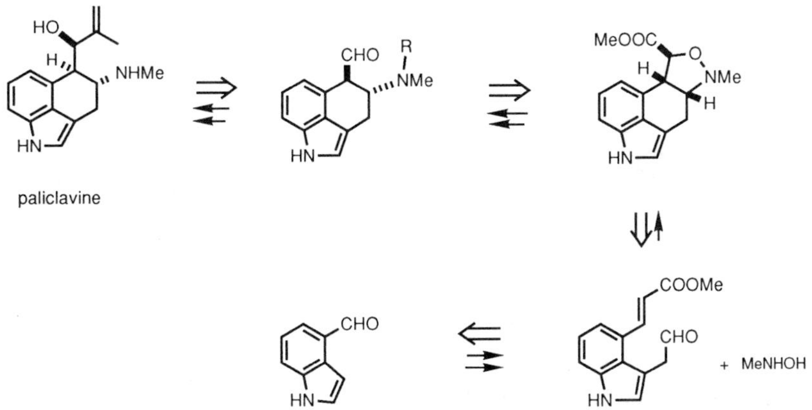

paliclavine

For maximum use of the potential intermediate, such as elaboration of other seco-ergot alkaloids, the secondary aldehyde is ideal. This aldehyde also allows for equilibration to establish the *vic–trans* substitution pattern [Oppolzer, 1983a; Kozikowski, 1984a].

It has been observed that the condensed ring system is the kinetic product. It is convertible to the bridged ring isomer at higher temperatures.

Undoubtedly the most elegant application of the cycloaddition involving nitrones is the approach to vitamin B_{12} [Stevens, 1986]. While it is not possible to discuss the intriguing assembly here, a glimpse of an advanced intermediate should give a hint to its underlying strategy.

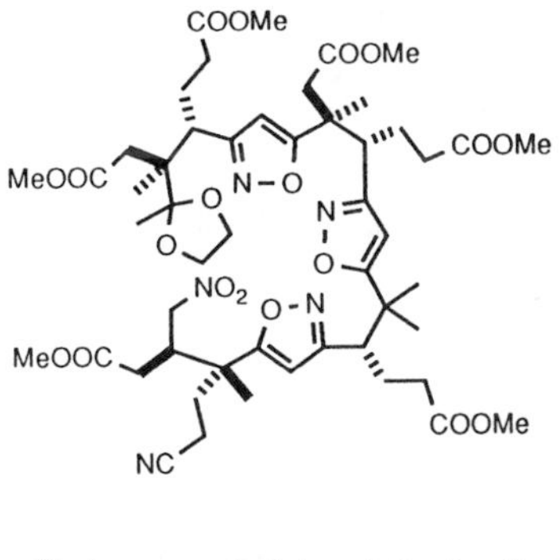

Finally, an extremely efficient assembly of 2-vinylindoles [Wilkens, 1987] which has enormous implications in the synthesis of *Aspidosperma* alkaloids, consists of tandem 1,3-dipolar cycloaddition, [3.3]sigmatropic rearrangement, retro-Michael elimination, and Schiff cyclodehydration.

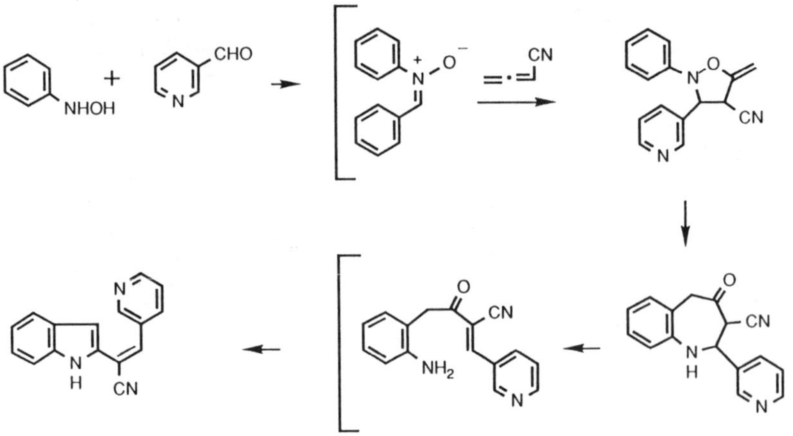

5.5.2. Cycloadditions of other Ylides

The pyrrolizidine alkaloids which contain a one-carbon pendant can be analyzed in the following manner. Thus, an acylic ester and pyrroliniuum methylides are the proper addends for their synthesis [Vedejs, 1980; Terao, 1982].

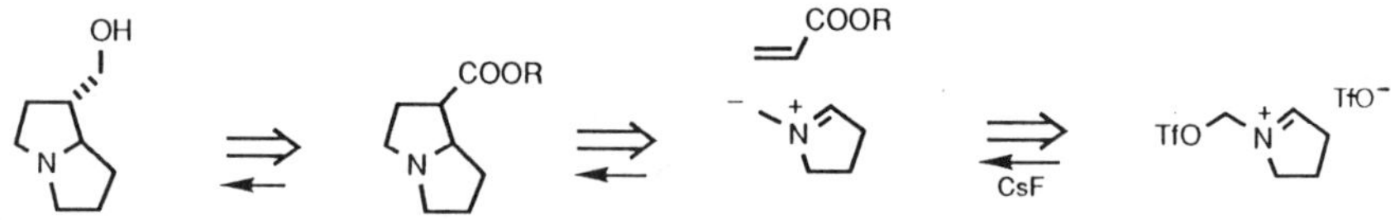

Aziridine-2-carboxylic esters undergo ring opening to give ylides which are obviously valuable in the synthesis of the kainic acids. Polarity matching indicates the requirement of an acceptor substituent at C-4 which carries an isopropenyl group in the target molecule(s). Accordingly, a 3-alken-2-one would be the suitable reaction partner [DeShong, 1986].

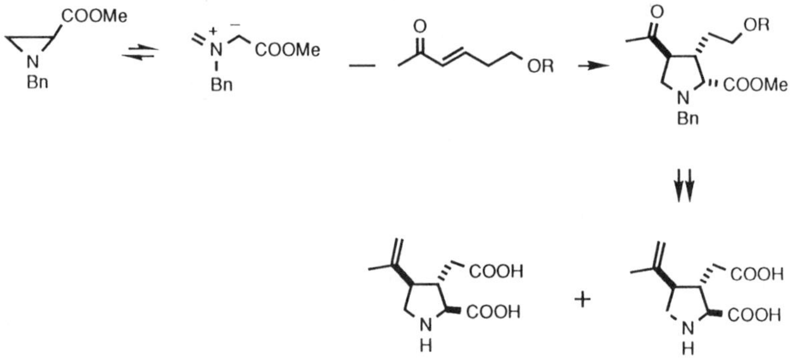

α-allokainic acid

An even more interesting work describes the utility of a thiazolium ylide as the 1,3-dipole [Kraus, 1985b]. Despite that a large portion of the original heterocycle had to be removed subsequent to the cycloaddition, the ready availability of the thiazolium salt offset the adverse economy of material utilization.

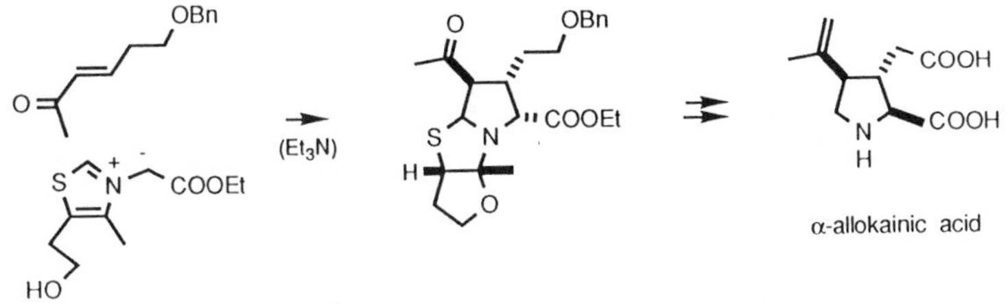

α-allokainic acid

A more elaborate 1,3-dipole has participated in a construction of the mitosane ring system [Rebek, 1984].

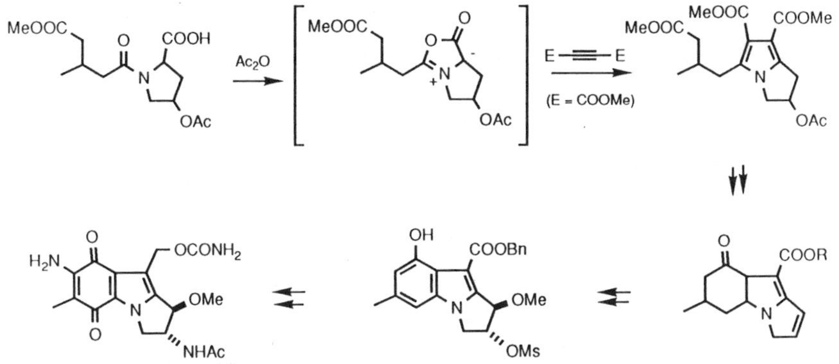

Valuable service of the 1,3-dipolar cycloaddition was obtained in an assembly of a quinocarcin intermediate [Allway, 1990].

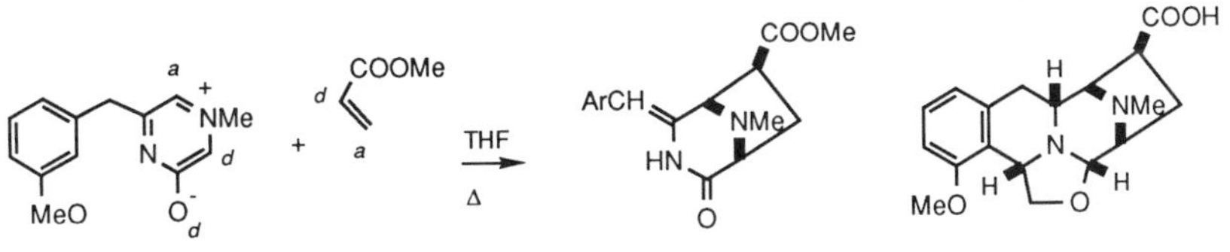

quinocarcin

The synthesis of biotin by a direct cycloaddition of a simple thiocarbonyl ylide with imidazolone is technically difficult in view of the instability of such 1,3-dipoles and their low reactivity towards electron-rich alkenes.

The scheme has been realized by modification of the synthons to accommodate the electronic requirements [Alcazar, 1990].

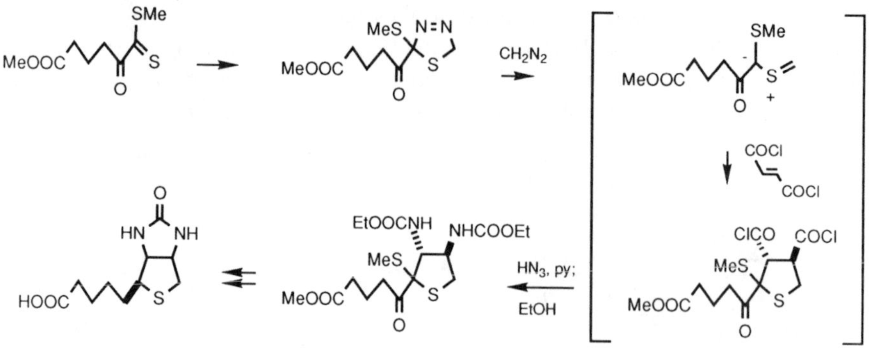

biotin

Generation of a spirocyclic β-amino ketone toward synthesis of perhydro-histrionicotoxin [Corey, 1976a] by an intramolecular 1,3-dipolar cycloaddition of an azide to an enone is an excellent concept. Rather unfortunately, the product is a diastereomeric mixture enriched in the wrong compound. Reductive C—N bond cleavage of the minor isomer also resulted in perhydrohistrionicotoxin in small yields.

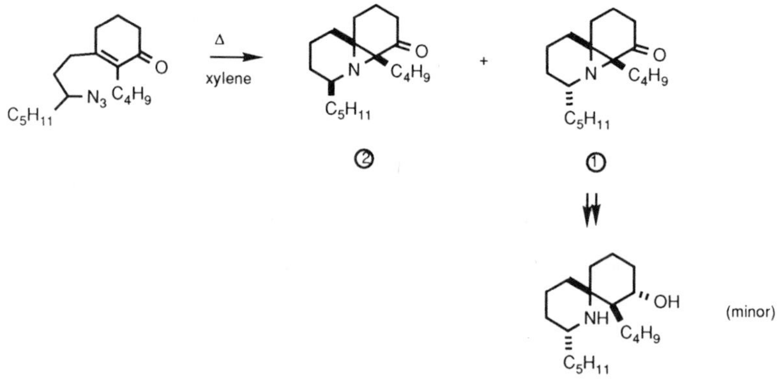

perhydrohistrionicotoxin

6

SYNTHESIS OF 3-OXY AMINO COMPOUNDS AND 1,3-DIAMINES

6.1. MANNICH REACTION

The Mannich reaction effects aminoalkylation (mainly aminomethylation) of substances possessing an active hydrogen [Tramontini, 1973]. It is a convenient method for the synthesis of β-amino carbonyl compounds (including esters), nitriles, nitro compounds, etc. Needless to say, the condensation products may serve as intermediates of other compounds by various functional group interconversion (FGI) and functional group removal (FGR) operations.

Generally, the Mannich reaction proceeds under acid catalysis. Formation of the iminium species (aldehyde + amine) is followed by attack of the nucleophile (donor) such as an enol. Preformed iminium salts such as $Me_2N^+=CH_2$ CF_3COO^- [Ahond, 1968] usually attach themselves to the more highly substituted α position of an unsymmetrical ketone. Dimethyl-(methylene)ammonium iodide which is known as the Eschenmoser salt [J. Schreiber, 1971] is a useful solid reagent for the Mannich reaction. Rather significantly, regioisomeric substitution can be accomplished by using iminium salts that are more hindered, e.g., $iPr_2N^+=CH_2$ ClO_4^- [Jasor, 1974].

The Mannich bases are conjoint. β-Amino ketones are structurally akin to the aldols. Their formation may be represented by a $\{d(NH) + a(CH=O)\}$ $+ d - a(CH_2C=O)$ combination.

Historically, the most celebrated Mannich reaction applied to natural product synthesis is Robinson's elaboration of tropinone [R. Robinson, 1917]. Referring to the following process he said: "By imaginary hydrolysis at the points indicated by the dotted lines, the substance may be resolved into succinaldehyde, methylamine, and acetone." This conclusion may be considered as the forerunner of the retrosynthetic analysis.

Improved yields of tropinone were obtained by replacing acetone with calcium acetonedicarboxylate, followed by acidification and decarboxylation. Extensive investigations by Schöpf [1937] led to optimized conditions by which the yield was increased to 90% by using a buffer solution. This is a case of polarity alternation accentuation (activation) at work; we know that nature employs this strategy in making various biomolecules, notably the polyketides.

Using glutaraldehyde in place of succinaldehyde, the Robinson–Schöpf condensation leads to the pomegranate alkaloid pseudopelletierine.

pseudopelletierine

The monoester of acetonedicarboxylic acid undergoes annulation analogously. This slight modification allows the synthesis of cocaine and its benzoyloxy diastereomer [Willstätter, 1923].

cocaine

The efficiency of this operationally simple reaction has not been forgotten in modern synthetic work on alkaloids. A pre-eminent case is an approach to the defense substances of the ladybug [Stevens, 1979]. These compounds which contain a perhydro[9b]azaphenalene skeleton can have up to five asymmetric centers (including the tertiary amine), several stereoisomers of which are known. It is significant and synthetically very helpful that the secondary methyl group in all these alkaloids is equatorial.

A synthesis based on the manifold Mannich or Robinson–Schöpf condensation is attractive as its reversibility could in principle be manipulated to produce certain stereoisomers. Accordingly, a scheme derived retrosynthetically calls for a condensation of acrolein with dimethyl malonate to reach an acetal ester of glutaraldehydic acid in a few steps. Conversion of the acetal ester to a symmetrical nine-carbon keto dialdehyde bisacetal is achievable by self-condensation (Claisen) and decarbomethoxylation. After reductive amination of the ketone, reaction of the amino dialdehyde (generated *in situ* hydrolytically) with dimethyl acetonedicarboxylate accomplishes the formation of four bonds. The well-characterized tricyclic product has been elaborated into precoccinelline.

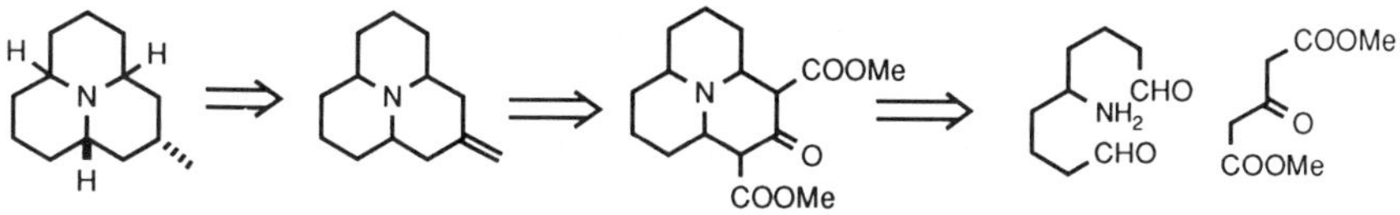

The facility of this multi-ring formation is the consequence of the perfect matching of various polar functionalities. In the preparation of the C_9 component all the reactions also benefit from the conjoint distribution of the reactive groups. But perhaps the most important facet of the annulation steps is the unraveling of stereoelectronic requirements. Thus, the first intermolecular C—C bond formation involves a chair-like transition state in which the attacking donor approaches from the opposite face of the pre-existing sidechain while maintaining maximum orbital overlap. The *syn* face approach to the alternative chair conformation of the protonated piperideine would encounter a severe 1,3-diaxial interaction.

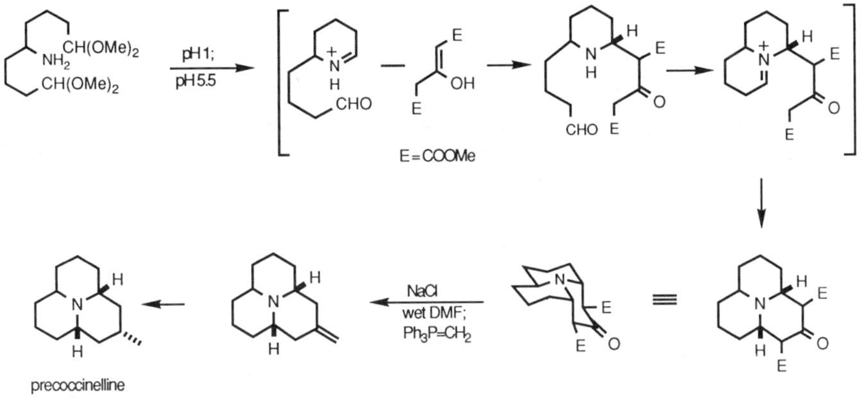

precoccinelline

The alkaloid luciduline is most amenable to assembly by the Mannich reaction as its structure yields directly a Mannich transform without functional group manipulation. A synthesis [Scott, 1972] indeed has been accomplished along this line. (A model study during an attempt at a luciduline synthesis by the same method terminated at the construction of N-methyl-3-azabicyclo[3.3.1]-nonan-6-one [Ho, 1968]).

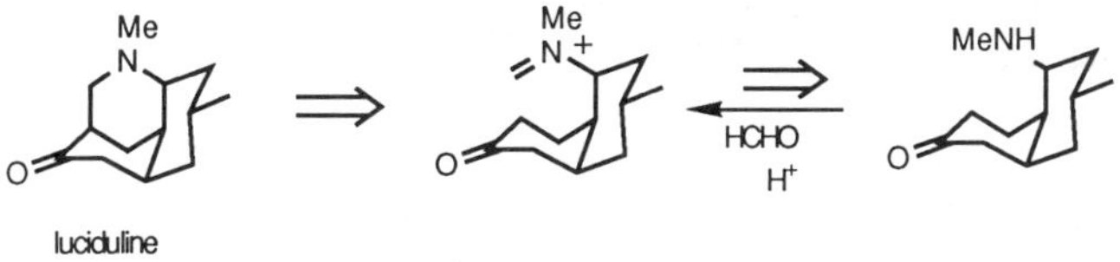

luciduline

When uleine is disconnected by removing the ethylidene portion, a 2-(3-indolyl)piperidine results. This disconnection is valid in view of the propensity of indole to undergo electrophilic substitution at the α-position when strain factors prohibit reaction at the β-site.

In synthesis, the C-4 of the piperidine ring must be functionalized in order to link up with the ethylidene moiety, and the most logical functional group is a ketone. This ketone possesses a Mannich retron and is most readily constructed by that reaction [Büchi, 1971].

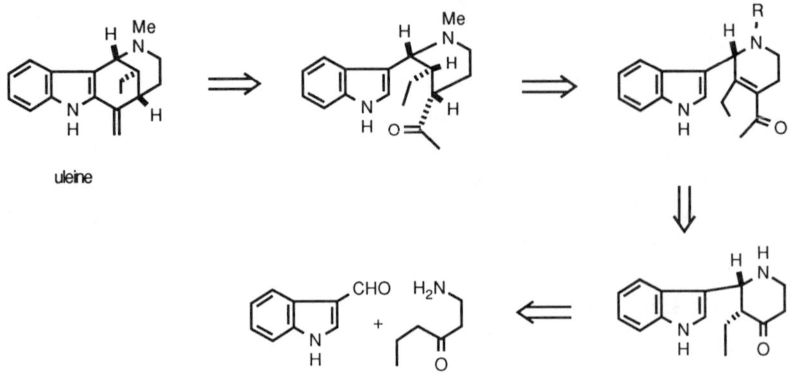

uleine

The tandem aza-Cope rearrangement and Mannich cyclization has been developed into a powerful method for the synthesis of azabicyclic and azapolycyclic compounds that otherwise require many more steps to prepare by conventional routes. The implication of this development in alkaloid synthesis is enormous [Overman, 1991]. While a few examples of its application will be discussed elsewhere, it seems appropriate to mention here an access to a pentacyclic compound serving in a potential synthesis of gelsemine [Earley, 1988a,b].

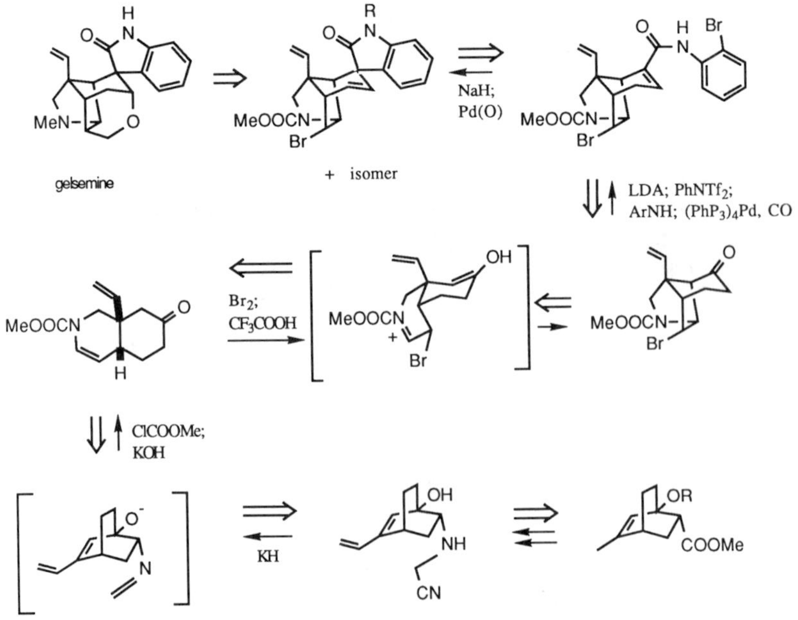

gelsemine   + isomer

In view of the ring strain, the reversibility of the Mannich reaction, and the desire to functionalize C-16 (gelsemine numbering) for eventual formation of the tetrahydropyran ring, the intramolecular Mannich reaction was performed independently, employing an N-acyliminium intermediate.

Nitramine and isonitramine are spirocyclic piperidine bases. A simple synthesis of these compounds [Carruthers, 1987] took into account the 1,3-relationship of the nitrogen and oxygen atoms. An intramolecular Mannich reaction is one of the most convenient ways to gain access to the molecular skeleton.

The intramolecular Mannich reaction has been entrusted a central role in a general scheme for alkaloid synthesis [Wenkert, 1968a, 1981a]. A case pertaining to an approach to epilupinine [Wenkert, 1968b] may suffice to illuminate the design. Here the iminium species was generated by protonation of the endocyclic enamine which also contains a ketal group. The Mannich reaction is favored by entropy factors, as the terminus of the enol tautomer of the ketal is separated by four atoms from the acceptor site of the iminium ion.

A synthesis of matrine [J. Chen, 1986] was based on the same key step; only that the electrophile is now an acyliminium ion. The application of acyliminium species in alkaloid synthesis has received considerable attention, and some more examples will be discussed later.

Iboxyphylline is likely derived *in vivo* from pandoline via oxidative fragmentation and Mannich cyclization. A biomimetic synthesis of this alkaloid [Kuehne, 1989] highlighted by a Mannich reaction has been achieved, although stereoisomers and structural isomers from a 5-endo-trig cyclization were also obtained.

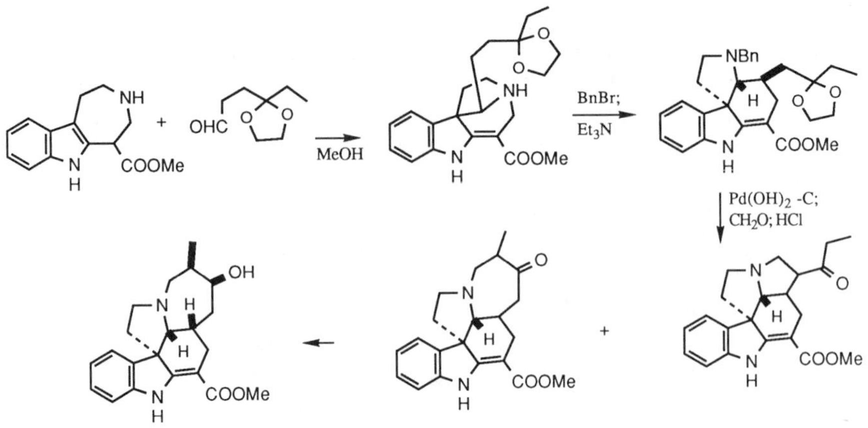

iboxyphylline

The first stage of the vitamin B_{12} synthesis [Woodward, 1968] is the preparation of a tricyclic ketone for the AD-ring segment by annulation of an indole with a $^+CH_2COCH_2^-$ synthon. This was done via propargylation, hydromethoxylation of the triple bond which was followed immediately by an intramolecular Mannich cyclization. Propargyl bromide is the umpoled reagent.

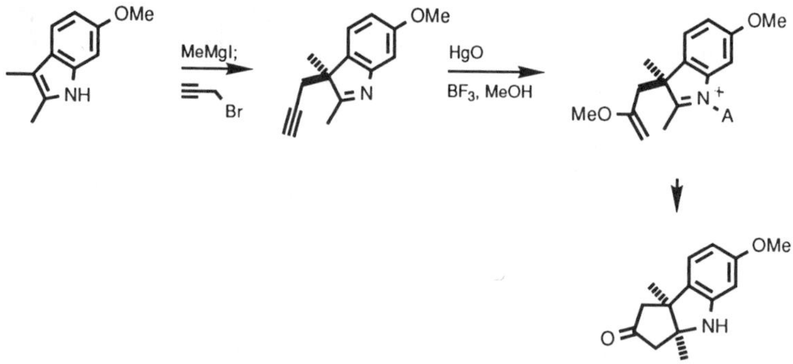

A careful retrosynthetic analysis of lycopodine [Heathcock, 1982a] has led to the conviction that the Mannich reaction holds promise to short routes, as the alkaloid is a β-amino ketone. The key observation was that the cyclohexane ring containing the secondary methyl group may act as an anchor to attach two properly functionalized carbon chains by which the Mannich reaction would assemble the bridged perhydroquinoline moiety. With a pendant on the ketone or the nitrogen atom the final steps could be a simple intramolecular alkylation.

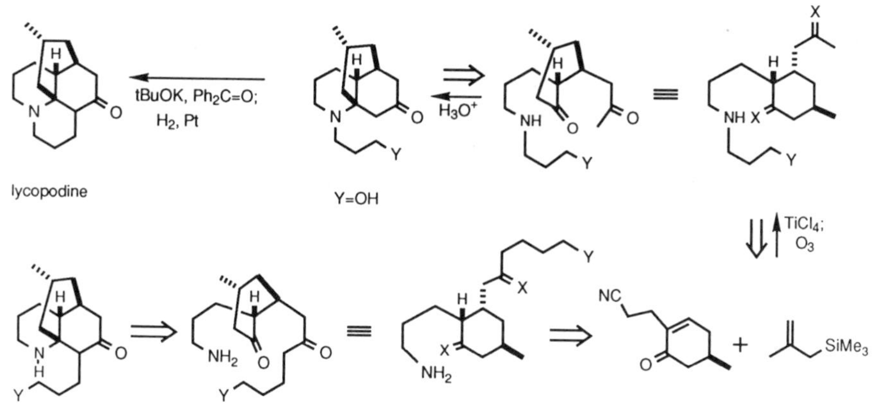

lycopodine

Y=OH

A C,E-seco intermediate for the synthesis of ajmalicine [van Tamelen, 1969] was constructed from tryptamine, formaldehyde and a keto triester. The three components were joined by a Mannich reaction which was followed by lactamization.

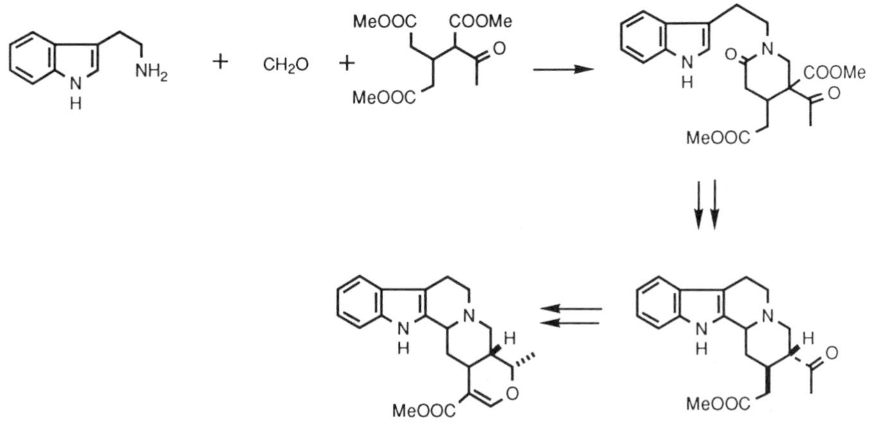

ajmalicine

The highly reactive β-keto ester subunit dictated the site of the Mannich reaction. Since the two ester groups of the adduct are equidistant from the nitrogen atom there are no inherent regiochemical problems attending the cyclization.

As far as stereochemistry is concerned, the remaining acetic ester sidechain played a crucial role in determining the stereocenters at C-3 and C-20. Thus, hydrogenation of the Bischler–Napieralski cyclization product occurred from the α-face which is opposite to the equatorial sidechain. At obliviation of the tertiary carbomethoxy group the unencumbered methyl ketone was subjected to equilibration to settle on a configuration *trans* to the vicinal substituent. The two carbon-chains were well juxtaposed for elaboration into the dihydropyran system of ajmalicine.

Conceivably in a synthesis of kopsinine the ethylene bridging is a more challenging task. It is also one of the most logical ways to complete the molecular framework by a Diels–Alder process as the cycloaddition involving a proper precursor is directed away from the β-face by the tryptamine link.

Construction of the requisite pentacyclic skeleton proved to be very efficient by a tandem enamine alkylation–anionic Mannich reaction of a 3-(2-piperidinyl)-indole, as all the reactive centers are conjoint. The keto ester is readily convertible to the homonuclear diene ester [M. Ogawa, 1987].

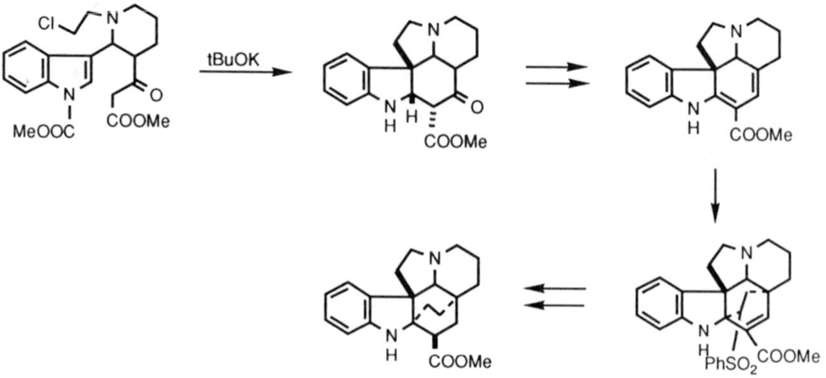

kopsinine

Four rings of the hexacyclic system of aspidofractinine can be assembled by two Mannich reactions [Cartier, 1989].

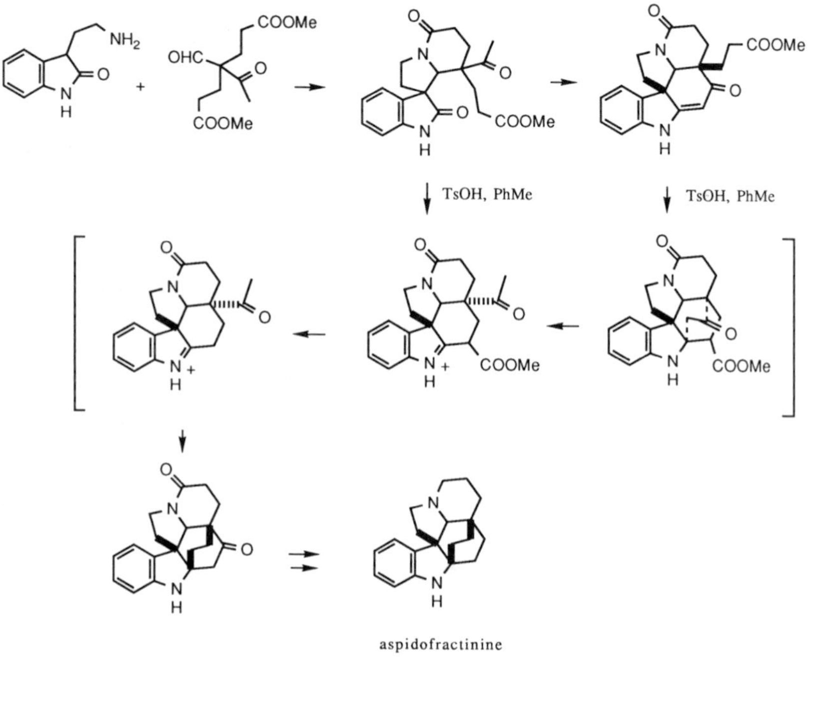

aspidofractinine

The spiro oxindole system is the most characteristic feature of rhynchophylline. The presence of an N—C—C—C=O segment in the molecule suggests a Mannich transform [Ban, 1972]. The facility with which the oxindole alkaloids undergo the Mannich (and retro-Mannich) reaction is shown by the isomerization of isorhynchophylline to rhynchophylline by dilute acetic acid.

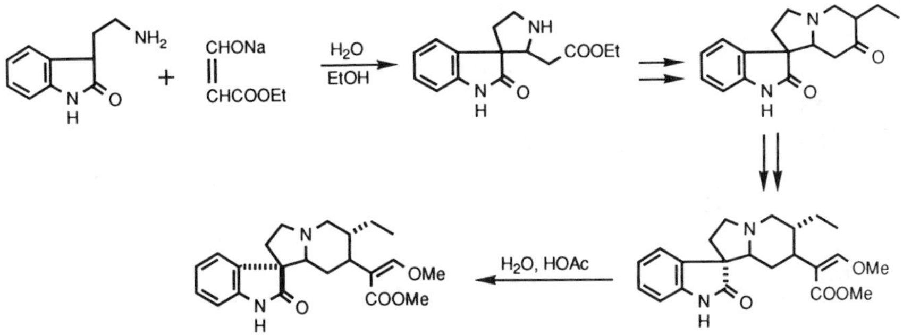

rhynchophylline   isorhynchophylline

The classic synthesis of strychnine [Woodward, 1963b] also employed a Mannich reaction. The donor in this case was an enamine of the indole nucleus. It required more drastic reagents (TsCl, pyridine) to render the reaction irreversible, the product being a more energetic indolenine derivative.

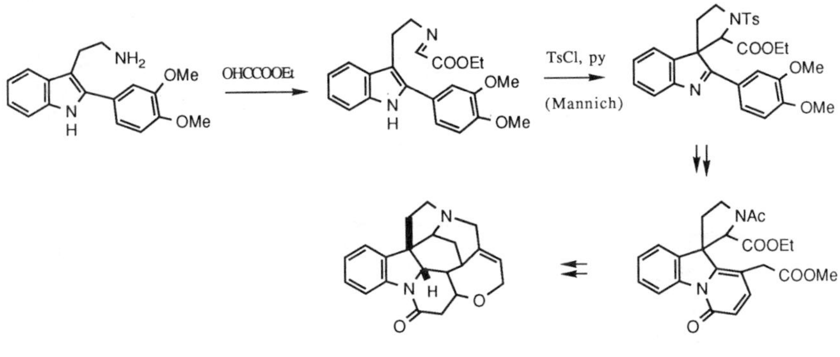

strychnine

An approach to *Strychnos* alkaloids such as tubifolidine can take advantage of an intramolecular Mannich reaction on a latent piperideine which contains an *o*-nitrobenzyl ketone subunit [Bosch, 1988]. The strong donor characteristics of this benzylic position cannot be overemphasized.

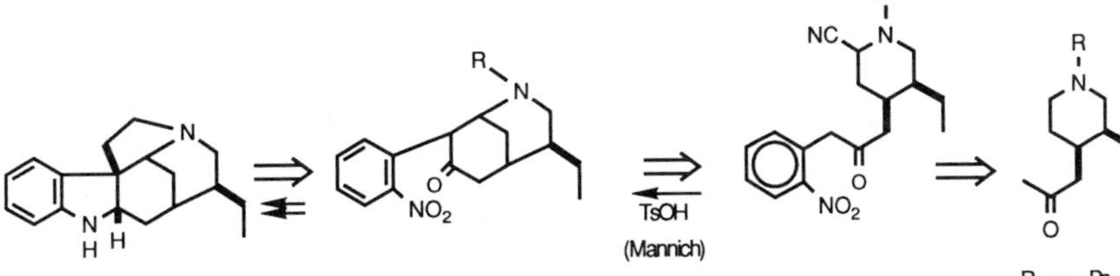

tubifolidine   R = Bn

An explicit demonstration of the retrosynthetic procedure for the formulation of efficient synthetic routes is the report on a porantherine synthesis [Corey, 1974]. From a retrosynthetic perspective porantherine contains a strategic bond between one of the sp^2-centers and an α-atom of the nitrogen. By FGI (replacing the double bond with a ketone at the far end) the molecule reveals a double Mannich retron. Eventually the synthetic target can be simplified to a symmetrical amino diketo aldehyde.

The realization of this scheme starting from such an acyclic compound must take into account the order of the condensation reactions. The more reactive aldehyde subunit cannot be set free before the first Mannich reaction. Actually it was introduced in the form of a vinyl group and released at the proper moment by an oxidative cleavage reaction.

Generation of both an imine and a ketone in a sterically mutually accessible state guarantees the success of a Mannich reaction. Photocycloaddition of vinylogous amides with alkenes leads to highly strained and fragmentable adducts. Often these adducts disintegrate into the δ-amino ketones which can then undergo cyclization at the alternative α-position of the ketone group, resulting in the β-aminocyclohexanones. This facile ring dismutation sequence is based on strain relief and a conjoint relationship of the functional groups. its synthetic potential is starting to show [Winkler, 1988].

Access to a Mannich reaction precursor via an aza-Cope rearrangement is instrumental to the facile annulation of the indolizidine nucleus onto a cyclohexane ring and it has enabled a rapid synthesis of perhydrogephyrotoxin [Overman, 1980]. Numerous elegant applications of the aza-Cope rearrangement–Mannich cyclization tandem have been reported.

perhydrogephyrotoxin

The preoccupation of biomimetic synthesis in the recent past has identified the Mannich reaction as the key to alkaloid synthesis. Whether the similarity of the laboratory version to the enzymatic process is close or remote, the guiding principle has been of tremendous value (cf. the Robinson–Schöpf reaction). In an elaboration of the sparteine alkaloids [van Tamelen, 1960] piperidine was condensed with formaldehyde and acetone. A double intramolecular Mannich process was required to lock up the bridged system, and it was performed by mercuric oxidation, admittedly a very non-biochemical event.

sparteine

In planning a synthesis of coronaridine and/or dihydrocatharanthine [Attur-ur-Rahman, 1980] one may focus on the N—C—C—COOMe segment. By a Mannich transform the molecular framework is simplified to a tetracycle. The iminium salt is available via oxidation (e.g., by mercuric acetate) of the corresponding tertiary amine which in turn may be derived from a C—N bond cyanolysis of an indoloindolizidinium species. It is now very easy to trace a pathway to this intermediate by starting from tryptamine.

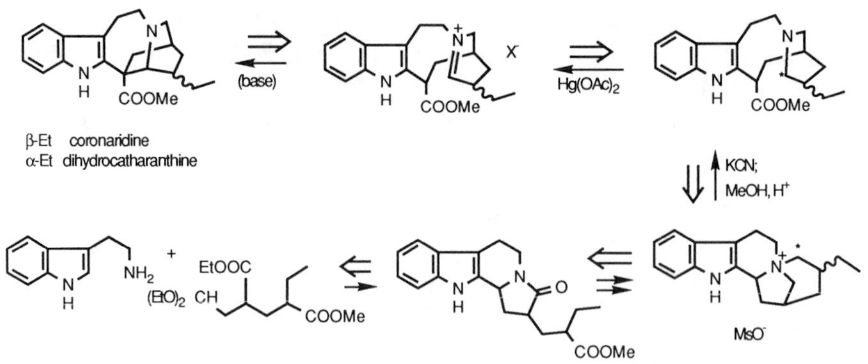

β-Et   coronaridine
α-Et   dihydrocatharanthine

As revealed in the synthesis of precocinelline (*vide supra*), stereoelectronic factors dominate the intermolecular Mannich reaction of piperideines. A route to porantherine [Ryckman, 1987] was formulated to test the limit of the control in an intramolecular environment.

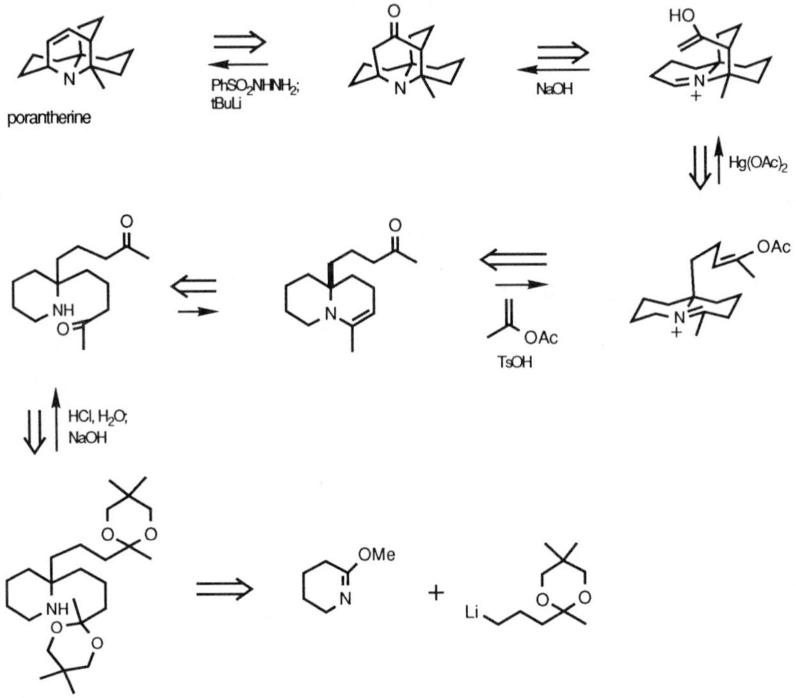

porantherine

While electronically the reaction sequence still conforms to polar complementarity, the closure of the last ring specified in this route requires the donor enol to be sterically interacting with an axial substituent at the α′-carbon of the iminium acceptor. The normally disfavored transition state actually becomes accessible because the precursor has already invested much of the activation energy in the tricyclic system. Furthermore, in the iminium salt, an interfering axial hydrogen has been removed, rendering the epimerization of the acetyl group energetically quite feasible.

Regiospecific generation of iminium ions has been achieved from degradation of α-amino acids. The α-carbon becomes a terminus of the double bond.

A synthesis of tropinone [Hess, 1972] showed the use of two separate Mannich reactions to accomplish the goal, the last stage involving N-chlorination and decarboxylation to give the iminium intermediate.

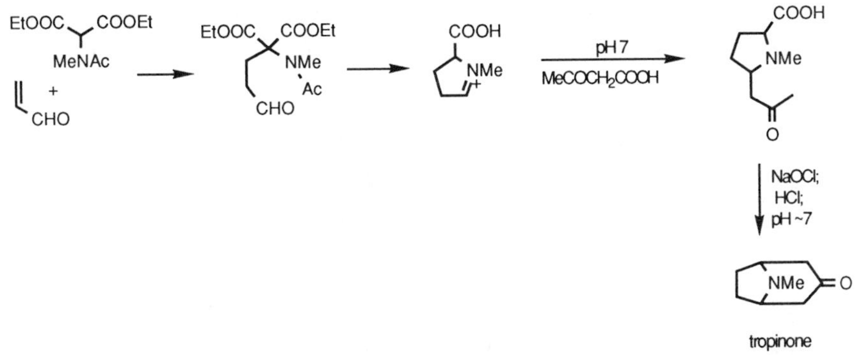

tropinone

An enantioselective approach to (+)-anatoxin-a [Sardina, 1989] started from glutamic acid to create a (2R,5R)-proline derivative. Exposing this N-benzyl amino acid to oxalyl chloride effected the C—C bond cleavage to afford the iminium chloride which was then trapped intramolecularly.

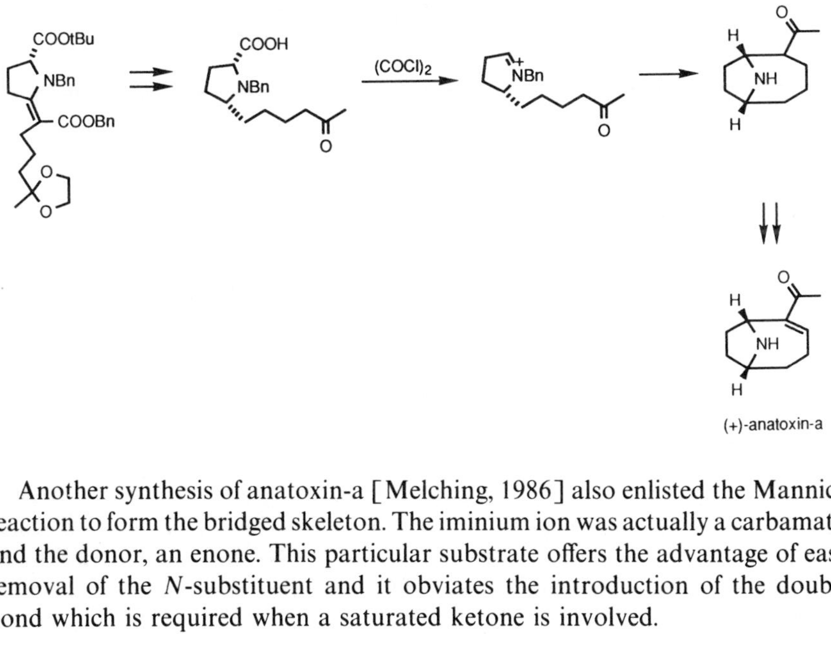

(+)-anatoxin-a

Another synthesis of anatoxin-a [Melching, 1986] also enlisted the Mannich reaction to form the bridged skeleton. The iminium ion was actually a carbamate, and the donor, an enone. This particular substrate offers the advantage of easy removal of the N-substituent and it obviates the introduction of the double bond which is required when a saturated ketone is involved.

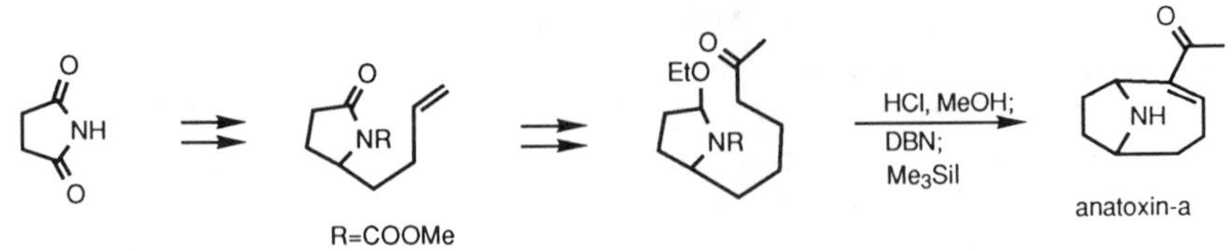

R=COOMe

anatoxin-a

The Mannich cyclization using an enone donor actually was mediated by hydrochlorination of the double bond. Base treatment of the cyclized product recovered the unsaturation. This variation is also found in a synthesis of elaeokanine B [Wijnberg, 1981].

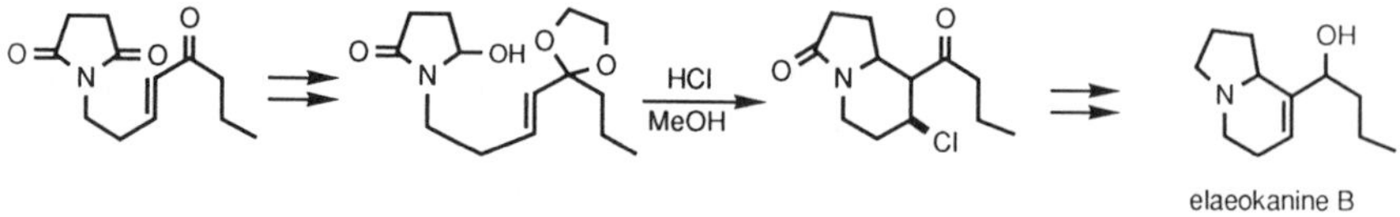

elaeokanine B

The variant of Mannich reaction based on acyliminium donors has been developed into a powerful synthetic protocol [Speckamp, 1985]. Two more examples that describe the assembly of mesembrine [Wijnberg, 1982] and a vindorosine intermediate [Veenstra, 1981] may be mentioned.

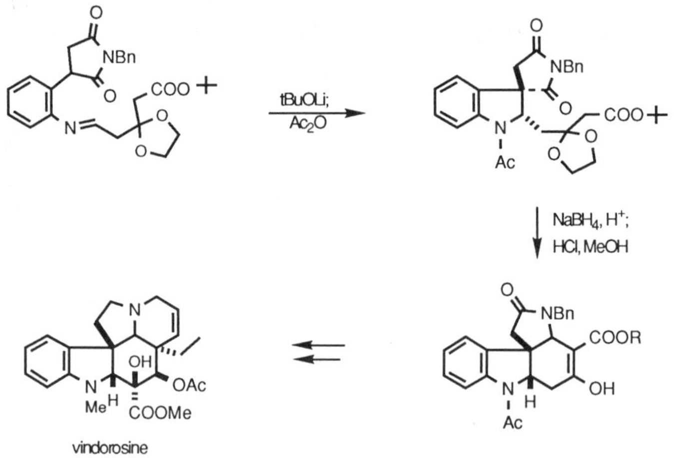

vindorosine

Not strictly belonging to the Mannich reaction are those processes featuring acyliminium trapping by C—C multiple bonds. It is interesting to note that many synthetic targets that have been elaborated using this strategy contain an N—C—C—C—O segment which is accessible by the classical Mannich reaction. Synthetic routes to isoretronecanol [Nossin, 1979], supinidine [Chamberlin, 1982], and elaeokanine A [Chamberlin, 1984] are a few representatives.

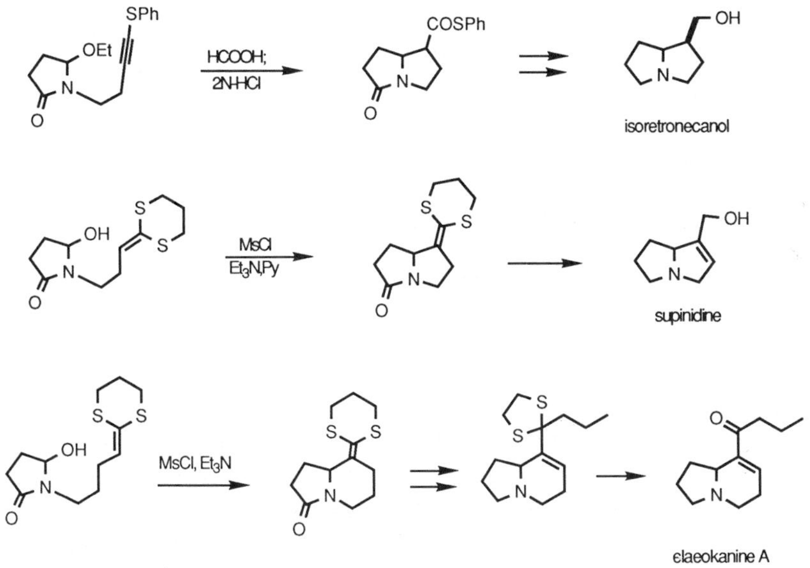

isoretronecanol

supinidine

elaeokanine A

Mannich cyclization at the reduced level, i.e., with an alkene replacing an enol donor, is a direct pathway to the 1,3-amino alcohol system present in perhydrohistrionicotoxin [Schoemaker, 1978; Evans, 1982c]. For acyliminium-ion-induced cyclization, the chair transition state, in which the butyl sidechain eclipses the trigonal portion of the existing ring, is expected to be preferred over that embodying an eclipse between methylene groups. Differences in hydrogen-bonding reorganization energies should also favor the transition state leading to the desired lactam. However, small amounts of cyclization products containing a cyclopentane ring cannot be excluded because the donor alkene is locally symmetrical.

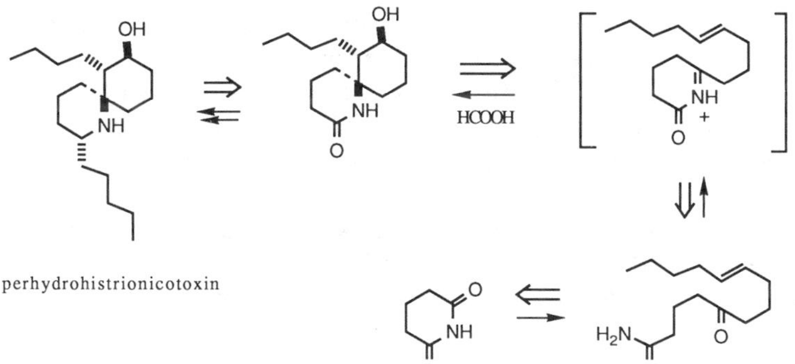

perhydrohistrionicotoxin

In previous paragraphs it is indicated that a ketal may serve as the donor in the Mannich reaction via its enol tautomer present in the reaction media. It is not surprising that a well-juxtaposed furan ring can act the same role. This notion has been realized in an approach to the spirocyclic skeleton of perhydrohistrionicotoxin [Tanis, 1987].

perhydrohistrionicotoxin

An excellent strategy for alkaloid synthesis [Stevens, 1977] is based on an extension of Stork's enamine annulation. By using endocyclic enamines, the reaction with methyl vinyl ketone (and homologs) furnishes cyclohexanoazacycle intermediates via the Michael–Mannich pathway.

The synthesis of perhydroindolones by this method requires Δ^2-pyrrolines. A convenient preparation is by thermal rearrangement of cyclopropyl imines [Stevens, 1968; Keely, 1968].

mesembrine

The powerful annulation reaction is subject to subtle stereoelectronic control. Assuming a stereobias favoring staggered transition states for the initial Michael addition, there are 12 available, and six of these transition states differ from the rest by the (E/Z)-geometry of the enolate. Those embodying severe 1,3-diaxial interactions between the enolate and the C-5 axial hydrogen atom of the piperideine ring can be rejected, and for the remaining ones, spatial charge interaction would become the determining factor, as closer association of the ions is definitely favored.

This rationale can be extended to interpret the failure of the annulation step in a projected synthesis of N-methyllycodine [Stevens, 1983a].

It is highly enlightening to consider the facile assembly in one operation of karachine [Stevens, 1983b] from berberine and a siloxydiene. For achieving the second annulation step after the initial vinylogous Mannich reaction and the enamine cyclization, a boat transition state is mandatory. Normally, such a reaction would be aborted, however, the boat conformation is already present in the intermediate and there is no need to spend additional energy otherwise required for a chair-to-boat change.

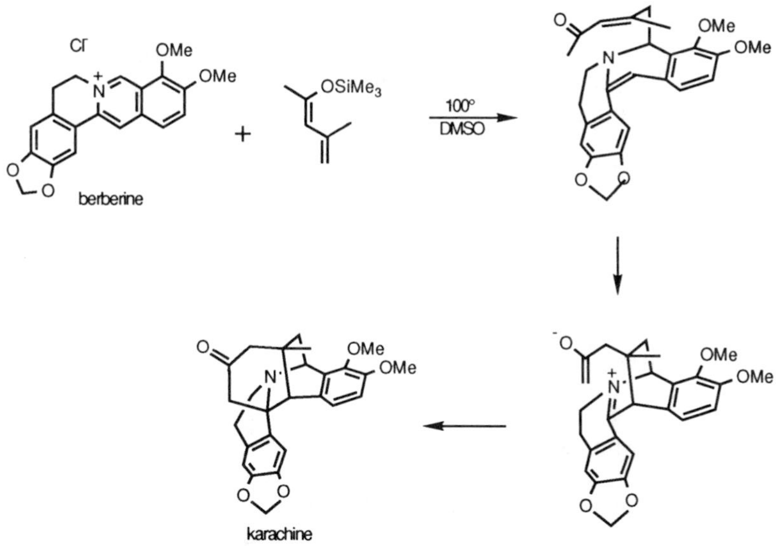

berberine

karachine

The vinylogous Mannich reaction in the above example is concerned with the donor part. It is possible to effect a Mannich reaction of vinyl imines (or 1-azadienes). Actually, a synthesis of yohimbone [Wenkert, 1976] involved an intramolecular C—C bond formation at C-4 of a pyridinium ring. On acidification, a dihydropyridinium species emerged, and it was immediately trapped by the indole moiety (Pictet–Spengler reaction). (Yohimbine has been synthesized [Wenkert, 1982] along the same lines, the major difference being that the vinylogous Mannich reaction was intermolecular, and the E-ring was closed immediately afterwards by an intramolecular aldol reaction.)

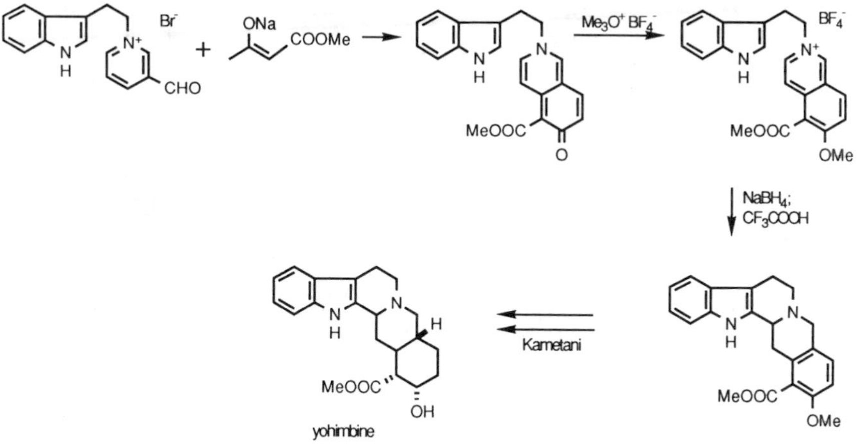

yohimbine

Condensation of a bicyclic conjugate azadiene with an enamide was the basis of an extremely concise approach to α-obscurine [Schumann, 1982]. The first step was a vinylogous Mannich reaction.

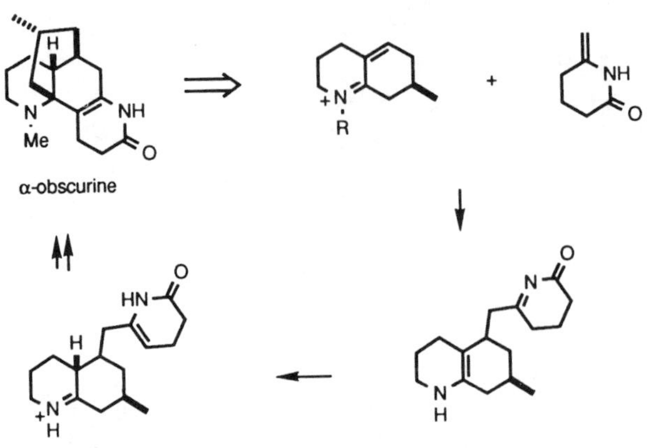

α-obscurine

This double Mannich process was employed earlier in a synthesis of lycopodine using acetonedicarboxylic acid as the donor [Schumann, 1982].

The synthetic plan for vindoline [Ando, 1975] is simplified by eliminating the reintroduceable hydroxyl and the ester groups, and replacing the acetoxylated site with a carbonyl. This latter compound is a β,γ-unsaturated ketone, and both nitrogen atoms are β to the carbonyl (conjoint arrays). Such an enone system can be set up by ethylation of the conjugate enone which in turn is an aldol product. The keto aldehyde precursor is an adduct of acrolein and a tetracyclic amine.

Disconnection of the tetracyclic amine generates a tryptamine derivative. The disposition of various functional groups makes it possible to employ a tandem Michael–Mannich or reflexive Mannich reaction sequence.

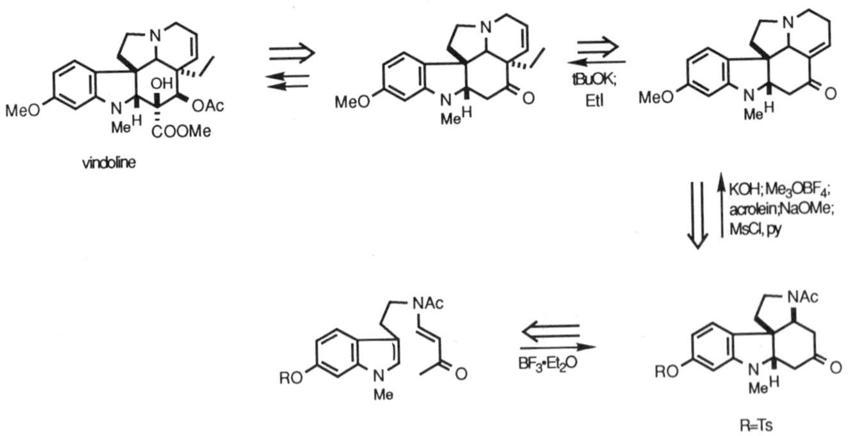

vindoline

Systematic disconnection with functional-group interchange of the emetine molecule results in a 3,4-dihydroisoquinoline and an enol derivative as the convenient building blocks to start a synthesis [Akiba, 1985]. All the relevant C—C bonds constitute compatible donor–acceptor pairs.

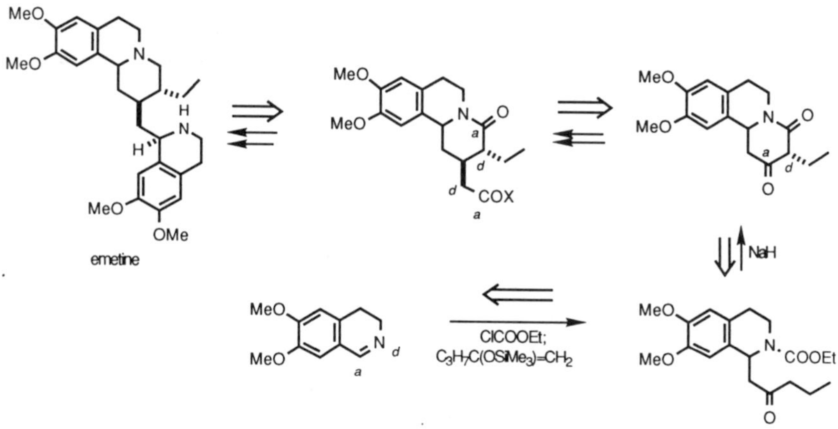

emetine

Quinoline alkaloids such as graveolinine contain a conjoint heterocycle. Retrosynthetic analysis points to a combination of Schiff bases and ketene acetals as a possible mode of ring closure [Kametani, 1986].

The idea has been put to the test. A low yield of graveolinine has been obtained, owing to the stability and hence inertness of the dimethoxycarbenium ion, and the intrinsic low nucleophilicity of the anilino group. Other reaction pathways intervened.

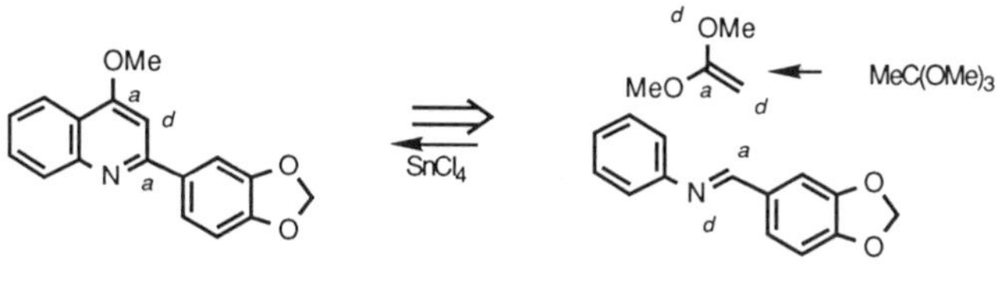

graveolinine

In brevicolline the two non-indolic nitrogen atoms are separated by three carbon atoms (conjoint situation), therefore normal condensation of a dihydro-β-carboline and a pyrrolenium ion leading to the complete skeleton is expected [Leete, 1979].

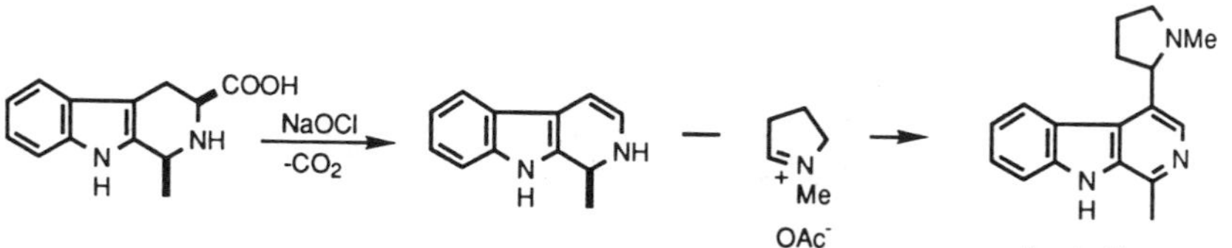

brevicolline

The use of tryptophan as starting material facilitated the synthesis in that the coupling with the pyrrolidine moiety totally depended on the reactivity of the non-indolic portion.

It is interesting that a synthesis of brevicarine [Kuchkova, 1971] was based on the Mannich reaction. Attack of the acceptor center attaching to the β-carbon by a donor species provided the remaining carbon elements of the pyridine moiety. A Beckmann rearrangement effected the introduction of the missing nitrogen atom and partition of the chain of the original donor.

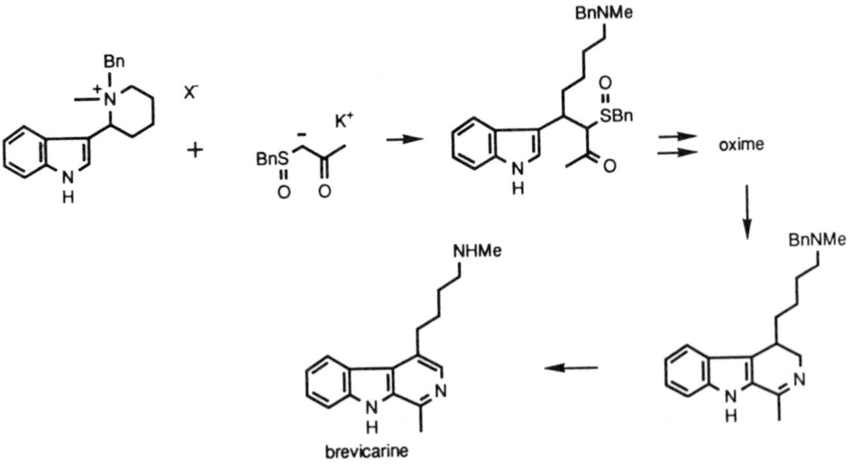

brevicarine

6.2. MICHAEL REACTION

The other general method for the preparation of β-amino carbonyl compounds is the Michael reaction of amines with conjugate carbonyl substrates. In a synthesis of complex alkaloids the Michael reaction often serves to provide such intermediates, not to limit its utility in obtaining structures with the obvious functionality.

The sidechain hydroxy group of perhydrogephyrotoxin is 1,3-related to the nitrogen atom. It is therefore conceivable that an intramolecular Michael reaction would find application in its synthesis [Overman, 1980], for a subsequent reduction is all that is required if the hydroquinoline intermediate is properly substituted. In fact, a synthesis completed by this method utilized a β,γ-unsaturated ester precursor which underwent conjugation and cyclization.

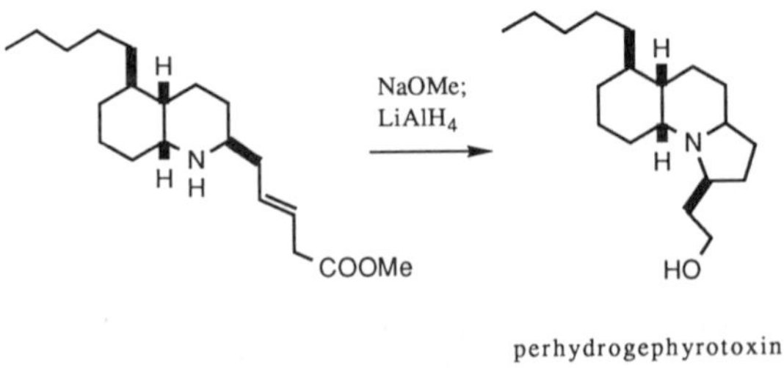

perhydrogephyrotoxin

The disconnection of lycopodine by an alkylation–Michael transform may seem tenuous because of the involvement of a bridgehead olefin. However, the demonstrated transient existence of such reactive species in recent years has at last made this approach possible, especially when the strained olefin is to be intercepted intramolecularly. A synthesis of lycopodine has been accomplished [Kraus, 1985a] according to this scheme and also one based on trapping of the corresponding bridgehead carbenium ion.

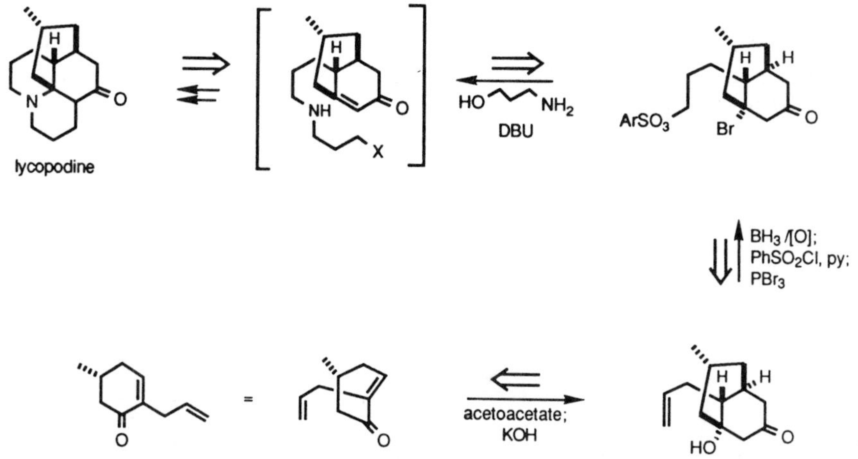

The 1,3-relationship of the nitrogen and oxygen atoms in perhydrohistrionicotoxin suggests a gainful application of an intramolecular Michael reaction to set up the spirocyclic system [Fukuyama, 1975]. The employment of a tethered amide as donor and an enone as acceptor indeed led to a successful construction of the ring skeleton, although the lack of stereocontrol resulted in the desired product as a minor component.

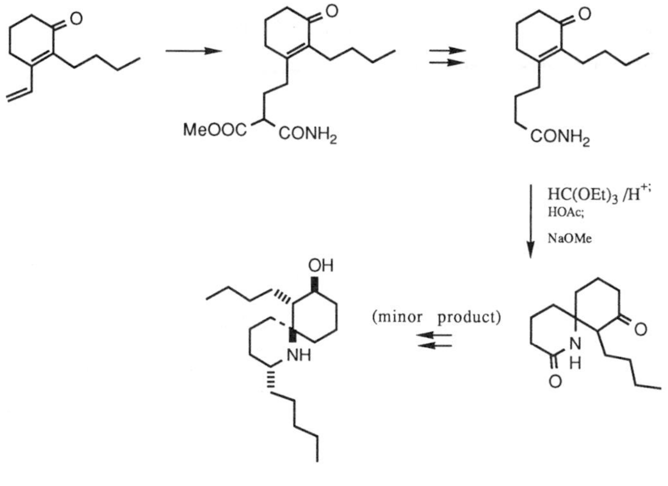

perhydrohistrionicotoxin

There are two separate chromophores in calicheamicin which interact on bioactivation to generate a 1,4-benzenediyl species from the enediyne moiety. In terms of synthesis of its aglycon the 4-(trithioethylidene)-2-cyclohexenone portion is more challenging because it requires exquisite timing in the installation of various functional groups. The presence of a carbamate substituent at C-2 of the enone system demands careful orchestration.

Several significant features in a synthesis of calicheamicinone [Cabal, 1990] include introduction of the trithioethylidene group via a lactone which was formed by an intramolecular Horner–Emmons reaction. Setting the carbamate function in place by an addition–elimination route via bromo and azido intermediates is rather remarkable, in view of the anticipated lack of stability of short-bridge enolate which must have intervened.

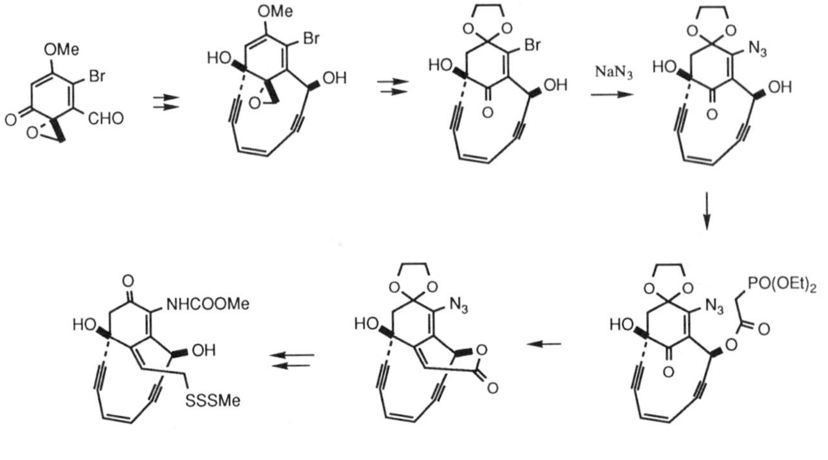

calicheamicinone

The addition–elimination route for establishment of an enaminone system also featured in a synthesis of crinine [Whitlock, 1967]. The donor was aziridine and the product was converted into a hydroindolone by a formal rearrangement (actually involving ring opening and intramolecular alkylation).

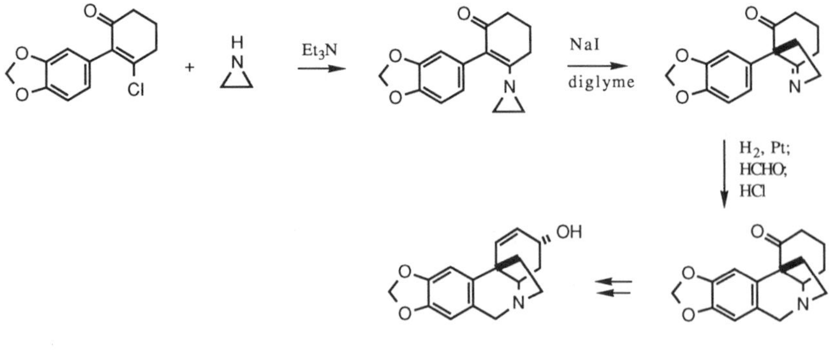

crinine

Formation of enaminones from 1,3-diketones probably does not proceed via a Michael reaction–dehydration pathway but is initiated by a Schiff condensation. However, a brief mention of this process here is not totally out of place in view of its kinship to the above syntheses. A notable application of this reaction is in a synthesis of gephyrotoxin [Fujimoto, 1980].

Disconnection of the pentakisnorketone of the alkaloid by an alkylation transform reveals a symmetrical *cis*-2,5-bis(hydroxyethyl)pyrrolidine precursor which is readily constructed. The reason for using the enaminone intermediate is twofold: to avoid retro-Michael fission during the alkylation step, and to be able to control the stereochemistry of the perhydroquinoline ring juncture.

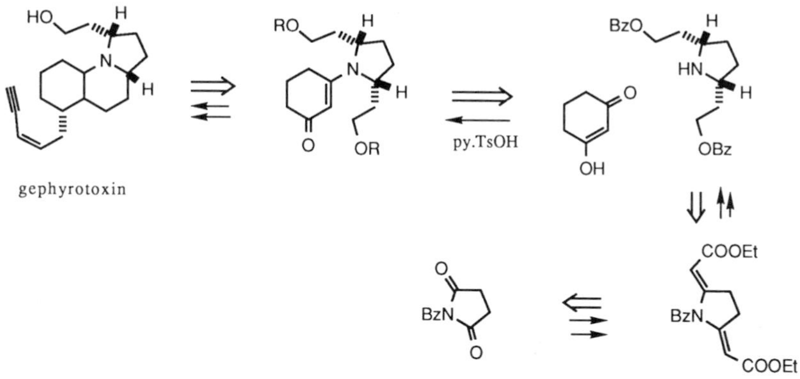

gephyrotoxin

The one-step reconstitution of tropinone [Bottini, 1971] from 2,6-cyclohepta-dienone and methylamine by a double Michael addition is the reversal of a degradation route.

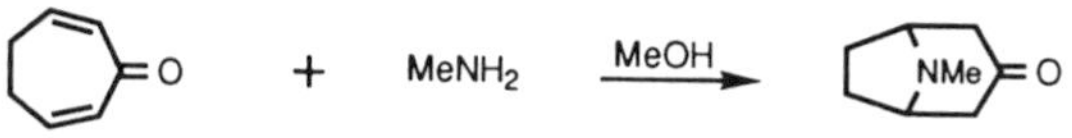

tropinone

Synthesis of 2,6-bridged piperidin-4-ones by exploiting the contrapolarizability of the sulfonyl group has been developed [Lansbury, 1990]. Thus, alkylation of dilithio-α,ω-bisphenylsulfonylalkanes with α,α'-methallyl diiodide to give the methylenecycloalkanes was followed by double-bond cleavage. Elimination of phenylsulfinic acid can then be effected by an amine. *In situ* Michael addition and a second-stage elimination–addition sequence led to the piperidinones. Alkaloids such as pseudopelletierine are available quite conveniently.

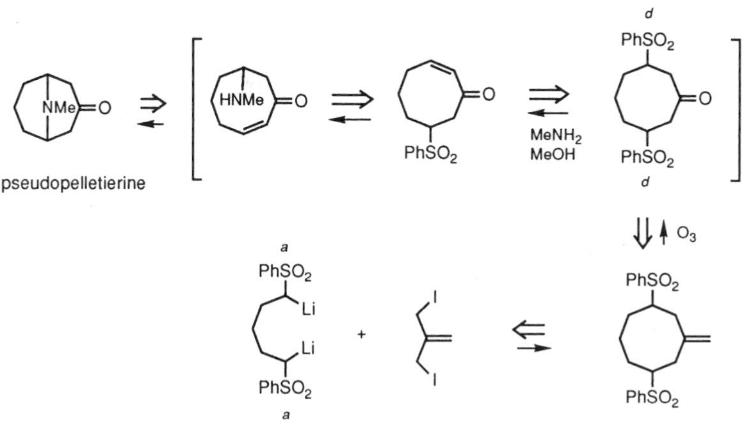

pseudopelletierine

Vindoline has been synthesized from a tetracyclic intermediate. A new elaboration of this intermediate involved a thermal reorganization of a benzocyclobutene encompassing electrocyclic and [3.3] sigmatropic reactions in tandem to generate a 4,4-disubstituted isochroman-3-one [Shishido, 1989]. The isochromanone was transformed into a cyclohexenone, and later a cyclohexadienone to accept the proper amino donors on two different occasions.

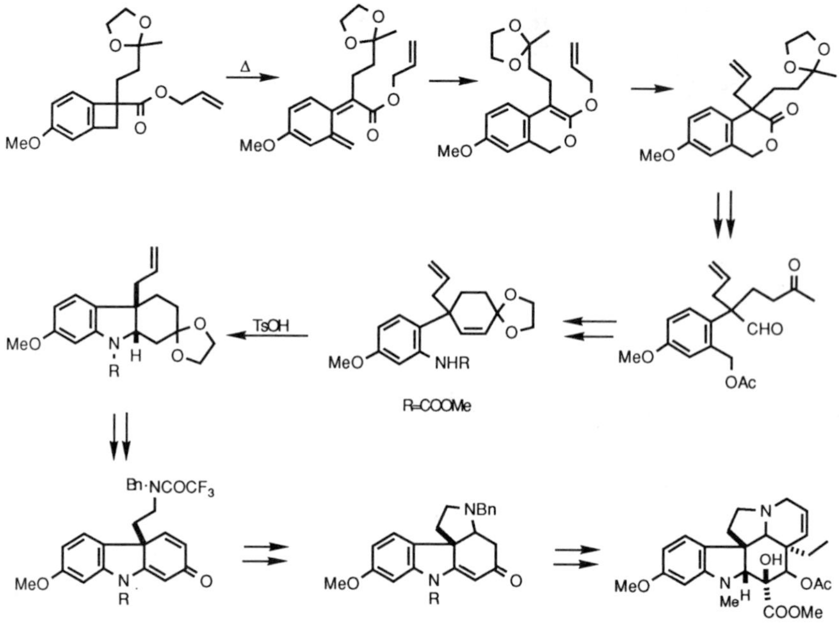

In dendrobine the nitrogen and the nearest oxygen atom are in a disjoint 1,2-relationship. After moving the oxygen one atom away and introducing a conjugate double bond, the pericondensed perhydroindole system can be closed by an intramolecular Michael addition. This strategy has actually been adopted in a synthesis [Inubushi, 1972].

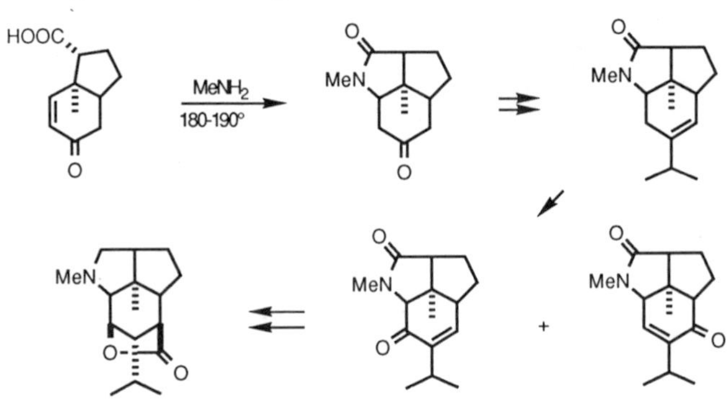

The key to a synthesis of aspidospermine [Stork, 1963b] was the preparation of a hexahydrolilolidinone in which the nitrogen atom is β to the ketone group. This tricyclic ketone was submitted to a Fischer indole synthesis to complete the skeleton of the alkaloid.

The amino ketone was assembled via the enamine method of cyclohexenone formation and subsequent transformations of the three-carbon sidechain at C-4 to a primary amine, which underwent ring closure to a perhydroquinolinone intermediate. The third ring was created by an intramolecular alkylation step.

aspidospermine

Dioscorine is a spirocyclic lactone with an isoquinuclidine nucleus. A key compound of its synthesis is obviously (+)-N-methyl-5-oxoisoquinuclidine, from which it is rather straightforward to reconstitute the lactone via a Reformatskii reaction.

Preparation of the isoquinuclidinone was relatively facile [Page, 1964], i.e., by a Mannich reaction on 2-cyclohexenone or acid treatment of the Birch reduction product from 2-methoxy-N-methylbenzylamine. An intramolecular Michael reaction took place.

dioscorine

The transform-based synthesis of muscopyridine [Utimoto, 1982] was keyed to the desilylative cycloacylation which was followed by a Michael reaction with ammonia. The presence of another ketone group, four bonds away from the β-carbon of the alkynone system, provoked a transannular Schiff condensation. The resulting heterocyclic ring was susceptible to aromatization, and the synthesis was concluded by Wolff–Kishner reduction of the remaining carbonyl, after its service in an actuating role.

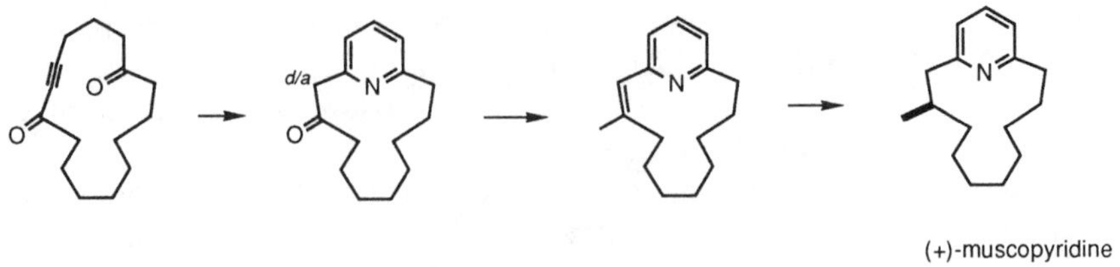

(+)-muscopyridine

An interesting route to both enantiomers of norsecurinine [Jacobi, 1989] involved a transannular bridging to the γ-position of the butenolide moiety in the final step. The procyclic intermediate was created from a facile Michael–Diels–Alder/retro-Diels–Alder reaction tandem. The reversible Michael reaction permitted equilibration of the unwanted isomer.

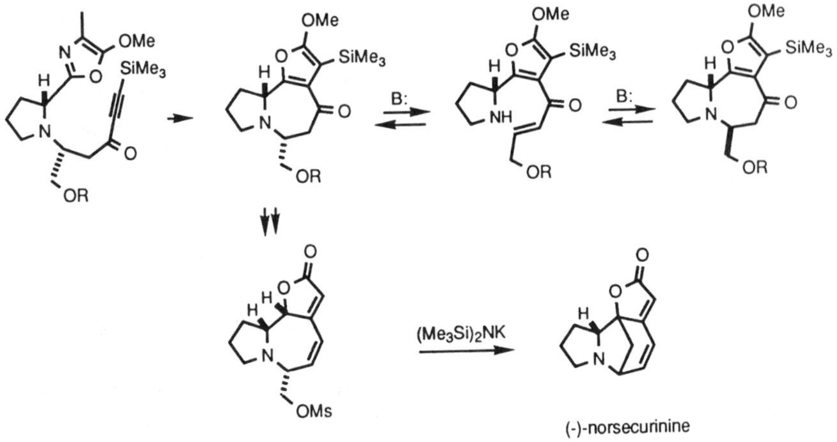

(-)-norsecurinine

For a description of a retro-Michael–Michael reaction sequence that converted a cephalotaxine skeleton into one with a quinolizidine moiety, see Dolby [1972].

It is easily recognized that a tetracyclic enamine would play a key role in synthesis of vincamine [Szantay, 1977]. As the γ-carbon of the enamine system is an acceptor site, the corresponding enaminone could be prepared readily by an intramolecular acylation. Thus the whole sequence of ring formation was initiated by a Michael reaction.

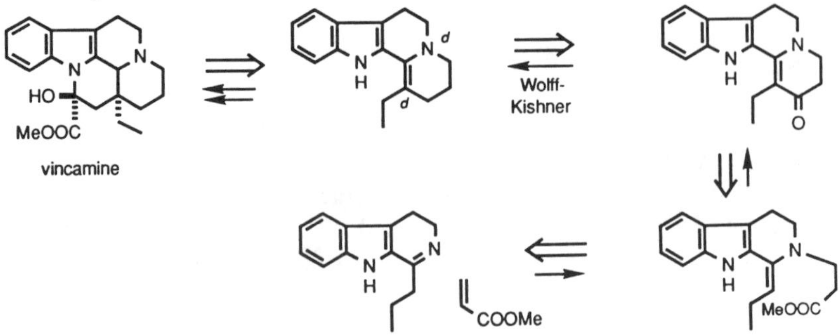

vincamine

The synthesis of hasubanonine [Ibuka, 1970] is mainly a matter of functionalization of the A-ring, once the tetracyclic skeleton is established. For the latter task it is advisable to choose a subgoal in which the carbonyl group is β to the aminated angular position since a Michael transform is readily perceived. Also such a substitution pattern indicates the Robinson annulation of a proper β-tetralone is most adequate to start the synthetic operation.

hasubanonine

A classical method for the synthesis of phenoxazones is by oxidation of o-aminophenols. The transformation was observed before the discovery of natural products (e.g., actinomycins, ommochromes, litmus dyestuffs) based on the ring system. A synthesis of xanthommatin [Butenandt, 1954] was simply accomplished by treatment of 3-hydroxykynurenine with potassium ferricyanide at pH 7.1.

xanthommatin

The dimerization followed from two consecutive Michael reactions which were also employed in the closure of the pyridine nucleus. The first Michael reaction involving the phenolic oxygen was perhaps under steric control.

The two nitrogen atoms in tryptamine are disjoint. Umpolung strategy is necessary to establish this structural segment from indole. The usefulness of the nitro function is apparent, as indole undergoes Vilsmeier formylation at the β-position, the Henry reaction gives β-(3-indolyl)-nitroethene which can be hydrogenated.

In a synthesis of brevicolline [Müller, 1977] the introduction of the pyrrolidine ring also took advantage of the special functional array of the nitroalkene. Michael addition with a pyrrole donor, reduction of the nitro group afterwards unveiled the substituted tryptamine.

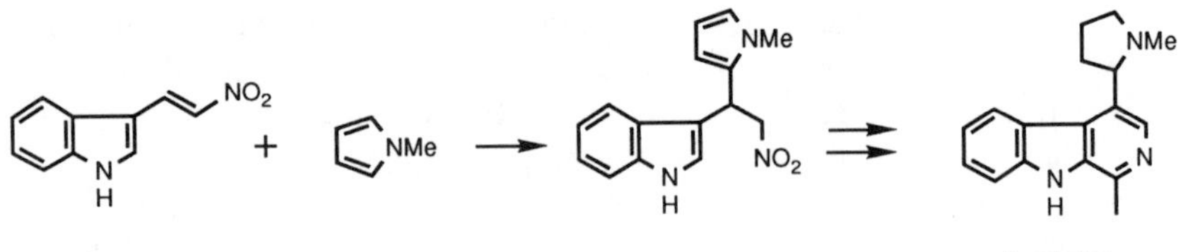

brevicolline

An internal Michael cyclization which is a formal 1,6-addition but resulting in a β-amino ester was embodied in a synthesis of lysergic acid [Cacchi, 1988]. The tricyclic precursor was assembled by a Heck-type coupling of a substituted acrylic ester with an enol triflate.

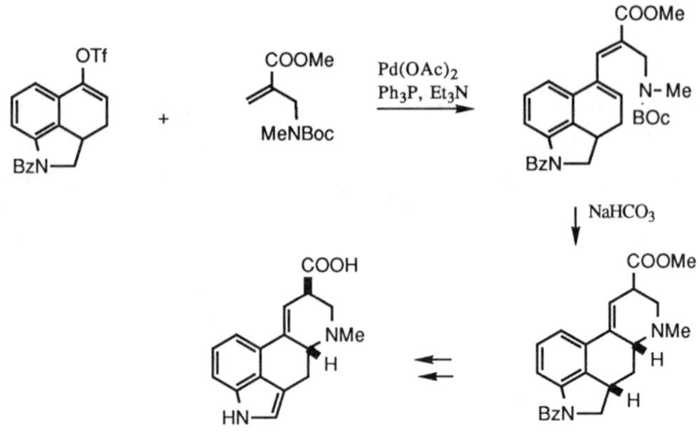

lysergic acid

As previously indicated, the strong support of the benzylic donor by the nitrogen atom of α-, γ-picolines, 2,6-lutidine, and *sym*-collidine makes 2- and 4-vinylpyridines excellent Michael acceptors. Of course, the vinylpyridines are internal imine derivatives of conjugate ketones. Exploiting this useful property is a synthesis of (−)-cytisine [van Tamelen, 1958a] in which N-benzyl-5-(α-pyridyl)-piperidine-3-carboxylic acid was formed by a Mannich reaction followed by the special Michael addition.

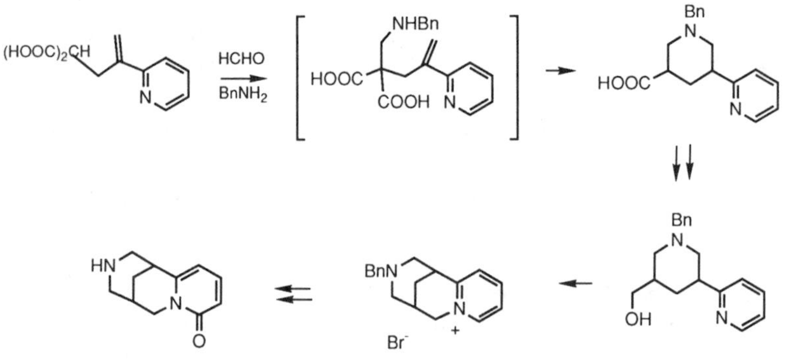

cytisine

An apparent intermediate for the synthesis of gelsedine is a tricyclic amino ketone. While the amino group is β to the ketone, undoubtedly accessible by the Michael approach, a construction of the synthon actually was effected by a photo Hofmann–Löffler–Freytag reaction [S.W. Baldwin, 1979]. In this case the choice of the method was dictated by convenience.

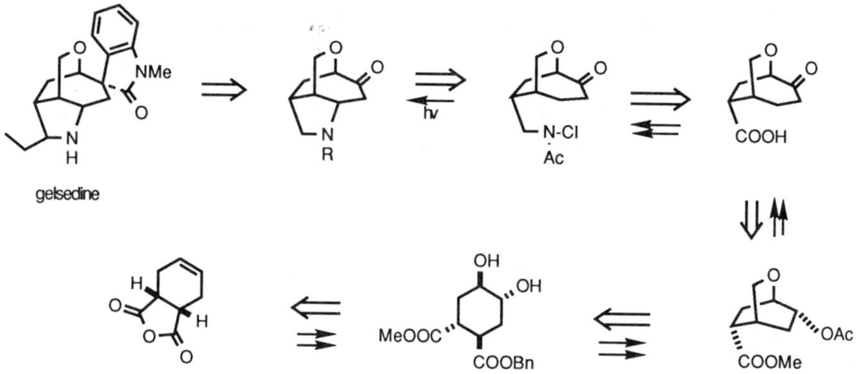

gelsedine

7

SYNTHESIS OF 1,4-DIFUNCTIONAL COMPOUNDS

7.1. MICHAEL REACTION WITH ACYL ANION EQUIVALENTS

The disjoint nature of 1,4-difunctional compounds of the same polarity necessitates the employment of umpolung techniques or disjoint precursors for their assembly. Thus, acyl anion equivalents may be persuaded to act as Michael donors, and the result is that the normal 1,5-separation of the hetero atoms is shortened by one atom.

The direct acyl anion equivalents are aldehydes. Such donors have been known for a long time in the form of the conjugate bases of cyanohydrins during benzoin condensation. In the presence of a Michael acceptor such a species may be captured. However, this method of 1,4-diketone formation is limited to the acyl donors in the aromatic series.

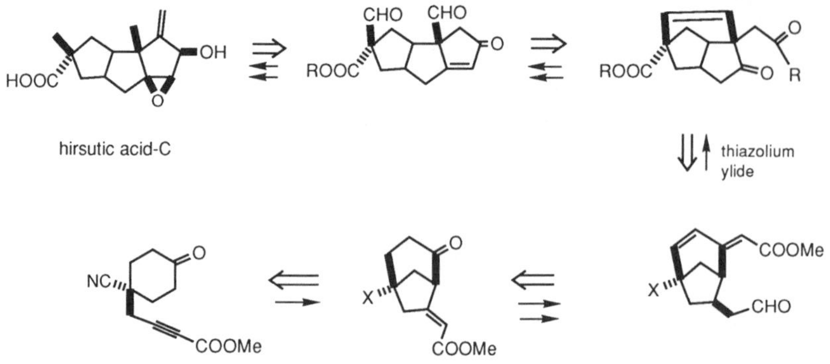

hirsutic acid-C

thiazolium ylide

X = CN

A more general procedure involves a thiazolium ylide as catalyst [Stetter, 1976]. This reaction is patterned after the vitamin B_1 coenzyme reactivity. While there have not been many examples of complex synthesis incorporating this reaction, an intramolecular version was highlighted in an approach to hirsutic acid C [Trost, 1979c].

An extension of the thiazolium-ylide-catalyzed condensation to the synthesis of symmetrical 1,4-diketones [Stetter, 1978] should be noted.

Monoketals of 1,4-diketones have been acquired by a photochemical method [Fraser-Reid, 1977]. Note that photoexcitation is generally associated with polarity inversion (for more of such reactions, see Chap. 14). Such polarity inversion can also be accomplished via electroreduction of oxazolinium salts and carboxylic anhydrides [Scheffold, 1983; Shono, 1987].

Aldehyde *t*-butylhydrazones undergo a thermal ene reaction with methyl acrylate or acrylonitrile. The resulting azo compounds may be reverted to the hydrazone forms on contact with trifluoroacetic acid and then hydrolyzed to give γ-keto esters [J.E. Baldwin, 1984]. This is a redox process.

Transient acylmetals effectively attack some Michael acceptors. These donors include acylnickels [Corey, 1969a], acylirons [Cooke, 1977], acylcobalts [Hegedus, 1985], and acylcoppers [Seyferth, 1985; Lipshutz, 1990]. Necessarily, the majority of these species are nonfunctionalized (elsewhere in the molecules).

A popular theme of current synthesis plays on the Michael addition and *in situ* trapping of the enolates with various electrophiles. On noting that the pyran ring of nanaomycin A is fused to a naphthoquinone and consequently two of the oxygenated carbon atoms are 1,4-related, a handsome synthetic scheme was devised [Semmelhack, 1985].

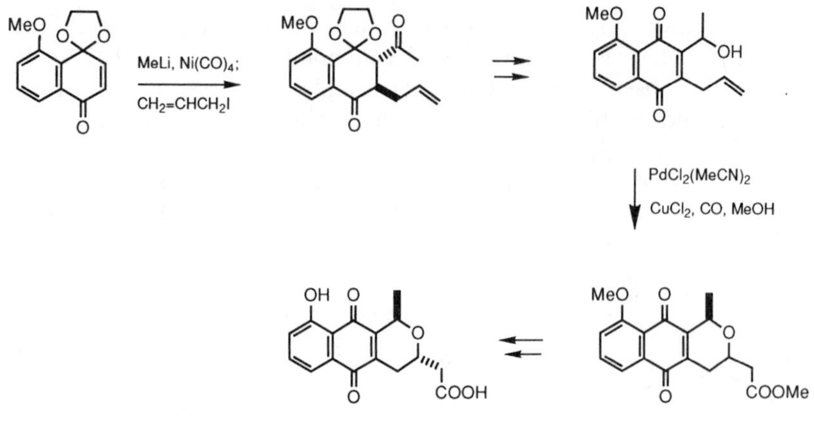

nanaomycin-A

γ-Keto acids in the spiroketal forms can be made from cyclic ethers, an acrylic ester, and (trialkylsilyl)manganese pentacarbonyl [DeShong, 1988]. The initial adducts are subjected to photodemetalation.

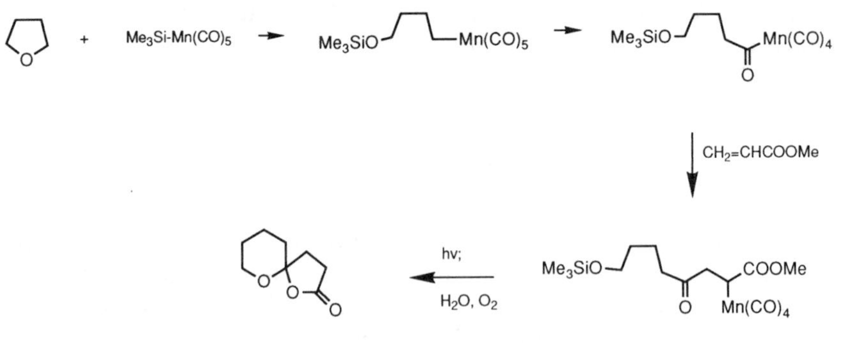

Biscarbomethoxylation of alkenes with palladium catalysts is a very useful functionalization reaction [Stille, 1979]. The β-palladoacrylate species undergoes CO insertion and methanolysis.

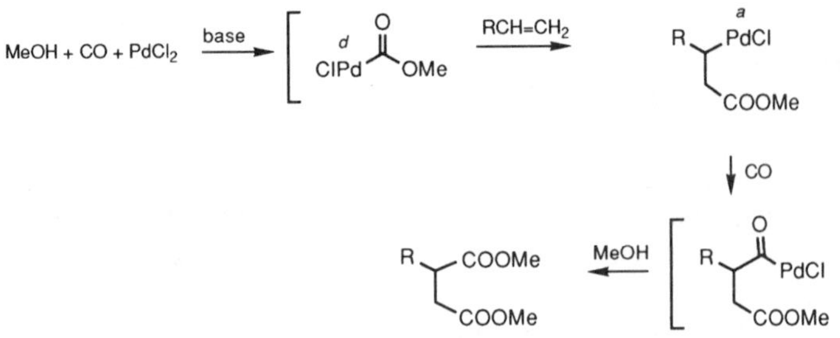

It has also been reported that reaction of iron carbonyl complexes of Michael acceptors with organometallic reagents results in 1,4-dioxo products [Rakshit, 1987; Kitihara, 1988].

dihydrojasmone

The most well-known acyl anion equivalents are probably the 2-lithio-1,3-dithianes [Gröbel, 1977]. These anions undergo Michael reactions in the presence of a cation-sequestering solvent such as hexamethylphosphoric triamide [C.A. Brown, 1979] and 1,3-dimethyl-2-oxo-hexahydropyrimidone [Mukhopadhyay, 1982]. Generation of the carbonyl group from the dithiane requires an individual operation.

β-Formylacyl compounds are Michael adducts of a formyl anion and the α,β-unsaturated acceptors. The conjugate base of bis(methylthio)methane monoxide is a useful synthon for this transformation which constituted a crucial step in a synthesis of carbomycin B aglycon [Tatsuta, 1977, 1980].

The asymmetric centers of the antibiotic are concentrated in the right-side domain. It is conceivable that a segment containing three consecutively oxygenated carbon atoms would map the same in a common carbohydrate molecule, therefore a bidirectional analysis is most profitable to the formulation of a synthetic plan. Accordingly, glucose diacetonide was identified and fashioned to a properly protected C_8 chain terminated with an α,β-unsaturated ester. The latter functionality was destined to receive the formyl anion equivalent. To complete the right-hand half of carbomycin B aglycon the saturated ester was converted into a lactol ether, the formyl group was unveiled, condensed with the butanone subunit, and the product reduced.

This reaction sequence involving two C—C bond-forming steps was predicated by the disjoint relationship between the ketone and the aldehyde groups in carbomycin B. It is pleasing that the Michael reaction was stereoselective, perhaps as a result of the chelation control by the methoxy substituent in the transition state.

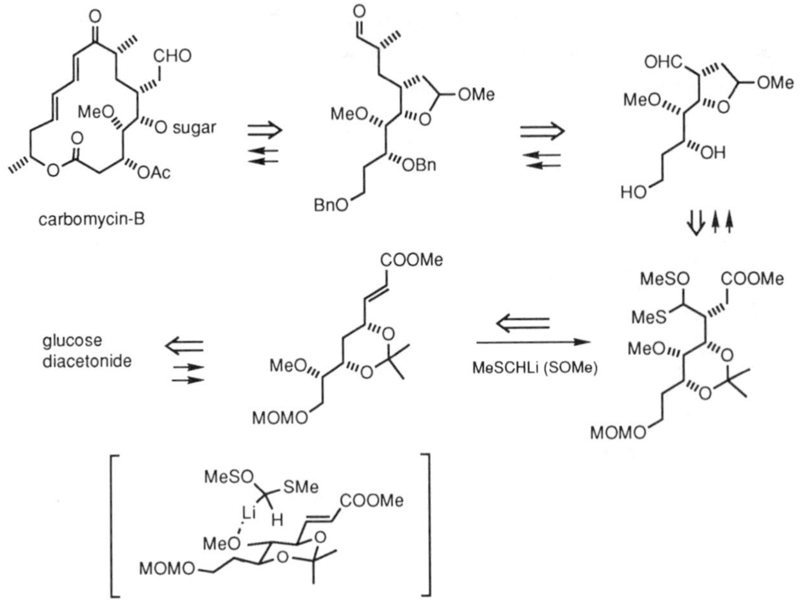

carbomycin-B

glucose
diacetonide

MeSCHLi (SOMe)

Tris(phenylthio)methyllithium functions in a similar manner as the cyanide ion toward enones [Manas, 1975]. Alcoholysis of the adducts leads to esters. It may be noted that cyano transfer has also been achieved using *t*-butyl isocyanide in the presence of a Lewis acid [Ito, 1982].

Trivalent phosphorus compounds are donors while most phosphorus(V) species are acceptors. Accordingly, the *ipso* carbon of an alkyl halide is converted from an acceptor site into a donor upon reaction with a phosphine. Deprotonation of the phosphonium ion is a prerequisite of the Wittig reaction. This contra-polarizability of a phorphosus(III) center has great significance in a synthesis of sarkomycin [Amri, 1989].

Generation of a β-phosphonopropanoate subunit from the dimer of an acrylic ester via a Michael reaction modified the reactivity of the phosphorus-bound carbon, enabling a Dieckmann-type cyclization to occur. The resulting cyclo-pentylidenephosphorane is delightfully equipped to be converted into the α-methylenecyclopentanone chromophore.

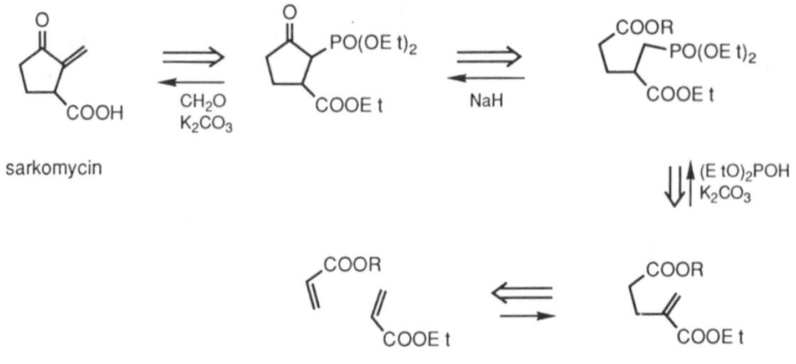

sarkomycin

7.2. FROM ALIPHATIC NITRO COMPOUNDS

The $CHNO_2$ group is an excellent acyl anion synthon, not in alkylation, but as a Michael donor. It seems that limitation of the reaction's utility is set only by the availability of the nitro compounds, as conversion of this group into an oxo function has been achieved in many ways: oxidation ($KMnO_4$, O_3, t-BuOOH,...), reduction ($TiCl_3$ [McMurry, 1974b],...) and Nef reaction (RO^-, H_3O^+). The Michael reaction of nitroalkanes has been effected in the presence of as mild a base as the fluoride ion. Such conditions allow the survival of many sensitive functionalities.

The γ-nitro oxoalkanes can actually be approached by switching the donor–acceptor moieties, i.e., Lewis-acid-catalyzed condensation of enol derivatives with conjugate nitroalkenes [Miyashita, 1976]. This latter mode of reaction is somewhat less convenient in the light of the instability of β-unsubstituted nitroalkenes which are usually prepared from other nitroalkanes. The choice of either approach also depends on the access of the oxo species.

Treatment of β-nitroalkyl acetates with NaCN leads to succinonitriles [Barton, 1983]. The occurrence of two sets of β-elimination and Michael reaction is evident. The contrapolarizability of the cyano group is responsible for the formation of the disjoint products.

At this juncture it seems appropriate to discuss briefly how to plan a typical synthesis in which there is no apparent correlation in functionality of a target molecule with early intermediates. Thus, the only tangible handle of isocomene is a trisubstituted double bond. However, the corresponding norketone is amenable to rapid disconnection by α-alkylation and Michael transforms, leaving a pentalenone skeleton. The latter is readily derived from a 2-(2-oxobutyl)cyclopentanone. The Michael–aldol approach is the logical proposal to access such an intermediate [Paquette, 1979].

isocomene

Of some significance is the first stage of the synthesis. Compared with the Robinson annulation, formation of 2-cyclopentenones is difficult to effect directly because special attention is required for the preparation of the disjoint 1,4-dioxo precursors. Another relevant fact is that odd-membered ring structures are also disjoint.

An application of the nitroalkane chemistry to the spiroketal synthesis [Rosini, 1989] may be mentioned.

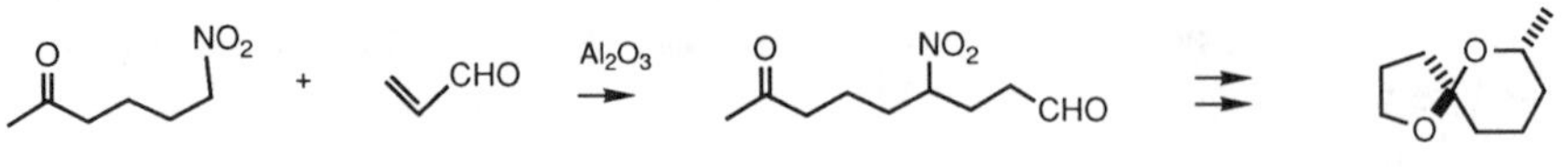

3-Nitropropanoic esters have been employed as acrylate β-anion equivalents. Thus, after condensation of the nitronate anion with an aldehyde, elimination of nitrous acid results in a 4-hydroxycrotonate [Bakuzis, 1978]. The value of this subunit assembly method may be illustrated by a synthesis of pyrenphorin.

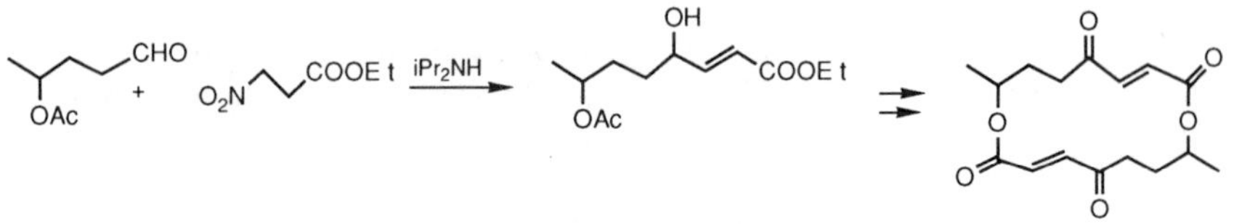

pyrenophorin

In the presence of a dehydrating agent, primary nitroalkanes from nitrile oxides. These 1,3-dipolar species react readily with acceptor-substituted alkenes in the Michael sense to give functionalized isoxazolines. A ramification of this reaction variant may be better appreciated on examining an approach to calythrone [Andersen, 1982].

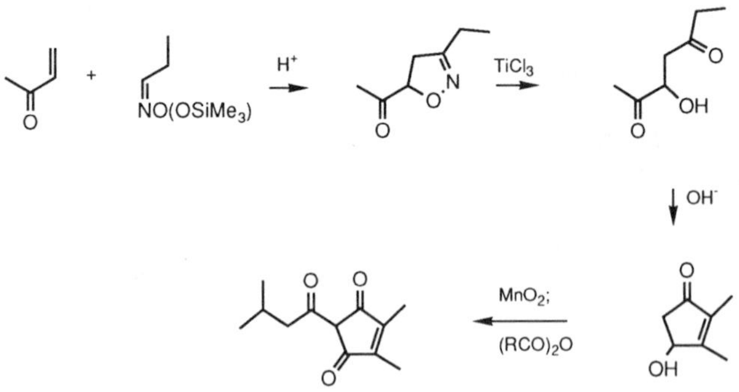

calythrone

A process for oxidative oxoalkylation of certain alkenes to give 1,4-diketones in three steps employs nitroalkenes as the key reaction partners [Denmark, 1986a]. A formal hetero Diels–Alder reaction initiates the coupling.

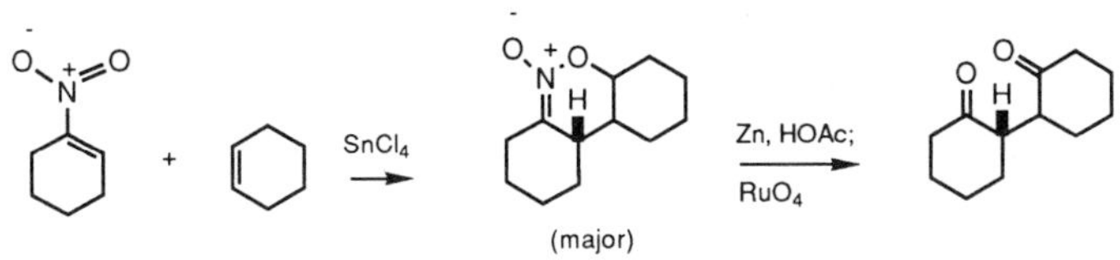

7.3. WITH UMPOLED ACCEPTORS

Besides nitroalkenes, there are a number of other Michael acceptors which are α-acyl cation synthons. These include α-amino acrylonitriles [Ahlbrecht, 1980] and ketene dithioacetals, their oxides [J.G. Miller, 1974; Kieczykowski, 1977], and onium ions [Oishi, 1974]. The advantage of these acceptors is that after the Michael addition, alkylation is possible. This capability magnifies the synthetic potential.

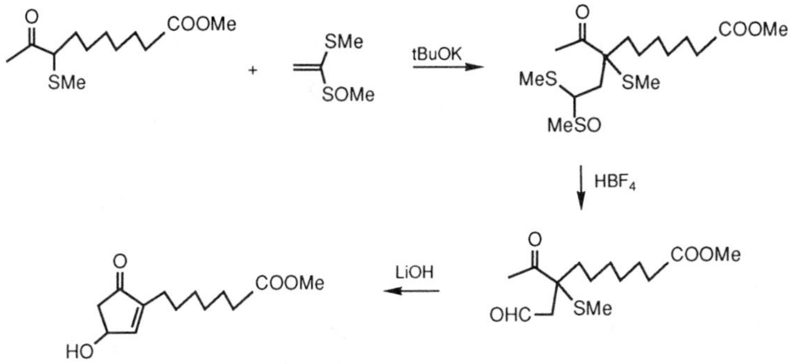

A classical method for preparation of 1,4-dioxy compounds is by alkylation of enol(ate) derivatives with α-halocarbonyl substrates or epoxides, these alkylating agents having an unnatural array of polar functionalities (i.e., $a-a$ in $^+CH_2C{=}O$, $^+CH_2C{-}O$). A most common practice in assembly of a γ-lactone is by alkylation of a ketone with an α-bromoalkanoic ester followed by reduction of the ketone group. An enamine is a useful donor (cf. synthesis of alantolactone [Marshall, 1966].)

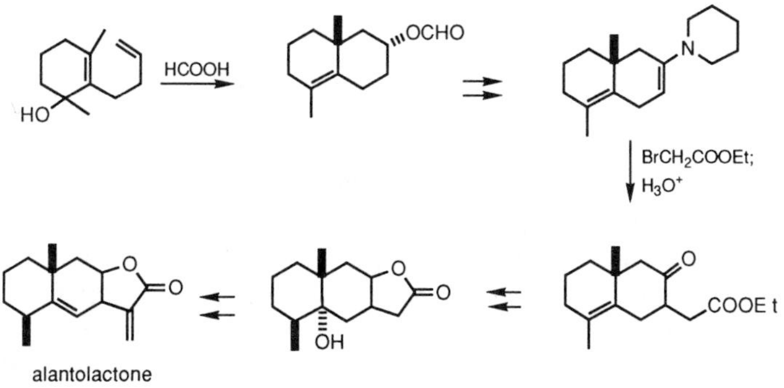

alantolactone

In a strigol synthesis [MacAlpine, 1976] assembly of the γ-lactone intermediate involved alkylation of a perhydroindanone. Interestingly, the precursor was prepared in one route by a reaction sequence starting from 2-cyclopentenone and 5-nitro-2-butanone and including a Michael reaction and aldol condensation.

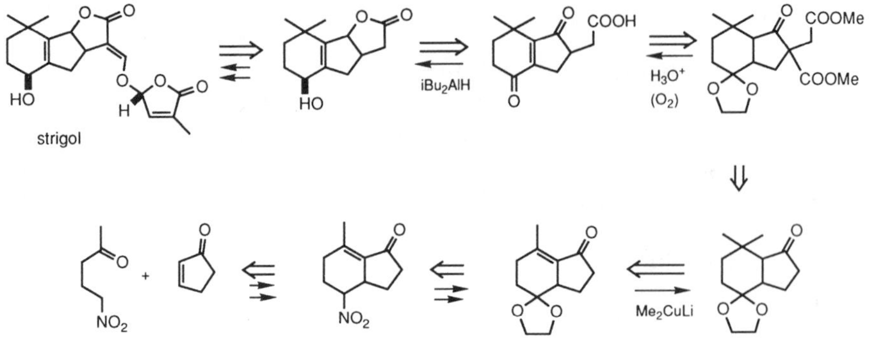

strigol

As seen above, control of alkylation at an α-carbon of a ketone may be facilitated by the enamine method or using an activated enolate (e.g., of a β-keto ester). The purpose of such a device is that a less acidic and more stable enolate (softer species) has a lower tendency to provoke side-reactions. It is more important when dealing with an α-haloketone as the alkylating agent because Favorskii rearrangement of which may be induced by stronger bases.

While in most instances the use of a β-keto ester is only for activation and regiocontrol, a synthesis of methylenomycin A [Jernow, 1979] is characterized by the preservation of the carboxyl group. Thus, a simple retrosynthetic analysis envisions a 2,3-dimethyl-2-cyclopentenone-4-carboxylic ester as an intermediate (aldol and epoxidation transforms). Derivation of this keto ester from acetoacetate and 1-bromo-2-butanone is evident.

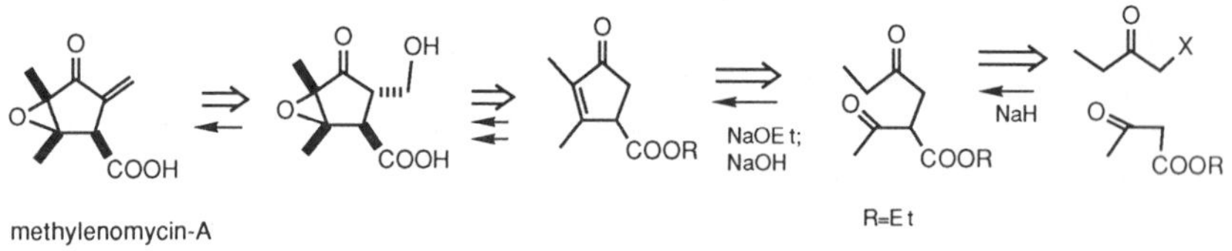

methylenomycin-A

2-Acylcyclopropanols, their ethers and esters are prone to fragment to relieve strain. The fragmentation is a retro-aldol reaction. Synthetically, the two-step process consists of the cyclopropanation (copper- or rhodium-catalyzed) of enol derivatives with α-diazo acyl compounds and the fragmentation accomplishes an alkylation in the umpolung sense. A biphilic carbenoid presents itself initially as an acceptor and then as a donor during the formation of the three-membered ring. (Concertedness of the cyclopropanation step notwithstanding, the electrophilic nature of such carbenoids indicates asynchronicity in the establishment of the two C—C bonds.)

Using this process many precursors of cyclopentanone natural products have been synthesized [McMurry, 1971; Wenkert, 1978b].

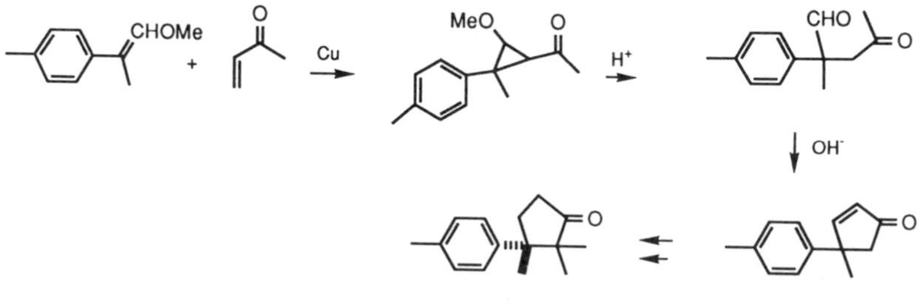

α-cuparenone

Extension of this particular alkylation method to the dihydropyran system allowed for the construction of a synthon for the *Eburna* alkaloids [Wenkert, 1978c].

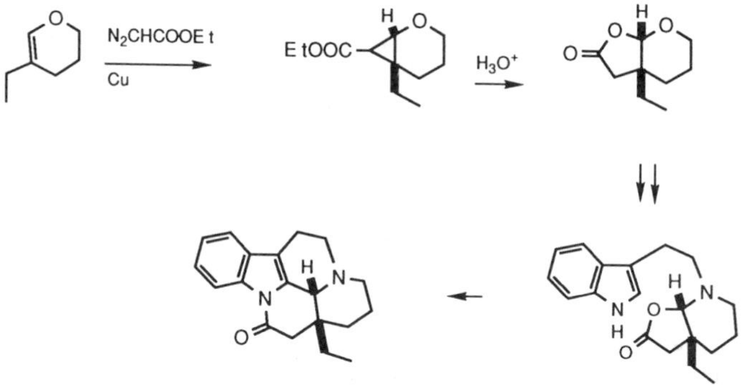

eburnamonine

The γ-keto ester α-anion arising from fragmentation of an alkyl 2-oxycyclo-propylcarboxylate may be captured by appropriate acceptors to broaden the synthetic utility of the process. Of course it is not possible to generate such anions directly from unprotected γ-keto esters themselves. Thus, a synthesis of pentalenolactone E methyl ester [Marino, 1987] was based on the reaction of a fragmented species with a vinylphosphonium salt. A Michael-type reaction was followed by an intramolecular Wittig olefination.

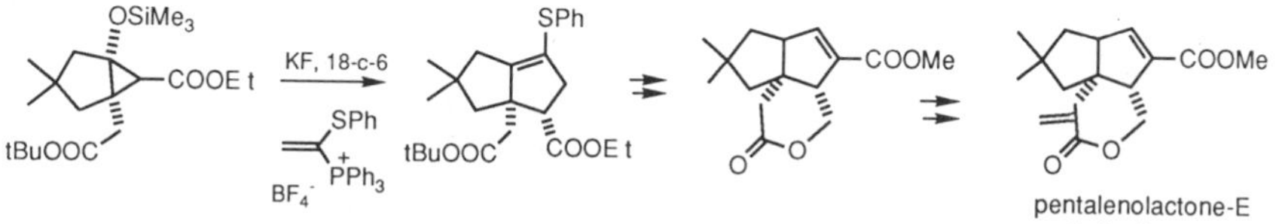

pentalenolactone-E
methyl ester

The formation of Δ³-butenolides by reaction of α-diazoacyl compounds with ketene [Ried, 1964] may involve 2-acylcyclopropanones as intermediates. Rearrangement of the latter entities must be facile.

7.4. FROM FURANS

Furans and 1,4-dioxo compounds are interconvertible. Consequently, synthesis of the latter class of substances may benefit from using precursors with a furan ring. Furthermore, the access of the enedicarbonyl chromophore on cleavage of the furan nucleus increases the value of such heterocycles.

A route to prostaglandin E_2 [Floyd, 1978] fully exploited the latent existence of a cyclopentenolone system in a furan derivative.

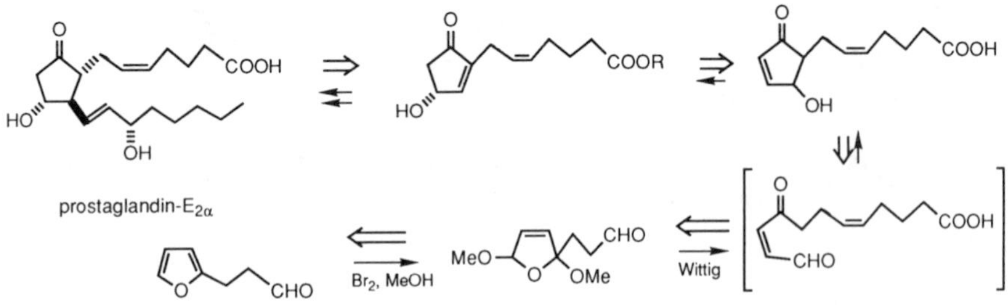

prostaglandin-$E_{2\alpha}$

Another advantage of the furan approach is that deprotonation of an α-hydrogen effectively creates the acyl anion equivalent, its utility is amply demonstrated in a synthesis of *cis*-jasmone [Büchi, 1966c].

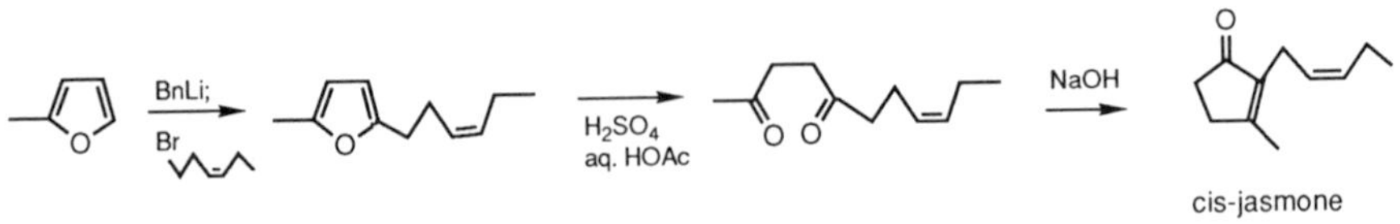

cis-jasmone

A formal synthesis of aphidicolin [Tanis, 1985] was based on a furan-terminated cationic polycyclization and the structural correlation of the product with a known 1,4-diketone precursor. This route may be slightly more lengthy than desired in view of the additional steps needed to introduce the methyl group to the furano-octalin.

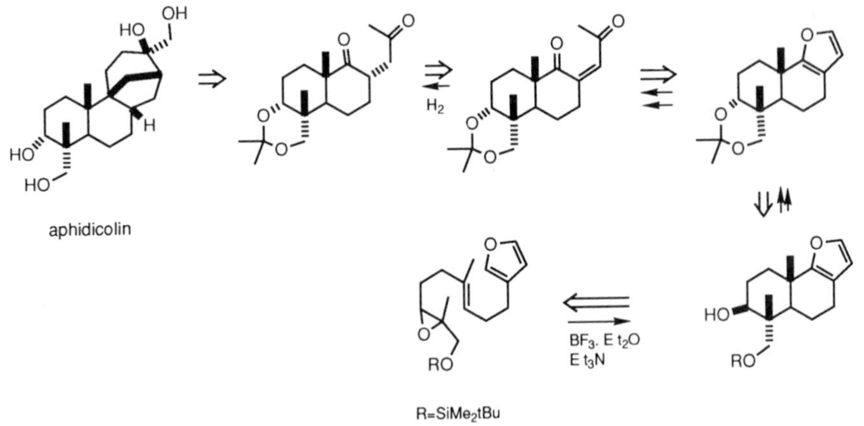

aphidicolin

R=SiMe₂tBu

Disconnection of camptothecin at the D-ring reveals a maleialdehydic acid portion (or butenolide equivalent) which is apparently derivable from a furan ring [Corey, 1975b]. The furan template allowed the construction of the pyrone system enantioselectively before its oxidative transformation into the appropriate electrophile. Thus, a synthesis was initiated by differentiating the two ester groups of dimethyl furan-3,4-dicarboxylate and elaborating them separately. Photooxygenation of the furanopyrone and converting the product into a pseudo acid chloride (as a mixture of regioisomers) led to the key intermediate.

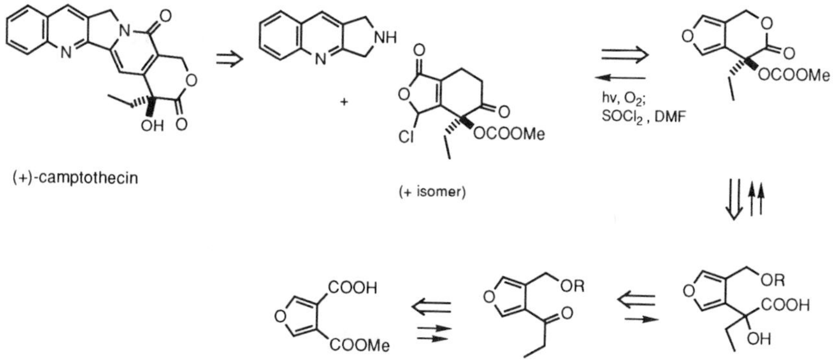

(+)-camptothecin

(+ isomer)

hv, O$_2$; SOCl$_2$, DMF

Oxidation of furan-2-carbinols leads to pyrone derivatives. Three consecutive asymmetric centers may then be created from the enone system of such products. When the furancarbinol sidechain contains additional hydroxyl groups, a bridged spiroketal may arise directly.

A clever manipulation of this chemistry has culminated in a stereoselective assembly of the C-3 to C-10 portion of erythronolide B [Martin, 1989b].

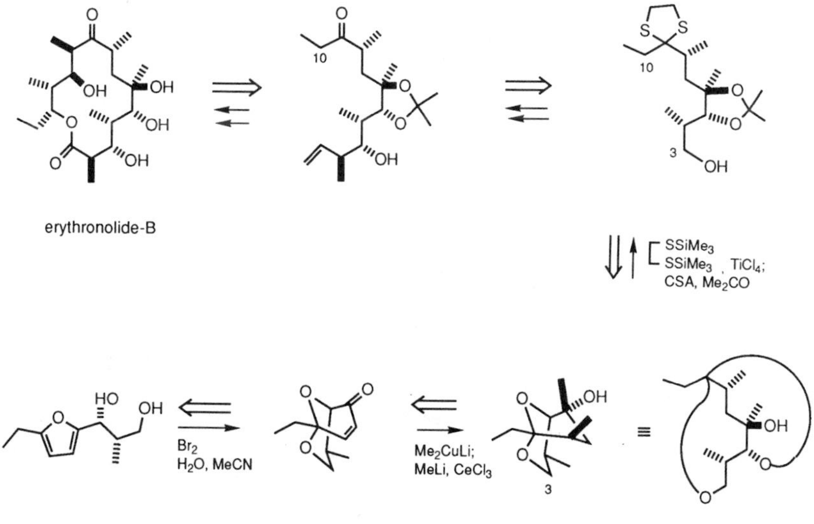

erythronolide-B

SSiMe$_3$
SSiMe$_3$, TiCl$_4$;
CSA, Me$_2$CO

Br$_2$
H$_2$O, MeCN

Me$_2$CuLi;
MeLi, CeCl$_3$

A great number of oxidants may be used to generate dioxo compounds from the furan precursors. Lead(IV) acetate was the choice reagent to convert 2-siloxyfurans into protected γ-keto-α-β-unsaturated carboxylic acids. Its application to a synthesis of pyrenophorin has been reported [Asaoka, 1980].

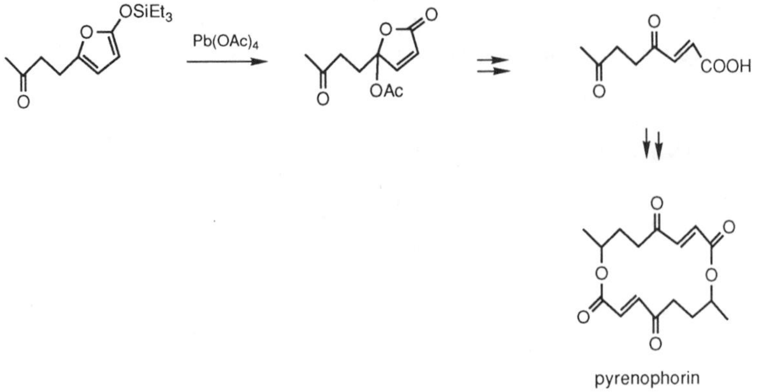

pyrenophorin

A less conventional use of a furan derivative to synthesize 1,4-diketones is exemplified by the condensation of α-angelicalactone with ynamines [Ficini, 1971a,b]. The enolate of the lactone is actually a furan; the adducts are readily hydrolyzed and decarboxylated.

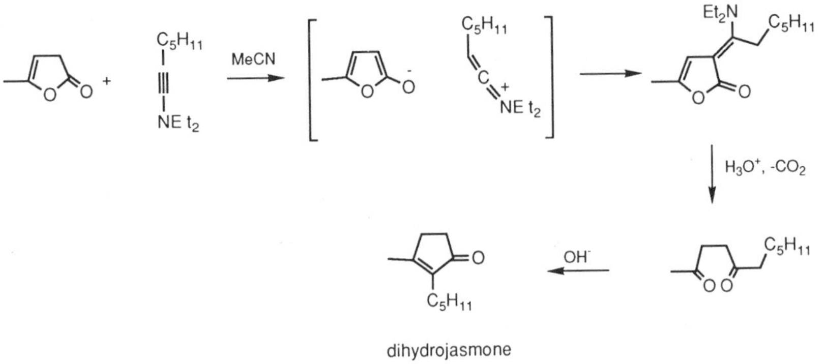

dihydrojasmone

7.5. FROM CYCLOBUTENES AND CYCLOBUTANOLS

Conversion of cyclobutenes to 1,4-dioxo compounds is straightforward. A synthesis of aldosterone [Miyano, 1981] took advantage of the proximity effect to oxygenate C-18 of the steroid skeleton via photoisomerization of a C-20 ketone group (via Norrish type II radical cyclization) and dehydration of the cyclobutene.

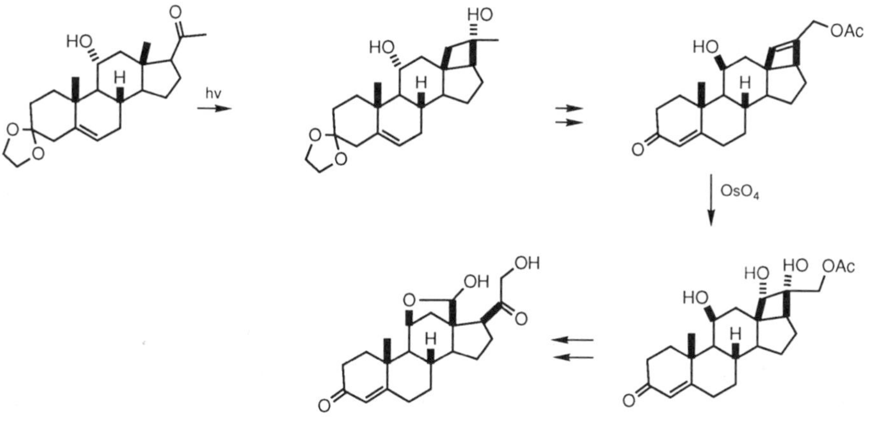

aldosterone

2,5-Alkanediones are accessible by cleavage of cyclobutanediols obtained from Grignard reaction of 2-hydroxy-2-methylcyclobutanone [Weinreb, 1972]. The latter compound is a photoproduct of 2,3-pentanedione.

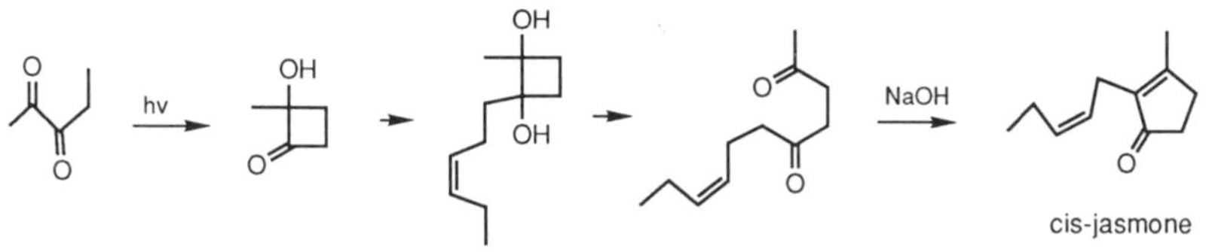

cis-jasmone

A synthesis of perhydrohistrionicotoxin precursor [Koft, 1984a] involved an intramolecular photocycloaddition of an enone double bond with an alkyne linkage and oxidative cleavage and decarboxylation. The same strategy was used in an approach to hibiscone C [Koft, 1984b], only that the oxidative cleavage step yielded a diketo aldehyde which was cyclized to a 3-acylfuran on acid treatment.

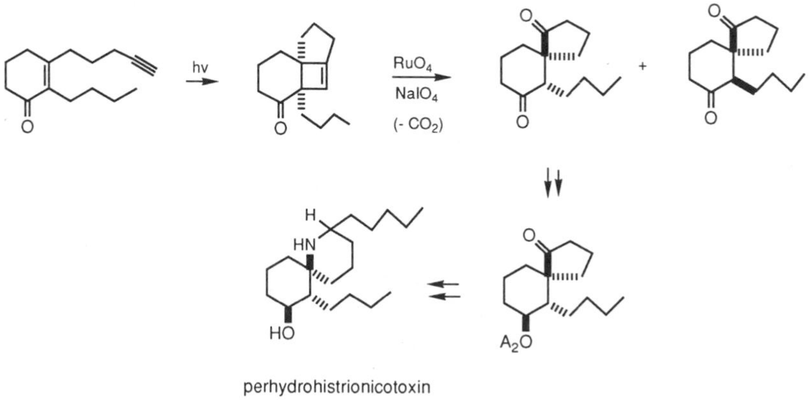

perhydrohistrionicotoxin

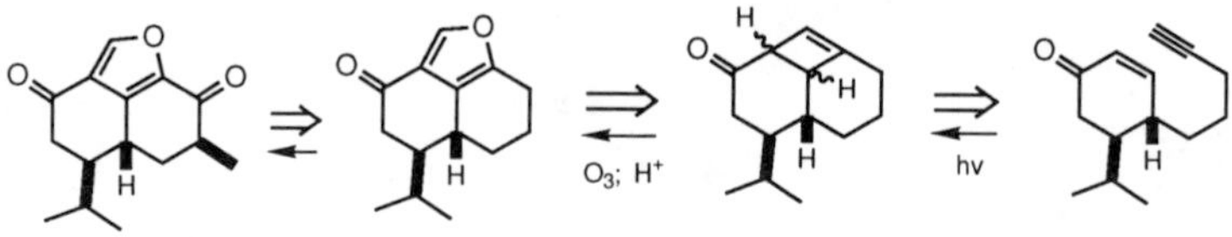

hibiscone-C

7.6. FROM ACETYLENES

1,4-Diols are available by stepwise condensation of acetylene with two carbonyl compounds and hydrogenation of the alkynediols. The triple bond in the alkynediols may be further converted into other functionalities. For example, isomeric 4,5-dihydro-3(2H)-furanones have been obtained from 1,1,4-trisubstituted 2-butyne-1,4-diols by selective hydration [Saimoto, 1983] and thereby a practical route to bullatenone was developed.

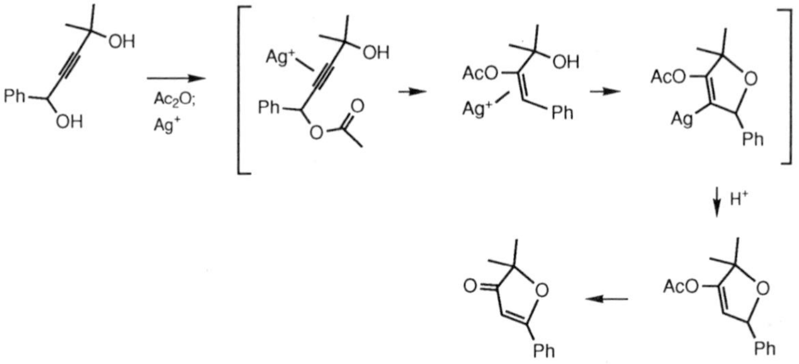

In trichodermin the anchoring atoms of the epoxide and the acetoxy group are both 1,3- and 1,4-related. This situation is of significance in its synthetic planning. Thus, an intramolecular aldol or Claisen condensation should serve well in the formation of the cyclopentane ring. The assembly of the 1,4-dioxo precursor may then be considered, and acetylene chemistry comes to the fore [Colvin, 1973].

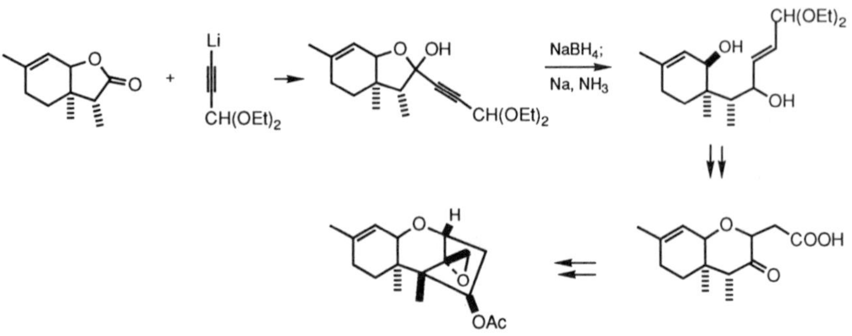

trichodermin

Greatly effective is the use of an acetylene link to construct the ribose segment of thymine polyoxin C [Garner, 1990].

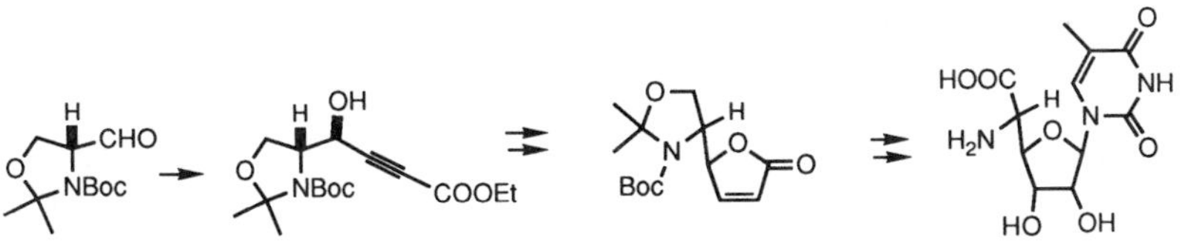

thymine polyoxin-C

In the spiroketal skeleton of chalcogran, all the oxygen functions are 1,4-related, therefore the use of acetylene in its synthesis constitutes a most convenient strategy. Outlined below is one route that has been completed [L.R. Smith, 1978], but it should be possible to approach it from the other direction.

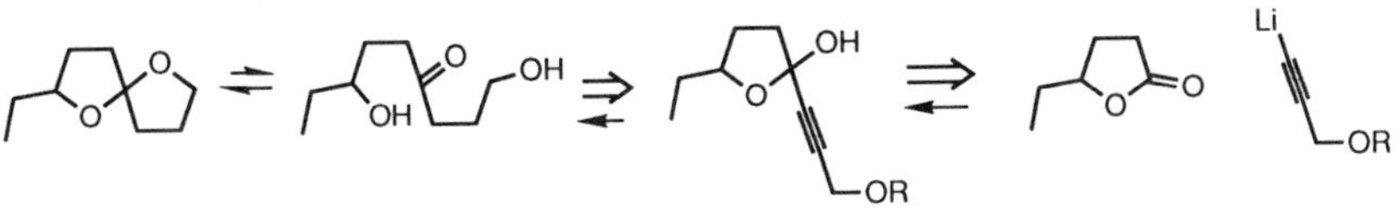

chalcogran

7.7. FROM DIELS–ALDER ADDUCTS OF HETERODIENOPHILES

A synthesis of tabtoxin [J.E. Baldwin, 1983] demonstrates the value of bond association in retrosynthetic analysis. Since the two amide carbonyls, the free amino group, and the tertiary hydroxyl are separated by four-carbon chain, pairwise association of the functionalities leads to an azoxabicyclo[2.2.2]octene intermediate, which enfolds a hetero Diels–Alder retron.

The regioselectivity of this Diels–Alder reaction arises from polarity matching at the reaction termini. Of course, the N—O linkage is intrinsically disjoint.

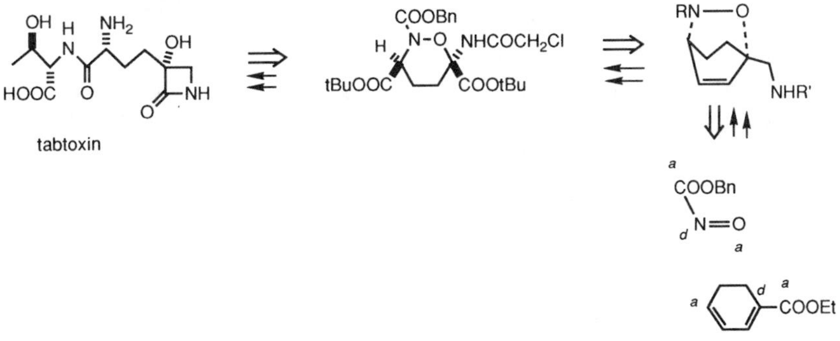

tabtoxin

7.8. REARRANGEMENT METHODS

Pinacol rearrangement of β,γ-dioxy ketones leads to 1,4-dioxo compounds. An outstanding application of the method is the construction of a potential precursor of quadrone [Monti, 1982].

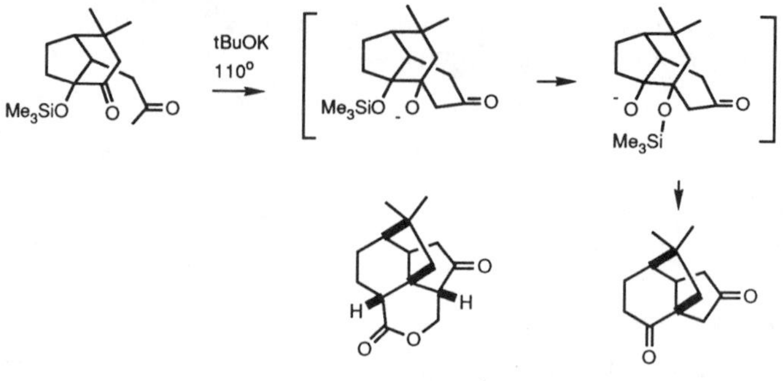

quadrone

The skeletal rearrangement was also an important feature of a synthetic plan for marasmic acid [Tobe, 1990] based on the photocycloaddition route to a bicyclo[3.2.0]heptane system.

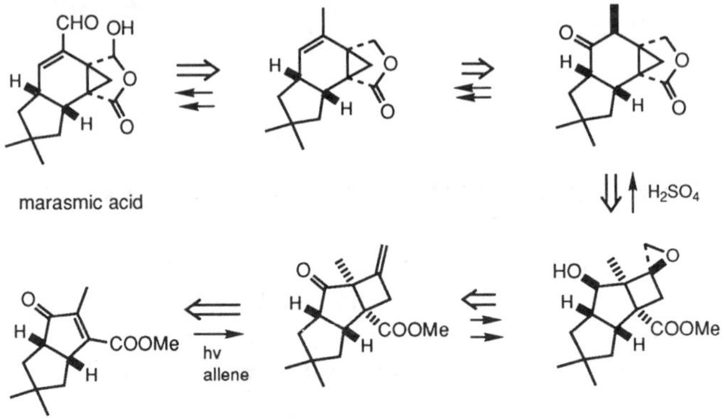

marasmic acid

The Barton reaction is useful for establishing 1,4-difunctional chains. Among the many applications of this stereoselective functionalization reaction is a synthesis of perhydrohistrionicotoxin [Corey, 1975a], although the target molecule is a *cis*-1,3-amino alcohol.

This route evolved from stereochemical considerations of the functional groups, the possible origin of the amine (Beckmann transform) and then the methods for generating the *syn*-γ-hydroxy oxime.

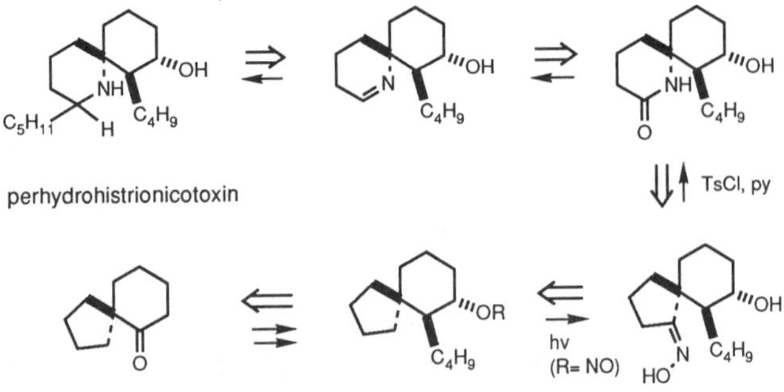

perhydrohistrionicotoxin

7.9. OXIDATIVE DEGRADATION

Strategically it is sometimes expedient to assemble a 1,5-difunctional compound and selectively degrade one of the terminal atoms to afford the lower homolog. For example, γ-keto esters may be prepared from direct autoxidation of the aldehyde enamine/acrylic ester adducts [Ho, 1974]. Similarly, *trans*-1,2-cyclopentanedialdehyde mono(ethylenedithio)acetal was acquired from an intramolecular Michael reaction of 2-octene-1,8-dialdehyde, selective protection and degradation. The cyclic aldehyde was used in a synthesis of brefeldin C [S.L. Schreiber, 1988b].

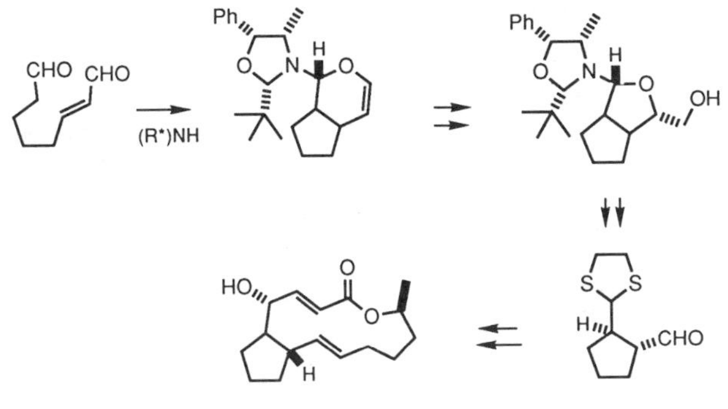

brefeldin-C

1,3-Dipolar cycloaddition of trimethylenemethane with acceptor-substituted alkenes generates methylenecyclopentanes from which 3-substituted cyclopentanones are readily unraveled by oxidative cleavage. Thus, a palladium(0)-catalyzed reaction of (2-acetoxymethyl)-2-propenyltrimethylsilane with a crotonic ester furnished the template for the elaboration of (+)-brefeldin A [Trost, 1986a].

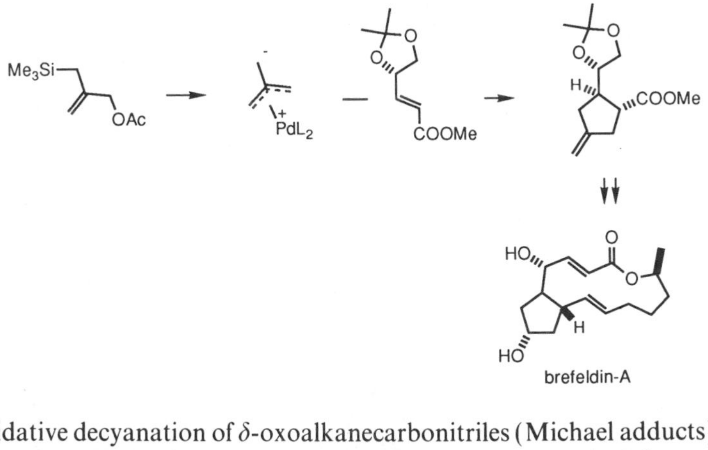

brefeldin-A

Oxidative decyanation of δ-oxoalkanecarbonitriles (Michael adducts) constitutes another method for the synthesis of 1,4-dioxo compounds. It is particularly pleasing that application of the reaction sequence to synthesis of anthraquinones offers much better regiocontrol than that based on the Friedel–Crafts reaction. The reaction conditions also allow the construction of acid sensitive functionalities such as those present in the methyl ether of aversin [Holmwood, 1971].

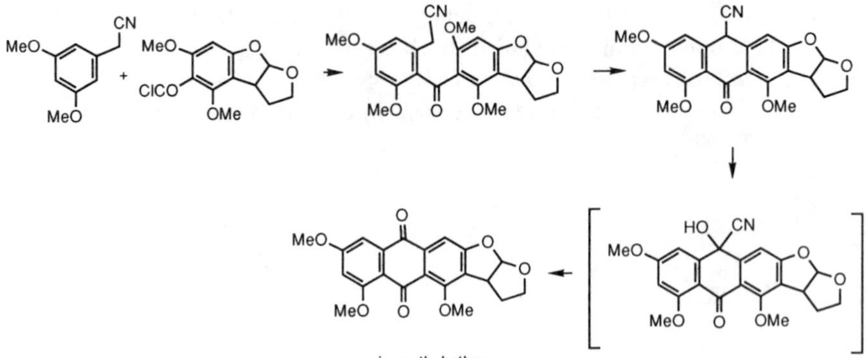

aversin methyl ether

8

SYNTHESIS OF 1,5-DIFUNCTIONAL COMPOUNDS: MICHAEL REACTION

8.1. MICHAEL REACTION

In the light of the pivotal role of the carbonyl group in organic chemistry, most 1,5-difunctional compounds are accessible from the corresponding dicarbonyl substances. This relationship helps affirm a pre-eminent position of the Michael reaction in synthesis.

The Michael reaction [Bergmann, 1959] and the aldol condensation are the two most popular nonpericyclic C—C bond-forming methods. As noted previously, these two reactions can occur in tandem to form a six-membered ring (Robinson annulation). The commonest version of the Michael reaction that features the union of a carbon donor (e.g., enolate) and an acceptor-substituted unsaturated compound (e.g., enone) provides a valuable 1,5-dicarbonyl product. The facility of the reaction is due to a harmonious electron transfer to create a conjoint system. And for exactly the same reason, the Michael reaction is reversible. Normally it can be driven to completion by using one reactant in excess (mass action law).

The Michael reaction has been generalized to include all sorts of addends. Reactive donors are nitriles, nitroalkanes, sulfones, and acidic hydrocarbons such as cyclopentadiene, indene, and fluorene, and RXH in which X is a heteroatom. While unsaturated carbonyl compounds are the most common acceptors, quinones, unsaturated nitriles, nitroalkenes, 2- and 4-vinylpyridines, vinyl sulfones, sulfonium salts and phosphonium salts are also effective.

The conjugate addition requires submolar amounts of a base (e.g., RONa), because the nascent carbanion would deprotonate the donor. Occasionally, a soft nucleophile such as triphenylphosphine may be employed as a catalyst. Such a base adds to the Michael acceptor to generate zwitterion species which then effect proton abstraction from the Michael donor to start the catalytic cycle.

To illustrate the latter process it is convenient to compare the hydrosilylation reactions of acrylonitrile [Pike, 1959] without any catalyst and in the presence of triphenylphosphine. It is obvious that in the phosphonium zwitterion is capable of deprotonating the silane and an S_N2 displacement leads to the product and also regenerates the catalyst.

It is interesting to note that the uncatalyzed reaction gives rise to an isomer in which the silyl substituent is conjoint with the cyano group.

Extension of the Michael reaction beyond assembly of a simple carbon chain has been constantly sought. For example, a reflexive Michael reaction brings about a ring formation different from the aldol process. Unlike the aldol condensation that must be preceded by enolate exchange, the reflexive Michael reaction uses the nascent enolate to attack the α',β'-enone system. The prerequisite for effecting this process is the kinetic enolate generation.

The intramolecular Michael reaction is facile when it leads to a common-sized ring product. Depending on the functional group arrangement, any of the three types of ring structure may be formed.

In the following we shall discuss some applications of the Michael reaction to synthesis. Some apparent targets are embellished with 1,5-dihetero substituents, while others are related to Michael retrons only after modification (e.g., FGA) of the molecular framework.

The 1,5-dioxygenated pattern that exists in zearalenone has a profound effect on the formulation of its synthetic plans. It is noted that the ketone group is also 1,5-related to the proximal carbon of the double bond. Convergent approaches to this molecule may be represented by two modes of assembly of the C_{10} chain [Vlattas, 1968; T. Takahashi, 1979] by means of a Michael reaction.

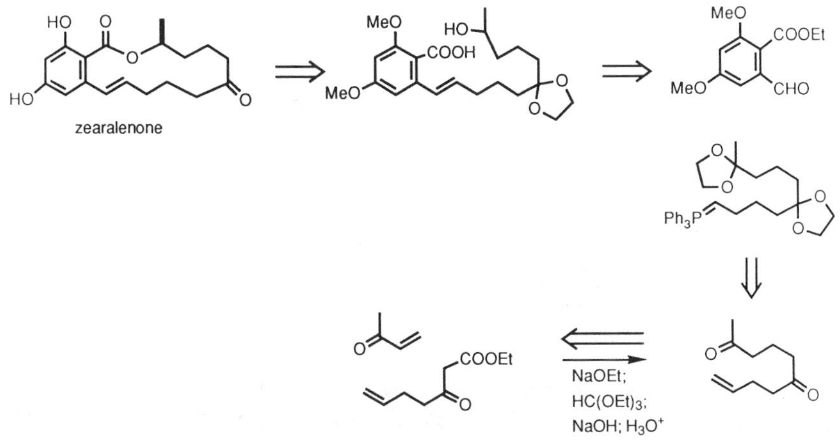

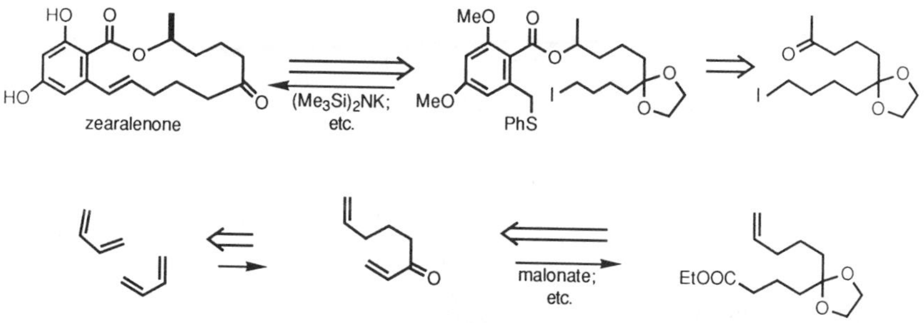

A tricyclic intermediate for the synthesis of gibberellic acid has been acquired from reductive cyclization of an angular allylated hydrindenedione [Stork, 1979]. As the latter compound is a 1,5-diketone, it is conveniently disconnected via a Michael reaction transform.

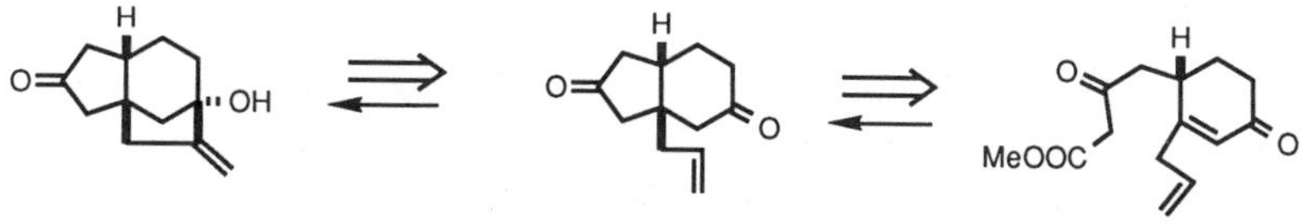

In a synthesis of morphine [Elad, 1954] a 2-arylcyclohexenone was identified as the starting material. The scheme is quite advantageous as the ketone provided an excellent site for the introduction of a two-carbon chain (as malonate) destined to become part of the B-ring via a Michael reaction. Furthermore, the benzylic position is donative, enabling creation of the azacycle by an intramolecular alkylation method.

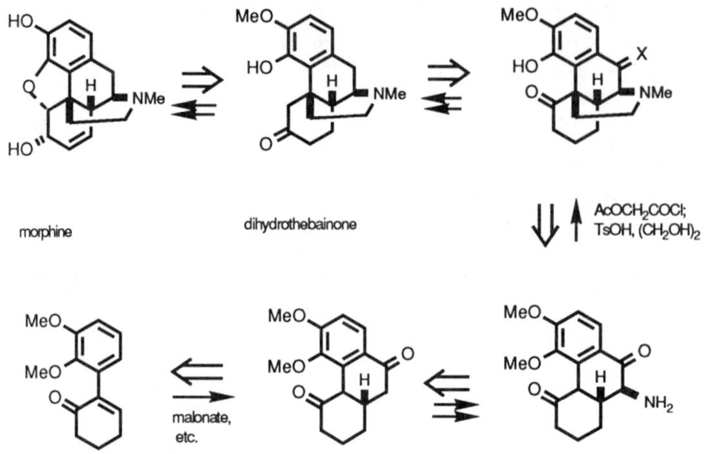

morphine
dihydrothebainone
$AcOCH_2COCl$; TsOH, $(CH_2OH)_2$
malonate, etc.

Piperidine and quinolizidines are dissectable into acyclic structures with conjoint functionality. A synthesis of castoramine [Bohlmann, 1963] fully exploited the structural features and it relied on two different Michael reaction steps to assemble the desired precursor.

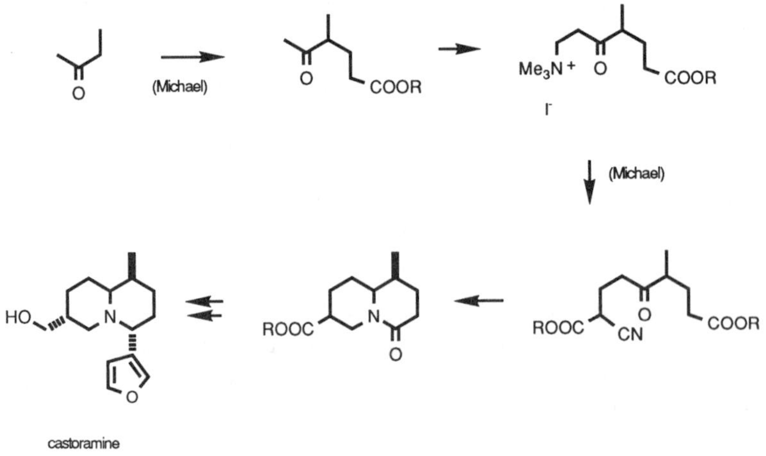

(Michael)
(Michael)
castoramine

The first synthesis of longifolene [Corey, 1961, 1964c] is significant since its full report harbingered the retrosynthetic analysis of complex molecules. The key step of this elegant work is an intramolecular Michael cyclization.

longifolene

ring expansion, etc.

Employment of this strategy hinged on the consideration that the ring juncture of the homooctalindione precursor did not have to be rigorously maintained in the reactive *cis* form, as it is equilibratable. Additionally, the carbonyl group of the acceptor moiety provided a handle for the introduction of the missing methyl group. (A logical variation of the scheme involving alkylation of a C_{14} homodecalone requires a *cis* ring juncture, and the displacement at a neopentyl center which is strongly unfavorable.)

An example of performing FGI prior to disconnection allows α-cedrene to reach a Michael transform [Horton, 1983]. Thus, converting the alkene linkage into a ketone affords many new opportunities to simplify the planning of the synthesis. Exchanging the remaining methylene group or the isopropylidene group of the bicyclo[3.2.1]octane system for another carbonyl and disconnecting an intervening C—C bond then creates several situations. However, only the longer interketone path of the bicyclo[3.2.1]octane-3,6-dione is conjoint (Michael transform) and therefore its closure is more direct.

α-cedrene

Analogously, the Wittig reaction transform of zizaene with stripping of the *gem*-dimethyl group (alkylation transform here is more obvious than in the cedrene approach which involves Simmons–Smith and hydrogenation transforms) leads to a tricyclic ketone. Addition of a hydroxyl group at the bridgehead allows an aldol transform to operate and the corresponding diketone is 1,5-related. A facile intramolecular Michael reaction can reach this ring system [Alexakis, 1978]. (In the synthesis, the Michael reaction was followed immediately by the aldol condensation because the tricyclic product suffered fewer nonbonding interactions.)

zizaene

It is expedient to replace the CHMe group of seychellene norketone with another carbonyl when a synthesis of the sesquiterpene is contemplated. The less hindered ketone is expected to undergo a Wittig olefination or Grignard–dehydration sequence to afford an unsaturated tricycle which likely undergoes stereospecific hydrogenation. The tricyclic 1,5-diketone involves a Michael transform [Jung, 1981].

seychellene

2-Oxoseychellene is a potential metabolite of seychellene. The presence of a carbonyl group renders the C-α/C-β bond strategic, and disconnection of this bond generates a bicyclo[2.2.2]octane system, and eventually a δ-keto ester. Although this is potentially a Diels–Alder adduct, experimentally the thermal reaction actually led to [2 + 2]cycloadducts [Spitzner, 1990]. On the other hand, the bridged ring skeleton with the proper functionalities has been constructed by a reflexive Michael reaction.

Vernolepin is adorned with two lactone rings, the closest oxygen atoms of the two being separated by four atoms. By shifting one of the oxygen functions by one atom, a conjoint system results, and this modified system is amenable to construction by the Michael reaction. An intramolecular cyclization also gives stereocontrol over the ring juncture.

Accordingly, a cyclohexenone constitutes the proper precursor. In fact, the transposed carbonyl is very beneficial to functionalization at the three consecutive carbon atoms C-6, 7, and 8 [Iio, 1979].

Interestingly, an extraneous cyclopropane ring was created during the synthesis from the oxadecalindione which served the purpose of regiocontrol.

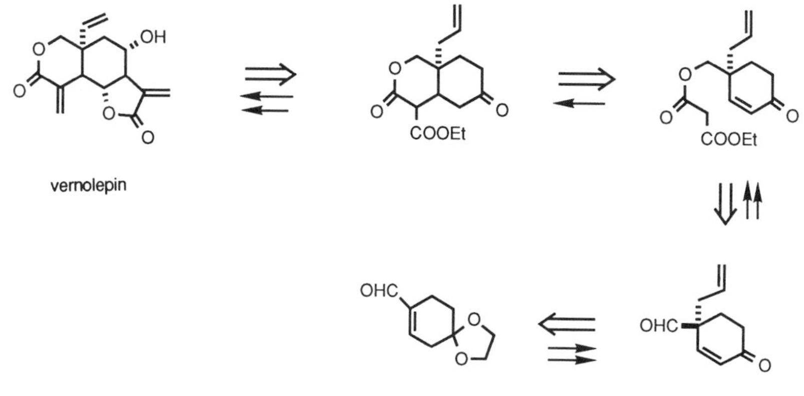

vernolepin

It is easy to imagine that annulation of menthone would pave the way to an entry to epizonarene. However, a Robinson annulation which normally involves a more highly substituted enolate would not be useful in this case, and the Stork annulation would also be difficult because enamine formation from menthone is unfavorable, due to allylic strain. (With an α-alkylcyclohexanone, enamine formation is accompanied by ring inversion to afford a conformer in which the alkyl group is axial, and it would place both the methyl and the isopropyl group in axial orientation upon enamination of menthone.) A synthesis of (+)-epizonarene [Belavadi, 1976] from (−)-menthone was conducted via a stepwise Michael reaction and aldol cyclization.

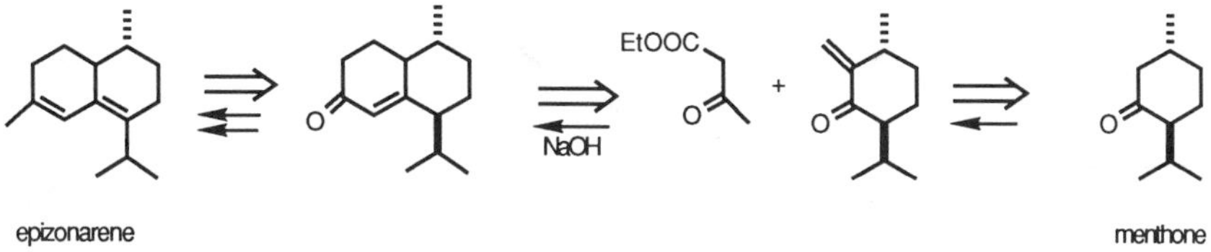

epizonarene   menthone

Annulation of a ketone with a fully substituted β-carbon by the Robinson method is militated against by steric hindrance. On the other hand, such a ketone undergoes α-methylenation without danger of isomzerization, and the annulation based on the 3C + 3C condensation (cf. 4C + 2C mode in the Robinson annulation) is viable. This version offers an entry to the hydrophenanthrenone precursor of jolkinolide E [Katsumura, 1982].

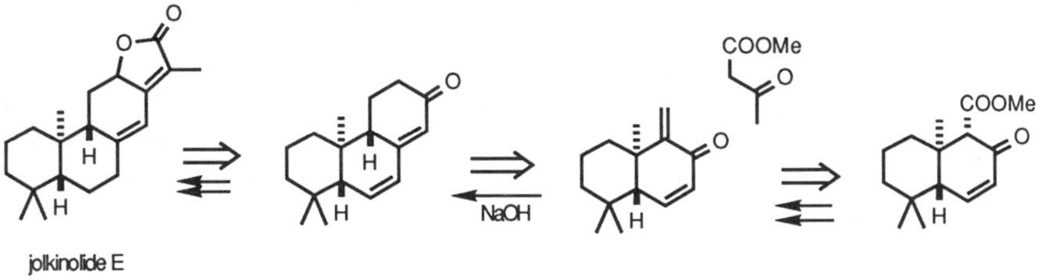

jolkinolide E

Reaction of an α-substituted cyclopentanone with a Mannich base under thermal conditions has been found to occur at the α′-position, presumably the Michael adduct at the more highly substituted carbon has a tendency to redissociate via the enamine. This regioselective Michael reaction was exploited in a route to guaiol [Buchanan, 1971].

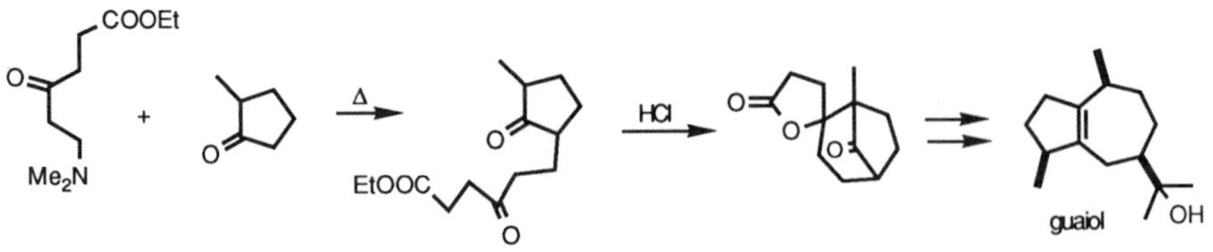

Completion of the C ring in an AB + D approach to estrone [Posner, 1981] via trapping the Michael adduct with vinyltriphenylphosphonium bromide was accomplished by an *in situ* Wittig reaction. Consecutive stitching of three C—C bonds was set up in each step by well-defined donor and acceptor centers, with the negative charge (reactive donor) successively transferred from one site to another to give ultimately a phosphonium ylide which was separated from the tetralone carbonyl by four atoms, an ideal situation for cyclization into a cyclohexene. This well-conceived route was marred only by a low yield.

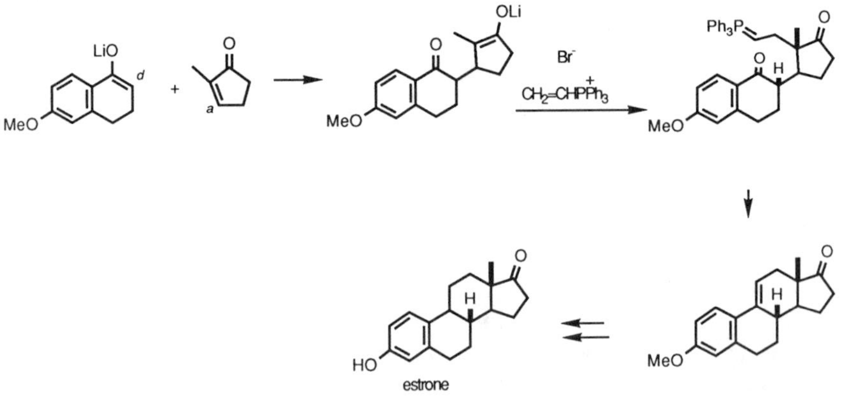

In camptothecin the carbonyl groups of the lactone and the lactam subunits are 1,5-related. Exploitation of this functional array has been extremely profitable and it has resulted in concise synthetic approaches. Considering the high acidity of the sidechain of quinaldine (and α-picoline), the disconnection of camptothecin can be simplified to a state that may involve a tricyclic Michael donor and an acceptor that carries a masked acceptor substituent at the γ-position [Meyers, 1973b].

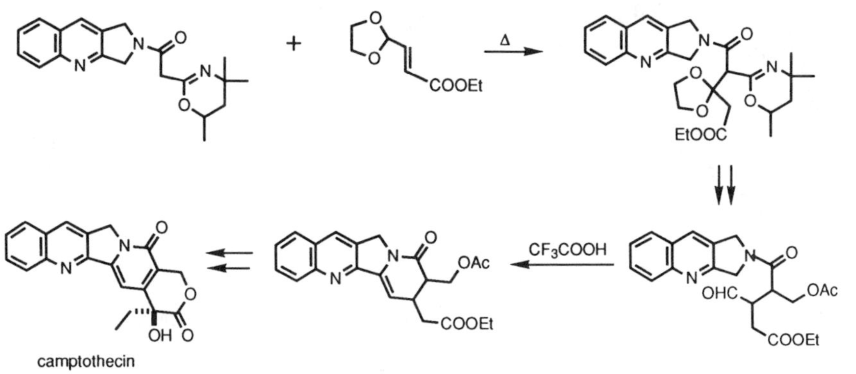

camptothecin

The dihydro-oxazine ring is an excellent masked carboxyl function and 3,4,4,6-tetramethyl-2-methylene-tetrahydro-1,3-oxazine serves well as an acetic ester synthon. Thus, a synthesis of methyl jasmonate [Meyers, 1973a] demonstrated its equivalence to the malonate α-anion.

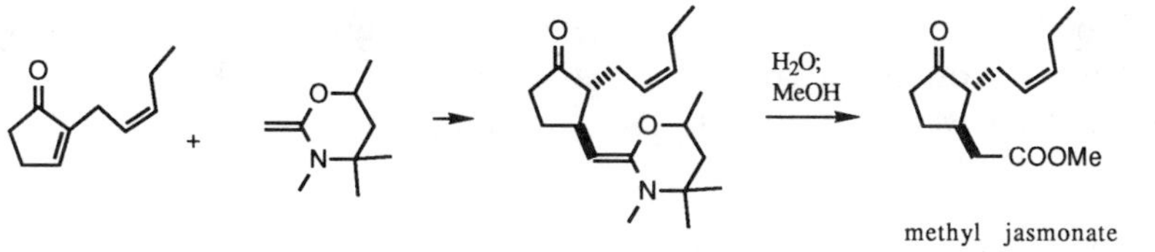

methyl jasmonate

The stereoselective approach to emetine hinges on the accessibility of a tricyclic ester. Since the ester group is conjoint with the nitrogen atom, an intercircuit C—C bond formation process is possible. Accentuation of the conjoint system calls for FGA in the heterocycle and an unsaturated lactam should serve well as a Michael acceptor. Further analysis of this acceptor indicates a convenient assembly of it by a Claisen condensation [Battersby, 1960].

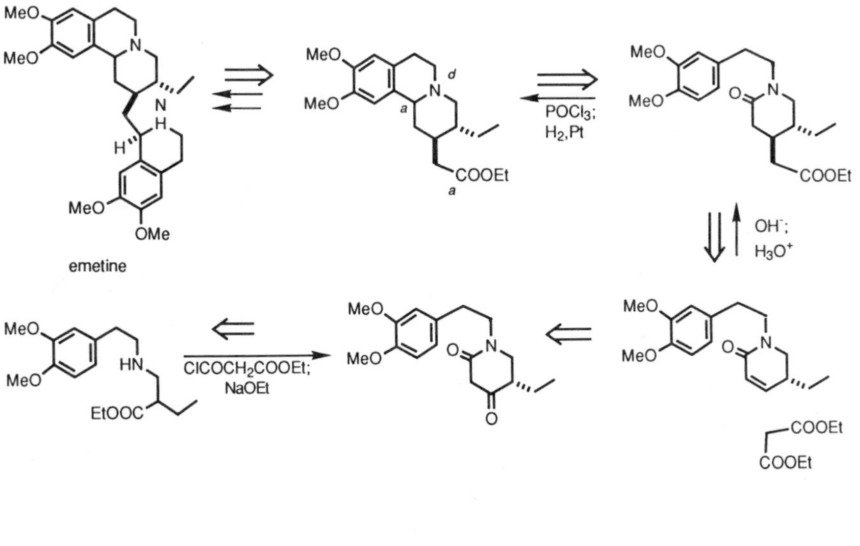

emetine

There are variations on this theme and one of them is shown by a route to akuammigine [Winterfeldt, 1968].

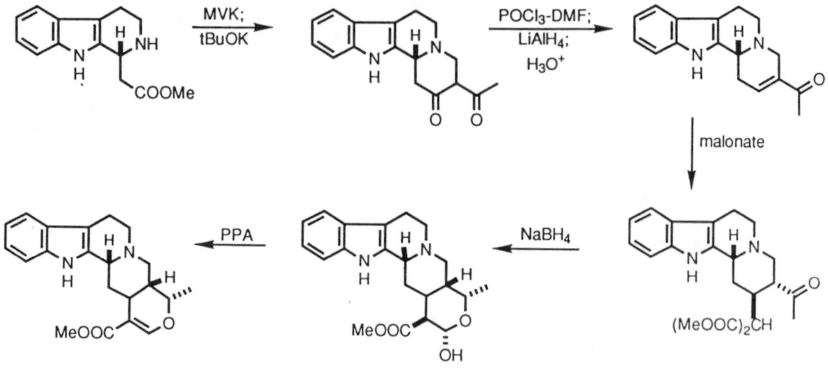

akuammigine

The aldol step that followed a Michael reaction between the kinetic enolate of 5-(methoxy)methoxymethyl-3-trimethylsiloxy-2-cyclohexenone and methyl crotonate was not desired during a synthesis of eriolanin [M.R. Roberts, 1981]. However, recovery of the initial Michael adduct was accomplished on base treatment.

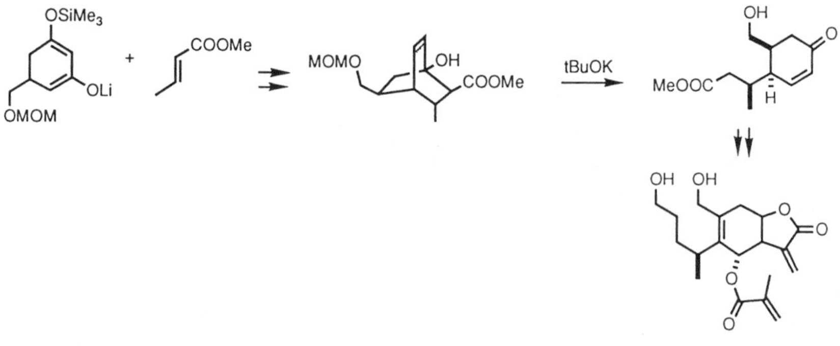

eriolanin

Less appparent applicability of the Michael reaction to the synthesis of natural products is exemplified by wedelolactone [Wanzlick, 1963]. In fact, *o*-benzoquinone has two acceptor sites and it reacts sequentially with a *C*-donor (an enolate of a 1,3-dicarbonyl compound) and then the enolate oxygen.

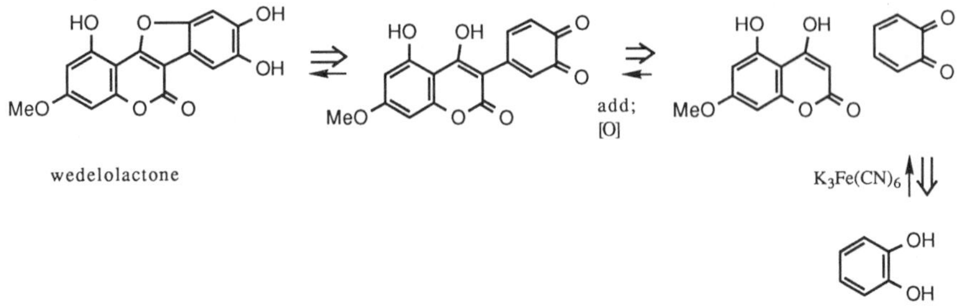

wedelolactone

8.2. INTRAMOLECULAR MICHAEL REACTION

The 1,5-dioxygenated pattern that prevails in many iridoid monoterpenes fits well into a Michael transform. Stereocontrol is the more difficult issue in their synthesis.

One solution to the problem is the use of a chiral amide α-anion to direct the Michael reaction [Yamaguchi, 1986]. It is most pleasing that the five-membered ring can be constructed immediately after the reaction, i.e. by an *in situ* Dieckmann cyclization.

(+)-dihydroiridodiol

The synthesis just described does not involve an intramolecular Michael reaction. However, it is interesting to compare it with the alternative approach in which the intramolecular reaction is employed. Dihydronepetalactone has been acquired by a Michael-initiated Michael cyclization [Uyehara, 1989].

dihydronepetalactone

Intramolecular Michael reaction accompanied by closure of the dihydropyran ring is also possible. Thus, a very short synthesis of nepetalactone [Denmark, 1986b; S.L. Schreiber, 1986] from citronellal has been reported. The cyclization was apparently initiated by enamination of the nonconjugated aldehyde.

nepetalactone

(+)-citronellal

A one-step closure of five rings from squalene-derived dialdehydes (*E*- and *Z*-isomers) occurred in the presence of ammonia, NH_4OAc, $Et_3N.HCl$, and acetic acid [Piettre, 1990]. The product, protodaphniphylline, is a plausible biogenetic parent of several alkaloids. The remarkable biomimetic cyclization is believed to be initiated by an intramolecular Michael reaction.

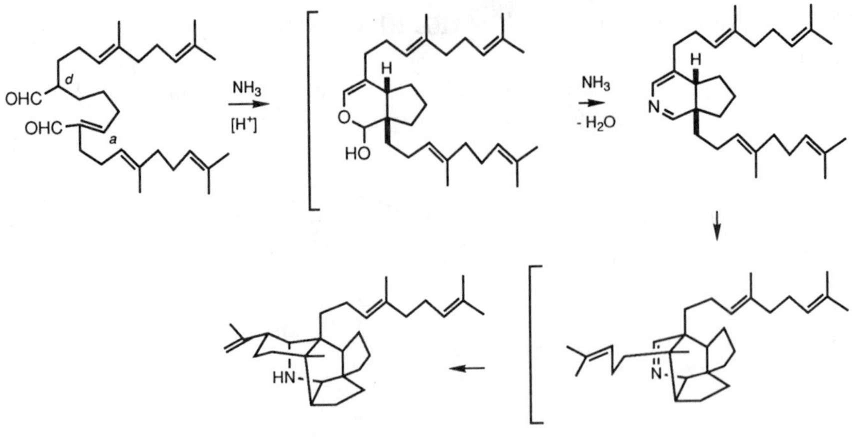

proto-daphniphylline

A synthesis of (−)-lycoserone from (+)-citronellal [Kuhnke, 1988] demonstrates the usefulness of tandem aldol–intramolecular Michael reactions.

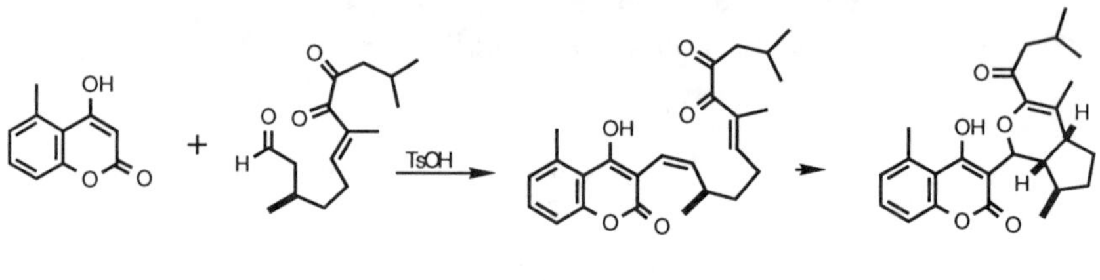

(-)-lycoserone

Cyclohexadienone formation from chromium-carbene complexes is very effective. In an attempt to synthesize taxodione [Gilbertson, 1989] the final step occurred in the Michael mode instead of the intended aldolization.

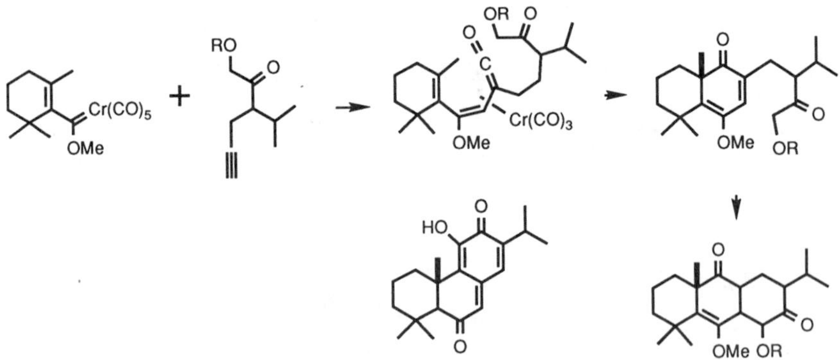

taxodione

The only functionality of isoclovene is a trisubstituted double bond. FGI of this olefin with a ketone opens a new vista for disconnection. For example, a hydrindenone may be identified as the subtarget, and a Michael transform relates the two structures [Baraldi, 1983].

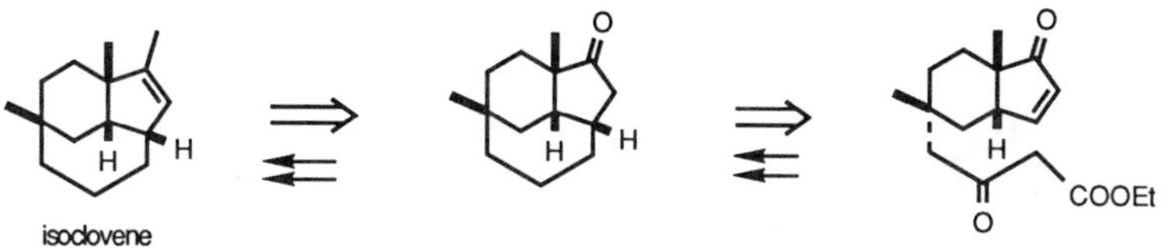

isodovene

Of any synthetic approach to cephalotaxine the carbocyclic moiety is always the focal point because it contains two oxygen functions. In the corresponding ketone a Michael transform is feasible [Auerbach, 1972]. The disconnection is also favored by the apparent accessibility of the cross-conjugated dienone in which the carbonyl group is part of an enaminone system, and acylation of the enamine represents a most reasonable method for its preparation.

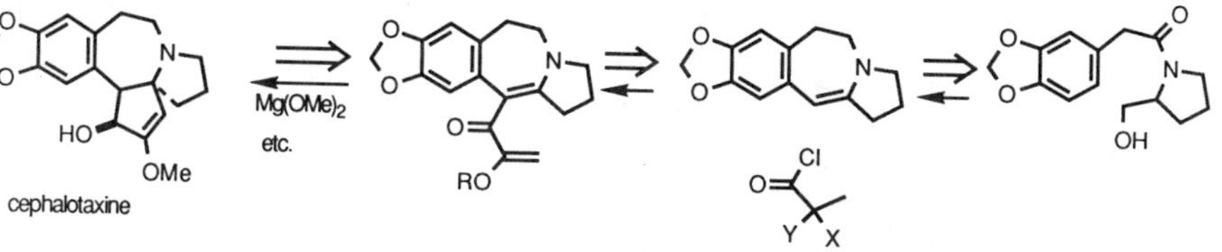

cephalotaxine

A route to gibberellic acid [Lombardo, 1980] featured efficient closure of the lactone and the A-ring by exploiting the conjoint relationship among various functionalities. Thus, the formation of the C-4/C-5 bond was done by a Michael reaction involving esters as activating groups, and the complete skeleton was consummated by an aldol condensation. The intramolecularity of these reactions contributed to their efficiency.

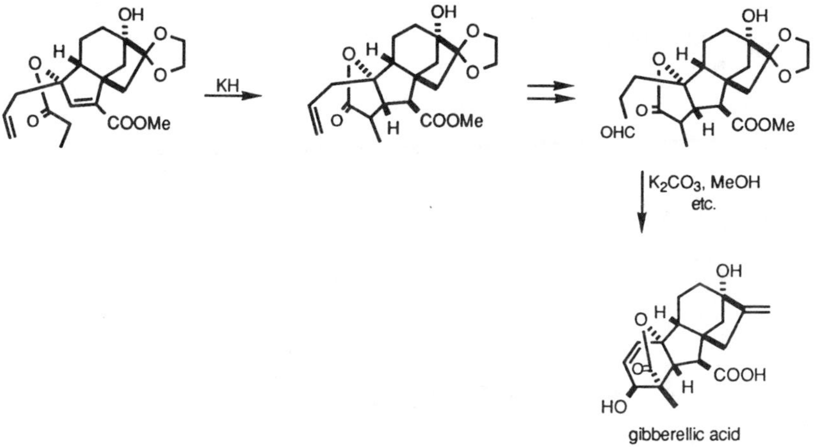

gibberellic acid

Because of the supernucleophilicity of the indole nucleus, disconnection of ibogamine into a C-seco derivative is meritorious. It is then crucial to devise synthetic routes to an azabicyclo[2.2.2]octane that allow for the incorporation of the C-ethyl group and an acceptor site for eventual linking to the indole. These two structural entities can be interrelated such that an intramolecular Michael reaction effects its formation [Imanishi, 1981]. Necessarily, the terminus of the ethyl chain is in a higher oxidation state.

ibogamine

A synthetic design for fawcettimine [Heathcock, 1989] was based on the recognition that the tricyclic amino ketone would automatically cyclize to the carbinolamine and the stereogenic center α to the cyclopentanone is self-adjustable. Thus, a hydrindenone should be created in the first stage of the synthesis.

Annulation onto a cyclohexenone should be stereoselective with respect to the ring juncture (*cis*), especially when a thermodynamically controlled process is employed. In principle, a hydrindenone containing the cyclopentanone subunit which is present in the alkaloid might represent a good subtarget, since the carbonyl group is conjoint to the nitrogen atom, and cyanoethylation is serviceable to assemble that chain. However, the alkylation site is more encumbered, leaving the α'-position a better donor (unless a blocking scheme is instituted). Consequently, an intramolecular Michael cyclization on a substrate that contains an existing chain is a more favorable solution. Such a condition demands a shortened conjoint sidechain, and a synthetic route according to this modification must involve an Arndt–Eistert homologation.

The synthesis consisted of a cycle of redox manipulation at the cyclohexanone site after it fulfilled its role of activating the α-position to effect the carbocyclization. The cyclopentanone subunit was introduced in a latent methylene form. In turn, a proper (α-oxymethyl)allyl anion equivalent was identified as a synthon to conjugate with a cyclohexenone. That the Hosomi–Sakurai reaction displayed a remarkably high facial selectivity was a rather unexpected bonus.

An excellent design for a synthesis of strychnine (yet to be accomplished) [Quesada, 1987] entails a Michael cyclization to build a perhydroindole system. using the nitrogen site to attach an acetylenic ester it was possible to assemble a tricycle stereospecifically, again by an intramolecular Michael reaction.

strychnine

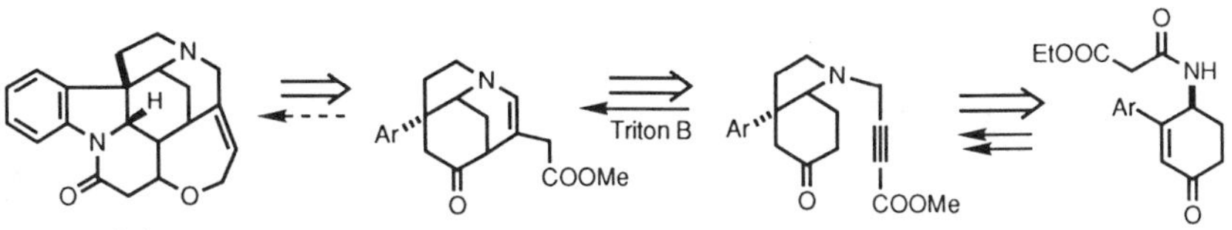

Inspired by biogenetic considerations, a retrosynthetic scheme for koumine [Magnus, 1990] focused on a vobasine-type intermediate which may be derived from fragmentation of a sarpagine skeleton.

Because the C_2 sidechain must be functionalized at its terminus with a donor substituent in order to participate in ring closure with the indole nucleus, the sp^2-carbon in the ring must be an acceptor. This electronic constraint renders the sidechain disjoint to the other oxygen function, therefore the assembly of the six-carbon subunit $O-CH_2-C-C-C=C-CH_2-X$ requires polarity inversion. However, skeletal construction of the conjoint norketone ($O=C-C-C=C-CH_2X$) would be much more facile, in view of the possible employment of an intramolecular Michael reaction. With this strategic modification the polarity inversion is deferred to the $C=O$ to $CHCH_2OH$ homologation step. Applying a combination of N-alkylation and Dieckmann transforms to the norketone intermediate results in a tetrahydro-β-carboline whose origin from tryptophan is apparent.

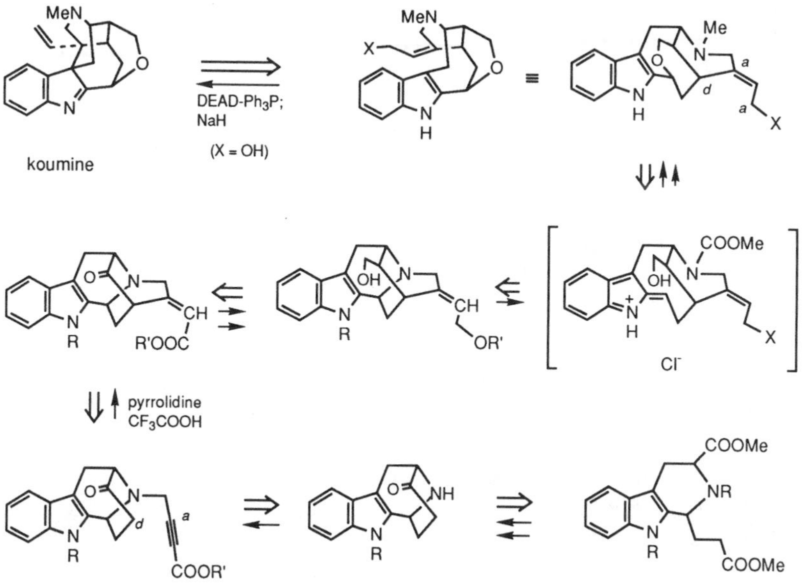

koumine

An intramolecular Michael reaction was the key step toward an assembly of alloyohimbone [Stork, 1987]. An enaminone served as the Michael donor in the formation of the C-15/C-20 bond leading to a useful precursor of heteroyohimbine alkaloids [Rosenmund, 1990].

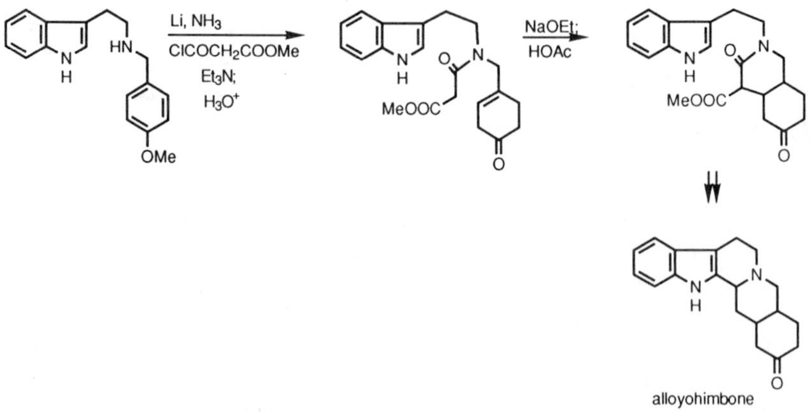

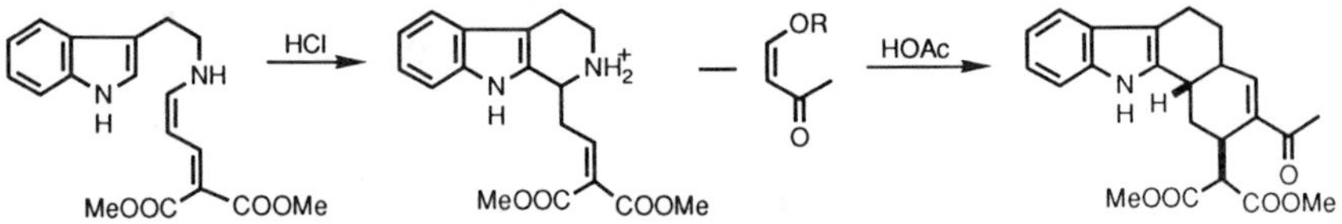

alloyohimbone

The donor–acceptor chains in this last process find a close correspondence to an intermediate for emetine [Hirai, 1986]. With an activated tetrahedral Michael donor site the *trans* stereochemistry of the two vicinal pendants can be ensured.

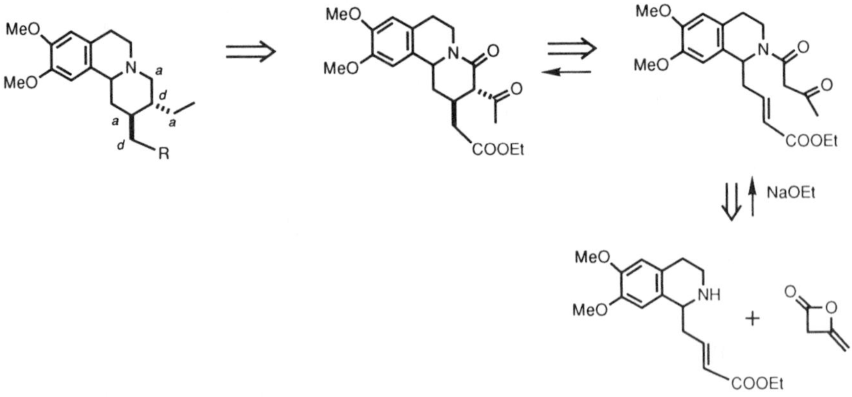

(+)-Methyl homodaphniphyllate can be disconnected at the axial C—C bond of the piperidine ring because its formation from a properly accentuated precursor should not be problematic. Note that conjoint circuits around the three six-membered rings are anchored by the nitrogen atom. Disconnection and accentuation by polar functionalities can result in an enone and a 4-piperidinone moiety [Heathcock, 1986]. Accordingly, this highly functionalized intermediate is amenable to stepwise disconnection to yield a useful synthetic scheme.

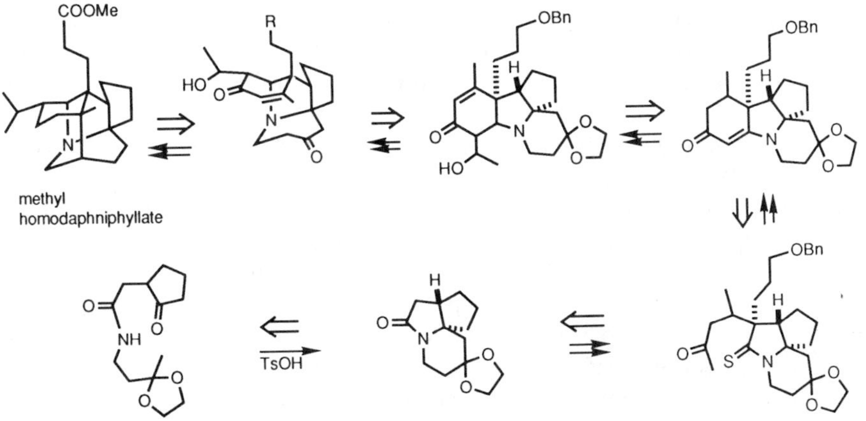

methyl
homodaphniphyllate

As shown in the alloyohimbone synthesis, decalone systems formed by an intramolecular Michael reaction (and many other reactions also) are usually *cis* fused. When the chain containing either the donor or the acceptor is constrained sterically, formation of a *trans*-decalone would occur. In connection with a synthetic study toward azadirachtin [Ley, 1989], such a Michael reaction played a prominent role.

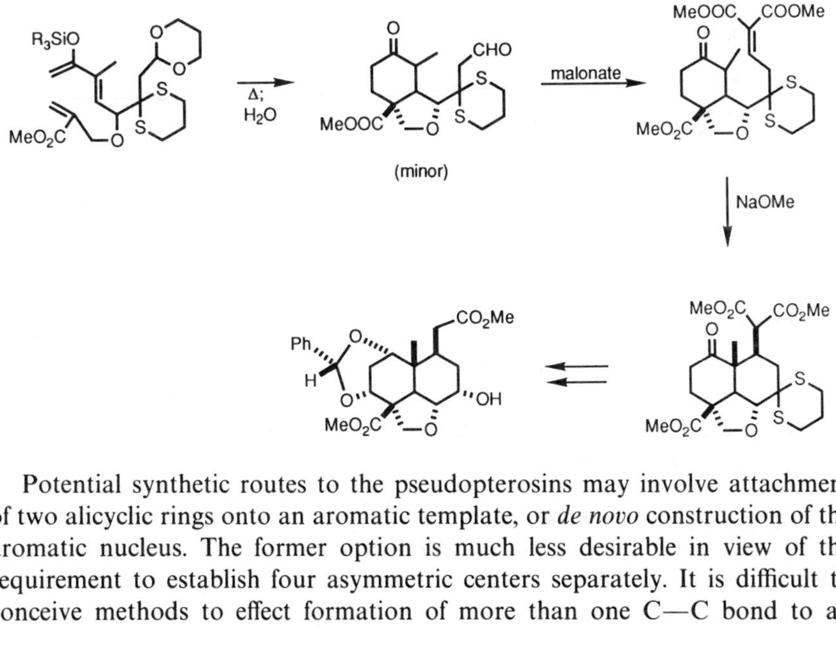

(minor)

Potential synthetic routes to the pseudopterosins may involve attachment of two alicyclic rings onto an aromatic template, or *de novo* construction of the aromatic nucleus. The former option is much less desirable in view of the requirement to establish four asymmetric centers separately. It is difficult to conceive methods to effect formation of more than one C—C bond to an aromatic system. On the other hand, the decalin skeleton is one of the most well known and well behaved with which stereocontrolled reactions may be executed with confidence. Furthermore, the two methyl substituents and the ring juncture comprise a structural array reminiscent of a menthane and the utilization of (+)-menthol to build the decalin intermediate would be most rewarding [Corey, 1989a]. (The greatest advantage of using abundant monoterpenes as building blocks in the elaboration of higher terpenes is probably the preservation of existing stereocenters, with which chiral induction may also be achieved.)

Identification of the structure goal serves to focus on a search for efficient transforms. Accordingly, the oxygen functions on the aromatic portion of the pseudopterosins become the prime loci of attention. 1,5-Interconnectivity of the peri-oxygenated carbon atom with the base of the isobutenyl sidechain divulges a Michael transform after FGI of the isobutenyl group with a carbonyl. This means that the forward synthetic operation would comprise coupling of an octalone with a 2-butynoic acid derivative, an intramolecular Michael cyclization, and oxidation of the resulting phenol to give a catechol subunit. The synthesis would then be culminated by chain extension at the benzylic carbonyl site and selective glycosylation at one of the phenolic oxygen atoms.

pseudopterosin-A

III

Regarding the execution of the synthesis, the third asymmetric center was established from equilibration of a *trans*-fused γ-lactone derived from (+)-menthol (Barton reaction, etc.), leaving the α-methyl group in an equatorial-like orientation. The fourth asymmetric carbon of natural pseudopterosins A and

E disposes the isobutenyl group in a pseudoaxial configuration, yet it could be created under thermodynamically controlled conditions of a Wittig reaction. The pseudoaxial chain is more stable because it experiences 1,3-diaxial interaction with only one hydrogen and it avoids a substantial in-plane peri repulsion with the aromatic methyl group.

A very efficient diquinane synthesis consists of two sets of tandem aldol–Michael reactions between an α-dicarbonyl substance and an activated ketone (e.g., acetonedicarboxylic ester). Particularly interesting is the access of staurane-2,6,8,12-tetraone [Mitschka, 1978] in only a few steps.

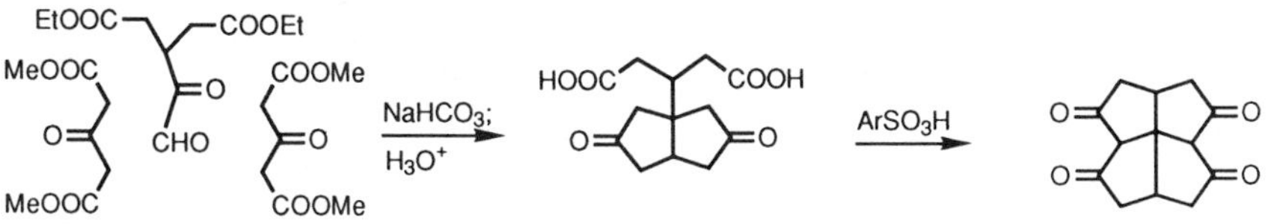

A key daunomycinone intermediate has been obtained from a quinizarin derivative [Krohn, 1986]. The cyclization occurred via the "outside" quinone tautomer.

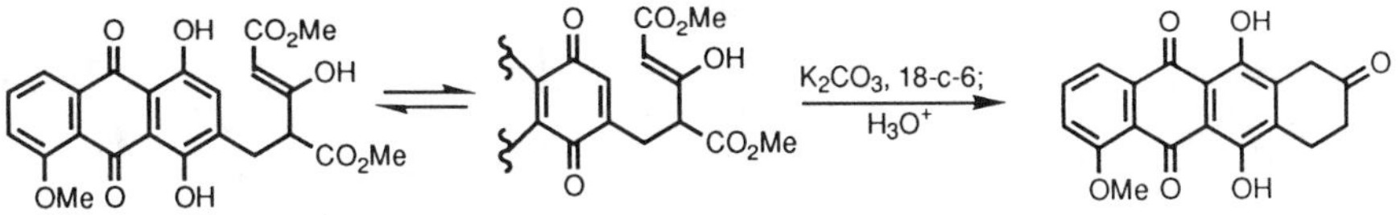

A double Michael reaction between *two* reaction partners always leads to a cyclic structure. The double Michael reaction of a 4-methylene-2-cyclohexenone with a ketone enolate constitutes an efficient protocol for generating a *cis*-decalin. An application of this method is found in a synthesis of occidentalol [Irie, 1978, Mizuno, 1980].

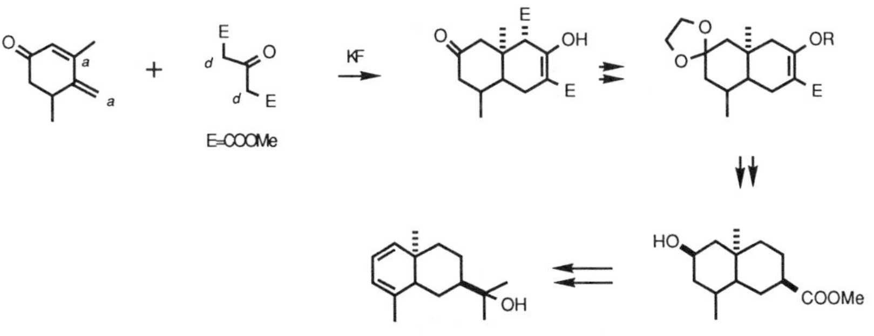

occidentalol

The bicyclo[3.3.1]nonane framework is very readily accessible because both ring components are conjoint. Disconnection of the system after placing a carbonyl at the single-carbon bridge realizes 1,3-dialkylation or Michael–aldol tandem transforms.

The clovane skeleton contains such a network which is further lateral-fused with a cyclopentane. This additional structural feature actually enables a combintion of three reactions to proceed in succession, that is a double Michael addition followed by a Dieckmann cyclization. Because of the thermodynamically controlled nature of this process, the carbon chain of the bicyclic intermediate is equatorial to the newly formed six-membered ring. Consequently the stereochemistry of the ring fusion is not the same as for clovane [Danishefsky, 1973].

With a cross-conjugated ketone instead of an extended conjugate acceptor, the double Michael reaction can give rise to a noncondensed ring. This reaction is the key to a concise synthesis of griseofulvin [Stork, 1962b].

8.3. REFLEXIVE MICHAEL CYCLIZATION

Reflexive Michael reaction forms a six-membered ring in one step. It is proposed that this term be used to denote two consecutive 1,4-additions in which the enolate of the initial adduct attacks the α',β'-enone system to form two bonds. The critical requirement of this process is the presence of a kinetic enolate of the enone at the start of the reaction. This process is different from the double Michael reaction in which a donor may react twice to a doubly unsaturated acceptor, using either one or two sites to form the new bonds.

The following delineates several examples of the reflexive Michael cyclization. First, the annulation of an acetylcyclohexene en route to khusilal [Hagiwara, 1989] has been delegated to this efficient reaction.

Pentalenic acid is 1,5-dioxygenated, but a Michael transform leads to a spirocyclic precursor. A reflexive Michael route [Ihara, 1987] retains the advantage of stereocontrol by virtue of its reversibility (at least in the second step), as well as the relative ease in the preparation of a useful substrate. As the synthetic target is a triquinane it is necessary to perform a ring contraction on the Michael product.

pentalenic acid

Rearrangement involving intracircuit atoms does not change the conjoint relationship of the two functional groups. Consequently, the Wolff rearrangement served to correct the structural features.

The elegant approach to pleuromutilin [Gibbons, 1982] was initiated by a reflexive Michael cyclization to build a tricyclic compound. When another six-membered ring was annexed, five of the eight asymmetric centers were erected stereoselectively. A C—C bond, which is part of the new six-membered ring and also connects to a bridgehead atom, was broken to unleash the framework of the terpene molecule.

pleuromutilin

The most intriguing application of the reflexive Michael cyclization to date may be that of a synthesis of atisine [Ihara, 1990]. The elaboration of the bridged piperidine from a double Mannich reaction into a molecule containing an acrylic ester chain and a cyclohexenone moiety set the stage for closing the BCD-ring system in one step. The remaining steps of the synthesis were conventional functional group adjustments.

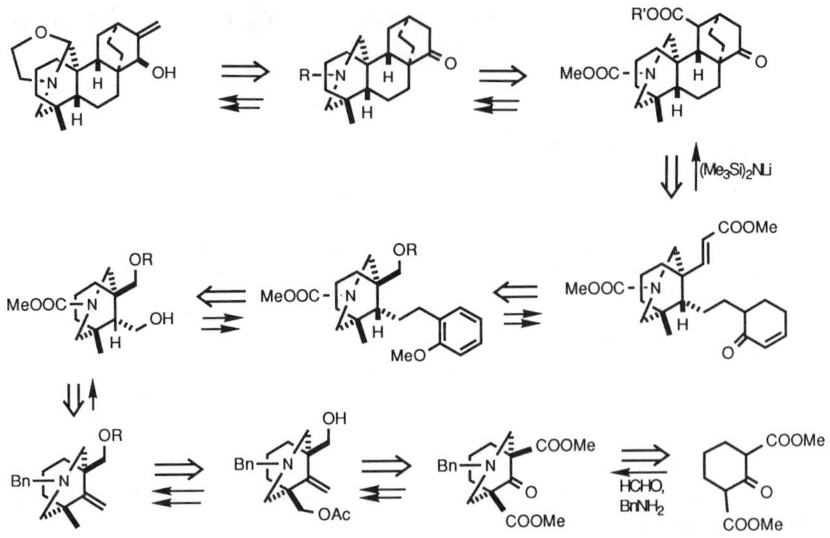

Lewis-acid-catalyzed condensation of a siloxycyclohexadiene and 1,4-penta-dien-3-one resulted in a bicyclic structure that is of obvious values for a synthesis of seychellene [Hagiwara, 1985]. The first ring was closed by a reflexive Michael reaction, and it was followed by another Michael step.

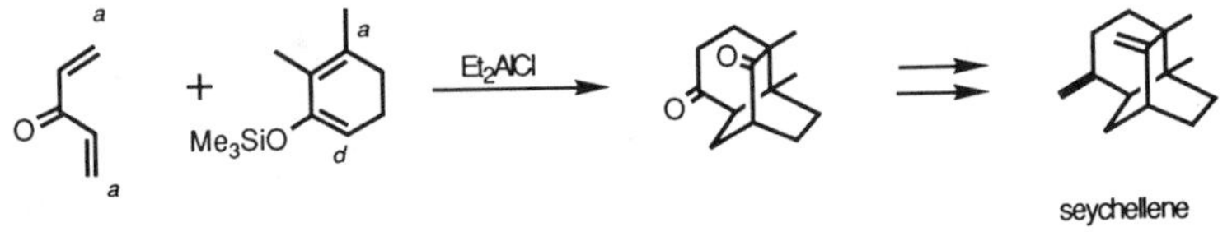

seychellene

Ishwarane is devoid of polar functionalities. Its synthesis has been greatly simplified by dismantling the three bridges and adding to the remaining decalin skeleton an enone system. The kinetic enolate of the dimethyloctalone can then be used to reconstruct the carbon network [Hagiwara, 1979]. With methyl α-bromoacrylate as the acceptor, the reaction conditions for the reflexive Michael cyclization also favored the intramolecular alkylation. The result is a one-step assembly of the intriguing ring system.

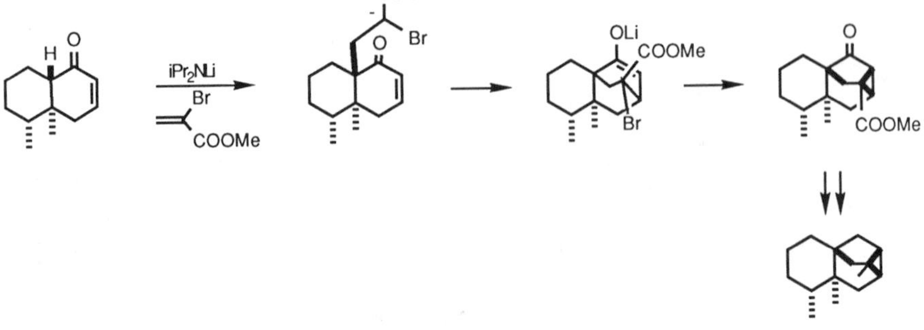

ishwarane

An enantioselective synthesis of aphidicolin [Holton, 1987] was based on the access of a previously known (racemic) synthetic intermediate [Corey, 1980] by a different route. Analysis of the spirotricarbocyclic intermediate by a partial intramolecular aldol transform leads to a drimane-type substance. The chain of atoms along the bottom periphery is conjoint and a Michael transform is indicated. However, a butanolide is not sufficiently acidic and a removable accentuating acceptor group must be attached to the α-carbon to help achieve the chemoselectivity. This requirement is mandatory because the intramolecularity of this Michael reaction dictates a higher activity of the donor precursor than the ketone of the acceptor moiety. The ideal accentuating group seems to be a sulfoxide, because it also facilitates the assembly of the intramolecular Michael substrate via another intermolecular Michael reaction. In fact, the reaction sequence represents an interrupted reflexive Michael cyclization.

aphidicolin

(Corey)

R = Me₂tBuSi

R = S(O)Tol
R = H

R = Me₂tBuSi

The interruption comes from the introduction of a vinyl group to generate a dienone system. The reaction with vinyllithium was uncomplicated by a mutual protection of the two functional groups habored in the α-sulfinyl lactone. Not the least important is that a sulfinyl group is chiral, enabling an enantioselective Michael reaction to be accomplished in the very first step of the synthesis.

Michael reactions involving more than two components have been extensively investigated [Posner, 1986]. In synthesis the economy associated with formation of more than one bond in a reaction is always coveted. In the multicomponent Michael reactions the selection of the reactants with compatible reactivities and reaction conditions is of utmost importance to the success of the controlled condensation. On the one hand, a reaction may stop at the formation of only one bond, while in the other extreme a reaction may give rise to polymers.

With proper attention, functionalized six-membered ring compounds can be synthesized in one step. A synthesis of juncunol [Posner, 1988] demonstrates the facile construction of an aromatic ring. One should be reminded that perfect alignment of polar sites for the various reactants is the key to the highly efficient ring formation.

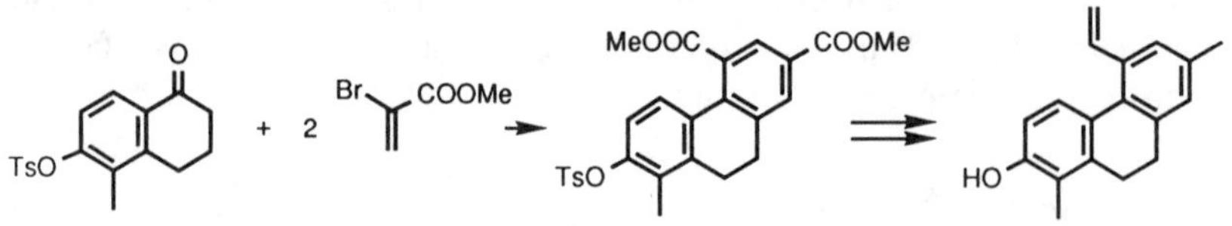

juncunol

A special kind of reflexive Michael cyclization is represented by a synthesis of α-allokainic acid [Barco, 1990]. The norketone is a δ-keto acid as well as a β-amino ketone. A Michael transform of the keto acid (ester) results in a seco species which can be further dissected into methyl vinyl ketone and a γ-aminocrotonic ester. Accordingly, this scheme was realized.

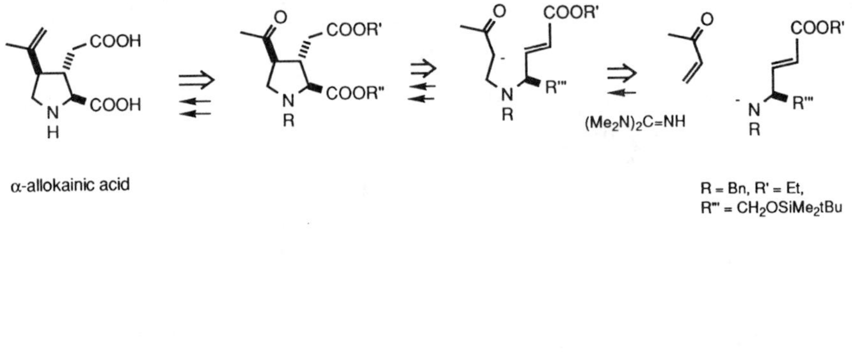

α-allokainic acid

R = Bn, R' = Et,
R''' = CH₂OSiMe₂tBu

9

OTHER METHODS FOR SYNTHESIS OF 1,5-DIFUNCTIONAL COMPOUNDS

9.1. ELABORATION FROM CYCLOBUTANE DERIVATIVES

9.1.1. deMayo Reaction

Two-carbon insertion of enolizable 1,3-dicarbonyl compounds can be effected via photocycloaddition with alkenes followed by retro-aldol fission. The combined steps are known as the deMayo reaction. When an unprotected enol component is used in the photocycloaddition, the retro-aldol fission occurs *in situ*. The driving force is the strain relief.

This homologation procedure has found numerous applications in the synthesis of natural products: loganin [Büchi, 1973, Partridge, 1973], longifolene [Oppolzer, 1978], daucene [Seto, 1985], and hirsutene [Disanayaka, 1985] are only a few that we can mention. The photocycloadduct may be subjected to structural modification before the fragmentation is induced.

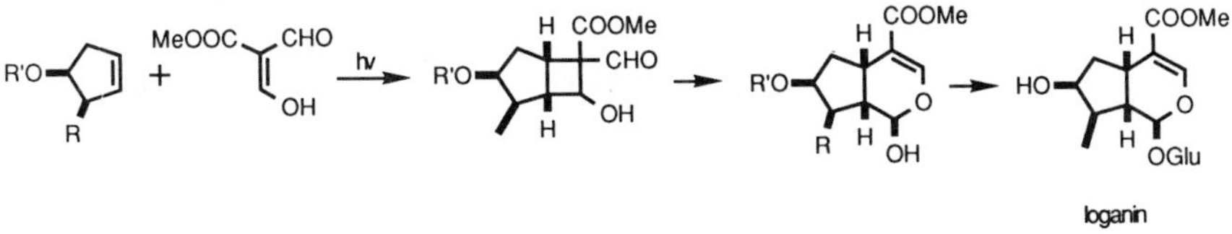

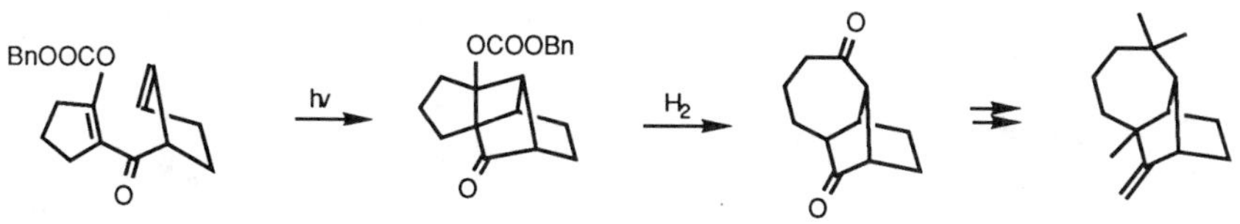

loganin

longifolene

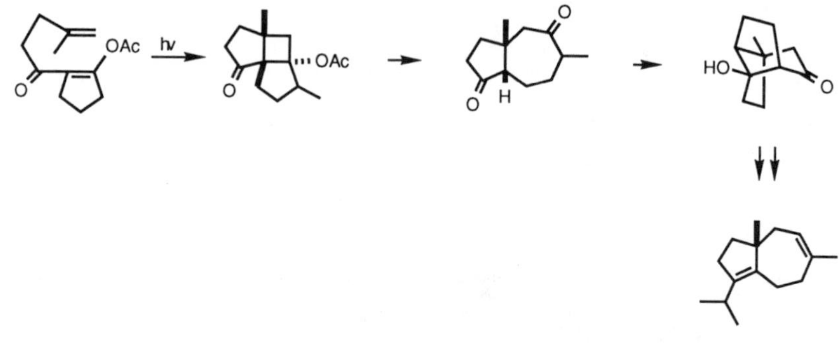

daucene

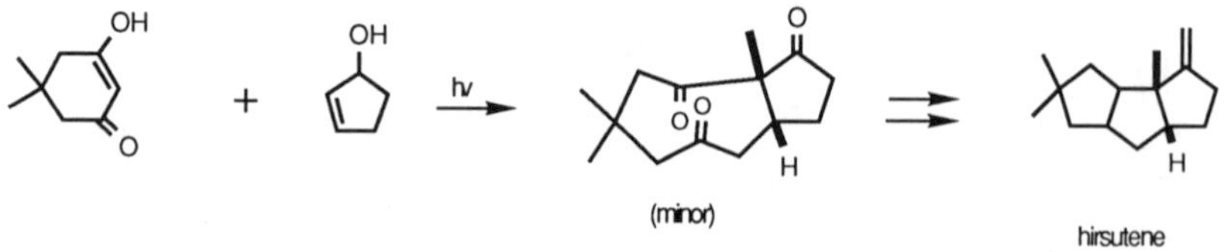

(minor)

hirsutene

The enolizable 1,3-dicarbonyl substrates may be strategically replaced by (5H)-5,5-dimethyl-3-furanone or (4H)-2,2,6-trimethyl-1,3-dioxin-4-one, which have been used in the synthesis of occidentalol [S.W. Baldwin, 1982] and *ent*-elemol [S.W. Baldwin, 1985], respectively.

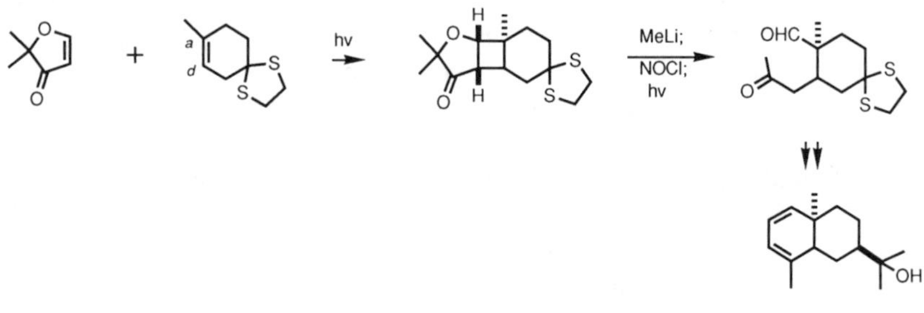

occidentalol

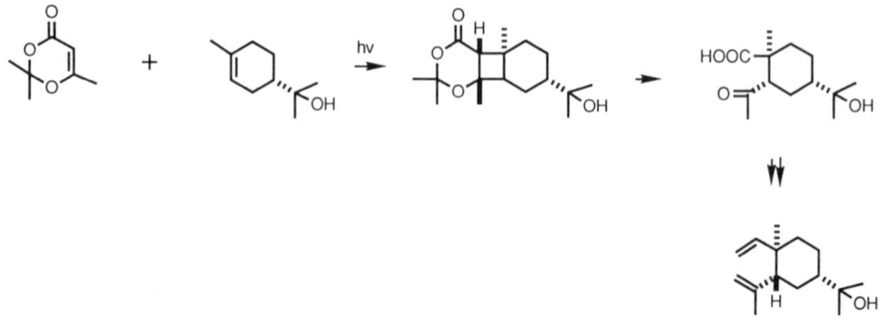

ent-elemol

The Corey lactone, a common intermediate for many families of prostaglandins [Corey, 1969b], is the anhydro form of a δ-aldehydo acid. Accordingly, its construction via the deMayo-type reaction has been studied [M. Sato, 1985]. But because the β-hydroxy aldehyde subunit is prone to elimination, only the unsaturated aldehydo lactone was obtained. Further decrease of its yield was due to the cycloaddition from the *syn*-face of the hydroxyl groups.

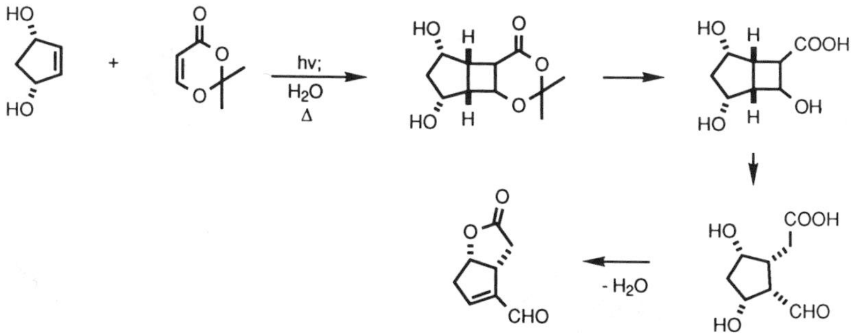

9.1.2. From Other [2 + 2] Photocycloadducts

The reaction sequence consisting of photocycloaddition of an enone with allene, oxidative cleavage of the exocyclic methylene group of the adduct, and any one of the various hydrolytic or reductive treatment gives rise to a 1,5-dioxygenated substance. If the cleaved product is a dicarbonyl compound, an intramolecular aldol condensation is expected to follow.

Many bridged ring systems have since been constructed according to this general scheme after the appearance of a report on its application to a synthesis of atisine [Guthrie, 1966].

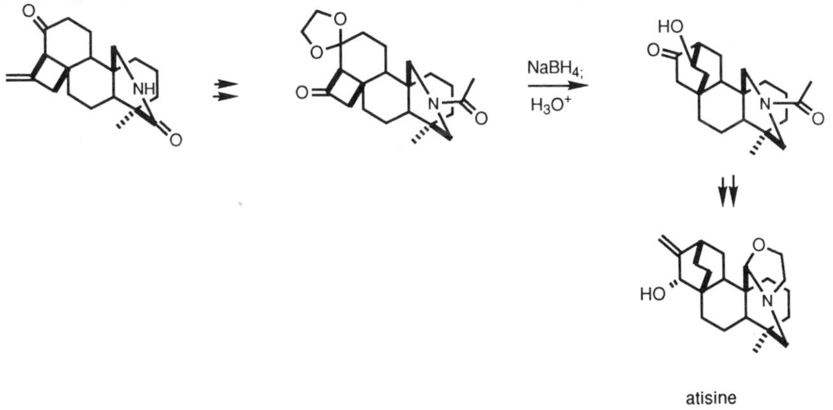

atisine

The introduction of the vinyl substituent to complete the skeleton of forskolin [Ziegler, 1987] also profited from this cyclobutane annulation–cleavage method.

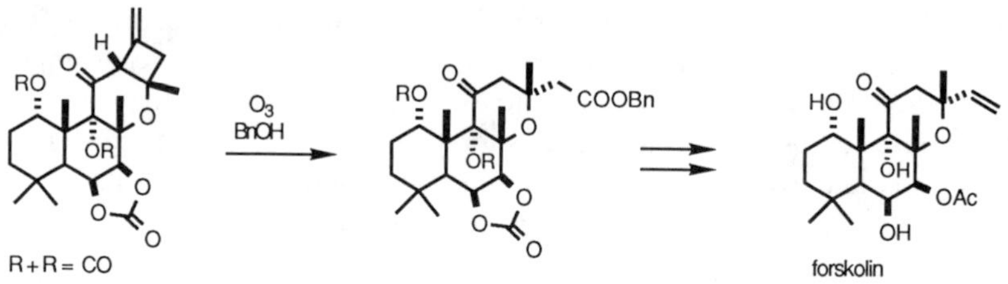

R+R = CO

forskolin

A synthesis of reserpine [Pearlman, 1979] featured an intramolecular photocycloaddition to assemble the DE-ring component. In this case it is more expedient to employ a substrate which contains a butenolide ring. However, the desired 1,5-dicarbonyl derivative could be obtained only after a Baeyer–Villiger degradation.

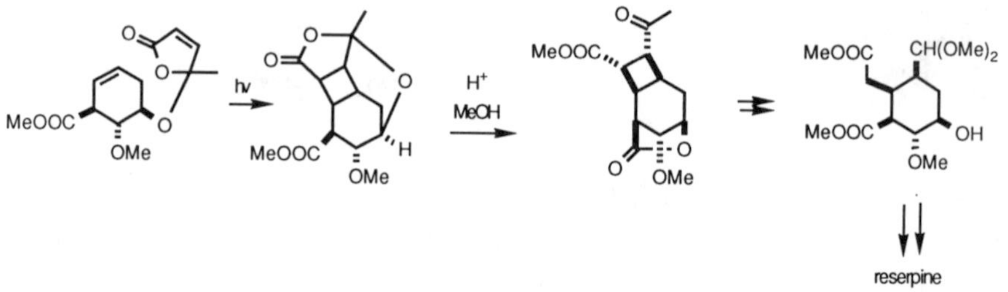

reserpine

There are a few nonphotochemical cycloadditions for the formation of four-membered ring compounds, most notably those involving ketenes. Ynamines also undergo cycloadditions with acceptor-substituted alkenes, the products are readily unraveled into the formal Michael adducts. The outstanding feature of this alternative is the stereocontrol at the two newly created tetrahedral carbon centers. An example of its application is a synthesis of dihydroantirhine [Ficini, 1979].

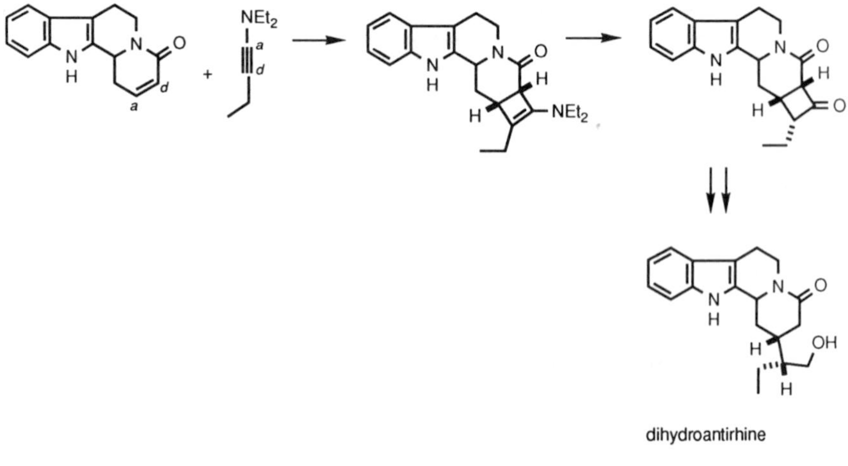

dihydroantirhine

9.2. MODIFICATION OF EXISTING 1,5-DICARBONYL SUBSTANCES

The most reliable synthesis of cyclohexenones has been the aldol condensation of 1,5-dicarbonyl compounds. A versatile method for the elaboration of latent 1,5-dicarbonyl compounds is based on bicyclic oxazolidinones which are obtainable from the condensation of δ-keto esters with *vic*-amino alcohols. The α-position of the lactams can be mono- or difunctionalized stereospecifically. The final stage of the reaction sequence involves an organometallic reaction on the lactam and subsequent release of the dicarbonyl product. Depending on the experimental conditions an aldolization may be promoted in the same step.

Using chiral amino alcohols which are readily available from α-amino acids to form the heterocycles, efficient asymmetric synthesis of cyclohexenones may be achieved. A synthesis of (−)-abscisic acid [Meyers, 1989] from isophorone and (S)-valinol is representative.

(-)-abscisic acid

Synthesis of (−)-agarospirol [Deighton, 1975] from (+)-3-methylcyclohexanone showed an exploitation of pseudosymmetry as the five-membered ring with a pendant can be created from a six-membered precursor. In turn, a single spiro[5.5]decanedione is derivable from a Dieckmann cyclization of the keto diester.

(-)agarospirol

The unfortunate aspect of this synthesis is the lack of regiocontrol during the conversion of the unhindered ketone into the trisubstituted olefin.

The spirocyclic diketone retains a 1,5-dioxygenated pattern which was introduced in the cyanoethylation–alcoholysis sequence.

9.3. DEGRADATION

A yohimbine synthesis [van Tamelen, 1958b] serves to highlight the evolution of a design that calls for elimination of one skeletal carbon atom from an intermediate.

For the nontryptamine portion of the molecule there are three contiguous carbon chains on the cyclohexane nucleus. A useful synthetic precursor must contain differentiable functional groups in these chains such that it would link up with tryptamine with a flanker pendant. It would be the best that the other two carbon chains are in some protected form when the reaction occurs. Connecting the two other chains into a ring would result in a *trans*-hydrindene nucleus. For steric reasons it is much easier to set up the *trans-vic* substituents in an octalin system.

This retrosynthetic reasoning can be further pushed back to reveal, ultimately, the butadiene–benzoquinone adduct. Homologation at one of the carbonyl sites of the adduct (after partial reduction) would give the required reactive connector. When the cyclohexene portion was oxidized, an extra carbon had to be degraded.

This work shows a strategy for vicinal attachment of two carbon chains to a ring structure via cycloaddition, double bond cleavage, and selective trimming.

The yohimbine synthesis also indicates the importance of recognizing the near-symmetry elements that exist in a synthetic target. These elements are not apparent and they may be revealed by stripping or adding functional groups which are to be reintroduced or eliminated at later stages of the synthetic operations.

In the molecular framework of ajmaline, C-17 is oxygenated and separated from each of the two nitrogen atoms by two-carbon units. The conjoint relationship indicates a possibility of effecting C—C bond formation by Mannich and enamine/aldehyde condensation reactions. This course of events probably occurs in the biosynthesis of the alkaloid.

A biomimetic synthesis [van Tamelen, 1970] started from an elaboration of a tetracyclic aldehydo carboxylic acid from $N_{(a)}$-methyltryptophan. The intermediary dialdehyde underwent spontaneous Pictet–Spengler cyclization. The tetracyclic amino acid was submitted to an oxidative decarboxylation, and the iminium ion was intercepted in the fashion of a Mannich reaction. The product with a *syn*-aldehyde group to the indole nucleus cyclized under reductive conditions, and the *anti*-isomer was also useful via its equilibration to the *syn*-aldehyde. Functionalization at C-21 via a C—N bond cleavage culminated in ajmaline.

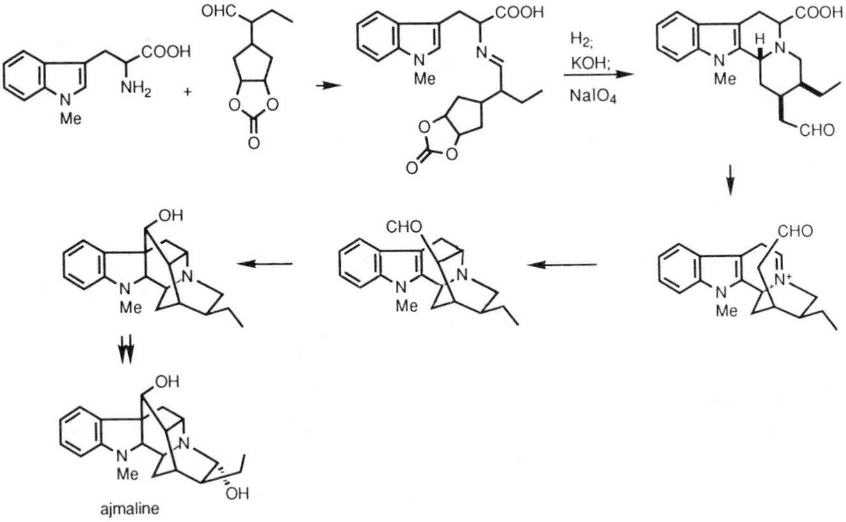

ajmaline

An approach to emetine [Burgstahler, 1959] from gallic acid was based on the same principle that a 3-substituted glutaraldehyde may be obtained and incorporated into different structural entities.

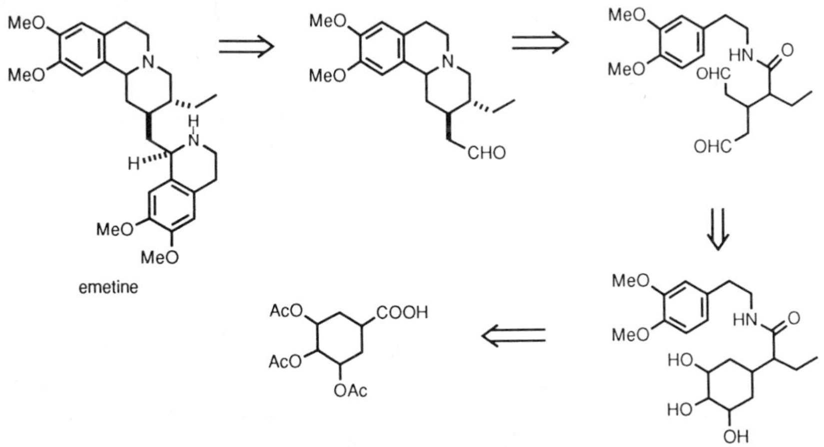

emetine

The peripheral segment of lycopodine extending between the nitrogen atom and the ketone is conjoint. The use of an aromatic moiety as the latent carbon chain would require redox manipulation that includes excision of a C_1 subunit.

Implicit in the overall transformation are two oxidation steps since the first oxidative ring cleavage would result in a six-carbon chain terminated by two acceptors.

With respect to the complete scheme for the synthesis [Stork, 1968] the identification of an aromatic intermediate also pointed to the possibility of utilizing a Pictet–Spengler-type cyclization, in view of the Ar—C—N structural subunit. The ultimate choice of an acylenamine acceptor was predicated by its adequate reactivity, the relative ease of its preparation, and the regiofixation of the reactive iminium center.

lycopodine

Consideration of a route to confertin and aromatin [Ziegler, 1982] in the retrosynthetic context is instructive. Disconnection of the lactone ring to a γ,δ-unsaturated carboxylic acid derivative immediately suggests a Claisen rearrangement transform. The subgoal is a 1,5-dioxygenated hydrazulene and an aldol–Michael transform scheme is revealed.

While undoubtedly a conventional reaction sequence might be developed, the propensity of 2-cyclopentenones to undergo conjunctive nucleophilic–electrophilic alkylation to afford 2,3-diorganyl cyclopentanones would indicate a more efficient access to the desired product. To this end, the conjugate addition of a lithio ketenedithioacetal with enolate capture by allyl bromide, and oxidative cleavage of the two double bonds in the adduct met the synthetic requirement. It is noted that an oxidation step returned a disjoint molecular segment which was assembled with an umpoled reagent (the dithiane) to a conjoint state.

aromatin

9.4. MISCELLANEOUS PATHWAYS TO 1,5-N,O AND 1,5-N,N SUBSTANCES

Morphine has a 1,5-N,O circuit. The application of a Michael reaction on a dienone has only appeared very recently as part of a synthetic route [Toth, 1987]. In this approach a major role was played by a sulfonyl substituent. First, it activated the conjugated double bond to participate in twofold C—C bond formation, and its subsequent elimination helped generate a dienone system to establish the piperidine ring by the Michael reaction.

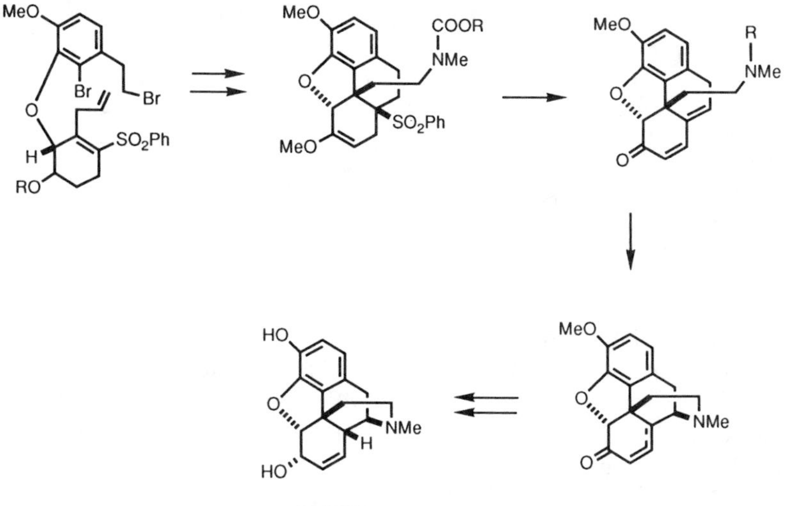

morphine

The Pictet–Spengler cyclization on activated aromatic compounds may be considered as a special version of the Mannich reaction in which the donor and the amine are of the same molecule, resulting in 1,5-difunctionalized heterocycles. This reaction has found thousands of applications in synthesis.

To elaborate ochotensimine [Irie, 1968] with the help of the Pictet–Spengler cyclization, an indanedione was condensed with a proper phenethylamine, and the remaining carbonyl of the spirocyclic product was submitted to Wittig methylenation. N-Methylation concluded the synthesis.

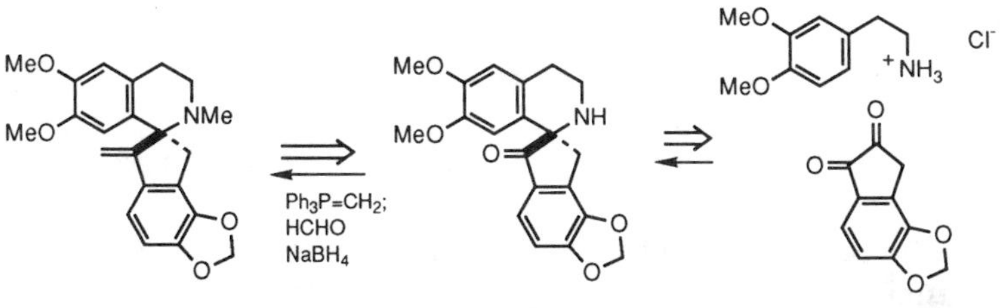

ochotensimine

Chemoselectivity in the Pictet–Spengler cyclization is noted. The nonbenzylic ketone is more reactive toward the amine.

The step leading to a tetracycle in a synthesis of ellipticine [Langlois, 1975] is a vinylogous Mannich reaction.

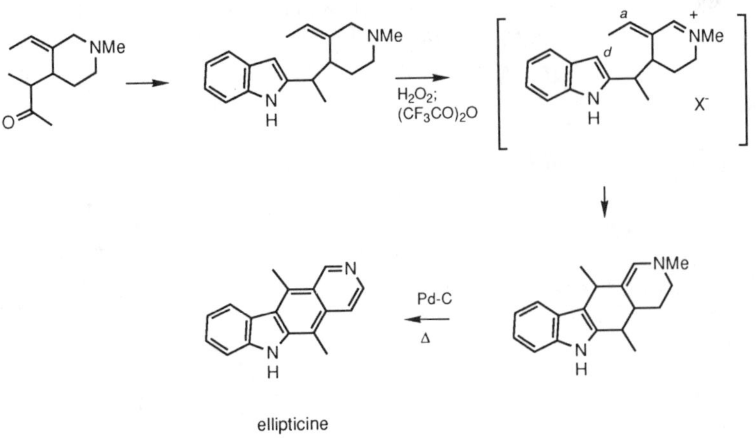

ellipticine

Enamine alkylation followed by a Pictet–Spengler cyclization on an *N*-tryptophyl-2-piperideine appears to be an efficient method for the assembly of an E-seco deoxyvincamine [Lounasmaa, 1986]. For the synthesis of vincamine the deletion of the keto group does not render the approach invalid, as the carbonyl can be reintroduced.

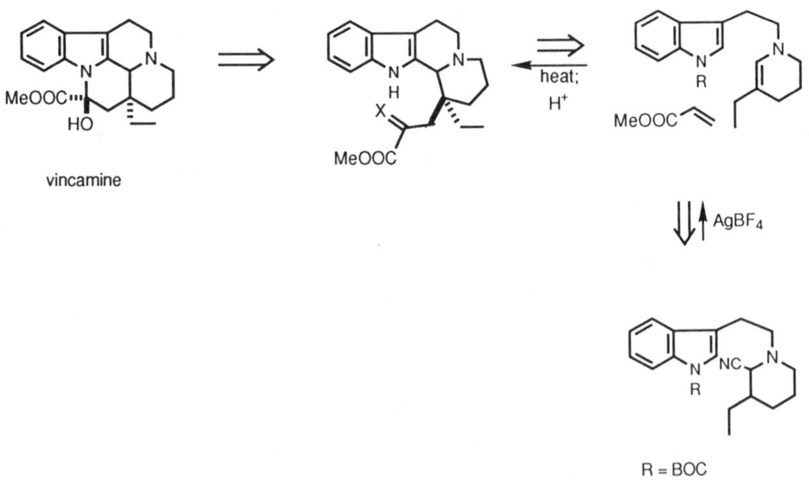

vincamine

R = BOC

A more direct coupling of the two cyclic moieties of emetine is by a Michael reaction using a lithiated imine as donor [Naito, 1986]. It only required a reduction step to conclude the synthesis.

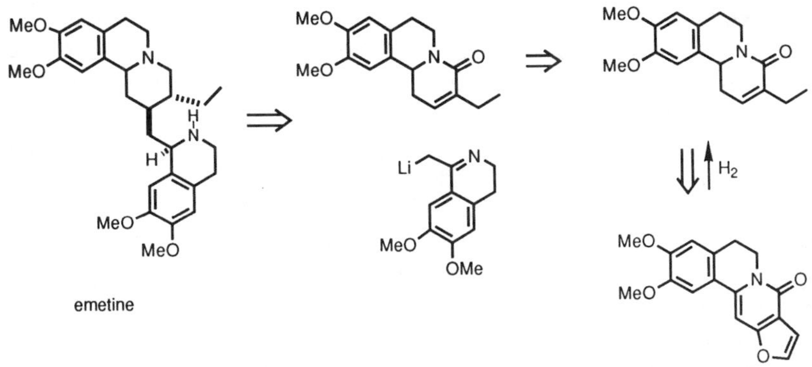

With a more reactive Michael acceptor, 3,4-dihydro-6,7-dimethoxy-1-methylisoquinoline is able to react in the absence of a strong base. A most convenient synthesis of protoemetinol [Kametani, 1979] was accomplished from the tricyclic adduct.

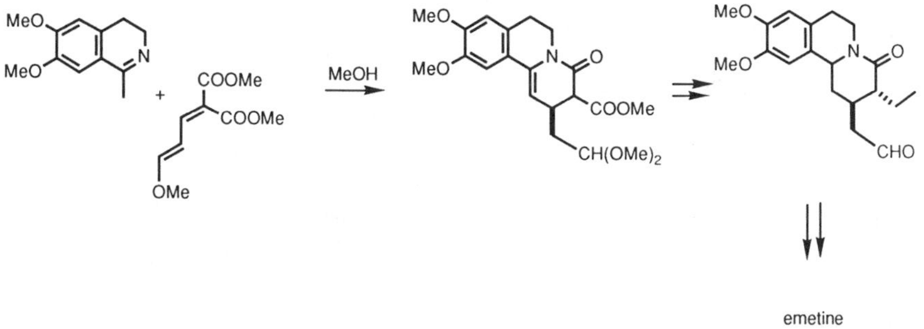

Special among 1,5-dihetero substances are the six-membered heterocycles which can be derived from the former compounds. Thus a twofold reductive azacyclization proved to be a most expedient method to establish stereoselectively the quinuclidine nucleus of matrine [Mandell, 1963].

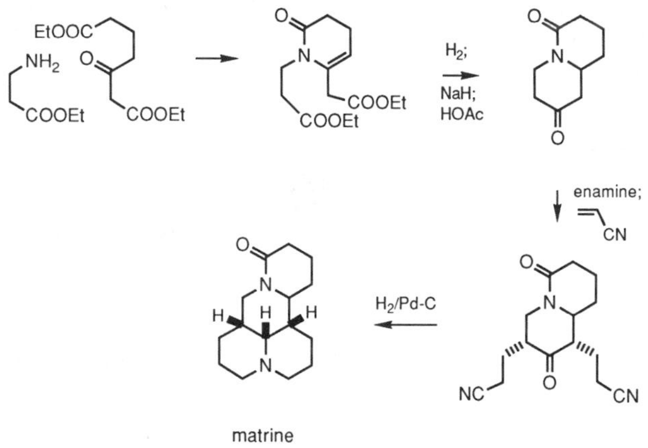

This synthesis also featured the preparation of a δ-lactam from a δ-keto ester and ethyl 3-aminopropionate. It is significant that all the starting materials contain fragments in total conformity to the polarity alternation rule. It means a most straightforward accessibility of them from simple chemicals.

A quite different strategy must be implemented for synthesis of cularimine [Kametani, 1964] regarding the nitrogen-containing portion of the molecule since the two functionalized chains to be linked together are separated by an aromatic ring. An intramolecular Friedel–Crafts acylation seems adequate as a means of establishing the 1,5-difunctional precursor. It may be envisioned that a diaryl ether bearing in each ring an acetic acid chain would serve admirably; the cyclization of which would proceed as desired because the only other possibility is the unlikely formation of benzocyclobutenone(s).

The identification of the diacid intermediate was a wonderful consequence of retrosynthetic analysis (perhaps subconsciously at the time). It enabled a simultaneous elaboration of both sidechains in the two separate aromatic rings.

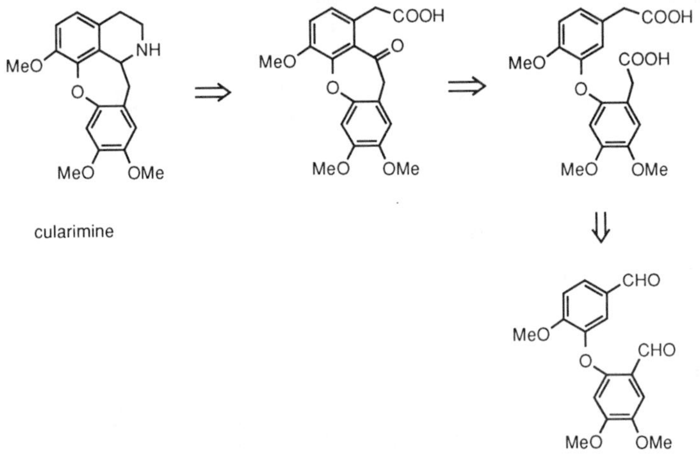

cularimine

The effectiveness of exploiting hidden symmetry of a complex synthetic target is also witnessed in approaches to ajmalicine [van Tamelen, 1969] and vincamine [Kuehne, 1964].

10

THE DIELS–ALDER REACTION

10.1. GENERAL FEATURES AND STEREOCHEMICAL ASPECTS

The Diels–Alder reaction [Sauer, 1980] is one of the most powerful methods for the construction of six-membered ring compounds including certain heterocycles by thermal induction. Its value lies in the stereospecificity and stereoselectivity which can most often be predicted on the basis of the highly ordered transition state it adopts. Up to four contiguous asymmetric centers can be created in the product by this [4 + 2]cycloaddition process.

In recent years developments in the intramolecular version [Brieger, 1980, Ciganek, 1984] and the use of chiral addends and Lewis acid catalysts have further magnified the versatility of the method. The intramolecular Diels–Alder reaction gives rise to polycyclic products, and because of the associated steric constraints, structures which are not easily accessible (e.g., via *exo* transition states) may be prepared. The catalytic reactions tend to heighten the regio- and stereoselectivity, and enable reactions to proceed under milder conditions. Chiral addends lead to chiral adducts. For very difficult reactions, applying high pressure [Dauben, 1977] is beneficial. The advantage is that reverse reactions that are favored at high temperatures are suppressed.

The reaction components of the Diels–Alder reaction are a conjugate diene and a dienophile. The diene must take a cisoid form during the reaction even if it is in equilibrium with the transoid conformation. A diene that cannot adopt the cisoid form is not reactive. The diene may also be introduced in a masked state (e.g., as a sulfolene) if it can be generated under the experimental conditions. Generally, donor-substituted dienes are more reactive toward (electron-deficient) dienophiles. For example, alkoxy-, siloxy-, and acylaminodienes have found many applications in natural product synthesis.

Cyclic dienes, especially cyclopentadiene and its derivatives, undergo the Diels–Alder reaction much more readily with dienophiles than similar acyclic dienes because the active cisoid coplanar form is already locked in place. This trend is not valid for medium-sized and large ring dienes as nonbonded interactions militate against coplanarity of the diene segment.

Most of the useful dienophiles are those containing a double or triple bond conjugated with an acceptor group such as COR, COOR, CN, and NO_2. Maleic anhydride, p-benzoquinone and their derivatives are exceptionally reactive dienophiles because of their cyclic structures. Steric repulsions in the stacked transition states are minimized.

The cyclic nature, as well as the high strain, makes benzynes and other dehydroaromatics in the pyridine and thiophen series unique. The [4 + 2]cycloaddition of benzyne to benzene! [Wittig, 1962] has been demonstrated.

As will be seen below, certain heterodienes and heterodienophiles have found special synthetic applications. Thus, dioxygen, nitroso compounds, imines, and aldehydes are frequently used as dienophiles.

The Diels–Alder reaction is characterized by high stereoselectivity, at least for the kinetically controlled process. Thus, the $(E-Z)$ configuration of the dienophile is translated to the *trans–cis* relationship in the adduct. The stereochemistry of the substituents at the 1- and 4-positions of the diene is also correlatively retained. The general validity of this "cis principle" suggests a concerted formation of the two new bonds.

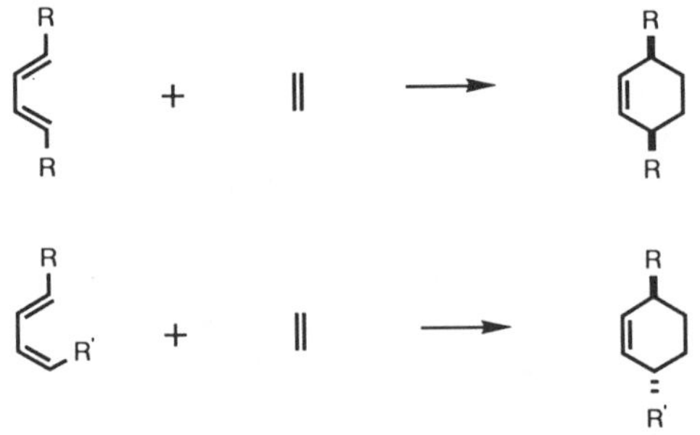

The celebrated *endo* addition rule applies well to reaction of a cyclic diene and cyclic dienophile. This is the consequence of secondary orbital overlap [Woodward, 1970] that stabilizes the *endo* transition state. However, other factors such as steric hindrances could severely undermine this rule. Nonbonding interactions in the *endo* transition state seem to be a major determinant in the adoption of the *exo* counterpart for many (especially intramolecular) reactions.

Although the Diels–Alder reaction is commonly regarded as a concerted pericyclic process and in the classical sense it is not subject to much electronic influence. However, the electronic compatibility between addends was early recognized as the normal reaction is facilitated by the combination of an electron-rich diene and an electron-deficient dienophile. The orientation of the major or exclusive adduct usually shows the donor–acceptor distribution according to the one favoured by its active interactions in the transition state. The Diels–Alder reaction of inverse electron demand involves an electron-deficient diene and an electron-rich dienophile. Common sense would have predicted such requirements.

Catalysis is also an aspect that indicates the accentuation of polarity alternation and the effect is felt at the terminal atoms participating in the bond formation. The phenomenon is definitely consistent with the favorableness of a somewhat polar transition state for the cycloaddition. Some reactions actually proceed by a concerted mechanism with two σ-bonds being formed at different rates. According to the frontier orbital theory the coordination of the dienophile with a Lewis acid lowers its LUMO (lowest unoccupied molecular orbital) energy substantially, thereby narrowing the energy gap of the $HOMO_{(diene)}-$ $LUMO_{(dienophile)}$ (HOMO is the highest occupied molecular orbital.) This change is manifested in a faster reaction. (In the Diels–Alder reaction of inverse electron demand the interaction is $LUMO_{(diene)} - HOMO_{(dienophile)}$.)

We shall illustrate the importance of polarity control for the Diels–Alder reaction in the following. However, like most chemical reactions, there are exceptions to the expected reaction courses. These arise mainly because the Diels–Alder reaction transition state is not planar. The stacking transition state that is favored by the secondary orbital overlap could have an opposite effect to the polarity alternation consequences. This a matter of through-space vs through-bond interactions [Hoffmann, 1971].

Although most of the anomalies in regioselectivity may be explained by the frontier molecular theory [Fleming, 1976], an empirical assessment of them is also possible based on the through-space interactions. For example, the dominance of a phenylthio group over a methoxy group as a regiodirector in the reaction of 2-methoxy-3-phenylthio-1,3-butadiene [Trost, 1980] may be due to the spatial interaction of the donor methoxy with the acceptor group in the dienophile (which leads to the "*meta*" adduct) reinforcing the donor effect of the sulfur substituent. In the absence of the sulfur substituent, the interaction of the reaction partners will be largely electronic which can be gauged by the polarity matching.

4 :1   carvone

Remarkably, frontier molecular orbital calculations fail to predict these results [Kahn, 1986].

The cycloaddition of 2-phenylsulfonyl-1,3-butadiene with methyl acrylate [Bäckvall, 1987] and 2-trimethylsilyl-1,3-butadiene with methyl vinyl ketone [Trost, 1986] each gave rise to a ca. 2:1 mixture of "*para*" and "*meta*" adducts. The condensation of 2-triethylsilyl-1,3-butadiene with ethyl acrylate resulted in a mixture of isomers favoring the *para* substitution pattern (7.3:1) [Batt, 1978]. Formation of the *para* adduct is electronically unfavorable, yet its appearance as a major product could be due to the avoidance of through-space interactions between the two acceptor substituents.

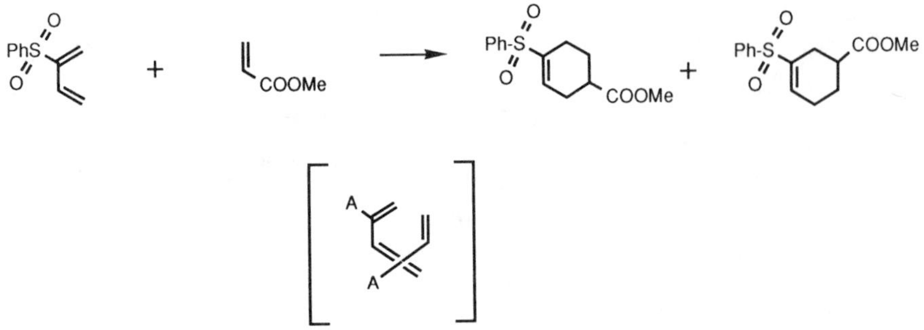

A similar rationalization may be given to the formation of one type of the dioxabicyclo[4.3.0]nonadienes by reaction of furan with allenyl ketones [Bertrand, 1967], in addition to the bridged bicyclics.

10.2. REGIOSELECTIVITY AND ITS MANIPULATIONS

Generally, a Diels–Alder reaction occurs with an orientation accountable by complementary interactions of polar substituents. In other words, a conjoint transition state is preferred. This leads to the so-called *ortho* and *para* cycloadducts as shown below. A synthesis that can exploit this reactivity pattern is rewarded by considerable abbreviation of steps.

Substituent manipulation (e.g., functional group addition) can be effective for inverting the regiochemistry of the Diels–Alder reaction. For example, 3-nitro-2-cyclohexenone has been employed to obtain the unattainable adducts of 2-cyclohexenone [Ono, 1982]. An extra step involved in the reaction sequence is the reductive removal of the nitro group.

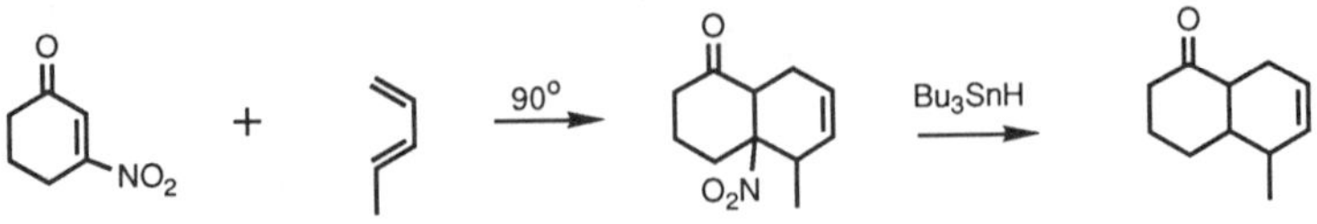

The difference in Diels–Alder reactivity of 2- and 3-methyl-2-cyclohexenone toward 1,3-butadiene is perhaps due to electronic effects more than steric effects. In the Lewis-acid-catalyzed condensation the complexed 2-methyl-2-cyclohexenone expresses a pronounced cationic character at C-3 which is intercepted readily by the diene. On the other hand, the corresponding complex derived from the 3-methyl isomer must be a more stable species which is more reluctant to react with the diene.

To circumvent the difficulty in securing Δ^6-4a-methyl-1-octalone by the convenient Diels–Alder approach, 4-acetoxy-2-methyl-2-cyclohexenone proved useful [Angell, 1989]. The existing carbonyl group was for enhancing the polarization and it was subsequently removed. The remaining acetoxy group of the adduct can then be processed accordingly to establish a new ketone.

Functional group interchange of the alkenes of β-chamigrene with carbonyl groups generates a conjoint spirocyclic diketone. In principle, a reflexive Michael addition is applicable to the synthesis of the diketone, but the polarity-matched Diels–Alder reaction is an excellent alternative method [A. Tanaka, 1967], since basic conditions which may be deleterious to the α-methylenecyclohexanone are avoided.

The regiochemistry of the Diels–Alder reaction between 1-methoxy-1,3-cyclohexadiene and α,β-enones is determined by the donor and acceptor substituents. This regiochemistry is crucial to a preparation of 4-substituted cyclohexenones with stereocontrol over two asymmetric centers. The synthetic potential was demonstrated in an approach to *threo*-juvabione [Birch, 1970]. Many natural product syntheses have since been achieved using this particular method.

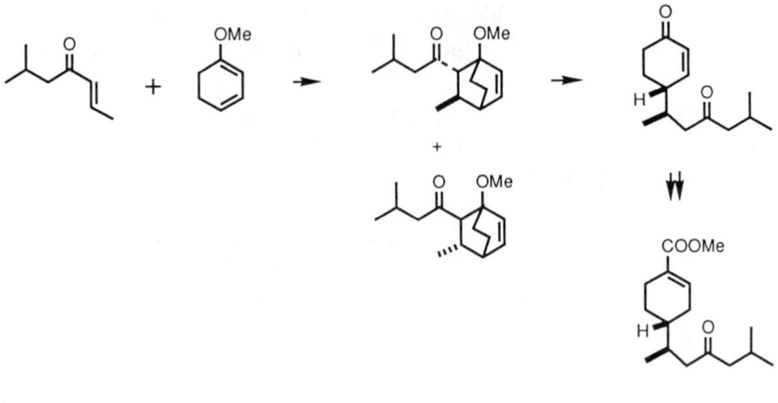

threo-juvabione

The angular fusion of three carbocycles of different sizes in illudol is quite remarkable. Besides the obvious synthetic essay based on the [2 + 2]cycloaddition involving a hydrindenone the skeleton may be assembled via a Diels–Alder route [Semmelhack, 1980] as practically all cyclohexane derivatives can be prepared in certain variations.

The only reasonable Diels–Alder transform on this sesquiterpene generates a cyclobutene dienophile and an (alkoxyvinyl)cyclopentene. Further considerations indicate a cyclobutenecarboxylic ester to be a more suitable reactant in terms of its reactivity and the stereo- and regiocontrol the ester function is able to confer, i.e., *endo* addition and polarity matching with the substituted diene. Of necessity reduction steps must be implemented to convert the ester to a methyl group.

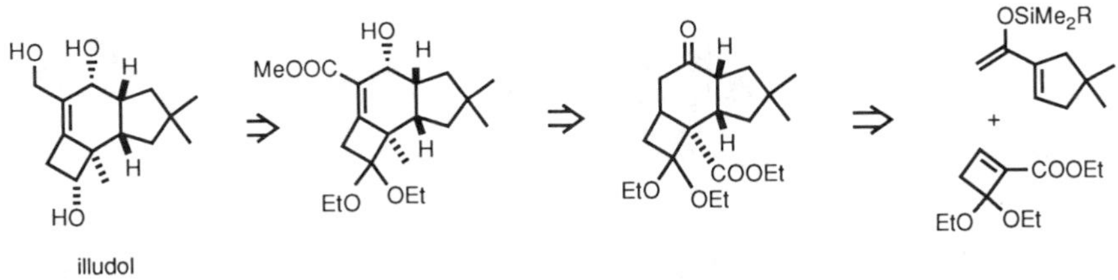

illudol

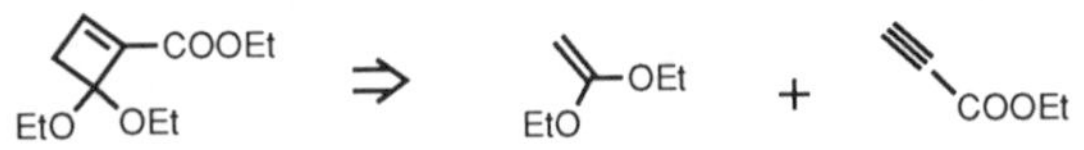

The secondary alcohol present in the four-membered ring of illudol also dictates an oxygenated cyclobutene. Fortunately, such a cyclobutenecarboxylic ester is also conjoint, therefore it is easily acquired. A thermal condensation of ethyl propiolate with 1,1-diethoxyethylene turned out to be a simple method for acquiring the dienophile.

The synthesis of phyllanthocin [Burke, 1985] using a Diels–Alder cycloaddition to build the cyclohexane ring is logical. The double bond of the adduct provides a handle for introduction of the ester group. The key feature is the *ortho* regioselectivity that places adjacently an oxygen function inherited from the diene component to a ketone group from the dienophile. The latter is to be converted into the epoxide ring and the spiroketal.

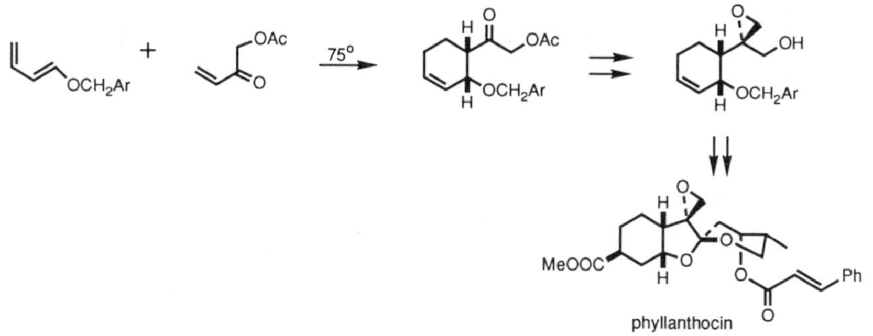

phyllanthocin

There is an even more obvious structural correlation between verrucarol and a Diels–Alder adduct [Trost, 1982a, Schlessinger, 1982].

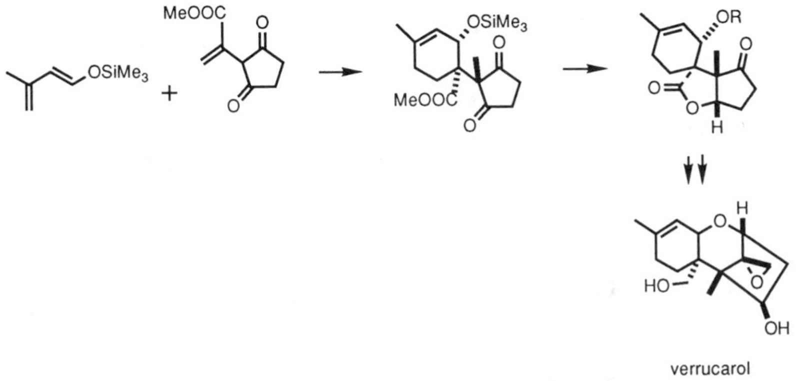

verrucarol

The presence of a trioxygenated cyclohexanecarboxylic acid residue in actinobolin is a sufficient hint for the Diels–Alder approach in its construction [Kozikowski, 1986]. A weakness of such approach is the lack of stereocontrol during the cyclization.

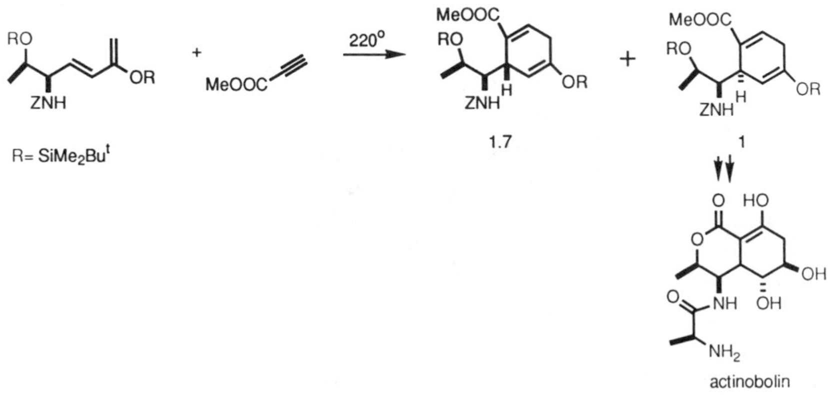

R= SiMe$_2$Bu^t

actinobolin

A novel synthetic route to isonitramine and nitramine [Wanner, 1989] involved a Diels–Alder reaction to construct the spirocyclic skeleton. The cyclohexanol moiety is a perfect partial retron for application of the transform. However, the dienophile must be activated.

N-Protected α-methyleneglutarimides fulfill the requirement, since the chromophore exerts regiocontrol over its reaction with a 1-oxygenated diene such that the hetero atoms to be preserved in the adduct(s) are correctly placed.

isonitramine

nitramine

An acylamino group attached to a diene is a powerful regiocontroller of the Diels–Alder reaction. It is not surprising that a short synthesis of pumiliotoxin C [Overman, 1978] could be executed by starting from the condensation of crotonaldehyde with 1-benzyloxycarbonylamino-1,3-butadiene.

pumiliotoxin C

The crucial thought process that generated a route to the pentacyclic intermediate for songorine [Wiesner, 1973] entailed the addition of a ketone group to the B-ring, and its assembly by an intramolecular aldol condensation and Michael addition. Ultimately this retrosynthetic reasoning led to a Diels–Alder reaction of a methoxylated benzyne with Thiele's ester.

The Diels–Alder reaction shows remarkable regioselectivity as arbitrated by the two polar groups. It is also important that the carbonyl group of the diene also served as a control element for the selective rearrangement–ring opening of an aziridine intermediate in a subsequent step.

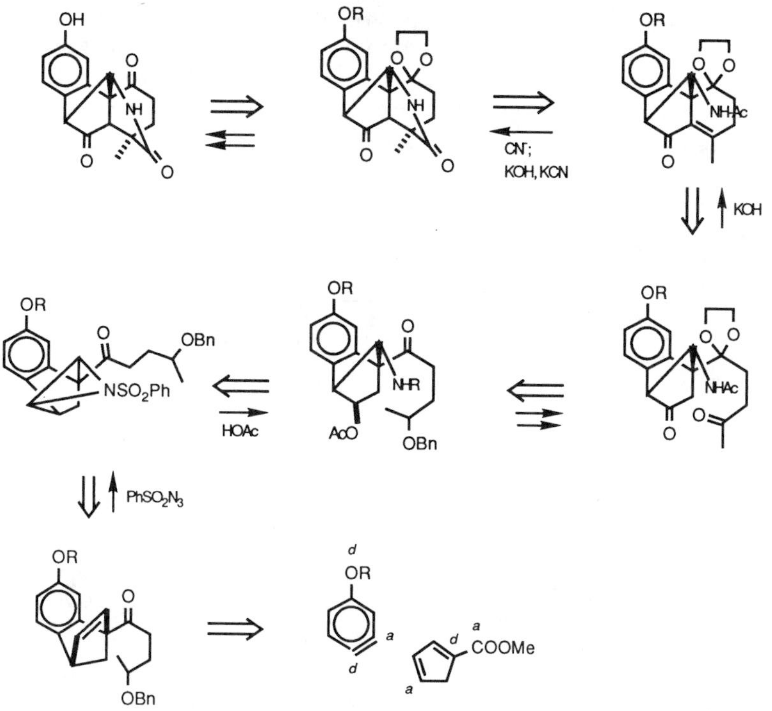

A less obvious electronic effect manifested in the Diels–Alder reaction is the following intramolecular process [Greuter, 1977; Fráter, 1974]. A more prominent electronic matching is that involved in the step leading to a sativene intermediate [Snowden, 1981].

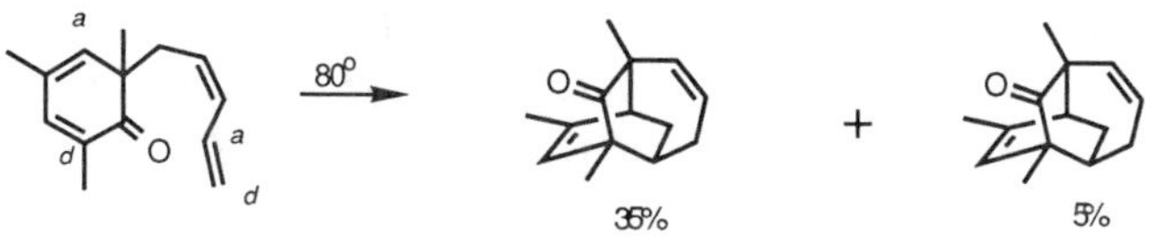

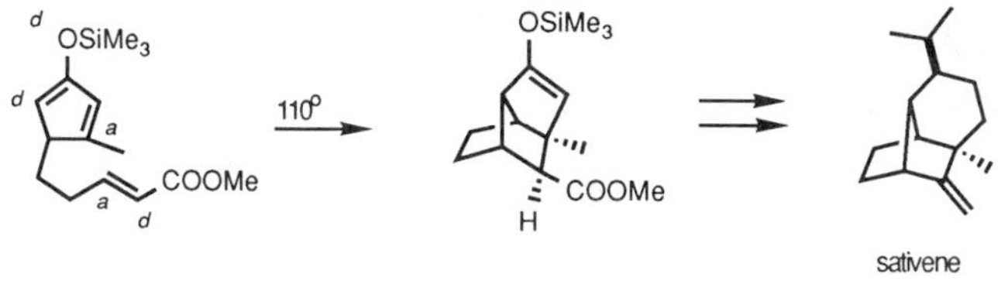

A 7:4 *meta:para* ratio of adducts has been observed in the Diels–Alder reaction of an ynone with a semicyclic diene in which the extracyclic double bond is substituted with a carbonate group and a terminal methyl. When the terminal methyl is trimethylsilylated, the *meta:para* adduct ratio is increased to 11:1 [Vedejs, 1983]. The unusual favoring of the *meta* adduct formation must be due to accentuation of the donor characteristic of the sp^2-carbon at the far end by the silicon atom. It is possible that in the *meta* cycloaddition transition state there was some through-space interaction between the oxygen atom(s) of the dienophile with the silicon.

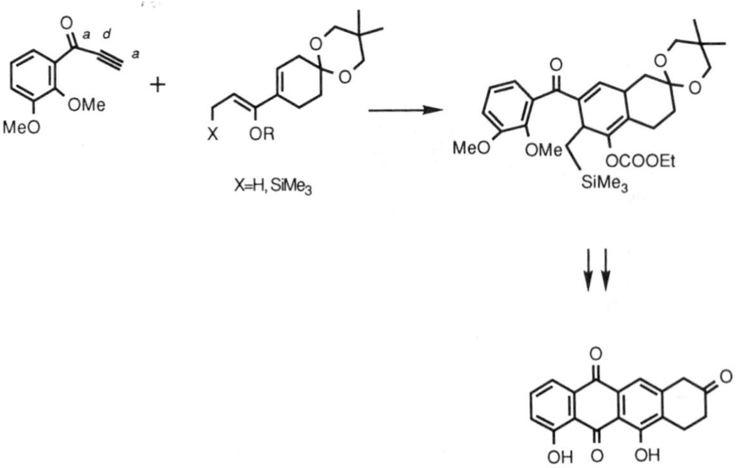

Enhancement of regioselectivity in the Diels–Alder reactions of isoprene and its derivatives by replacing the methyl group with a trimethylsilylmethyl or trimethylstannylmethyl substituent [Wilson, 1979; Hosomi, 1980]. This accentuation effect has significant implications in a synthetic approach to the cytochalasans [Vedejs, 1982].

There are subtle electronic influences by remote substituents on the regiochemistry of the Diels–Alder reactions involving 2,3-bis(methylene) bicyclo[2.2.1]heptanes as the dienes [Tamariz, 1983]. The effect of a methoxy group which is seven bonds away from the nearest terminus of the diene system appears to dominate the mode of cycloaddition via a combination of through-bond and through-space interactions.

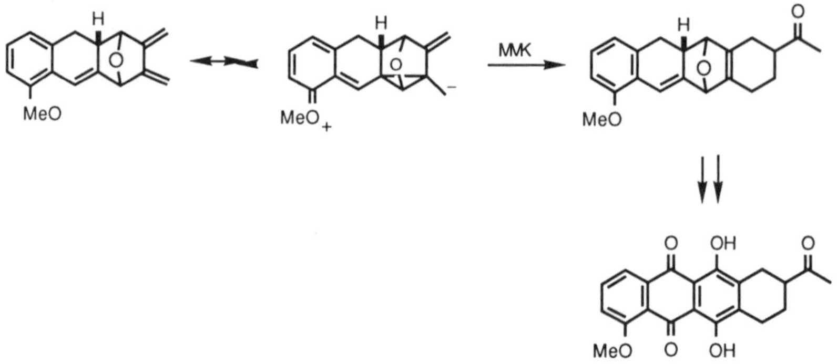

Although speculative, the significant "para" directing effect of the carbonyl group in 5,6-bis(methylene)-2-norbornanone may be the result of a favorable dipole alignment in space.

favorable   unfavorable

At first sight, the application of the Diels–Alder reaction to the synthesis of 2,5-cyclohexadienones of the prephenic acid type would not be particularly rewarding. There are not too many highly reactive acetylene-equivalent dienophiles and conceivable predicaments in unraveling the double bond from such adducts which contain delicate functionalities. However, the development of polyoxygenated dienes [Danishefsky, 1981] has paved the way to new disconnection leading to alternative Diels–Alder retrons and thereby meeting the synthetic challenge. A clear illustration of this approach is a synthetic embroidery of disodium prephenate [Danishefsky, 1977a].

disodium prephenate

The presence in the adduct of two potential leaving groups at the β-positions of the masked ketone is the key to the rapid deconvolution of the desired ring structure. The overriding polarization of the dienophile by the lactone carbonyl (vs sulfinyl group) is also critical.

A method for the synthesis of 4,5-disubstituted 2-cyclohexenones has been developed [Paquette, 1982] which involves the Diels–Alder reaction of alkenyl sulfones and 1-methoxy-3-trimethylsilyloxy-1,3-butadiene and subsequent introduction of the C-4 substituent with the aid of the sulfone.

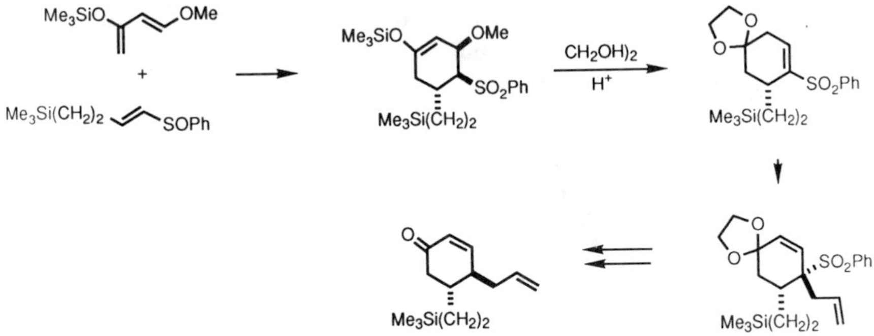

As far as the carbocyclic portion of forskolin is concerned, its most direct assembly by the Diels–Alder method consists of an intramolecular reaction of a terminal methoxylated diene moiety with a butenolide [Hashimoto, 1988]. The adduct contains a double bond in the 6,7-position and the methoxy group can serve as a handle for the necessary transformations. The reaction appeared to proceed asynchronously via an *exo* transition state. The internal bond formation advanced ahead of the terminal bond.

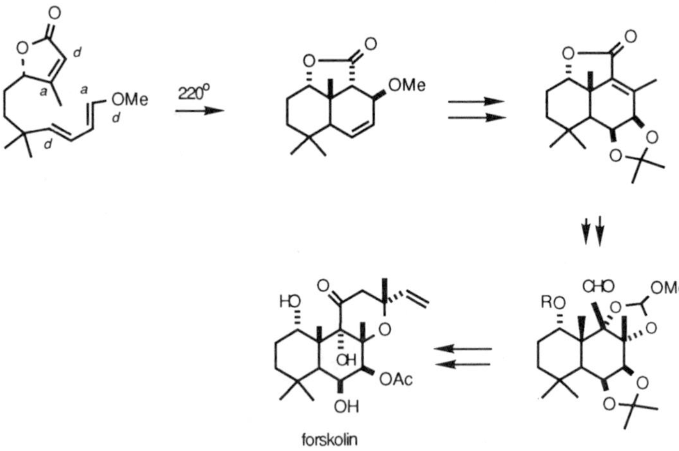

Two syntheses of marasmic acid have been reported. Both employed an intramolecular Diels–Alder reaction to construct the hydrindene portion of the molecule. Interestingly, the conditions required are vastly different, as one proceeded at room temperature via an *endo* transition state [Greenlee, 1976], whereas the other was conducted at 200°C [Boeckman, 1980, 1982] which produced a 1:1 mixture of two isomers. These observations indicate comparable energetics in the *exo* and the *endo* transitions states in the latter case.

It should be noted that in the substrate for the high-temperature reaction there are acceptor groups on the proximal termini of the diene and the dienophile, hence a mismatched dipolar interaction exists.

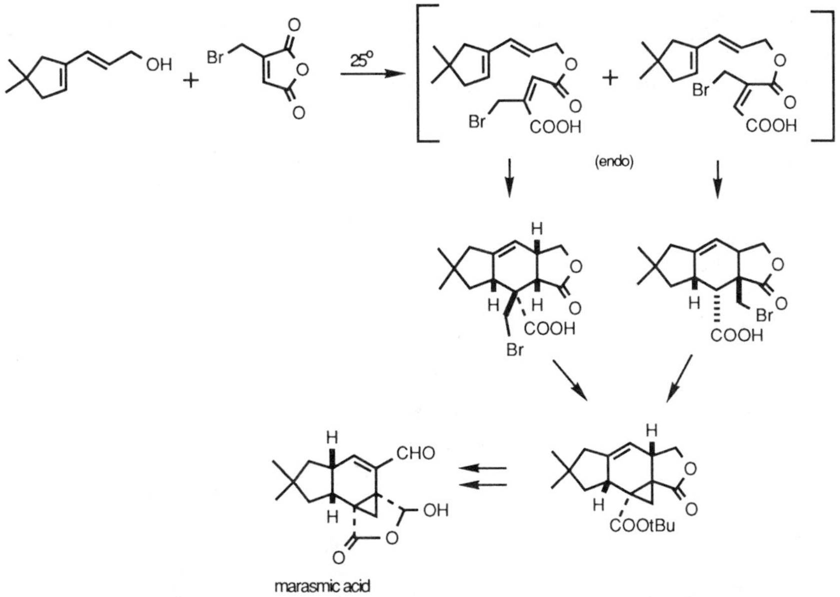

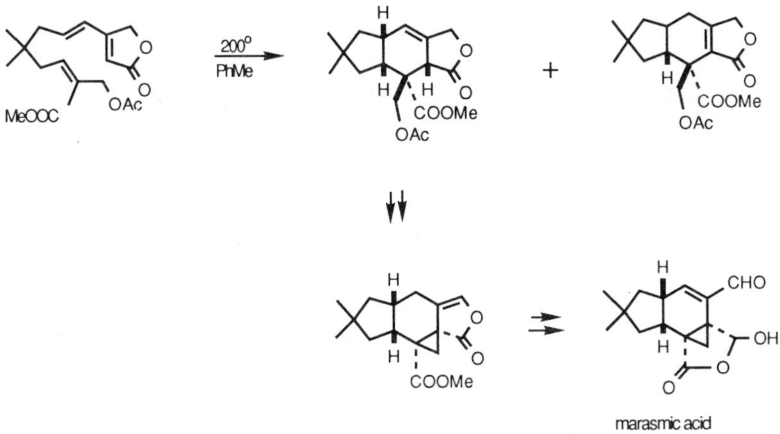

marasmic acid

marasmic acid

In connection with this work a synthesis of lycorine [Boeckman, 1988] must also be assessed.

Lycorine embodies a hydrololilidine framework which is fused to the aromatic portion. This alkaloid with a single double bond in the six-membered carbocycle presents itself as an obvious candidate for a synthesis based on a Diels–Alder reaction. The critical aspects are the construction of a precursor with stereochemically correct substituents. As both the *vic-diol* subunit and the *para*-dihetero substitution pattern in lycorine are disjoint, the Diels–Alder reaction cannot profit from a polarity-matched transition state. Furthermore, the fact that substituents in both the diene and the dienophile components are donors means they do not contribute to a lowering of the activation energy, and this scenario may be reflected in the relatively harsh conditions for the cycloaddition. Countering the electronic malaise is the intramolecularity.

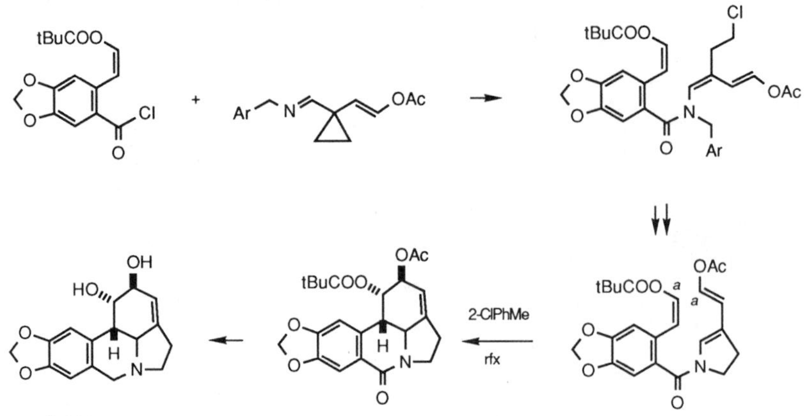

lycorine

Certain marine sponges elaborate amphilectane diterpenoids. The challenging synthesis of one of its members was met by adopting a Diels–Alder approach [Piers, 1989]. This had the merit of placing a double bond between C-12/C-13 that allowed allylic oxidation and eventual introduction of an exocyclic methylene group, as well as stereochemical adjustment at these sites via controlled enone reduction. By choosing acrolein as the dienophile a latent equatorial methyl group was set in the form of the aldehyde.

The pleasing result is that unwanted regioisomers were not isolated. The allylic ether might have exerted a certain influence on the cycloaddition (see polarity matching).

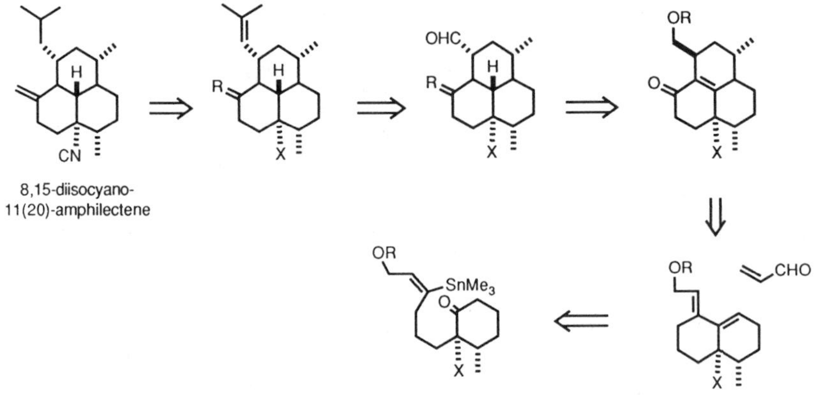

8,15-diisocyano-
11(20)-amphilectene

An earlier synthesis of (+)-7,20-diisocyanoadociane [Corey, 1987a] incorporated two intramolecular Diels–Alder steps. An extra carbon atom in the diene subunit for the assembly of the tetracyclic system had to be excised to provide a means of inverting a ring juncture.

(+)-7,20-diisocyanoadociane

A potential intermediate for aldosterone has been acquired from an intramolecular Diels–Alder reaction of a nascent *o*-quinodimethane [Nemoto, 1985b]. The methoxy substituent in the aromatic ring must have had certain effects on the regioselectivity of the cycloaddition. It is also interesting to note that the 1,3-dithiane moiety served to block the transition state leading to the *cis* C/D isomer, whose formation was found to be favored in the thermolysis of the substrate lacking the dithiane moiety.

10.3. SYNTHESIS OF AROMATIC RINGS

Occasionally it is expedient to create aromatic rings *de novo* during synthesis. The main reason is that certain polysubstituted aromatic substances are not readily available as building blocks. Three examples are discussed here, and a few others will appear in the section dealing with α-pyrones as dienes.

Lunularic acid is a 6-alkylsalicylic acid. The synthesis of this compound is more intriguing when it avoids the use of aromatic precursors. Thus, the route involving *in situ* generation of the salicylic acid moiety via a Diels–Alder/retro-Diels-Alder reaction sequence [Arai, 1972] is intellectually more satisfying.

It is known that the cycloadducts of cyclohexadienes and alkynes lose ethylene at moderate temperatures, the driving force being the aromatization. Accordingly, the use of 1-methoxy-1,3-cyclohexadiene and a proper dienophile leads to a substituted anisole, and with an unsaturated carboxylic acid derivative, the methyl ether of the salicylic acid would result (note the orientation governed by the polar groups of the reaction partners). For the synthesis of lunularic acid the dienophile chosen was dimethyl 2,3-pentadienedioioate. The choice was a tactical one as this substance is more readily accesible than the acetylenic diester. The adduct eliminated ethylene and then rapidly isomerized to dimethyl 3-methoxyhomophthalate.

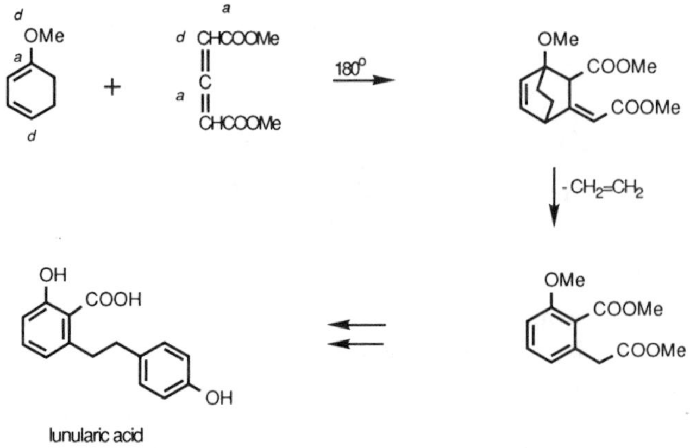

lunularic acid

Reaction of another allene dienophile was featured in an imaginative synthesis of lasiodiplodin [Fink, 1982]. In this case a retro-Diels–Alder elimination is not a prerequisite for aromatization.

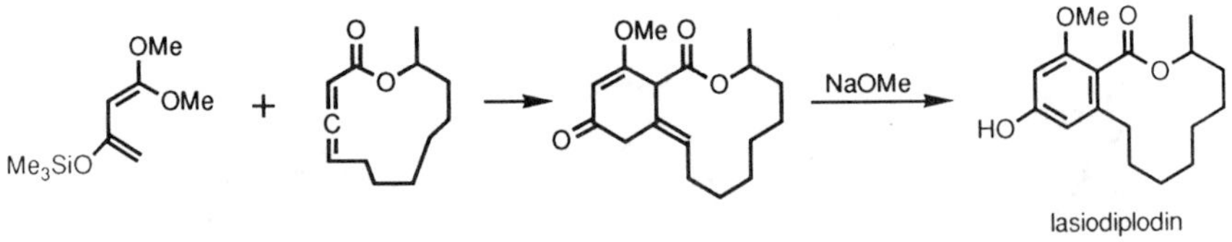

lasiodiplodin

A convenient route to xanthones is via cycloaddition of 1-methoxy-3-trimethylsiloxy-1,3-butadiene with a 3-arylsulfinylbenzo-4-pyrone [Cremins, 1987]. The electronic complementarity of the addends is responsible for the occurrence of the reaction and the facile elimination of methanol from the tricyclic product after the thermal decomposition of the sulfoxide.

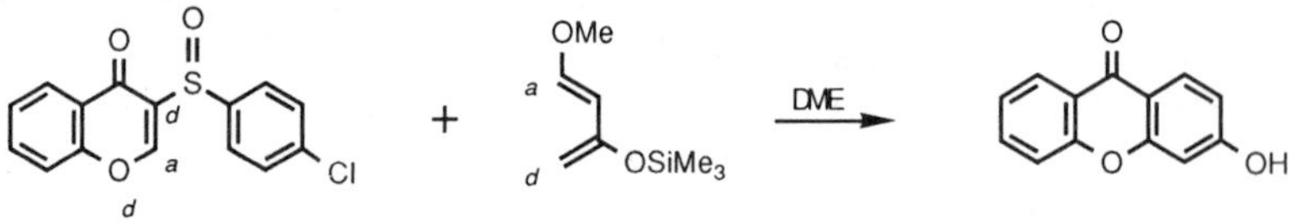

10.4. QUINONES AS DIENOPHILES

The effectiveness of quinones as dienophiles owing to their planar structures and the availability of functionalities for secondary orbital overlap in the *endo* transition state has been indicated above. The regiochemistry of the Diels–Alder reactions for the unsymmetrically substituted quinones depends on the substitution pattern and the electronic nature of the substituent. For example, it is easily understood that 2-acyl-1,4-naphthoquinones are strongly polarized. An excellent exploitation of this arrangement has culminated in a short synthesis of nanaomycin A [Kraus, 1987].

A Diels–Alder reaction of 2-acetyl-8-methoxy-1,4-naphthoquinone with 1-ethoxy-1-(*t*-butyldimethylsilyloxy)-1,3-butadiene was conducted at < 0°C, and the adduct underwent a retro-Claisen fragmentation on treatment with fluoride ion. In this simple manner all the skeletal atoms and functionalities required for the elaboration of the antibiotic molecule were assembled.

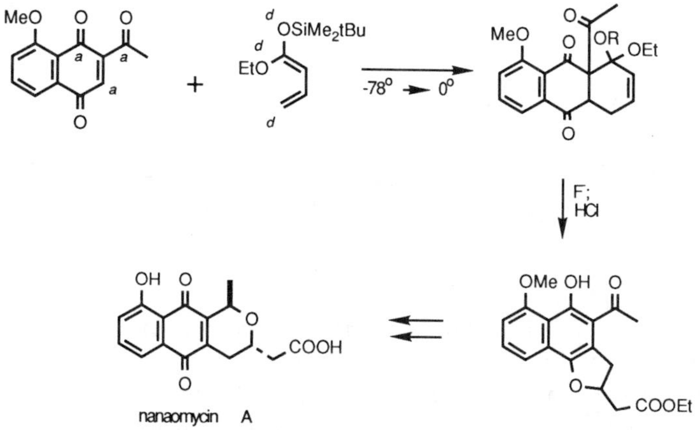

nanaomycin A

Regiocontrol in the Diels–Alder reaction of furanobenzoquinones is exerted mainly by the ethereal oxygen atom. There are essentially two resonance systems. The enolone circuit takes its carbonyl out of a dominant influence on the dienophilic double bond, leaving the other to play the major role in decreeing the orientation. This phenomenon has profound and beneficial consequences in a synthetic approach to ligularone [Bohlmann, 1976; Yamakawa, 1977].

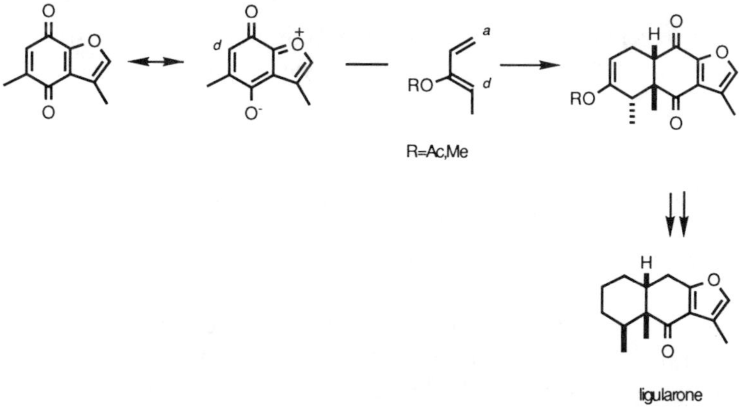

R=Ac,Me

ligularone

Juglone and its derivatives are capable of forming Diels–Alder adducts with electron-rich dienes with high regioselectivity. The internal hydrogen bonding apparently polarizes the juglones and elevates the peri carbonyl to a dominant role [T.R. Kelly, 1977]. The property opens up simple routes to unsymmetrical anthraquinones [Krohn, 1980] and anthracyclinones.

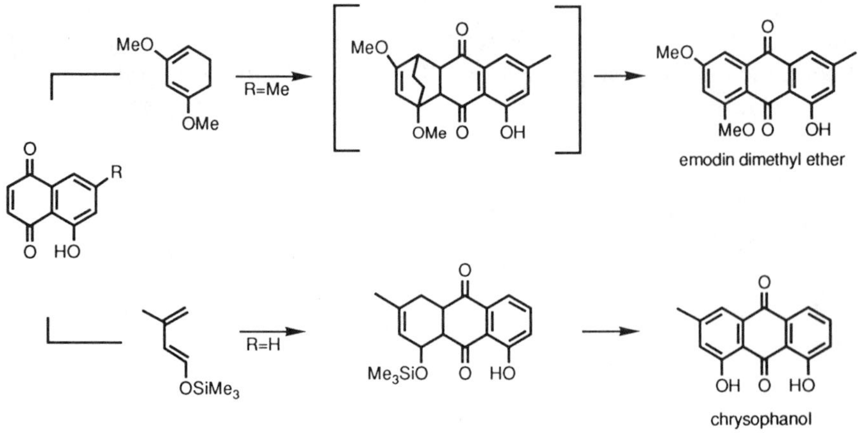

emodin dimethyl ether

chrysophanol

The reactions of naphthopurpurin monomethyl ether with dienes are fascinating [T.R. Kelly, 1978a]. Both carbonyl functions participate in hydrogen bonding, and the control element is consigned to the remote methoxy group. Its effect is manifested in a 3:1 regiopreference in the reaction with 1-methoxy-3-methyl-1,3-butadiene.

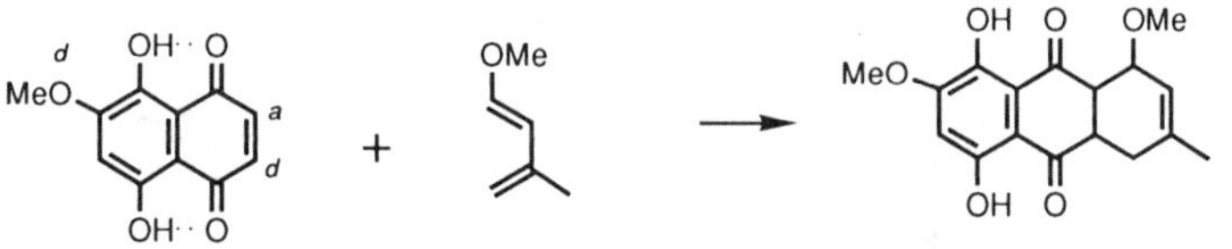

A complementary regiocontrol for the anthraquinone synthesis is possible by using a "neutralizing tactic", e.g., by removing one of the hydrogen bonds.

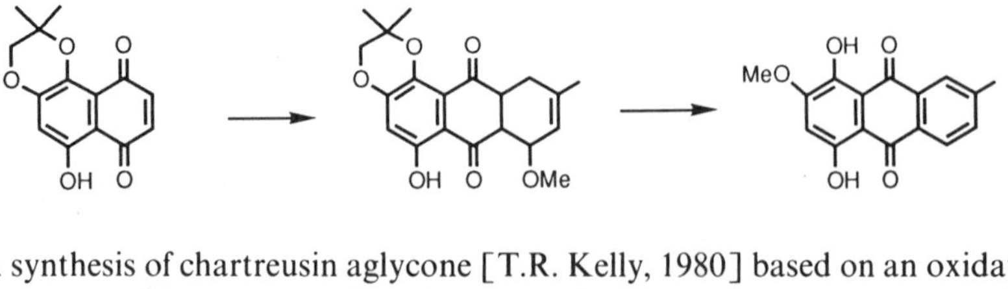

A synthesis of chartreusin aglycone [T.R. Kelly, 1980] based on an oxidative cleavage transform suggests the advantage of a benzanthracene precursor. The latter compound is a potential adduct of juglone and a substituted styrene.

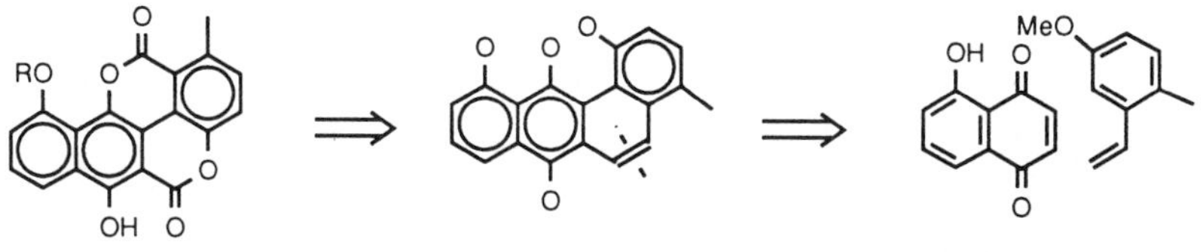

10.5. CATALYZED REACTIONS

The effect of Lewis acids on the Diels–Alder reaction can be quite dramatic. Most often such catalysts exert their influences by complexing the dienophile and thereby increasing the reactivity of the latter species. These reactions are also characterized by an increased stereoselectivity and regioselectivity. Frontier molecular orbital theory offers an explanation of the phenomenon.

While the thermal reaction of 2-methoxy-5-methyl-1,4-benzoquinone with isoprene gives a 1:1 mixture of two adducts (both of which have the methyl group at an angular position), the BF_3-catalyzed and the $SnCl_4$-catalyzed reactions are much more regioselective [Tou, 1980]. Interestingly, the two reactions exhibit opposite results, the reason being that the tetracoordinate boron tends to complex with the least hindered carbonyl of the dienophile, and the tin atom prefers chelation with a bidentate ligand. The coordination polarizes the C=C bond in one direction or the other.

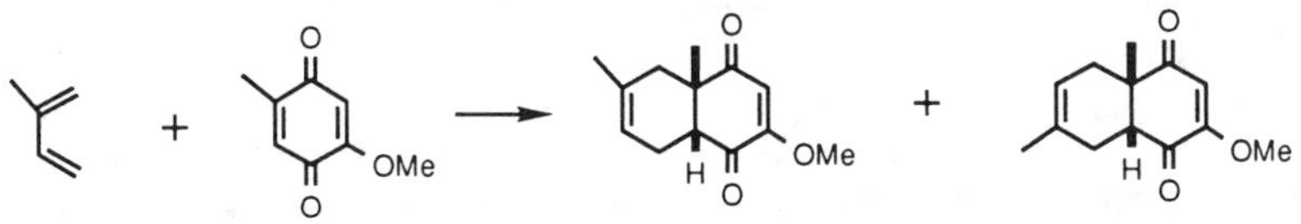

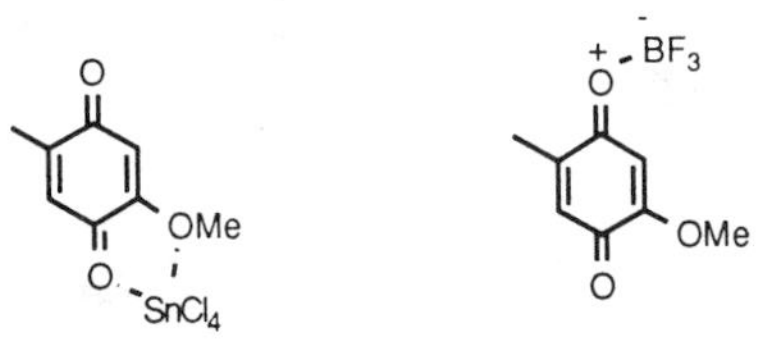

A specific application of these very useful observations is a synthesis of butyrospermol [Kolaczkowski, 1985]. However, the synthesis of estrone [Dickinson, 1972] starting from a BF_3-catalyzed reaction of 2,6-dimethyl-benzoquinone and 6-methoxy-1-vinyl-2,4-dihydrotetralin defies explanation regarding its regioselectivity.

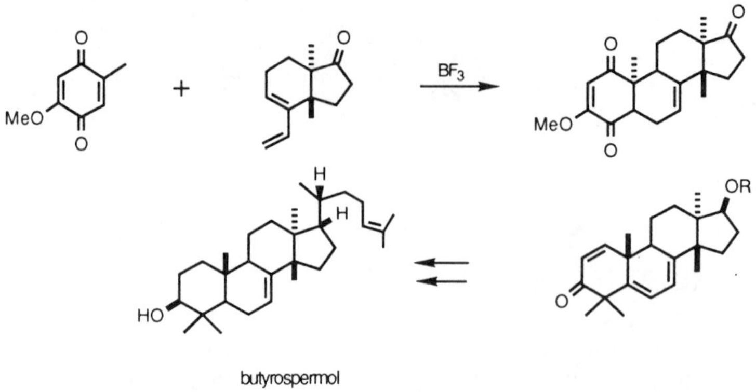

butyrospermol

On the other hand, the formation of a precursor of altersolanol B [T.R. Kelly, 1978b] by a triacetoxyborane-catalyzed Diels–Alder reaction is readily understood.

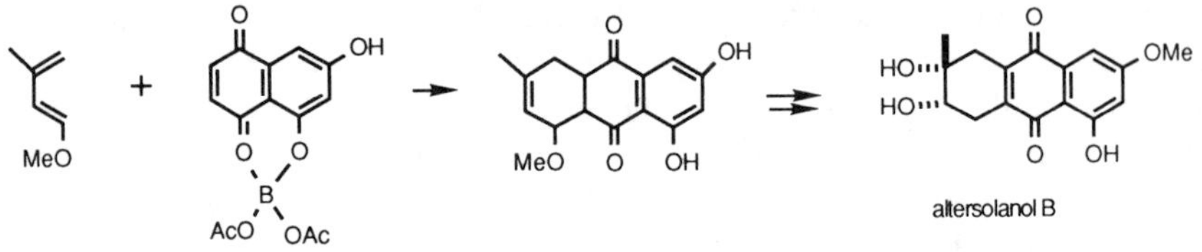

altersolanol B

The catalyzed cycloaddition of 2-carbomethoxy-4,4-dimethyl-2-cyclohexenone with isoprene apparently proceeds from a transition state in which the diene is *endo* to the ester group [Liu, 1988]. The alternative transition state is disfavored sterically. It is significant that in the corresponding reaction with the dehydro analogue of the dienophile, i.e., the cross-conjugated dienone, the regioselectivity is decreased. The preference for the ketone-*endo* transition state leads to a sacrifice in regiocontrol, because the *meta* cycloaddition suffers less steric repulsions, the methyl groups of the reaction partners being farther apart.

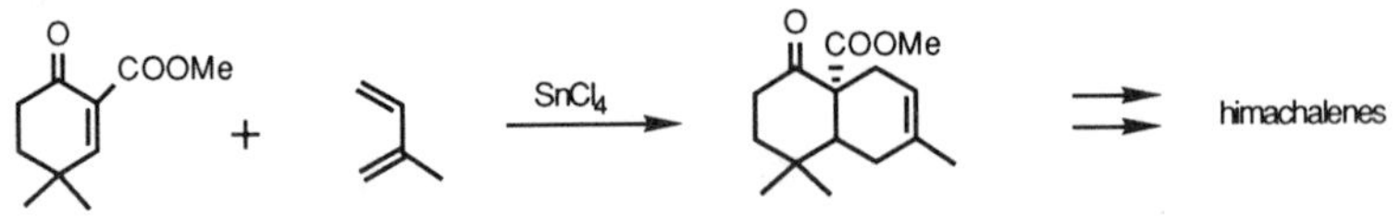

himachalenes

10.6. HIGHLY POLARIZED CYCLOADDITIONS

10.6.1. Cycloadditions Involving α-Pyrones and Related Dienes

The Diels–Alder adducts of α-pyrones often lose carbon dioxide, resulting in cyclohexadienes. An astute application of the process is in a synthesis of occidentalol [Watt, 1972]. The most interesting aspect of the reaction involving the methyl 2-oxopyran-3-carboxylate and 4-methyl-3-cyclohexenone is its regioselectivity. It has been suggested that the carbonyl group of the dienophile exerts its effect (perhaps via the enol form) on the double bond. This electronic transmission designates C-3 as an acceptor site and C-4 as a donor.

occidentalol

In a colchicine synthesis [J. Schreiber, 1961] featuring a Diels–Alder reaction between an α-pyrone and α-chloromethylmaleic anhydride to build the tropolone ring, strong preference for one isomer was observed in the cycloaddition.

It seems that the major adduct should be the less favorable one, considering the severer steric interactions in its transition state. Perhaps the chloromethyl substituent of the dienophile exerted a sufficiently strong electronic effect on the orientation of the two addends.

colchicine

Recent developments in the α-pyrone cycloaddition study have included the use of α-sulfinylpyrones with electron-rich dienophiles. The matching cycloaddition mode leads to versatile synthetic intermediates such as chorismic acid [Posner, 1987]. In this case phenyl vinyl sulfide was employed as the dienophile. Most notably the valuable oxygen functions of the lactone ring were retained to become part of the target molecule, and the pre-existing sulfoxide group of the adduct served to introduce stereospecifically the allylic oxygen function and the conjugate double bond. The *endo* sulfide remained as a latent double bond until the latter was needed.

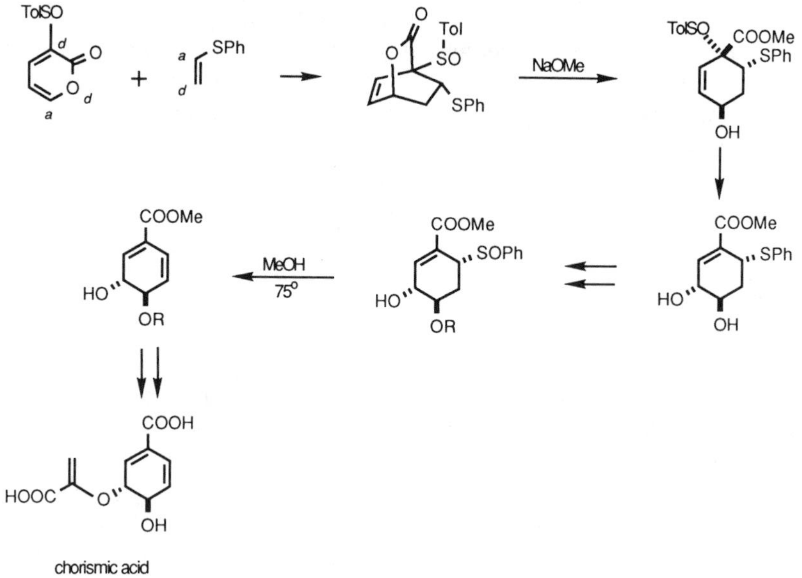

A fully functionalized precursor for the bottom half of ivermectin has been assembled from a pyrone and 3,4-dibenzyloxyfuran [Jung, 1987]. The direct Diels–Alder reaction of an alkoxy furanone would not be fruitful, as at best a mixture of regioisomers would result. While the effects of the two ethereal oxygen atoms may cancel one another, the net polar influence of the dienophile reactivity is the ketone, and the "wrong" isomer will predominate. Masking the ketone group in the form of an enol ether removes its adverse effect.

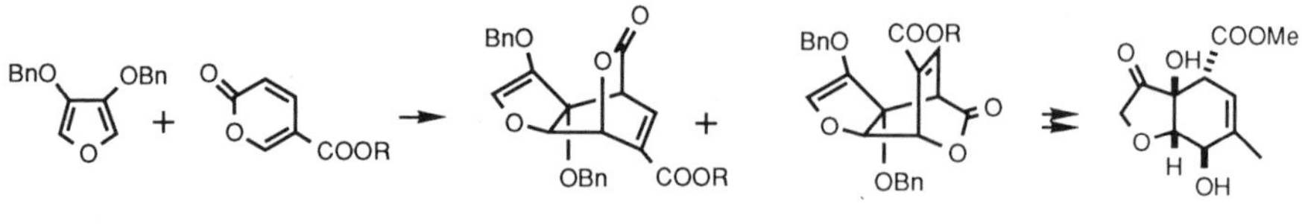

(88%:exo:endo = 47:53)

Most frequently cycloadducts derived from α-pyrones are used as predecessors of aromatic componds. For example, in an approach to lasalocid [Ireland, 1980c] an aldehyde was synthesized according to this reaction sequence.

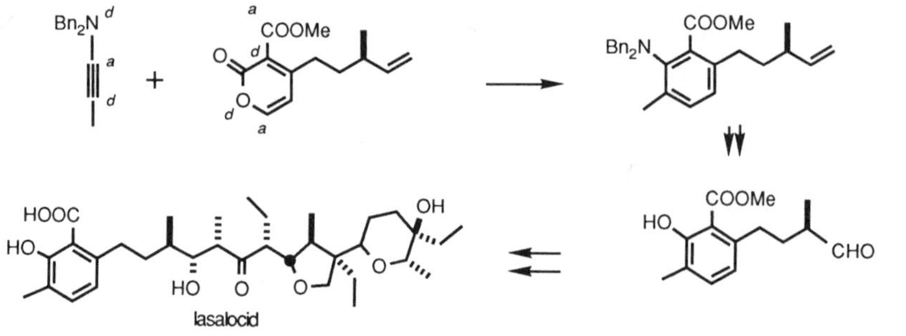

lasalocid

An annulated α-pyrone has also been used in a synthesis of juncusol [Boger, 1982]. Its merit is again the regioselectivity associated with the cycloaddition step.

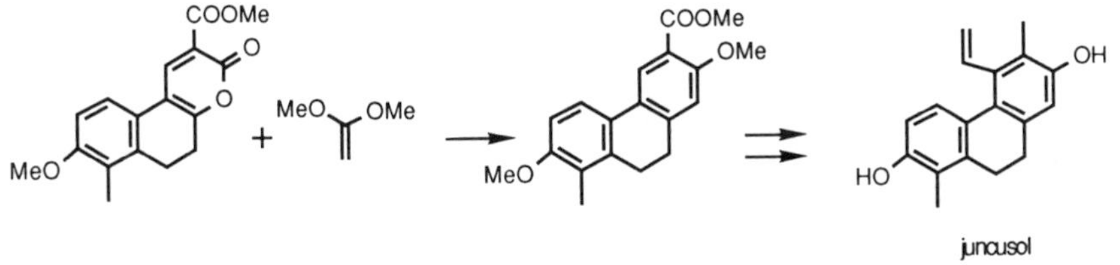

Dihydroaromatics can be obtained from the reaction of the Fischer carbene complex analogs of α-pyrones [Wang, 1990] owing to the exceptionally facile cycloaddition and extrusion of the metal carbonyl fragment (vs CO_2 elimination) from the initial cycloadducts.

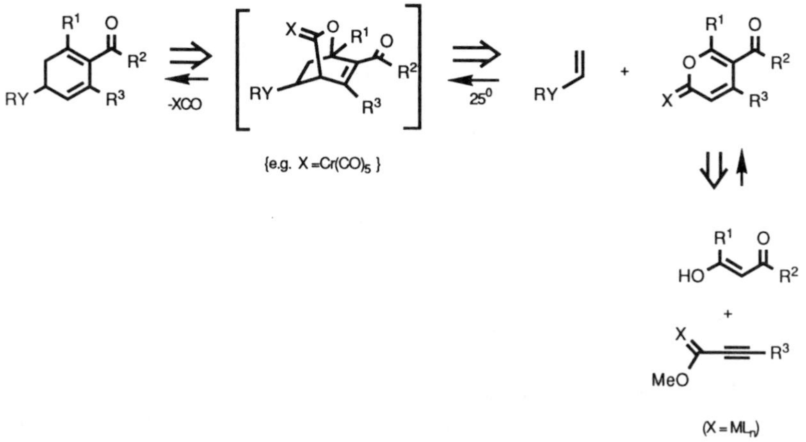

Previously the regioselective Diels–Alder reactions of juglone were discussed. The desirable feature can be coupled to the chemistry of α-pyrones. Thus, a route to chrysophanol [Jung, 1978] demonstrated an electronic control which is very beneficial to synthesis.

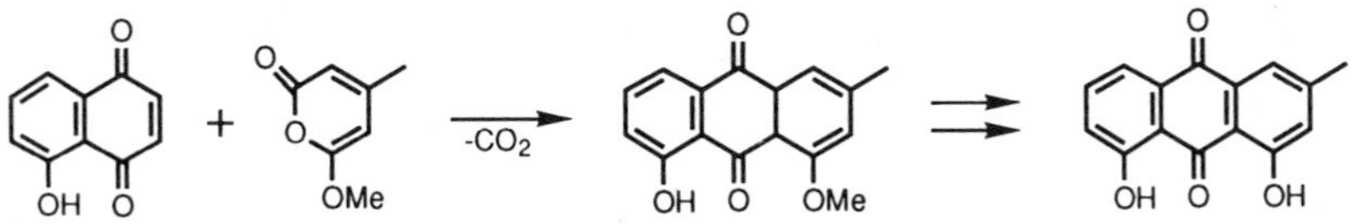

Pyrano[3,4-b]indol-3-ones undergo Diels–Alder reactions with ease. However, the regiochemistry of the reaction is hard to predict because the nitrogen atom and the carbonyl group are disjoint, i.e., there is a competition between the enamine system and the pyrone. Frontier molecular orbital theory predictions are not always valid in this area.

The reaction of the 1-methyl derivative with diethyl mesoxalate gives one type of product [Pindur, 1989], whereas its reaction with α-chloroacrylonitrile leads to a product with an opposite orientation [Narasimhan, 1985]. The latter is a carbazole that has been converted into olivacine.

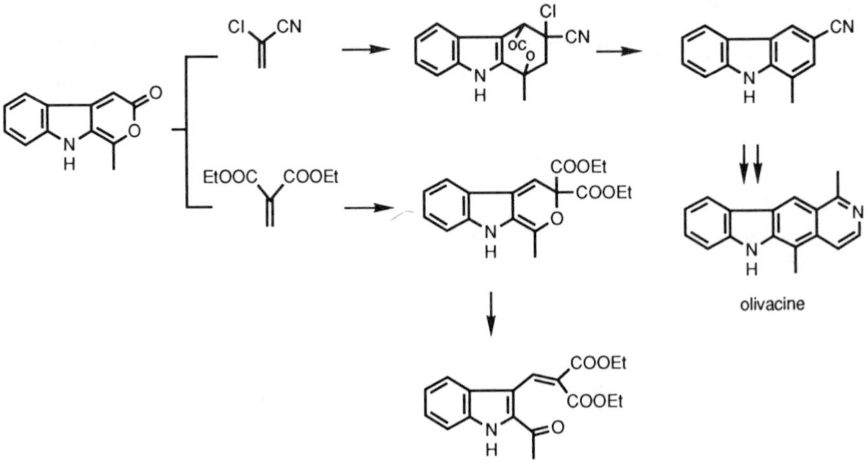

olivacine

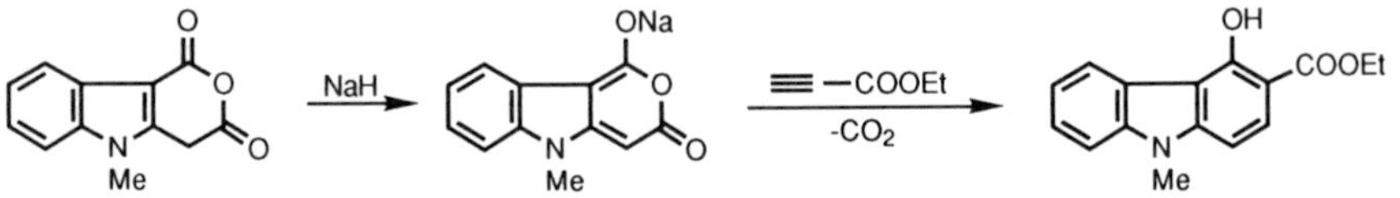

Cycloaddition routes to ellipticine generally suffer from a lack of regiocontrol. An apparent solution to this problem is the use of a lactam as the dienophile [Davis, 1990]. When polarized with a strong silylating agent the lactam elicited attack by 1,3-dimethyl-4-(phenylsulfonyl)-4H-furo[3,4-b]indole with its "bottom carbon" which is also part of a sulfonyl enamine system.

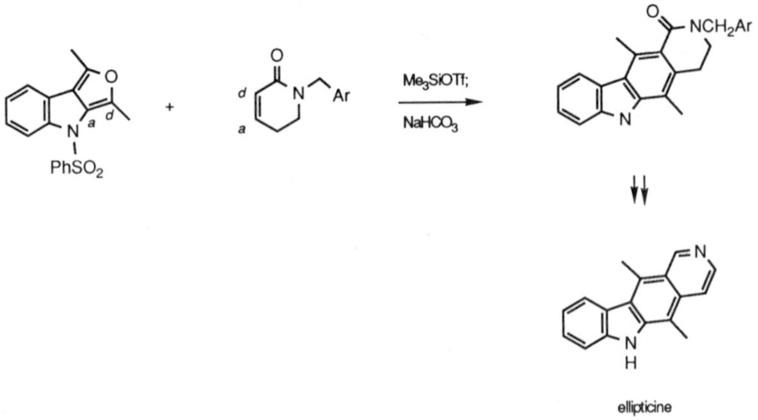

ellipticine

Although the regioselective [4 + 2] cycloaddition of 2-phenylsulfonyl-1,3-dienes to the magnesium salt of indole is not really a Diels–Alder reaction, its discussion here is quite appropriate. As a result of polarity control this reaction provides an entrance key to the carbazoles of appointed substitution patterns. The usefulness of the reaction is seen in its application to a direct approach to ellipticine and olivacine [Bäckvall, 1990].

ellipticine (R = Me, R' = H)
olivacine   (R = H, R' = Me)

X = SO₂Ph

X = SO$_2$Ph

10.6.2. α-(2-Indolyl)acrylic Esters as Dienes

α-(2-Indolyl)acrylic esters are highly reactive toward such dienophiles as enamines. The two termini of the diene system have prominent donor and acceptor characteristics owing to the individual conjugation with the nitrogen atom and the ester group, respectively. Under certain circumstances the cycloaddition could become nonconcerted.

Synthesis of *Aspidosperma* alkaloids by this efficient mode of cycloaddition has been inspired by biosynthetic considerations. Current theories suggest secodine as an *in vivo* precursor, whose intramolecular Diels–Alder-type reaction is the key process for the elaboration of such alkaloids.

A synthesis of minovine [Ziegler, 1973b] was pursued along this line, but using an intermolecular Diels–Alder reaction which proceeded according to polarity matching. The isolated product has the indole framework, because this isomer is aromatic and easily accessible by double bond migration.

minovine

A biomimetic approach to the pentacyclic alkaloid skeleton by an intramolecular version has also been realized [Kalaus, 1985].

A formal retro-Diels–Alder/Diels–Alder process transformed Δ^{18}-tabersonine into andranginine [Andriamialisoa, 1975]. However, the reaction must be stepwise as 21-epiandranginine was also obtained, together with other products.

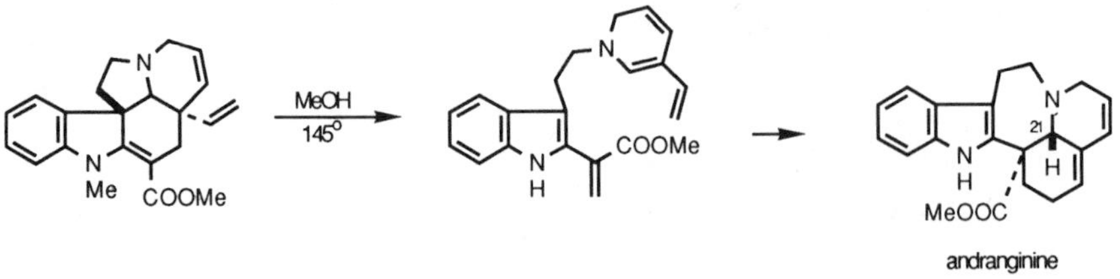

andranginine

The indolylacrylic ester may act as a dienophile when the 3-position of the indole nucleus is fully substituted. This behavior has enabled a synthesis of catharanthine [Marazano, 1981] from a Diels–Alder adduct of a dihydropyridine.

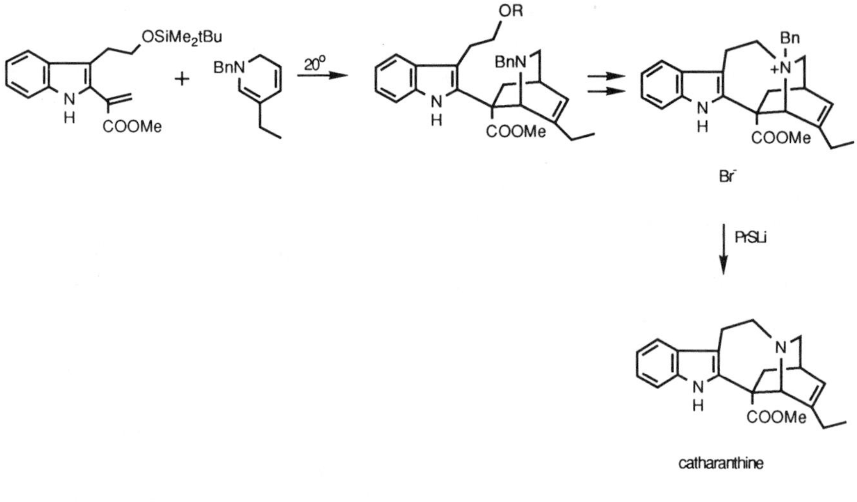

catharanthine

10.7. REACTIONS OF AZA DIENES AND DIENOPHILES

The early use of dihydropyridines as dienes in the synthesis of indole alkaloids may be traced to a synthesis of ibogamine [Büchi, 1966a]. A dihydronicotinonitrile was condensed with methyl vinyl ketone to provide an azabicyclo[2.2.2]octene which was then linked to a 2-(β-indolyl)ethyl chain after proper modification. The orientation of the cycloaddition was of course controlled by the polar groups of the reactants. The late stages of the synthesis were devoted to cyclization toward the α-position of the indole ring and to functionality adjustment.

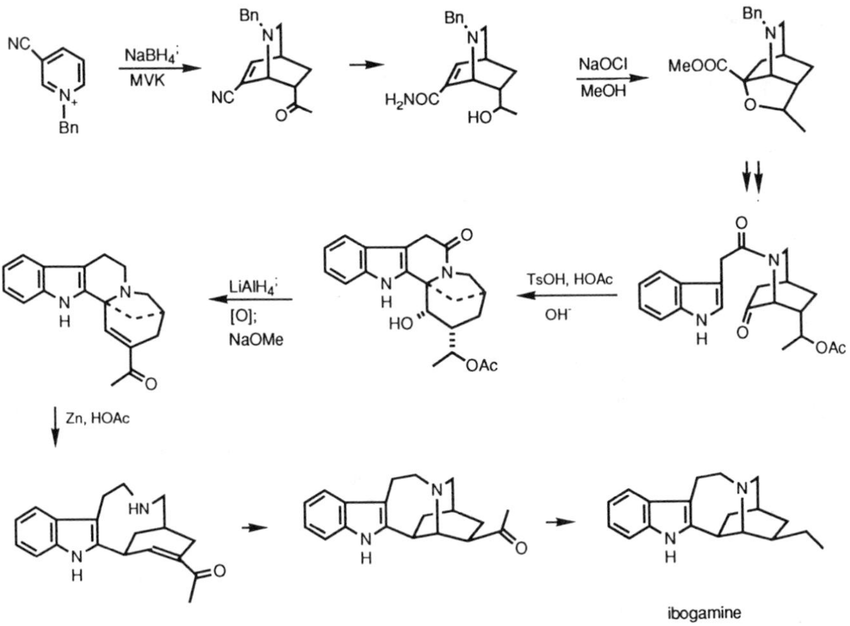

ibogamine

Dihydropyridines such as those involved in the above synthesis are activated dienes. Azadienes in which the nitrogen atom is part of the conjugate system are also known to participate in cycloadditions. Such reactions have great potential in the synthesis of piperidine derivatives and their intramolecular versions are particularly useful for assembling polycyclic substances. While the following example of gyphyrotoxin synthesis [Ito, 1983] has no implication in polar substituent effects, its presentation may demonstrate the power of the reaction. It is expected that such a reaction is also subject to polarity control and efficient approaches to more highly functionalized heterocycles may be tailored accordingly.

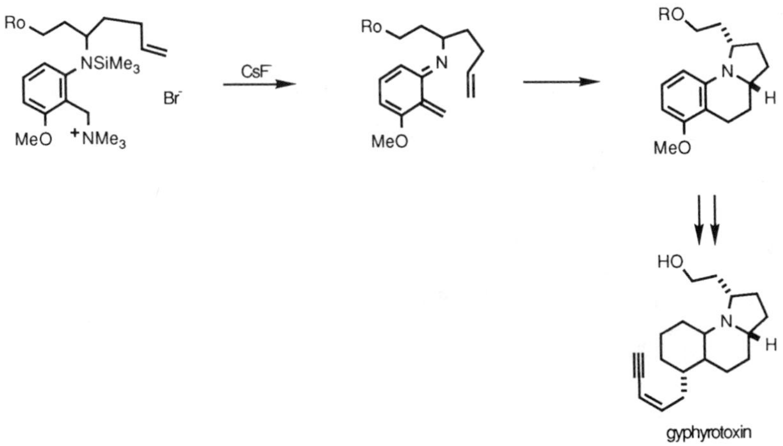

gyphyrotoxin

A simple construction of epilupinine [Ihara, 1985] serves to illustrate the principle. Note that the siloxy-azadiene matches the electron demand of the unsaturated ester moiety.

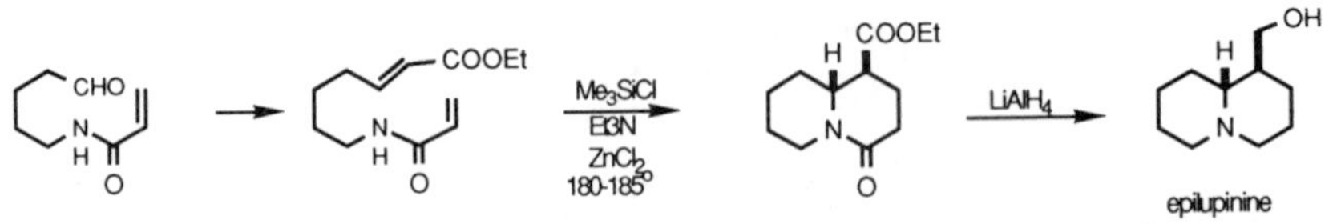

Isoquinolinium and acridinium species are reactive toward electron-rich dienophiles. A convegent synthesis of cryptosporin [Gupta, 1989] that employed a glycal is fully matched with the 5-alkoxyisoquinolium salt. The adduct was hydrolyzed to an amino aldehyde and rational degradation and aromatization of the latter proceeded well. The enantiomer of the natural naphthoquinone that emerged from this work was predicated on the absolute configuration of the glycal.

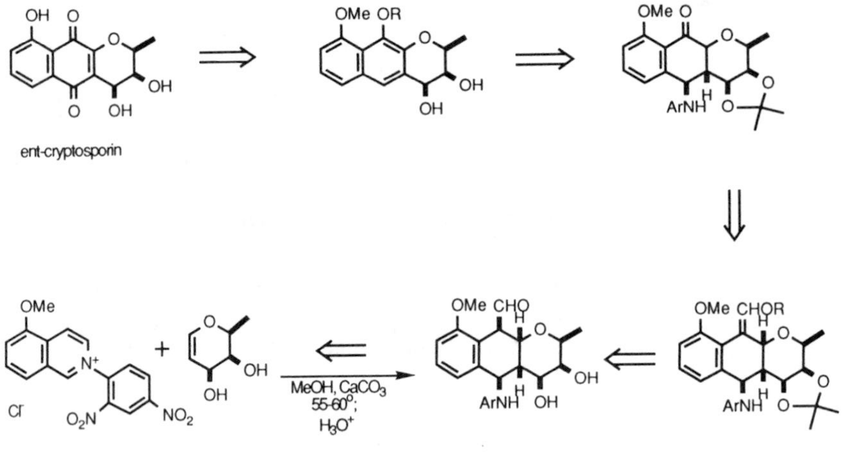

ent-cryptosporin

Imines including 3,4-dihydroisoquinolines and 3,4-dihydro-β-carbolines react with electron-poor dienes with the expected regiochemistry. For example, a synthesis of xylopinine [Kametani, 1975] involved the interception of a cyano-o-xylylene. The cyano group of the adduct was removed reductively.

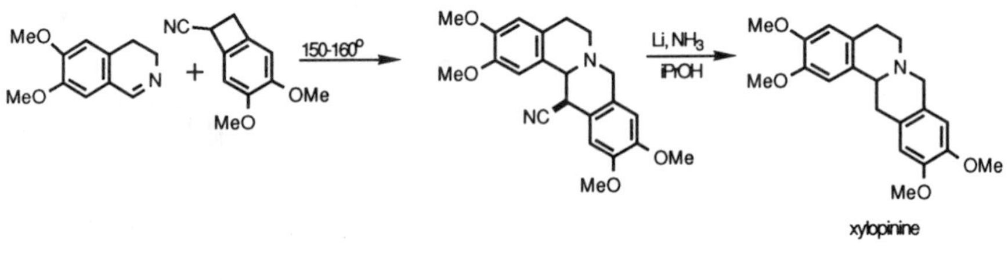

xylopinine

A most straightforward assembly of the tetracyclic intermediate of vincamine [Langlois, 1983] called for the condensation of a dihydro-β-carboline with 2,4-pentadienonitrile. The cyano group of the adduct was necessary for the subsequent ethylation and chain extension to build the final ring.

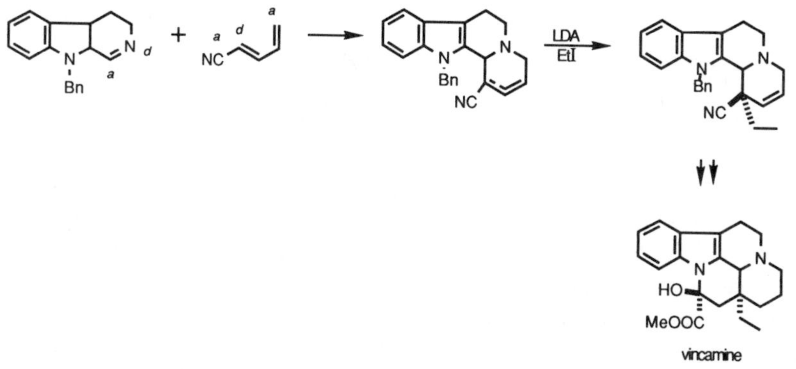

Methyl 2,4-pentadienoate may be used instead of the nitrile. It is noted that this cycloadduct contains a double bond in a position corresponding to that of vindorosine. Advantage of this process was therefore taken to gain access to the alkaloid [Andriamialisoa, 1985]. The skeletal rearrangement of the fused ring system to that incorporating a spirocyclic array has been documented.

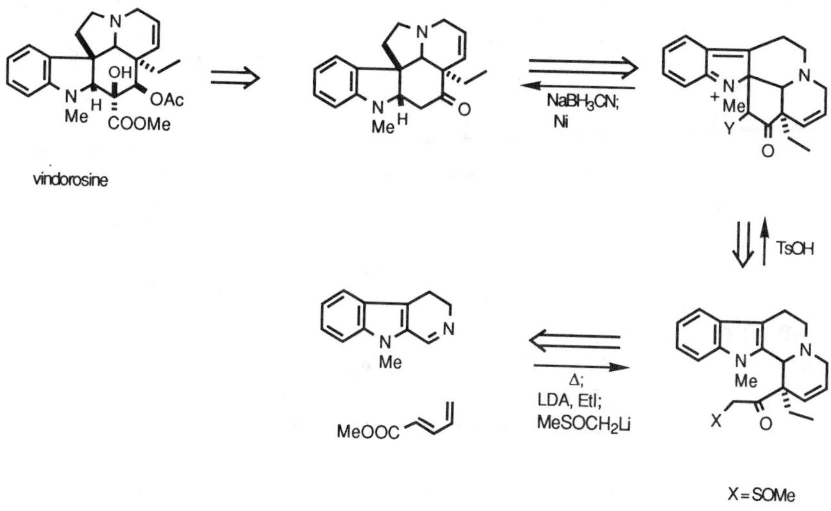

Among other imino Diels–Alder approaches to alkaloids may be mentioned the syntheses of lupinine [Bohlmann, 1967] and rutaecarpine [Kametani, 1976], although in the latter case the cycloaddition might be stepwise (not involving the imino ketene).

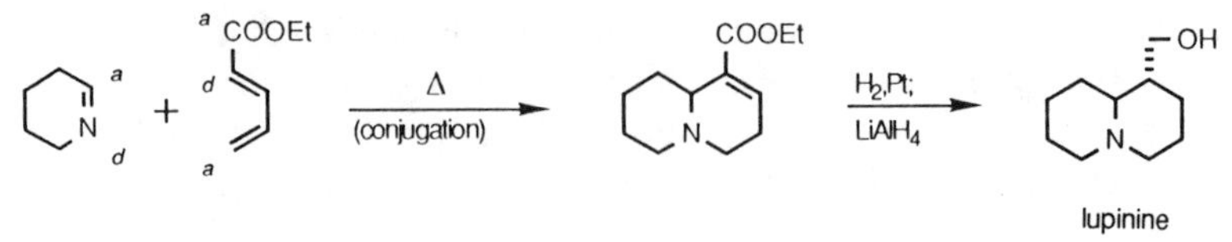

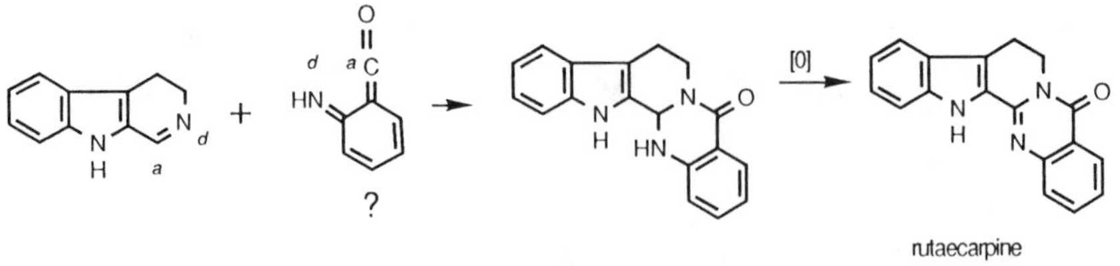

lupinine

?

rutaecarpine

Although there is a full Diels–Alder retron in lysergic acid the direct approach does not seem to be meritorious for reactivity reasons. An intramolecular version would generate a double-bond isomer but the isomerization to the styrenic chromophore is not difficult. Such a intramolecular approach [Oppolzer, 1981] is even more attractive in view of the simultaneous formation of two rings and the simplified preparation of the substrate.

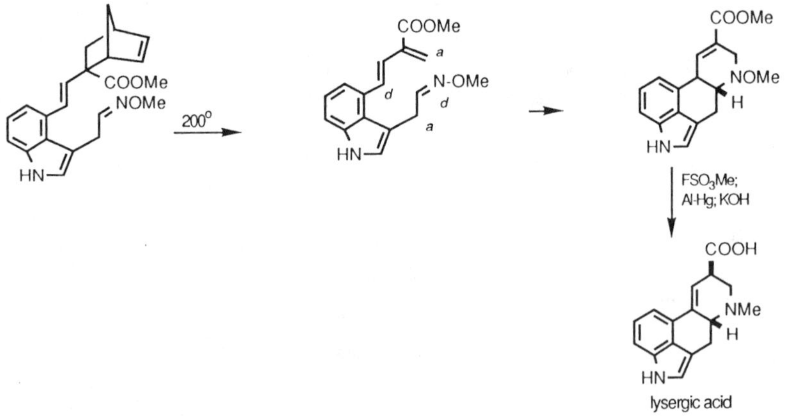

lysergic acid

In connection with a total synthesis of streptonigrin the Diels–Alder reaction of 1,2,4-triazines was extensively investigated as a method of the pyridine C-ring construction. The regioselectivity for the cycloadditions with enamines is strikingly different between the parent heterocycle and the 3-ester [Boger, 1989]. Due to the inverse electron demand nature of the cycloaddition the reactivity of the 3,5,6-tricarboxylic ester is greater than the parent triazine.

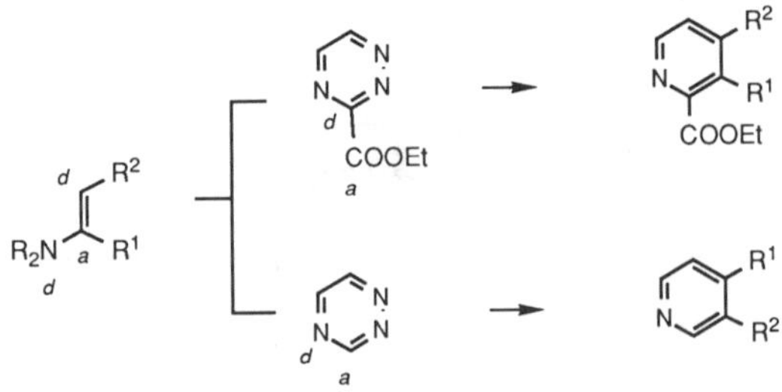

There has been intense research in the asymmetric Diels–Alder reaction. A piperidine synthesis that deserves mention is that of (S)-anabasine [Pflengle, 1989]. Here, an imine derived from N-(tetra-O-pivaloyl-β-D-galactopyranosyl)amine and 3-pyridinealdehyde was coupled with 1-methoxy-3-trimethylsiloxy-1,3-diene in the presence of zinc chloride to give an optically active adduct. A simple reaction sequence elaborated the alkaloid with good recovery of the chiral auxiliary.

(S)-anabasine

The heterocycle in the final product is devoid of any other functionality and therefore the regiochemistry of the cycloaddition is of no consequence. Nevertheless, the efficiency in asymmetric induction has everything to do with an ordered, polarity-matched transition state.

10.8. REACTIONS OF OTHER HETERO DIENES AND DIENOPHILES

Oxomalonic ester react with dienes to form dihydropyrans, therefore it is not really surprising that an oxalylpyrimidone opts for this mode of reaction although the dienophile contains a C=C bond which could also participate in the cycloaddition [Keana, 1976]. It was with the normal Diels–Alder reaction in mind as the first step toward a synthesis of tetrodotoxin that the unexpected reaction was identified and it forced the abandonment of this approach.

One of the most exciting advances in the realm of the Diels-Alder reaction is the catalyzed condensation of nonactivated aldehydes with siloxydienes [Danishefsky, 1989a]. Contrary to the all-carbon version, these dienes escape destruction by the Lewis acids, probably because the aldehydes are more effective sequesterers of the catalysts.

The versatility of the method in synthesizing a variety of pyrans has been ascertained. Many aldehydes have been submitted to the condensation which proves to be stereoselective. Besides common Lewis acids such as BF_3, $SnCl_4$, and $TiCl_4$, the lanthanides are also effective. Moreover, the milder conditions with the latter catalysts enable preservation of all the sensitive functionality in the adducts.

Notable chemoselectivity was witnessed in a synthesis of β-methyl lincosamide [Danishefsky, 1985]. The cyclocondensation with the aldehyde group took precedence over the all-carbon mode which predominated in the absence of Lewis acids.

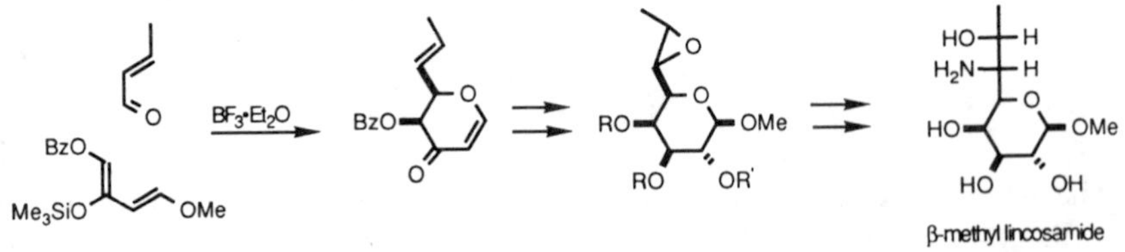

β-methyl lincosamide

Asymmetric induction by α-substituents of the aldehyde is routinely observed. Impressive remote asymmetric induction has been discovered during a total synthesis of compactin [Danishefsky, 1988]. (An even higher selectivity (10:1) was reported in an analogous approach to mevinolin [Wovkulich, 1989].)

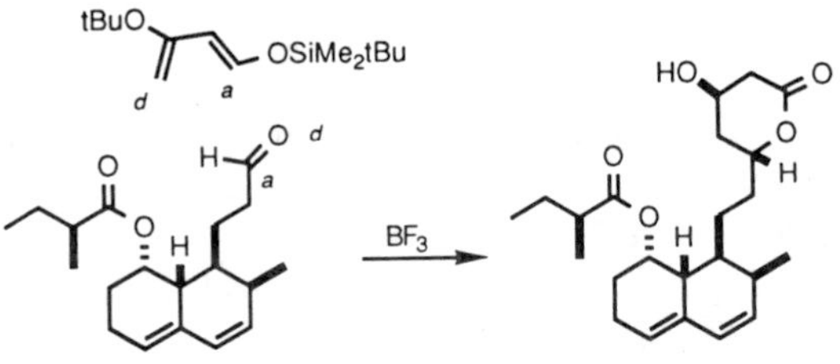

compactin

The less obvious applications are exemplified by polypropionate synthesis. Thus, the access to zincophorin [Danishefsky, 1987] using two different cyclocondensation steps is very enlightening.

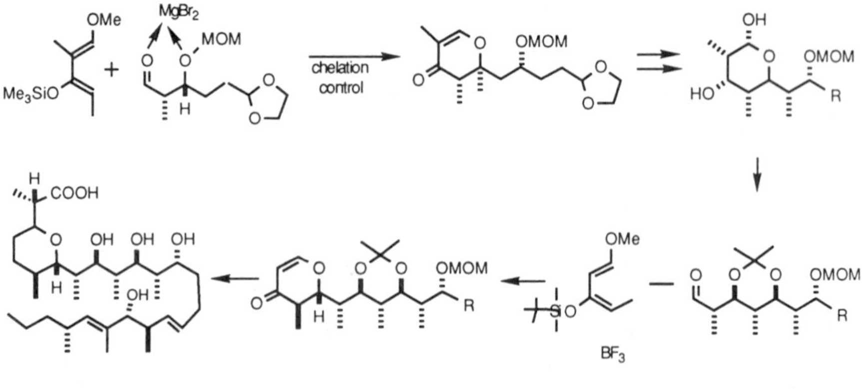

zincophorin

There are quite a few examples of heterodiene/heterodienophile condensations. A simple synthesis of the Norway spruce beetle aggregation pheromone chalcogran [Ireland, 1980b] started from the preparation of a spiroketal from acrolein and 2-methylenetetrahydrofuan.

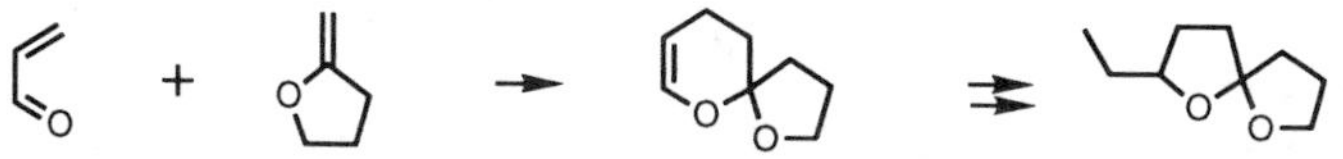

The cross-condensation of α-methylenecycloalkanones with acrylic esters appears to proceed via a misaligned transition state, as far as polar substituent matchings are concerned. These reactions may be controlled by frontier orbitals; through-space interactions of the heteroatoms may also contribute to the observed results. Nevertheless, very significant applications of this process have been reported, among them is a synthesis of aphidicolin [Ireland, 1984].

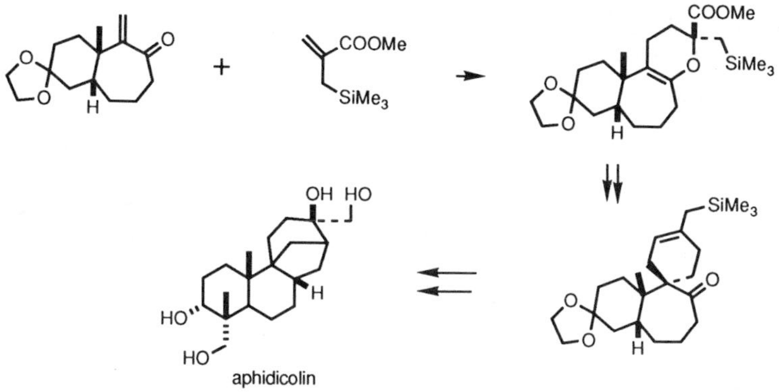

aphidicolin

Recently, macrolide and polyether antibiotics have become popular targets for synthesis. The polyethers often contain spiroketal subunits whose formation is subject to stereoelectronic control. This latter aspect can be exploited in synthesis. For example, stereorational introduction and manipulation of substituents in a carbon chain is rendered possible when the chain can fold into a rigid spiroketal.

In this context a synthesis of methynolide [Ireland, 1983], the aglycone of a macrolide antibiotic, from a carbohydrate derivative which passes through spiroketal intermediates is very engaging.

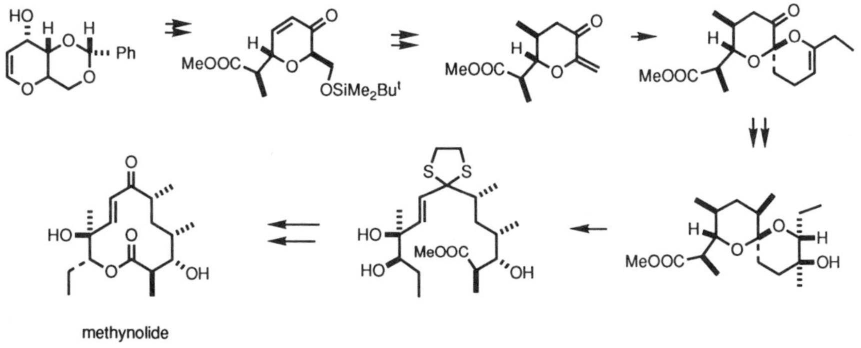

methynolide

The first spiroketal in this synthesis emerged from a Diels–Alder reaction as the major diastereoisomer (ratio 74:26). Other regioisomers apparently did not form.

Although this Diels–Alder reaction does not enjoy an ideal electronic match of the two components, the ethereal oxygen of the dienophile definitely helps to counterbalance the influence of the carbonyl group.

Acylketenes have a highly polarized structure and they undergo Diels–Alder reactions with enol ethers to provide promising precursors of deoxy sugars [Coleman, 1990].

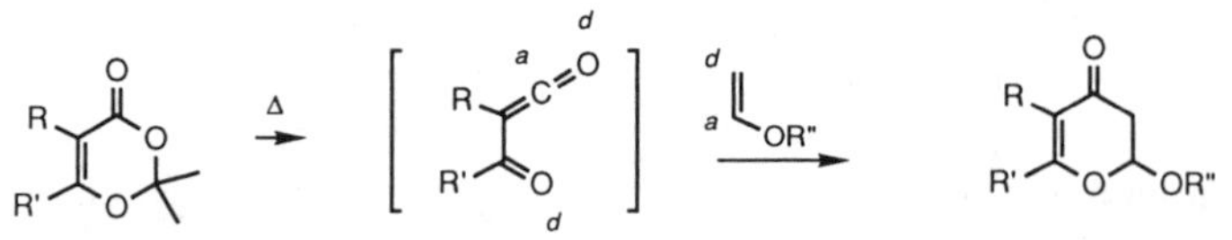

Finally, it should be noted that the remarkable synthesis of carpanone [Chapman, 1971] by an oxidative dimerization of an *o*-(1-propenyl)phenol involved an intramolecular Diels–Alder reaction.

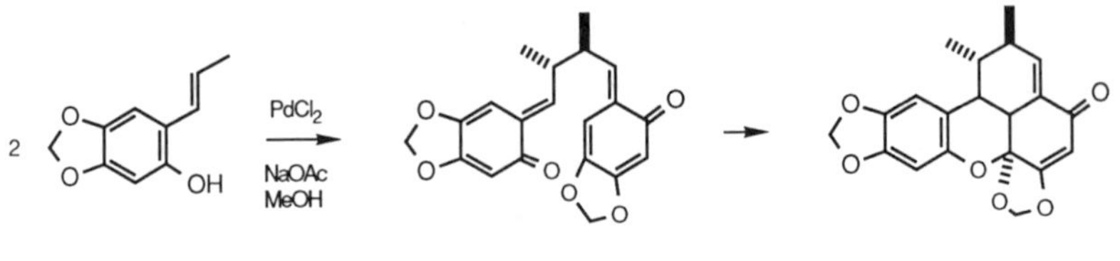

carpanone

11

OTHER CYCLOADDITION REACTIONS

11.1. CYCLOADDITION OF KETENES AND ISOCYANATES

Ketenes are capable of participation in a thermal [2 + 2] cycloaddition in an antarafacial fashion. In view of the scarcity of any methodology in the formation of functionalized cyclobutanes, and the simplicity of ketene generation by dehalogenation of α-halocarboxylic chlorides or dehydrohalogenation of acid chlorides with tertiary amine bases, this cycloaddition is very valuable. It is fortunate that the regiochemistry of such cycloaddition is subject to control by polar substituents. Thus, the regioselectivity of the ketene cycloaddition with simple allylsilanes [Brady, 1977] is in excellent agreement with the prediction on the basis of the polarity alternation rule. The inertness of vinylsilanes to undergo a similar cycloaddition is also understandable.

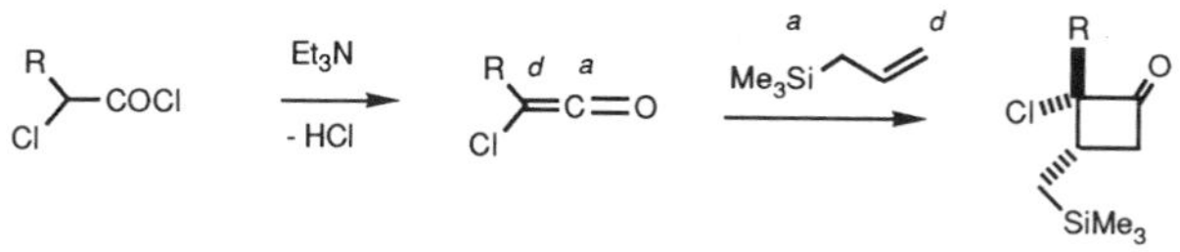

A silyl substituent at the α-position of a ketene tends to favor its formation through stabilization by polarity alternation. A dramatic increase in the yield of a bicyclic cyclobutanone from a thermal rearrangement [A. Alder, 1983] has been witnessed.

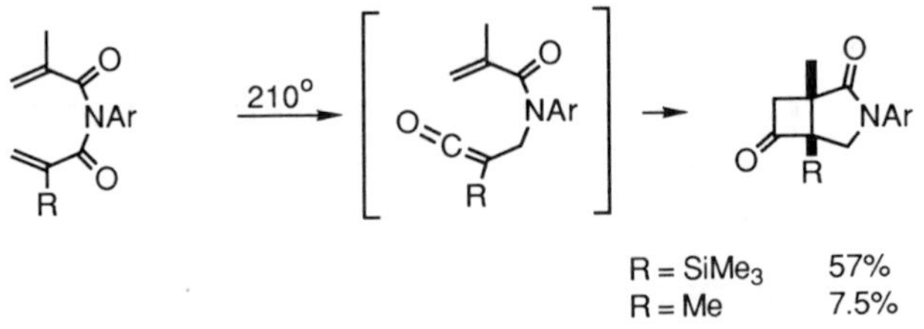

R = SiMe₃   57%
R = Me   7.5%

$R = SiMe_3$ 57%
$R = Me$ 7.5%

Two syntheses of eriolanin relied on annulation with dichloroketene for the establishment of the γ-lactone subunit. Interestingly, due to the difference in the substitution pattern of the cycloalkene addends, totally opposite regioselectivity was observed.

From a cyclohexadiene the cyclobutanone adduct contains a direct Baeyer–Villiger/dechlorination retron of the γ-lactone ring [Grieco, 1978]. On the other hand, an allylic methoxy group directed the cycloaddition in a contrasting manner [Wakamatsu, 1988]. As indicated in the formulas, the donor–acceptor characteristics of the two trigonal carbon atoms of the cycloaddends are determined by the nature of the allylic atom.

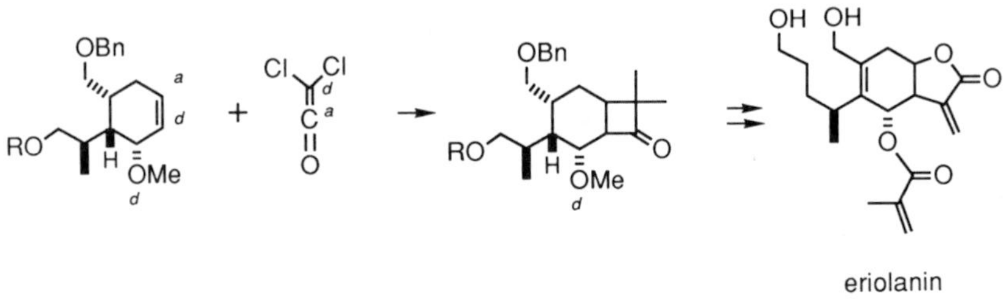

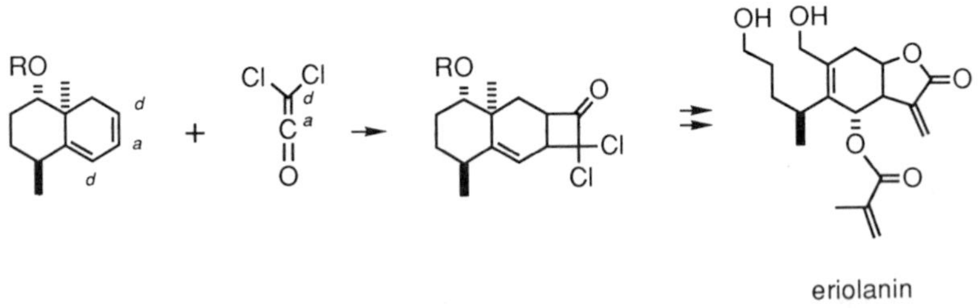

eriolanin

eriolanin

A *de novo* access of resorcinol derivatives which exploits a series of pericyclic reactions may be illustrated by a synthesis of maesanin [Danheiser, 1990]. The first step of the sequence is a [2 + 2] cycloaddition on an α-vinylketene and an alkoxyalkyne according to a polarity-matched transition state. The adduct underwent electrocycloreversion and then a six-electron electrocyloaddition to give, in one operation, the aromatic percursor of the natural product.

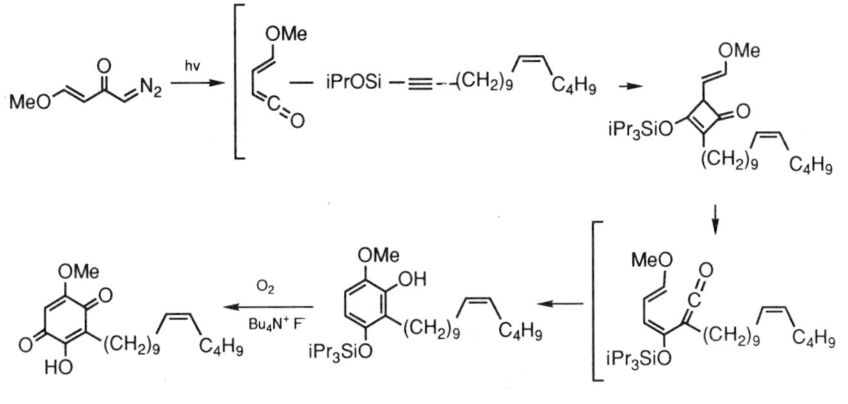

maesanin

Regiocontrol by alkyl groups of the alkene in the intramolecular [2 + 2] cycloadditions is felicitous, as attested by the synthesis of β-*trans*-bergamotene [Corey, 1985b; Kulkarni, 1985] and methyl dehydrojasmonate [S.Y. Lee, 1988].

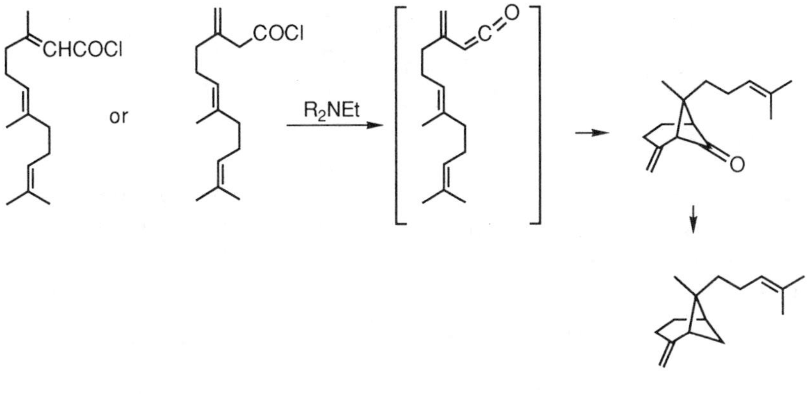

β-trans-bergamotene

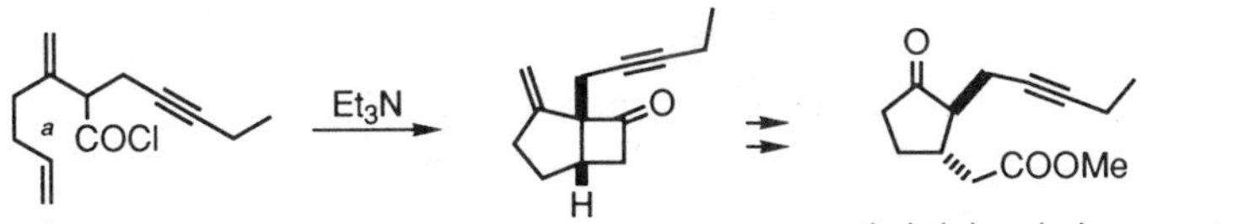

methyl dehgydrojasmonate

A cyclopentenecarboxylic acid moiety is present in retigeranic acid and naturally it is the prime focus for retrosynthetic analysis. While this full retron for a vinylcyclopropane/cyclopentene rearrangement transform is evident [Hudlicky, 1989], an alternative approach [Corey, 1985c] based on aldol and C=C oxidative cleavage transform is intellectually more challenging. The subgoal is now a cyclohexene and the Diels–Alder simplification can be implemented. Necessarily the dienophile must be modified (actuated).

retigeranic
acid

(acid chloride)
Et₃N

The propensity of alkylated dienes and dienophiles for the *ortho* condensation mode is well suited for the required regiochemistry. An *endo* transition state should lead to an adduct with two stereogenic centers in the correct relative configuration. It appears that 2-methyl-2-cyclopentene-1,3-dione is a good candidate for the dienophile, in view of the differentiable environment about the two carbonyl groups.

Disconnection of a bond pair in the diquinane portion opens up the new opportunity of applying the ketene cycloaddition. This consideration can be equated to a Diels–Alder transform involving derivative of citraconic acid.

The intramolecular cycloaddition of ketene and an alkene ensures a correct stereochemistry of the remaining asymmetric centers. The reaction also places a carbonyl group at a position which facilitates ring expansion and elaboration of the secondary methyl substituent. (An intramolecular [3 + 2] cycloaddition involving an ionic trimethylenemethane synthon may provide yet another pathway for the annulation of the DE-ring component, although such a reaction is favored by the presence of acceptor substituents in the dipolarophile.)

The intramolecular ketene/alkene cycloaddition was featured prominently in the awe-inspiring construction of ginkgolide B [Corey, 1988b,c]. This reaction created the spirocarbocycles stereoselectively (as controlled by the *t*-butyl group) with the cyclobutanone subunit serving as a latent lactone. The substitution pattern of the alkene partner for the ketene was fully cooperative (polarity matching) to furnish the intended intermediate.

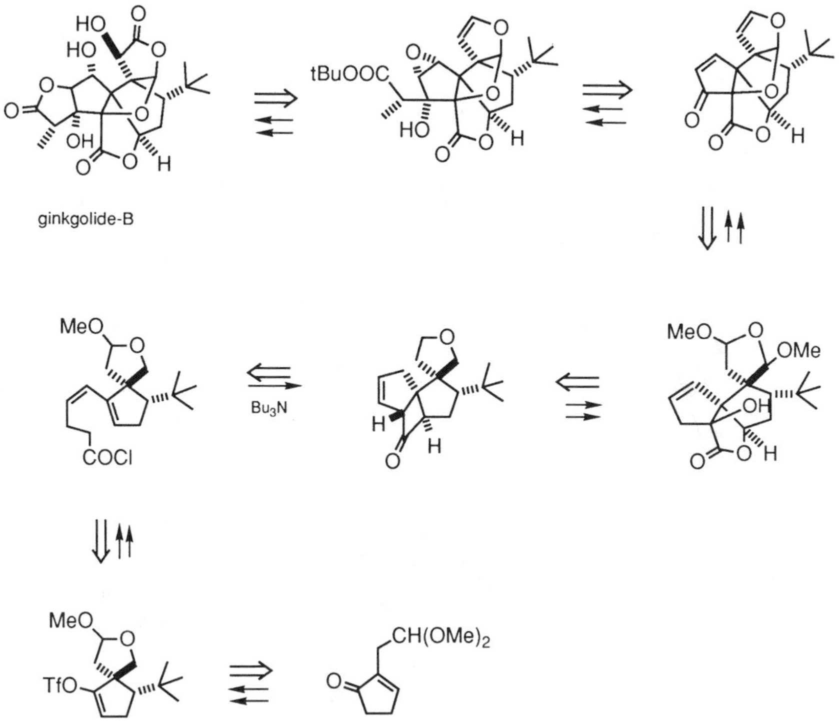

ginkgolide-B

The preference in the formation of either a 2-aroylbenzofuran or an isoflavone from an intramolecular [2 + 2] ketene cycloaddition to a benzil carbonyl depends on the substitution of the remote benzene ring [Brady, 1988]. Accentuation by a conjoint methoxy group on the distal carbonyl makes it a more reactive donor. Otherwise formation of the 5:4 fused intermediate is favored.

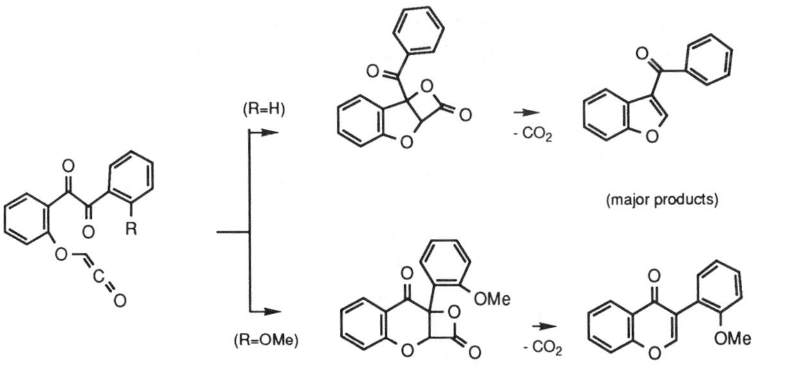

(major products)

Cycloaddition of allylsilanes to *N*-chlorosulfonylisocyanate also proceeds regioselectivity in the predicted manner to give β-lactams [Colvin, 1990].

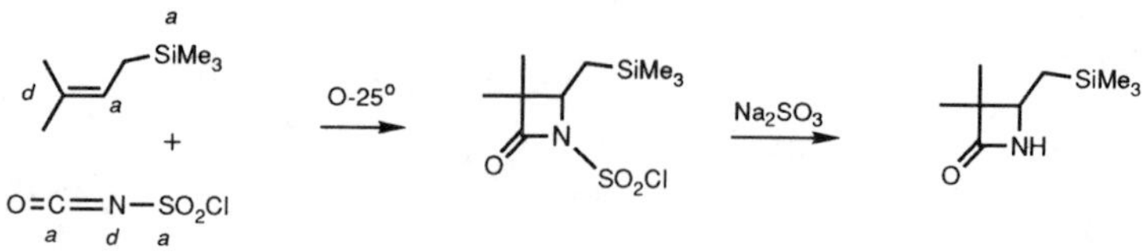

Although the regiochemistry of the cycloaddition between chromene and chlorosulfonyl isocyanate would not have affected the conversion of the adduct into biotin [Fliri, 1980], it is pleasing to observe that the reaction was under polarity control.

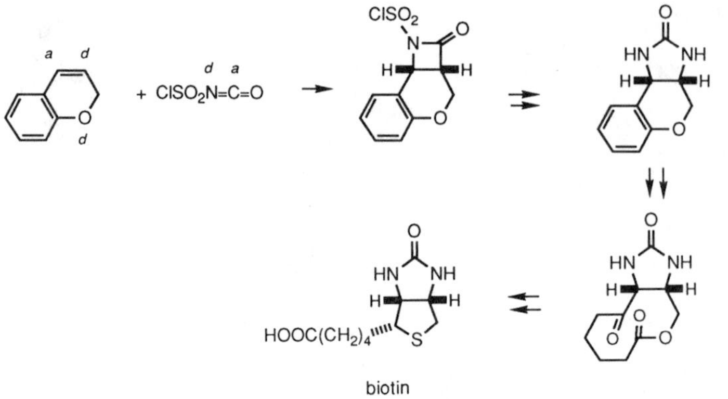

biotin

11.2. CYCLOADDITIONS INVOLVING ACETYLENES

Enamines react with activated alkynes to give cyclobutene derivatives. Electrocyclic opening of these adducts is favored by the polar substituents (push–pull mechanism) such that frequently the cyclobutenes cannot be isolated. However, discussion of these processes here seems appropriate.

The enamine reaction constitutes a useful method for ring expansion of ketones by two carbon atoms, and many synthetic operations, including an approach to vellerolactone [Froborg, 1978], rely on it.

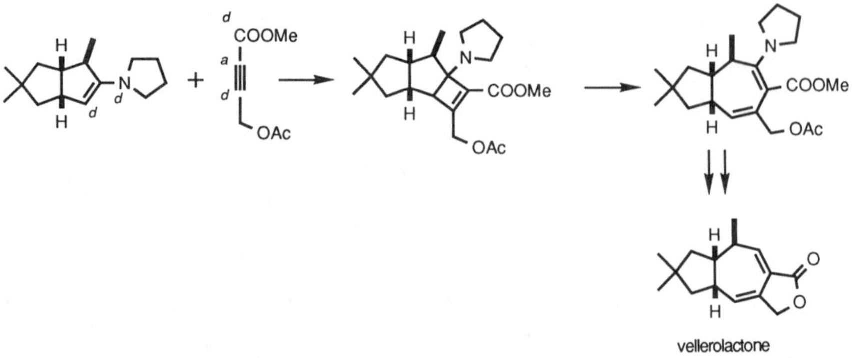

vellerolactone

A wide range of synthetic applications of ynamine cycloadditions to acceptor-substituted cycloalkenes [Ficini, 1976] has been developed. Because of the polarity matching in these reactions, the cyclobutanones which are readily formed upon hydrolysis of the aminocyclobutene adducts undergo facile fragmentation to afford compounds with 1,5-acceptor sites. The most important aspect of this process is the stereocontrol at the two newly created stereogenic centers. Thus, fragmentation of the kinetically (*exo*) protonated product of the aminocyclobutene provides one diastereomer, whereas equilibration prior to fragmentation leads eventually to the other.

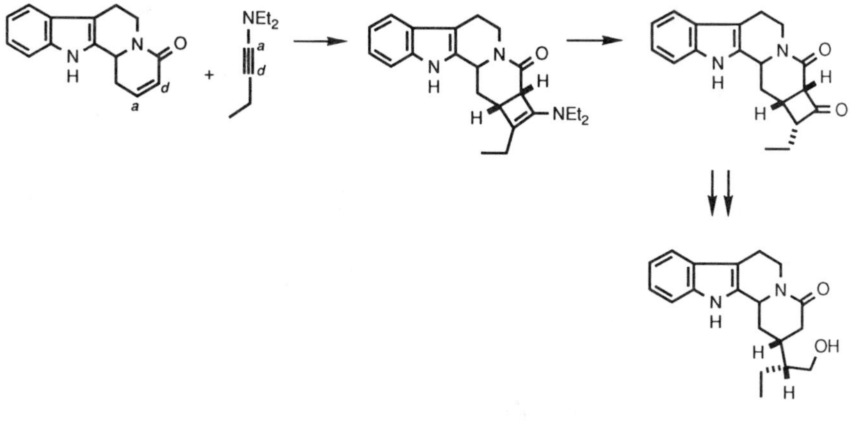

dihydroantirhine

An interesting observation pertaining to the reaction of methyl alkynoates with quadricyclane derivatives is the site-dependence on the nature of the nonbridgehead atom [Prinzbach, 1968]. A donor-atom bridge favors new bond formation at the α-carbons (acceptors).

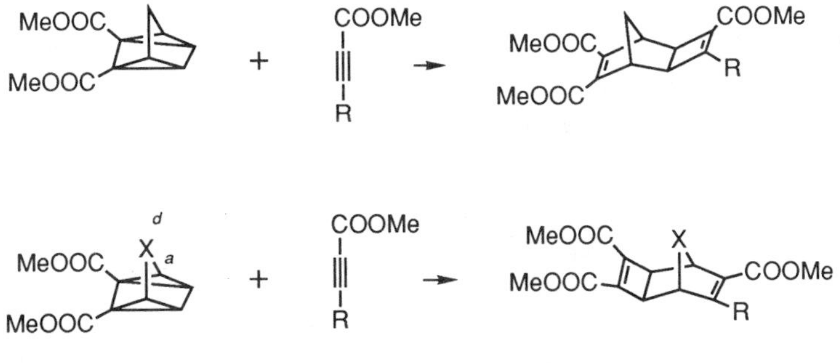

X = O, NR'

Addition to benzynes can be regioselective. Cycloaddition with electron-rich alkenes leads to benzocyclobutenes of synthetic value. In a synthesis of taxodione [Stevens, 1982a] the nearest methoxy group to the benzyne controlled the regiochemistry of the cycloaddition with 1,1-dimethoxyethylene. It is interesting that the same substituent was responsible for a selective cleavage of the tricyclic benzocyclobutenol intermediate in unraveling an enone system.

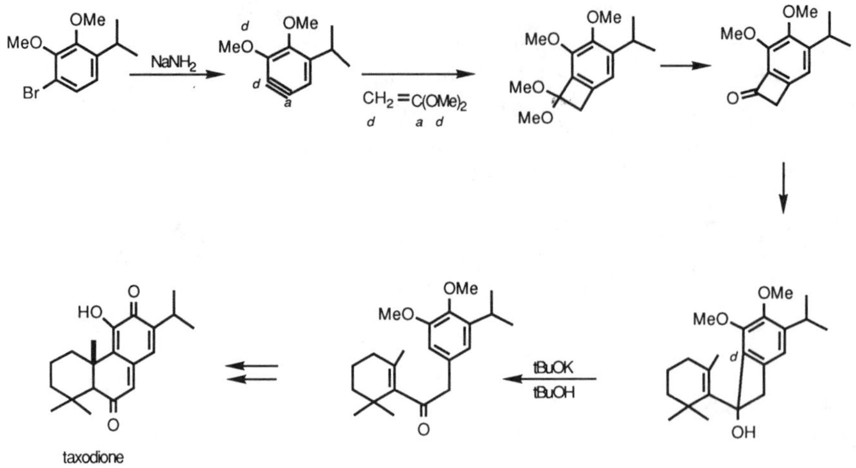

taxodione

12

SYNTHESIS OF 1,6- AND 1,7-DIHETEROSUBSTITUTED COMPOUNDS

12.1. 1,6-DIOXO COMPOUNDS

12.1.1. Oxidative Cleavage of Cyclohexenes

The Diels–Alder reaction gives cyclohexene derivatives, and these adducts are an excellent source of 1,6-dicarbonyl compounds. Innumerable examples are found in the chemical literature describing the oxidative cleavage of such adducts as a synthetic step.

An elegant application of the two-step sequence formed the key operation of an approach to 11-oxygenated steroids [Stork, 1981, 1982a]. The perceptive recognition of a 1,6-relationship between C-3 and C-11 of the steroid skeleton led to the formulation of the route featuring an intramolecular Diels–Alder reaction.

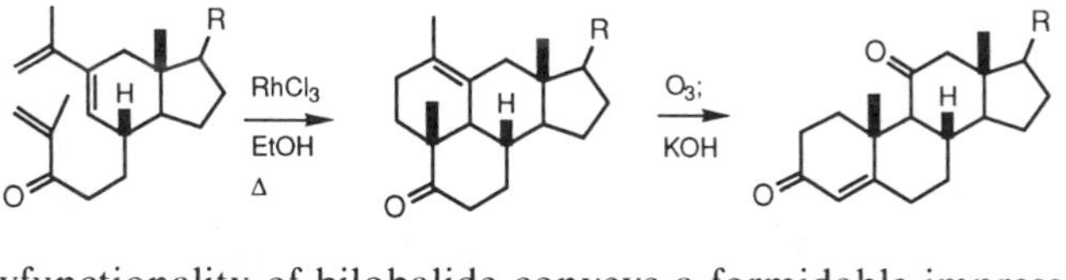

The polyfunctionality of bilobalide conveys a formidable impression to the chemist who endeavors to synthesize it. However, retrosynthetic analysis based on functional-group reconnection has enabled a simplification to a very manageable hydrindene derivative as the precursor [Corey, 1987b, 1988d]. Specifically, the two nonadjacent lactone carbonyls in a 1,6-relationship with their *cis* orientation on the cyclopentane core were noted. Recombination of these carbonyl groups then displayed the role of the angular substituents for the element of the central lactol ring. Further reward for choosing the hydrindene precursor is the stereocontrol in establishing the secondary oxy substituent in the cyclopentane via reduction. In turn, this substituent helped direct the entry of the tertiary alcohol (via bishydroxylation and selective deoxygenation) after its incorporation into a γ-lactone.

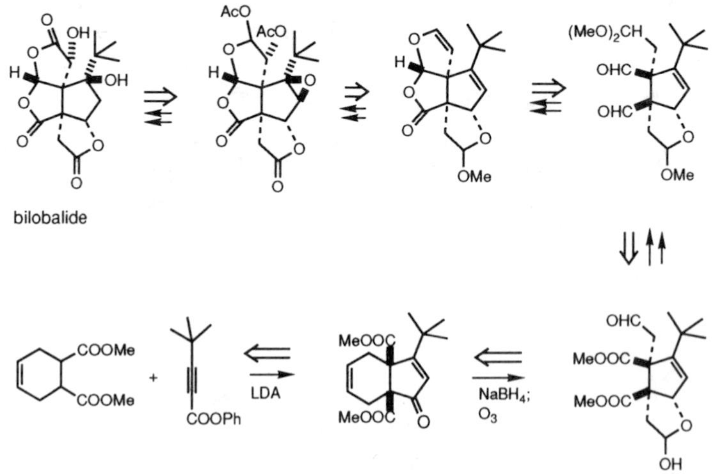

bilobalide

The combination of a Diels–Alder reaction, a $C=C$ oxidative cleavage, and an aldol condensation constitutes an important concept in a synthesis of retigeranic acid [Corey, 1985c].

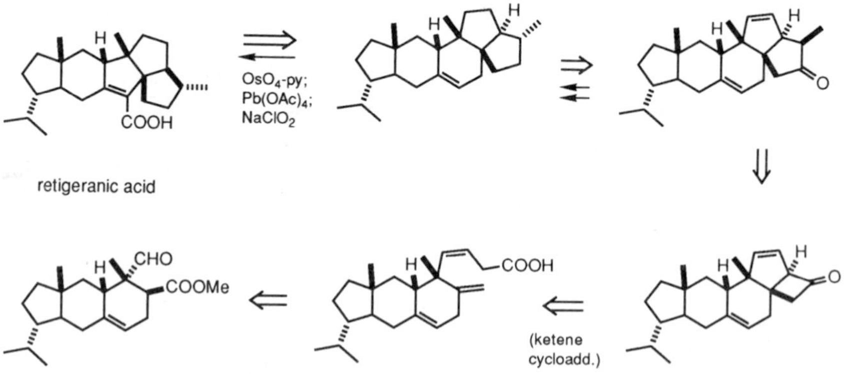

retigeranic acid

The cyclohexene precursors of 1,6-dicarbonyl compounds are not limited to those obtained from the Diels–Alder reactions. A synthesis of vermiculine [Y. Fukuyama, 1977] also took advantage of the 1,6-related diketone in each of the half-molecules and accordingly that functionality was generated from the cyclohexene moieties via the diols. A distinguished aspect of this synthesis is the deliberate avoidance of the dimerization strategy.

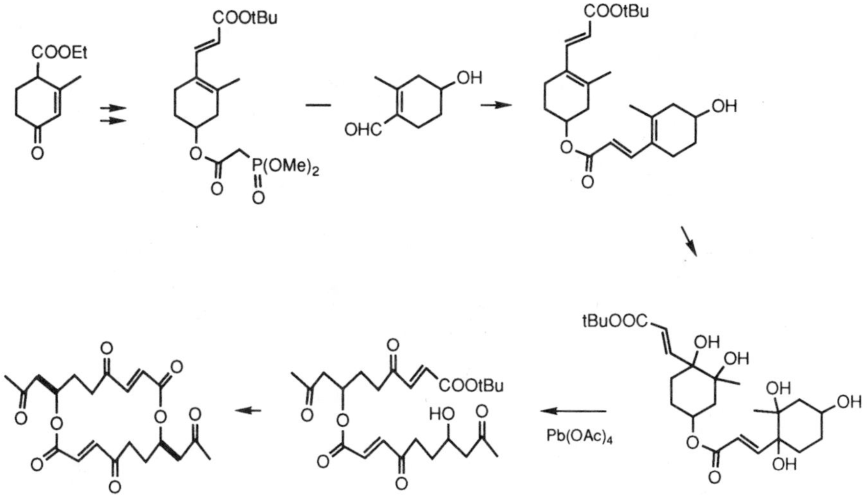

vermiculine

It may be noted that anisole derivatives are also excellent sources of 1,6-dicarbonyl compounds. Birch reduction of anisole gives 1-methoxy-1,4-cyclohexadiene which can be selectively cleaved at the enol ether linkage. In an elegant synthesis of the cecropia juvenile hormone JH-1 [Corey, 1968b] the focus was the stereospecific elaboration of a trisubstituted double bond, and the oxy groups served other functions. The concept is exemplary of synthetic expediency gained by paying attention to hidden structural correlation. Of course, the synthesis of strychnine [Woodward, 1963b] is an earlier masterpiece featuring the oxidative cleavage of a veratrole ring at one site, and without affecting the indoline moiety.

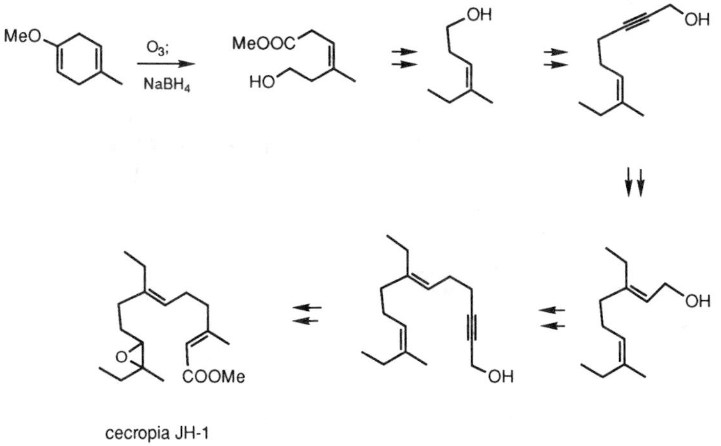

cecropia JH-1

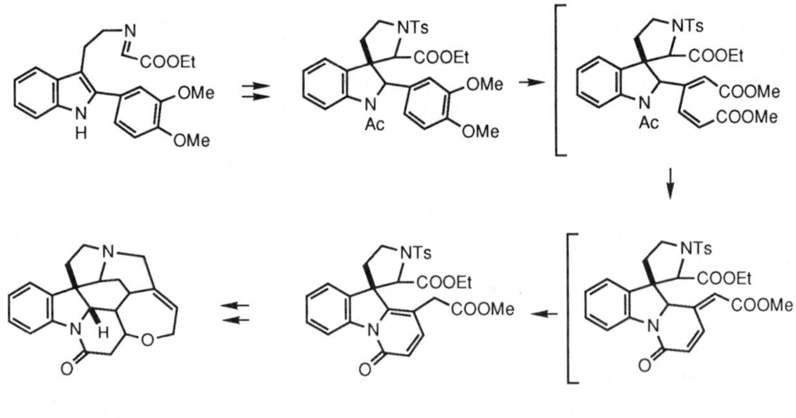

strychnine

Although the Baeyer–Villiger reaction of cyclic ketones is a quite general method to acquire 1,n-dioxy alkanes with differentiable functional groups, its use in generating a 1,6-dioxygenated molecule in combination with other reactions in the context of E-ring construction during a reserpine synthesis [Stork, 1989] was particularly enlightening.

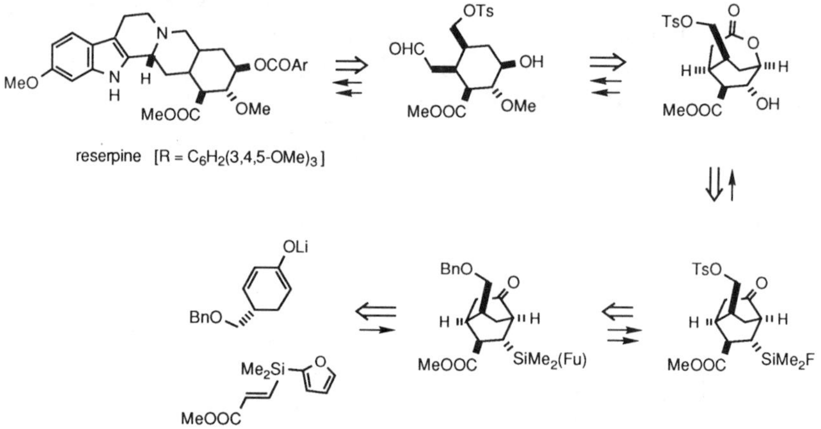

reserpine   [R = C6H2(3,4,5-OMe)3]

Accordingly, a reflexive Michael annulation led to a bicyclic ketone containing five asymmetric centers. Fluorination of the silicon substituent activated it toward an oxidative C—Si bond cleavage in the same step as the Baeyer–Villiger reaction. It should be noted that the oxidative maneuver was predicated by reactivity considerations such that the Michael adduct would not lose the β-substituent of the Michael acceptor under the basic conditions (see also Sec. 16.1.).

12.1.2. Reductive Cleavage of 1,2-Diacylcyclobutanes

Reductive cleavage of 1,2-diacylcyclobutanes furnishes 1,6-dicarbonyl compounds. From the synthetic viewpoint the access of the cyclobutanes is the crux.

An example of this process is found in a synthesis of pentalenic acid [Crimmins, 1984]: the photocycloaddition-reductive cleavage sequence led to a spirocyclic ester which was elaborated quite readily into the target molecule via Claisen condensation and ancillary reactions.

pentalenic acid

(Reductive cleavage of *vic*-diacylcycloalkanes of other sizes has also been reported. Notable among these are 1,7-diacyl intermediates for dodecahedrane [Ternansky, 1982], cocaine [Krapcho, 1985], and a 4-octenedicarboxylic ester, actually a tetraester was generated initially, for the AD-component of a corrin [Bertele, 1964]).

A steroid building block containing the D-ring and the asymmetric centers C-13, 17, 20 has been acquired from a bicyclo[2.2.1]octan-6-one-2-*endo*-carboxylic ester by reductive cleavage [Shimizu, 1990]. The merits of this particular approach include the uncomplicated and predictable stereoselective creation of the three asymmetric carbon atoms within the bridged ring framework.

12.2. 1,7-DIHETEROSUBSTITUTED COMPOUNDS: MICHAEL-TYPE REACTIONS

Substances in which two heteroatoms are six bonds apart are more readily assembled in a stepwise manner and by functional group relay processes. The strategy of FGA at an internal carbon atom profits from more synthon options but the synthetic route is less convergent and it requires steps for removal of the extra functional group. On the other hand, while there are not many syntheses featuring direct assembly of the long atomic segment, those few that have taken advantage of the structural characteristics which permit such a mode of synthetic operation, have been well rewarded.

An early synthesis of α-santonin [Abe, 1956] recognized the functionality distribution that a Michael reaction between a malonate anion and a dienone would be an efficient method for bringing together all the framework carbons.

α-santonin

It has been repeatedly shown that FGA and FGI are of great value in the formulation of synthetic pathways. Strategic insertion of a heteroatom to an "apparent" monofunctional molecule could extend the possibilities of its construction. With respect to hidden 1,7-dihetero compounds, a discussion of a synthesis of the lactonic sesquiterpene frullanolide [Kido, 1979] is instructive.

Customarily, fused γ-lactones have been prepared by alkylation of cyclo-alkanones with an α-bromoacetic ester, followed by reductive lactonization. Novel approaches to these compounds involve annulation of an existing lactone ring.

A strategic bond of frullanolide is identified as C-5/C-6 because C-5 is trigonal. Thus, an FGI operation that places a tertiary alcohol at C-5 demands C-6 (γ-carbon of the lactone moiety) to be a donor site. Without introducing elaborate activating groups to the system the demand can be met by using the corresponding butenolide subunit. Examination of the resulting seco intermediate reveals the applicability of a Michael transform to it, the Michael acceptor being 3-vinylbutenolide.

The synthesis that followed this logical analysis actually employed methyl 3-methyl-2-oxocyclohexanecarboxylate as the Michael donor. The reaction led directly to the desired decalin system.

frullanolide

(R= COOMe)

It is significant that the aldol condensation proceeded as shown, instead of cyclization at the α-position of the lactone. The reversibility of the reaction enables equilibration of the two possible products, and the observed tricyclic lactone should be more stable in view of the presence of a conjugate lactone.

An alternative mode of lactone annulation is depicted in a synthesis of dihydrocallistrisin [Schultz, 1976].

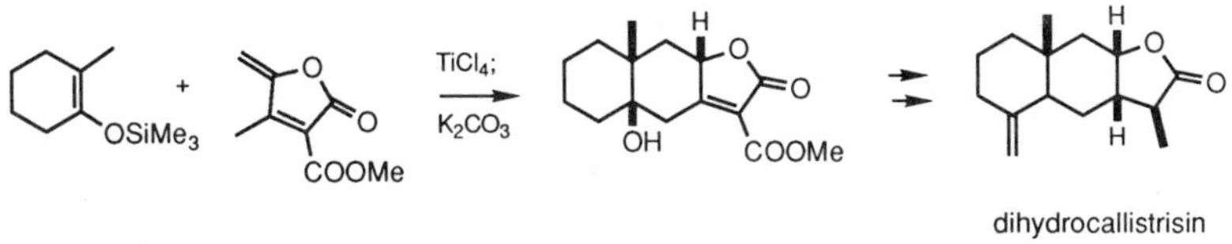
OSiMe3
COOMe
TiCl4;
K2CO3
OH
COOMe
dihydrocallistrisin

Of relevance to the above is an elegant enantioselective synthesis of aphidicolin [Holton, 1987] based on a retrosynthetic analysis of a spiro-tricarbocyclic intermediate. The analysis uncovered a Michael transform.

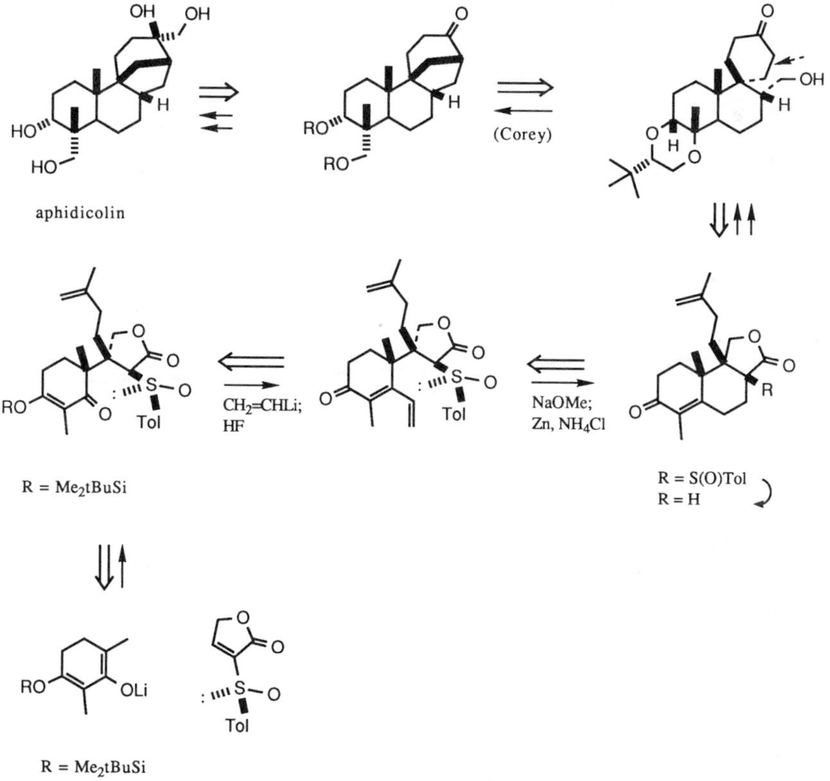
aphidicolin
(Corey)
CH2=CHLi;
HF
NaOMe;
Zn, NH4Cl
R = Me2tBuSi
R = S(O)Tol
R = H
OLi
Tol
R = Me2tBuSi

Finally, it should be mentioned that a short synthesis of nootkatone [Dastur, 1974] belongs to the same category. This synthesis hinged on an alternative cyclization of a trienone derived from fragmentation of a bicyclo[2.2.2]octene to give a 7-formoxy-2-en-1-one embedded in a decalin skeleton.

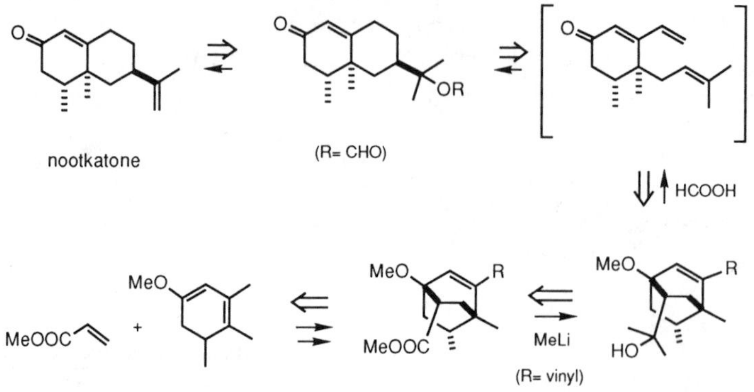
nootkatone
(R= CHO)
OR
HCOOH
MeO
R
MeOOC
HO
MeLi
(R= vinyl)
+

13

REACTIONS AND SYNTHESIS OF ODD-MEMBERED RING COMPOUNDS

13.1. CYCLOPROPANES: SOME PREPARATIVE METHODS AND REACTIONS AFFECTING THE RING STRUCTURE

Odd-membered ring compounds, either carbocycles or heterocycles, are disjoint. This electronic nature determines both the modes of ring formation and the functional-group distribution of the ring cleavage and bond insertion products. For example, cyclopropanes are most readily created by the cycloaddition of carbenes/carbenoids to alkenes, the 1,1-dipolar species are of course disjoint.

Let us consider a way to synthesize trinoranastreptene [Kang, 1988]. This terpene molecule, embedded with a three-membered ring, suggests an intramolecular acylcarbenoid capture pathway as the key step of its construction. Routes involving substrates with cyclopentadiene and cyclohexene subunits of their latent forms (i.e., substituted cyclopentene and cyclohexene derivatives) also require an unproductive removal of a carbonyl group from the products, and the transannular version has the problem of its precursor accessibility, although the ketone group would be gainfully converted into a C=C bond.

The best option seems to be the approach generated by disconnection of the cyclohexene portion. A carbonyl substituent of the cyclopropane definitely facilitates the formation of a carbenoid. Considering the length of the chain and functionality required of the acyclic precursor for this approach, alkylation of methyl acetoacetate dianion with a geranyl halide emerges as an excellent choice to start the synthesis.

trinoranastreptene

The umpolung action by organometallic intervention is shown in a synthesis of cyclopropanols from α,β-enecarbonyl compounds [T. Sato, 1988]. The transformation involving reaction with trimethylstannyllithium and treatment of the adducts with $TiCl_4$ is attended by an $a \rightarrow d$ change at the original β-carbon.

Regarding the consequence of cyclopropane cleavage, the situation is more easily visualized from reactions involving polarized substrates. In this context it is useful to analyze the synthetic operation of introducing a β-methyl group to a ketone.

Beside the method via cuprate addition to the corresponding α,β-enone, it is possible to generate the product from a cyclopropyl ketone by Li/NH_3 reduction. The cycopropyl ketone may be obtained from the enone ($+ Me_2S{=}CH_2$), allylic alcohol (Simmons–Smith reaction and oxidation), or β-tosyloxy ketone (internal alkylation).

Following the internal alkylation–reduction sequence one witnesses polarity inversion in several atoms. Thus, the tosyloxylated carbon of the procyclic precursor is an acceptor, yet the same atom must become a donor during its conversion into the methyl group. As this role change is strongly opposed by the carbonyl group, the bond cleavage necessitates umpolung maneuver, e.g., reduction. The end result is that a disjoint array present in a γ-oxy ketone is transformed into a conjoint species by incorporating a step that inverts atomic polarity. (Note that $C{-}O \rightarrow C{-}H$ conversion may be considered as an $a{-}d \rightarrow d{-}a$ process.)

An acylcyclopropane gives rise to a disjoint chain when it is attacked by a donor reagent on the ring carbon. On the other hand, its reaction with a donor which is contropolarizable (e.g., CN) will lead to a conjoint molecule.

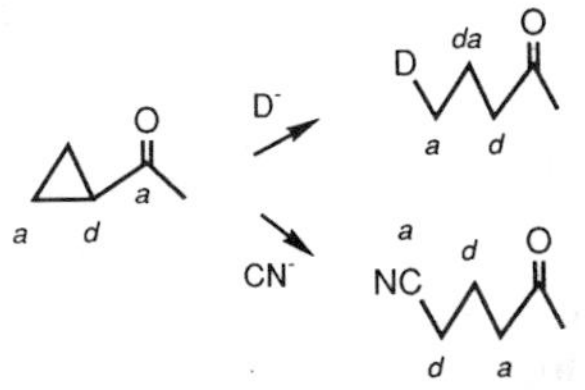

(Contrapolarizable reagents have also been exploited in cyclopropane formation, e.g., in a synthesis of ethyl *trans*-chrysanthemate [Martel, 1967].)

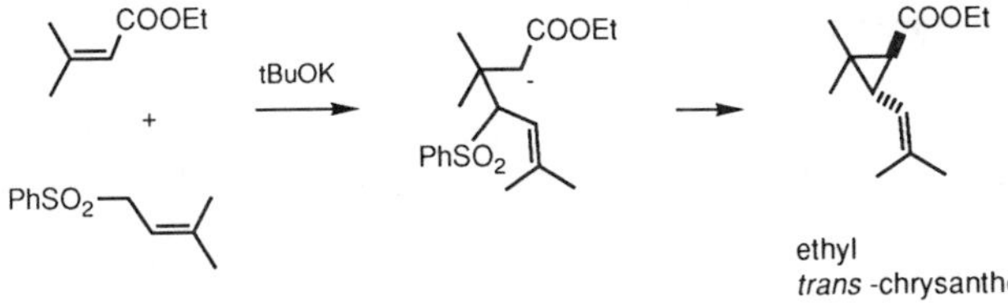

ethyl
trans -chrysanthemate

The ring expansion of cyclopropanone on exposure to diazomethane follows the same trend.

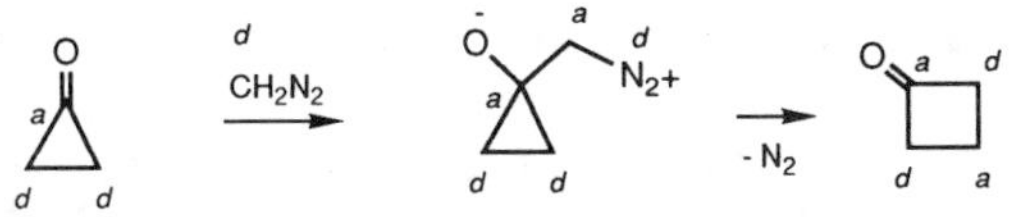

Note the pairwise contrapolarization during the rearrangement step, which may also be witnessed in the following stepwise reaction.

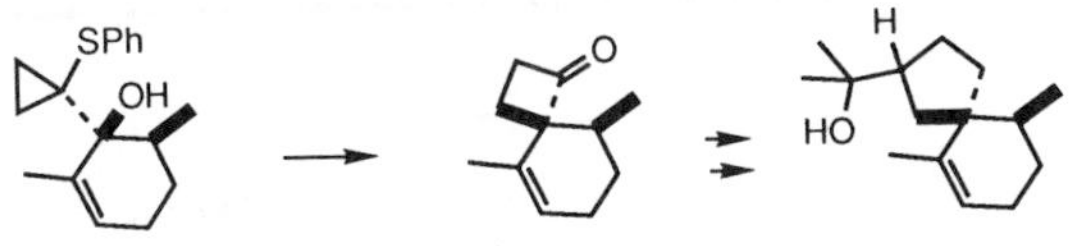

hinesol

The formal 1,3-rearrangement of 1-(2-oxoalkyl)cyclopropanols on base treatment [Carey, 1986; Narasimhan, 1986] may actually involve an incipient carbanion (homoenolate) to effect an intramolecular aldolization to afford 3-hydroxycyclopentanones. It should be emphasized that the long path between the two oxygen functions is a disjoint one.

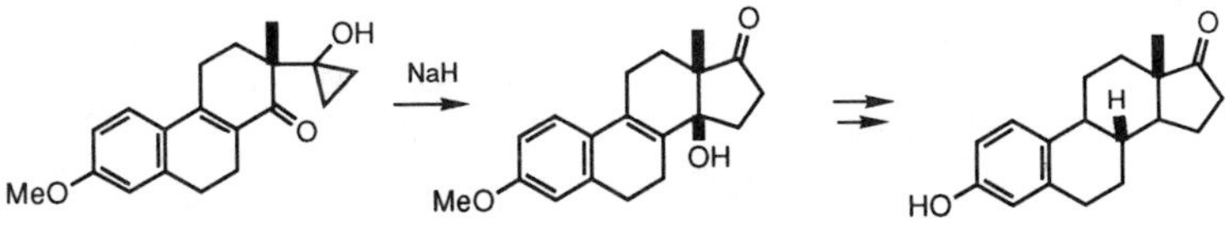

estrone

Acylation of latent homoenolates results in the disjoint 1,4-diketones [Aoki, 1989].

Evidently the effect of polar substituents on a cyclopropane is similar to that on a C=C bond. In terms of synthetic application of such an activation one may compare the synthesis of 3-cyclopentenols [Danheiser, 1981b] with a two-step annulation of conjugate dienes via 2-alkenylcycloproyl ethers and release of the cyclopropanoxides from the latter species. The stereoselective rearrangement to give the five-membered ring products occurs at room temperature instead of ca. 500°C for the conventional vinylcyclopropane to cyclopentene reorganization.

The preparation of 1,4-dicarbonyl compounds by an alkylation method is often fraught with complications due to proton exchange and self-condensation of the alkylating agents. An indirect route is via a copper acylcarbenoid addition to enol derivatives in which the donor and the acceptor components are modified (e.g., enolate ion to enol ether, haloketone to diazoketone). The resulting acylcyclopropane derivatives readily undergo ring cleavage which is formally a retro-Michael reaction. Because such a 1,3-dioxygenated system is skipped, a disjoint 1,4-dicarbonyl product is obtained.

It is apparent that many cyclopentenones and hence cyclopentanones can be synthesized by this method (Wenkert, 1978b; McMurry, 1971].

dehydrojasmone

The use of an enamine or N-acylenamine instead of an enol derivative as the donor in the cyclopropanation may be of some advantage if the nitrogen atom is to be retained for other synthetic purposes. A synthesis of eburnamonine [Wenkert, 1978c] from a tetrahydropyridine derivative illustrates this point well.

eburnamonine

2-Siloxycyclopropanecarboxylic esters undergo ring cleavage readily in the presence of fluoride ion. The nascent enolate of the ester may participate in C—C bond formation with an appropriate acceptor partner present in the reaction vessel. For example, a Michael–Wittig reaction tandem has been achieved in the presence of a vinylphosphonium salt, and a diquinane precursor of pentalenolactone E methyl ester has been acquired in one operation [Marino, 1987].

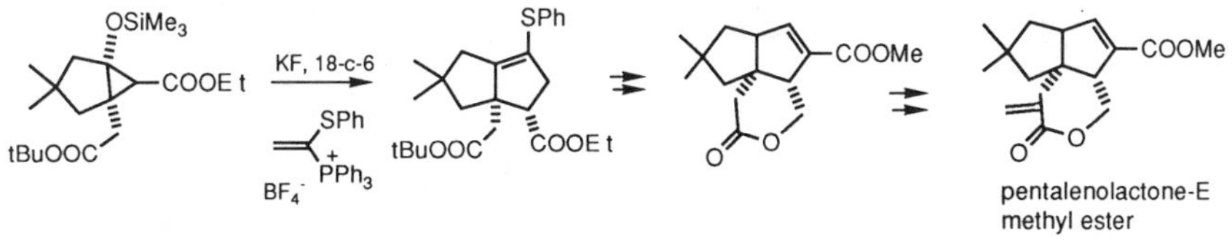

The generation of a vinyl ketone and its intramolecular interception by a remote diene moiety has been accomplished [Zschiesche, 1986].

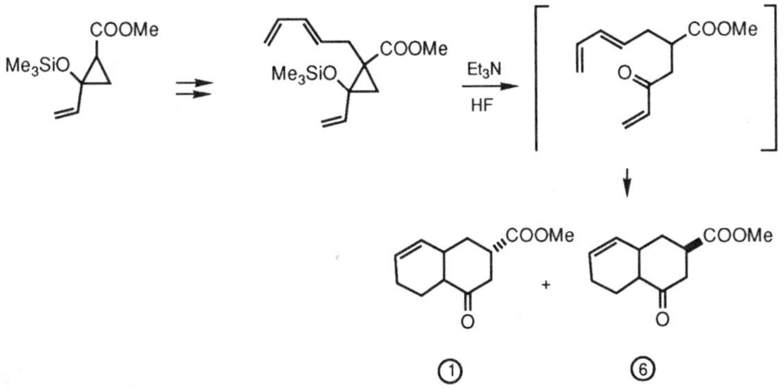

Rather unexpected is the efficient and virtually stereospecific [3 + 2]cyclo-addition of a 2-methylenecyclopropanone ketal with electron-deficient alkenes [Yamago, 1989]. A concerted reaction is implicated.

In situ fragmentation further increases the efficiency of directed C—C bond formation via cyclopropanation. A synthesis of 4-oxo-β-ionone [Wenkert, 1989] from a furan derivative was predicated upon such a process.

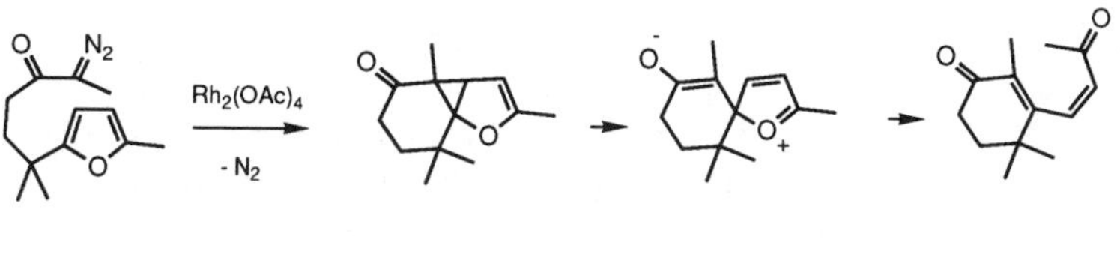

The oxadi-π-methane rearrangement of bicyclo[2.2.2]octenones on irradiation is an excellent method for entry into fused ring systems such as shown in an approach to pentalenolactone G [Demuth, 1984]. In this case the presence of a strategic methoxy group enabled the cleavage of the cyclopropane ring to generate a diquinane with the most desirable functionality in the carbon network.

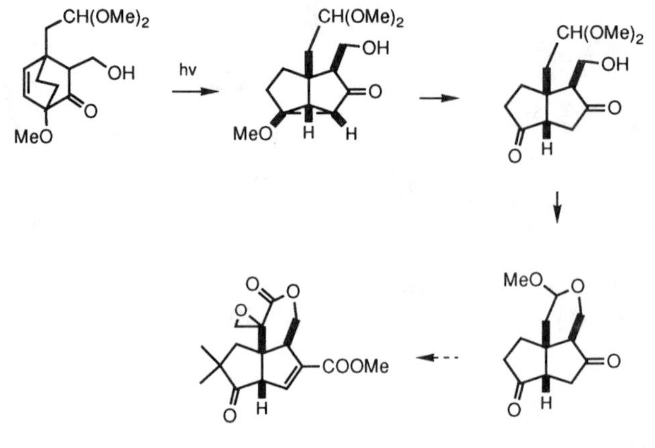

pentalenolactone-G

In daunomycinone the benzylic hydroxyl and the sidechain carbonyl are 1,4-related. This structural subunit can be devolved from a solvolytic cyclopropane scission [Reddy, 1988].

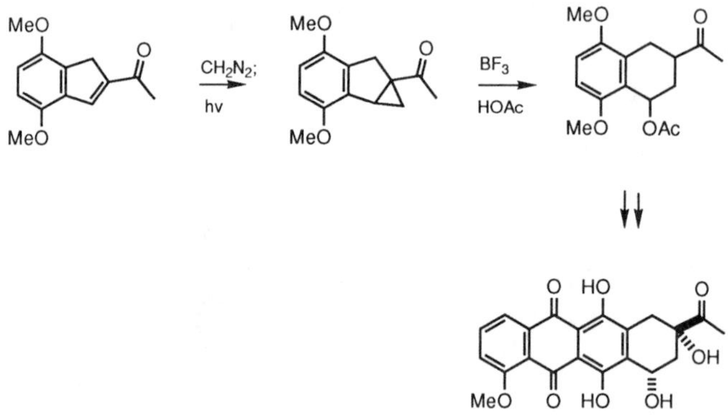

daunomycinone

In grosshemin there is a disjoint array spanning the A-ring ketone and the γ-lactone which is fused to the B-ring. To construct this array *de novo* an umpolung process must be involved. A reaction sequence of intramolecular cyclopropanation and a regio- and stereoselective opening of the cyclopropane ring provided a solution to the synthetic problem [Rigby, 1987]. Trapping the emerging enolate also permitted methylation at C-4.

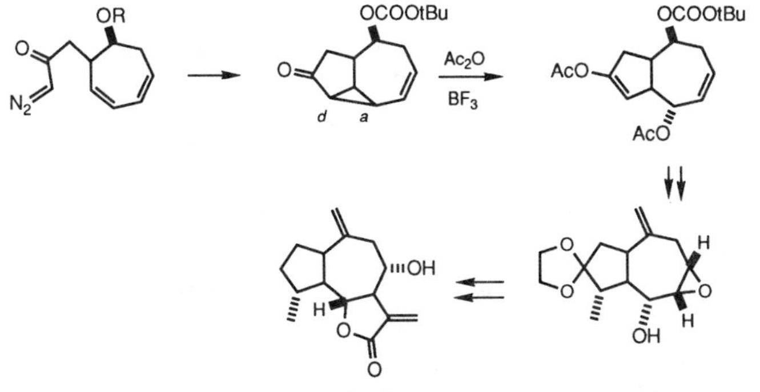

grosshemin

It should be noted that an oxatricycle which arose from a formal hetero-Diels–Alder reaction of a cycloheptadiene proved useless in the elaboration of the terpene as the downstream intermediate is a β-oxy ketone which is extremely prone to elimination.

Ring expansion with allylic to homoallylic hydroxyl transposition constituted a key step in a synthesis of confertin [Marshall, 1976]. With respect to the original donor–acceptor sites in the vicinity of the cyclopropane ring, the action of the solvolytic lactonization caused a wholesale inversion,

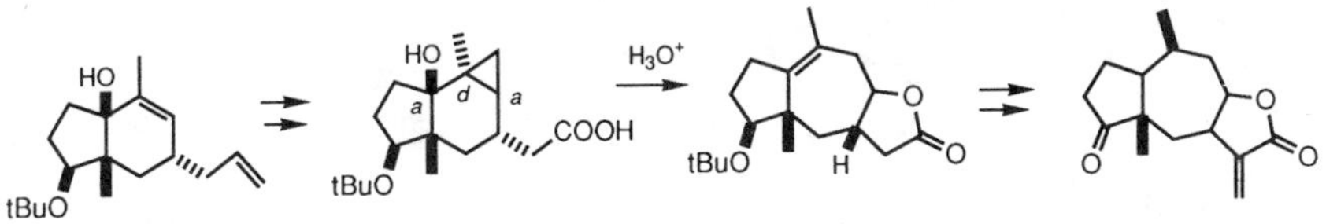

confertin

In ketonic prostaglandins the oxygen atoms of the C-15 hydroxyl group and the C-9 carbonyl are six carbons apart. This disjoint array requires an umpolung reaction or a building block already containing an umpoled functionality. Synthesis passing through a vinylcyclopropyl ketone intermediate would resolve the problem as ring opening with a nucleophile leads to a C-13 functionalized substance, which may undergo allylic transposition to introduce the desired oxygenation pattern [Kondo, 1977; Taber, 1977]. For a reactivity reason and based on the well-defined [2.3]sigmatropy of allylic sulfoxides, the cyclopropane opening was achieved with the benzenethiolate anion.

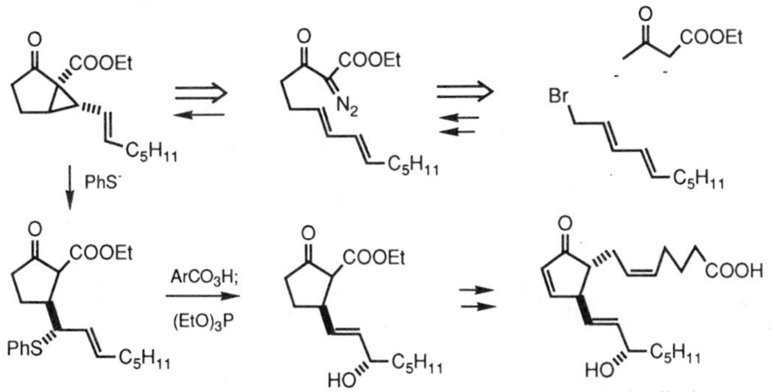

prostaglandin-A₂

An early synthesis of prostaglandin E_1 [Just, 1967; Schneider, 1968] addressed the ene-1,5-diol subunit by generating it from a disjoint cyclopropyl *vic*-diol. Noteworthy is the transposition of the hydroxyl group proximal to the small ring, upon opening of the latter, which led to a conjoint molecule.

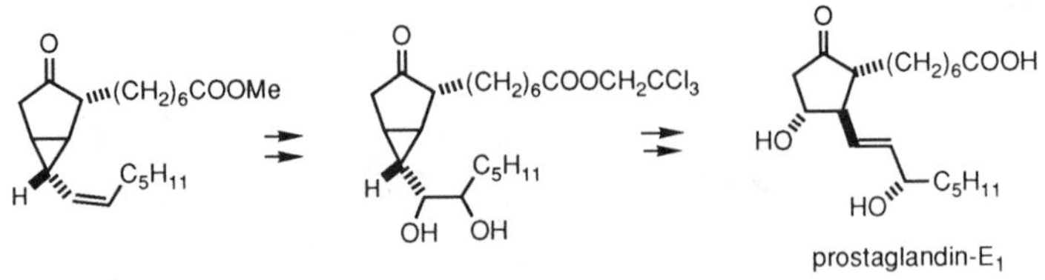

prostaglandin-E₁

An intramolecular carbenoid addition served to establish the oxadecalone ring system of vernolepin [Zutterman, 1979]. It also permitted a stereospecific functionalization of C-6 and subsequently the remaining asymmetric centers in the B-ring.

It is especially interesting that another approach [Isobe, 1978; Iio,1979] in which the same ring system was assembled by an intramolecular Michael reaction actually moved forward by a separate cyclopropanation (via bromination–dehydrobromination) in order to introduce a functional group at C-6.

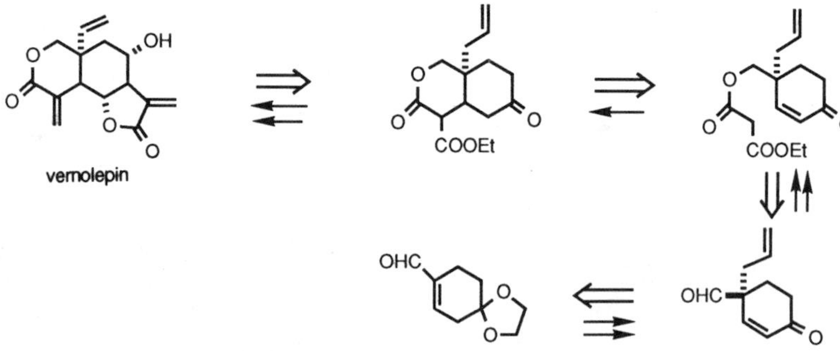

vernolepin

Isoretronecanol is liable to disconnection in many ways. One reasonable retrosynthetic pathway involves a homo-Michael transform.

Based on this analysis a synthesis [Celerier, 1987] was carried out starting from a 1-succinoyl-1-cyclopropanecarboxylic ester. Its reaction with benzylamine led to an enamine which possesses all the skeletal atoms of the alkaloid.

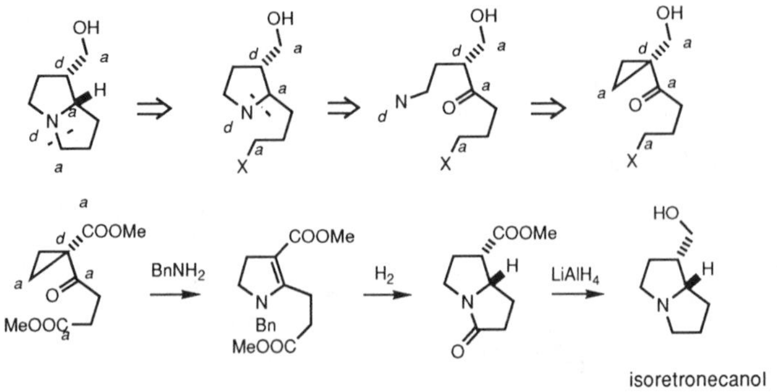

isoretronecanol

A very elegant scheme for alkaloid synthesis consists of the condensation of endocyclic enamines with methyl vinyl ketone as the key operation [Stevens, 1977]. Many of the 2-pyrrolines required for this approach have been acquired from cyclopropyl ketones. A route to an intermediate of cepharamine [Keely, 1970] is shown below.

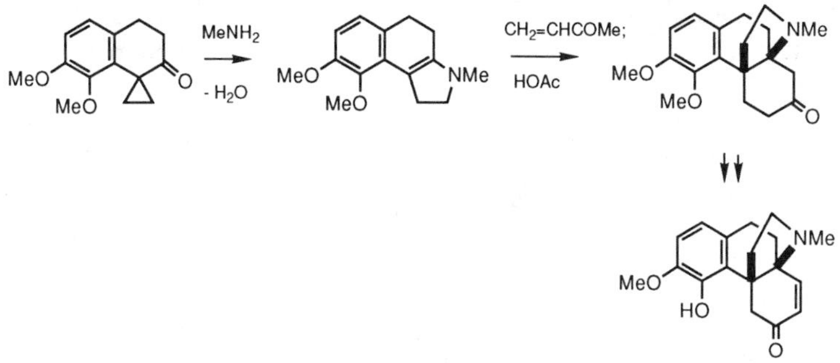

3-Acyl-2-pyrrolines are particularly valuable for synthesis of various alkaloids. While the N—C$_3$—O segment is conjoint, the longer span is a disjoint circuit. Accordingly, ring closure involving a connection within the latter circuit must engage an umpolung action or use an umpoled (disjoint) precursor. The thermal reorganization of 1-iminocyclopropanecarboxylic esters fulfills this need (cf. the vinylcyclopropane to cyclopentene rearrangement). This process has implications in the synthesis of isoretronecanol [Pinnick, 1979] and many other more complicated azacycles including an intermediate of aspidospermine [Stevens, 1971].

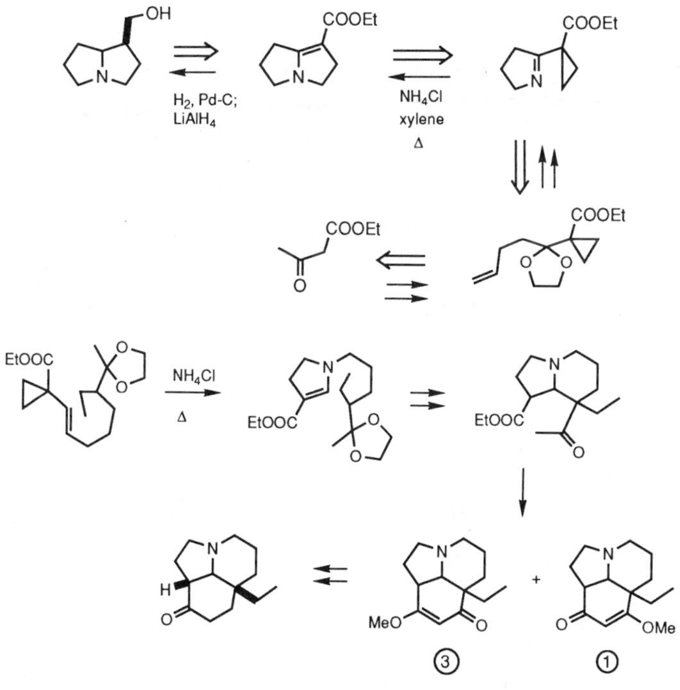

(A synthesis of shihunine [Breuer, 1975] may have also involved imine formation and rearrangement.)

A variant of the necine base synthesis embodies a ring dismutation [Danishefsky, 1977b]. A pyrrolidine ring was formed at the expense of a cyclopropane rupture.

In this synthesis two chirality centers were established in the cyclopropanation step. These in turn were transferred from the geometry of the alkene precursor.

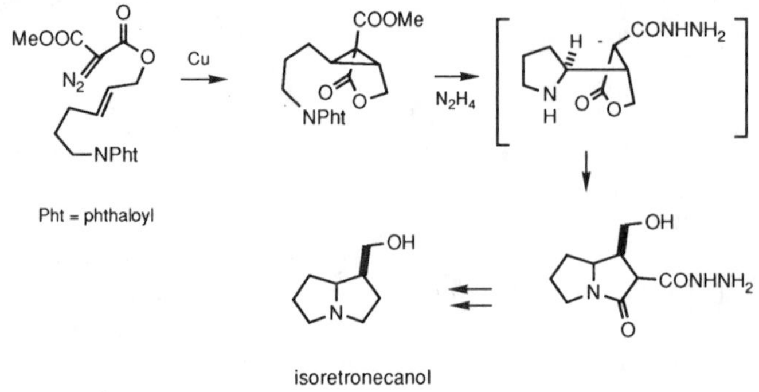

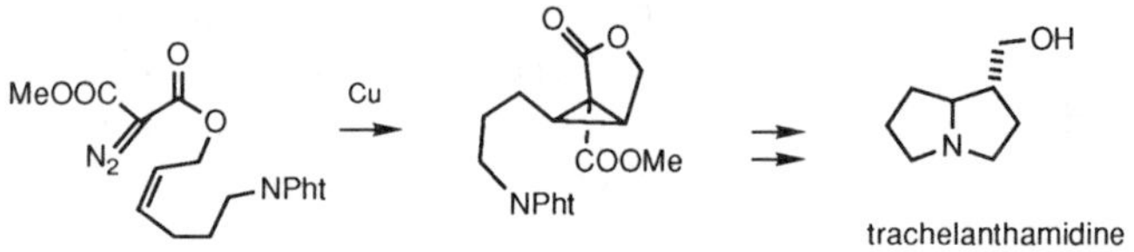

Pht = phthaloyl

isoretronecanol

trachelanthamidine

Opening of activated (acceptor-substituted) cyclopropanes with carbon nucleophiles is also possible. In an intriguing synthesis of eburnamonine [Klatte, 1977] the condensation of tryptamine with a functionalized cyclopropane set the stage for ring cleavage by methyl lithiocyanoacetate, Dieckmann cyclization, and reorganized lactamization.

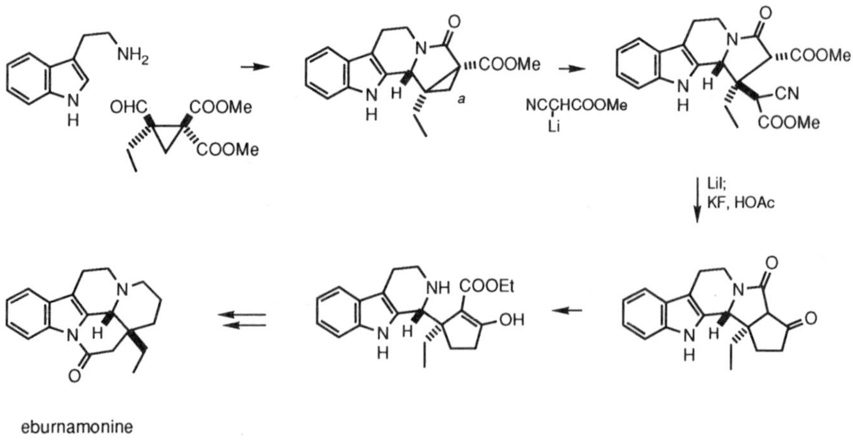

eburnamonine

The D-ring for an estrone synthesis [Bryson, 1980] was constructed from 2,4-pentanedione and (1-carbethoxy)cyclopropyltriphenylphosphonium tetrafluoroborate. The condensation is a homo-Michael reaction in tandem with a Wittig reaction. Both reactants contain highly accentuated polarity alternation sequences that were responsible for the efficient process.

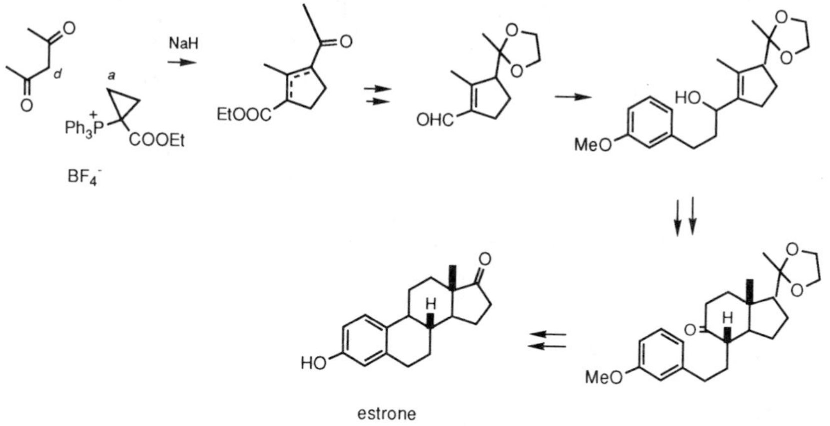

estrone

The opening of a cyclopropane ring with nucleophiles is an S_N2 displacement. For an approach to (+)-estrone [Quinkert, 1982] dimethyl (*S*)-2-vinylcyclopropane-1,1-dicarboxylate was reacted with the sodium salt of dimethyl methylmalonate to form a cyclopentanone. Under the basic conditions a Dieckmann cyclization also occurred. A Michael reaction combined the optically active 2-methyl-3-vinylcyclopentanone with an aryl vinyl ketone, and the product was photoenolized. An ensuing intramolecular Diels–Alder reaction afforded 9α-hydroxyestrone methyl ether.

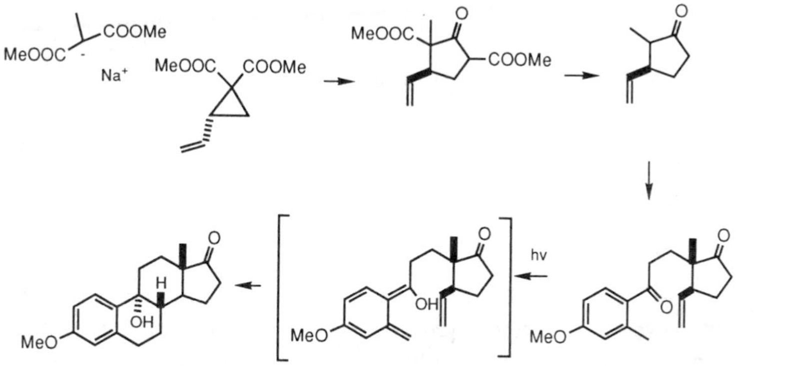

The sidechains of steroids and certain tetracyclic triterpenes usually contain an asymmetric center which is directly attached to a cyclopentane ring (D-ring). Stereocontrol at the point of attachment and the adjacent exocyclic carbon atom is the most important issue in the assembly of this portion of such a synthetic target.

This challenge has been met by the translation of the double bond geometry of an acyclic precursor to the chirality centers by means of a stereoselective alkylative cleavage of cyclopropyl ketones derived therefrom [Trost, 1976b].

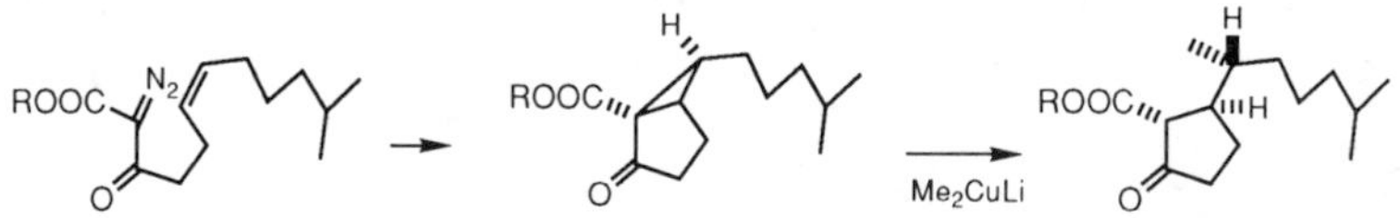

A novel idea to gain access to a properly substituted bicyclo[2.2.1]heptanone which can act as a synthetic intermediate for the prostaglandins consists of a prior cyclization of a bicyclo[3.2.0]heptan-6-one [Newton, 1980]. The increased strain of the resulting tricyclic ketone enables the attack of a nucleophile in the desired fashion. (Note the alternative mode of ring opening to regenerate the original ring system is less favorable, since the targeted bicyclo[2.2.1]heptanone is less strained.)

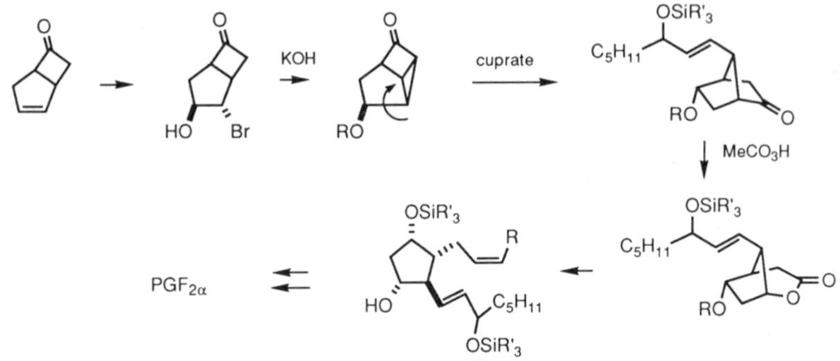

An intramolecular cycloaddition (likely nonconcerted) was devised as the key step of a cedrol synthesis [Corey, 1973]. Ionization of the bicyclo[3.1.0]hexanone with acetyl mesylate was assisted by double-bond participation. The unsymmetrically substituted double bond favored the desired mode of cycloaddition both electronically and sterically.

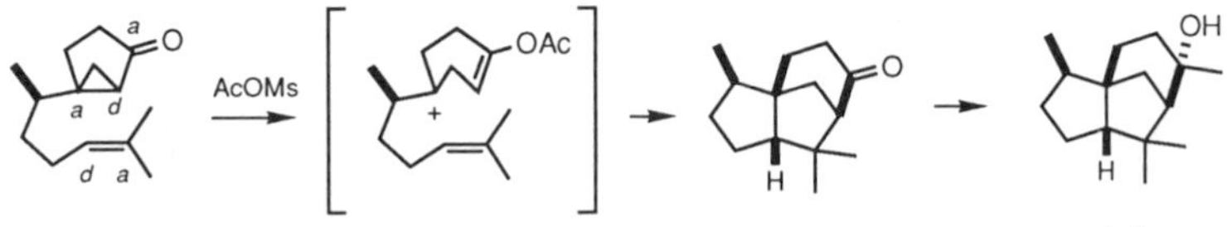

cedrol

A well-positioned aromatic ring may act as an intramolecular donor to assist opening of a cyclopropyl ketone [Stork, 1969]. These substrates can be prepared by an intramolecular carbenoid addition to double bonds [Burke, 1979].

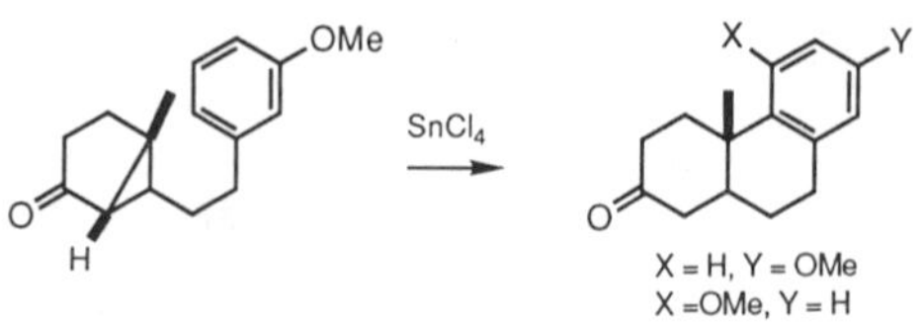

X = H, Y = OMe
X = OMe, Y = H

When the carbenoid addition involves a phenolic partner, de-cyclopropanation may follow. The overall transformation represents an alternative intramolecular alkylation of phenols with an α-haloketone chain under essentially neutral conditions. This spiroannulation technique has been applied to a synthesis of α-chamigrene [Iwata, 1979], among other substances.

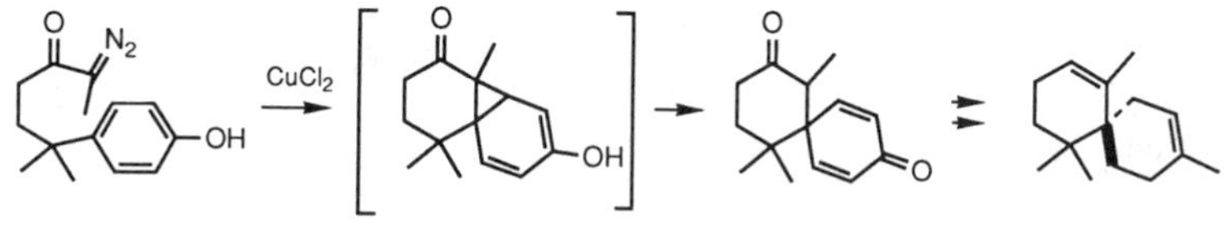

α-chamigrene

The regioselective de-cyclopropanation may have to be carried out separately when the substrate does not possess a donor trigger which can be revealed under the reaction conditions [Beames, 1974].

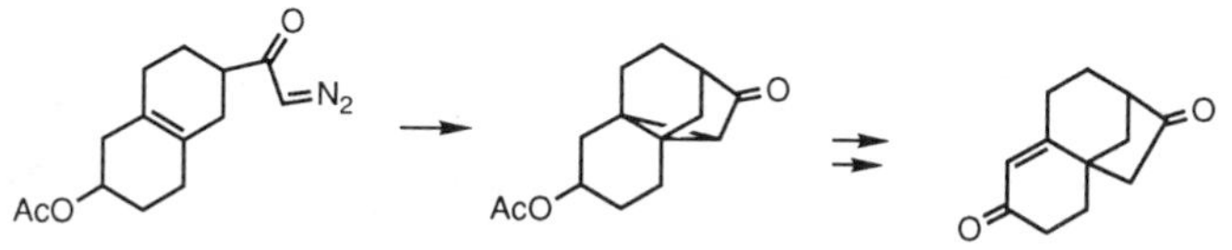

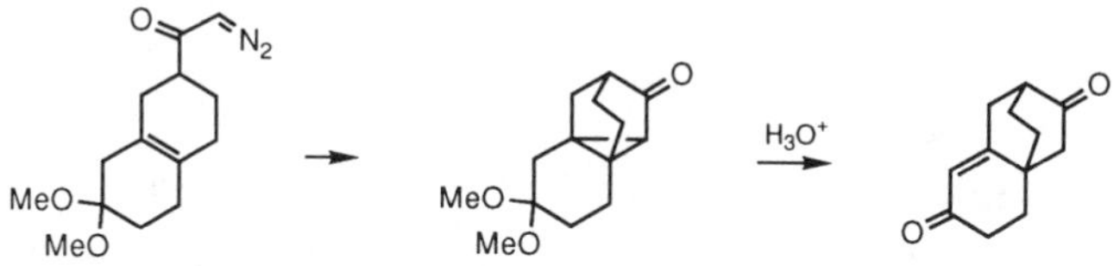

The initial stage of a brefeldin A synthesis [Corey, 1976c] was the preparation of a bicyclo[3.1.0]hexanonedicarboxylic ester and its de-cyclopropanation by treatment with a base. The resulting enone then accepted an alkenyl residue from a cuprate reagent on being elaborated into the macrolide.

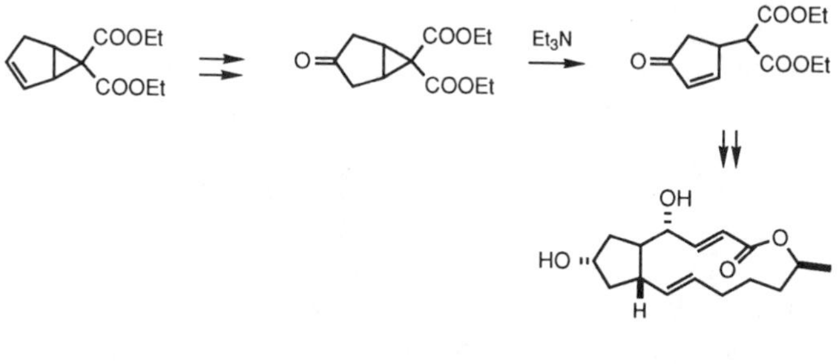

brefeldin-A

The retro-Michael fission of cyclopropanecarboxylic esters has been known for a long time. The transformation of dimethyl α-tanacetonedicarboxylate to tanacetophorone [Wallach, 1912] is interesting in that a Dieckmann cyclization followed the ring opening.

tanacetophenone

An intriguing synthetic route for cafestol [Corey, 1987c] relates the diol subunit to the furan ring at the two opposite corners of the molecule. Apparently, the efficient polyene cyclization [Johnson, 1981] motivated the design. Such cyclization of a B-seco intermediate would resolve the relative stereochemistry at C-9, 10. With an intact furan ring as charge terminator a double bond would be generated between C-5 and C-6, and its reduction (Li, NH$_3$, ROH) may then be controlled to give the *trans-anti* ring system.

The retrosynthetic reasoning must now address the cationic initiator end. While it is certainly possible to construct a CD-ring portion, embedded with an ionizable group (e.g., allylic alcohol) in the C-ring to induce the cyclization, there are potential complications from rearrangement of the bicyclo[3.2.1]octane subunit into a [2.2.2] isomer. An even more important consideration is that setting up a complex substrate to create only one C—C bond is not an attractive proposition.

The idea underlying the brilliant synthesis evolved from the cognizance that a bicyclo[3.2.1]octane can be unveiled from a tricyclo[3.2.1.0$^{2.7}$]octane derivative by solvolytic cleavage of a cyclopropyl carbinol. The latter process may be directed by a double bond, as the formation of an incipient allyl cation is much more favorable than a nonconjugated carbenium species. At this point the further disconnection by a ketocabenoid addition transform became clear and a 1-substituted 2,4-cyclohexadiene-1-carboxylic acid emerged as an easily accessible subtarget molecule. A minor operation inherent in this plan is the removal of the extra oxygen function in the D-ring.

The angular strain of a cyclopropane is high enough to cause bond scission on protonation, and only a simple polar substituent nearby is sufficient to control the regiochemistry of the ring opening. This process is the basis of a conversion of estrone into androst-4-ene-3,17-dione [Birch, 1964].

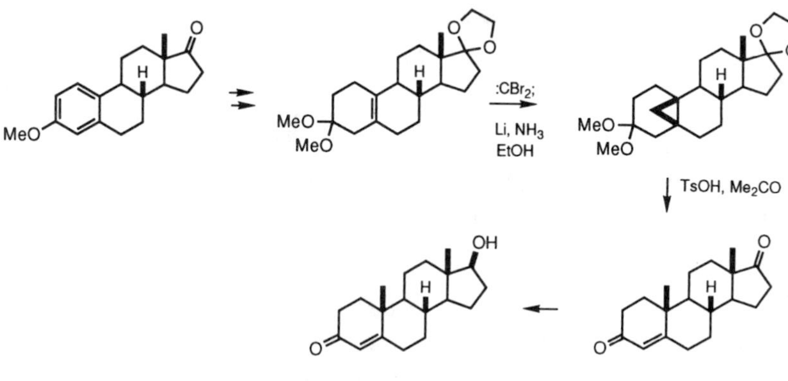

testosterone

Birch reduction of the aromatic ring, ketalization, cyclopropanation at the C-5/C-10 double bond using dibromocarbene and subsequent debromination, and acidolysis constitute the steps of this transformation. When the C-3 ketone was unmasked, it defined C-5 as an acceptor site, and the 3-enol displaced the protonated C-5/C-19 bond.

13.2. RING OPENING OF AZIRIDINES AND EPOXIDES

Corydolactam contains an α-substituted α,β-unsaturated carbonyl subunit. It is logical to conduct its synthesis by an alkylation route with an acetoacetic ester as donor. The acceptor synthon is a β-aminoethyl cation and its synthetic equivalent is an aziridine [Kametani, 1971].

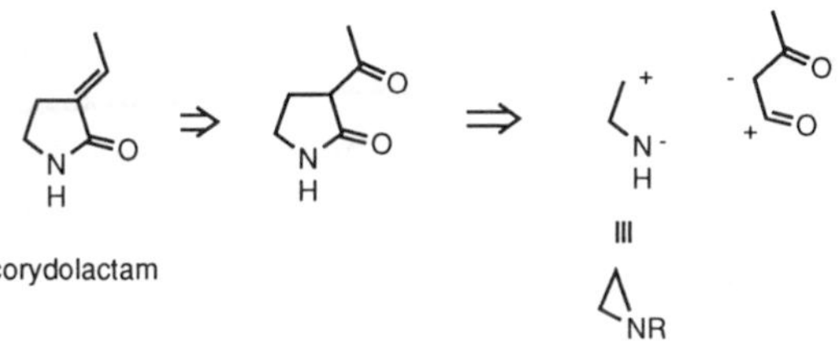

corydolactam

Because aziridines are highly strained, they are susceptible to attack by nucleophiles if active N—H is absent. Generally, quaternization or N-sulfonylation facilitates the ring opening.

There are two asymmetric centers in pseudoconhydrine, and since a β-amino alcohol can be derived from hydration of an aziridine a synthesis of the alkaloid based on such a strategy [Harding, 1984a] is worthwhile.

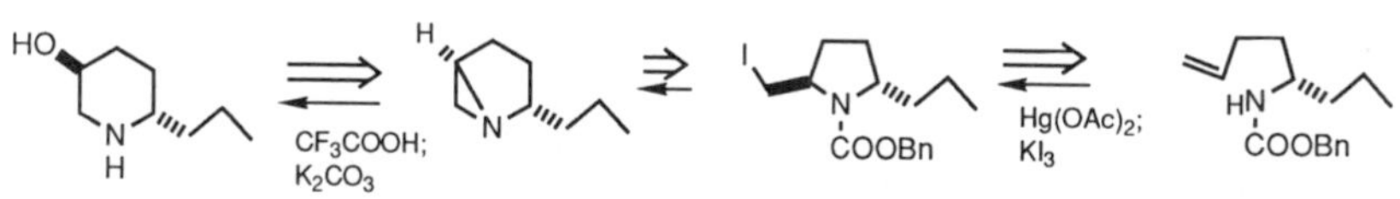

pseudoconhydrine

Aziridines can be prepared via amidomercuration, halogen/mercury exchange and internal alkylation of the corresponding amines. Opening of a 1-azabicyclo[3.1.0]hexane can lead to a 3-substituted piperidine and a 2-substituted pyrrolidine. Adjustment of reaction conditions may favor one product over the other.

The benzhydryl center in cherylline acquires a strong acceptor character from its conjointment with two *p*-hydroxy groups. Advantage can be taken of this situation in the synthesis of the alkaloid [Kametani, 1982] via a C(ar)—bond formation.

An interesting precursor for the skeletal assembly is an aziridinium ion which is easy to prepare from the corresponding benzaldimine. The transient quinonemethide species which is expected to be in equilibrium with the aziridinium ion is conducive to cyclization.

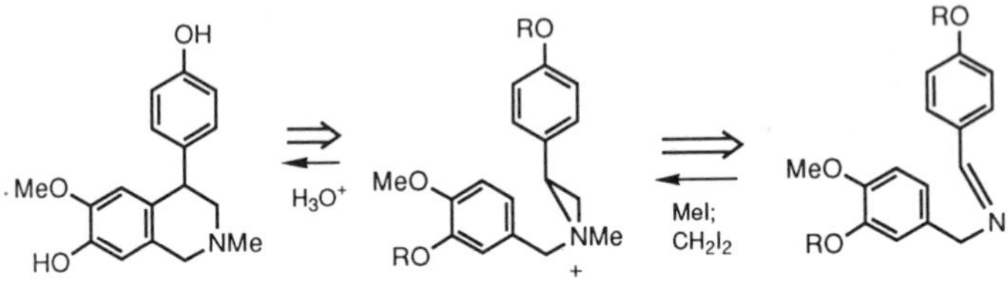

cherylline

A bond disconnection of ibogamine exposes 2-(3-indolyl)ethyl and quinuclidin-6-yl synthons. The latter species is disjoint as the acceptor center is *β* to the basic nitrogen atom. Consequently, an aziridine would serve well in the acceptor role.

The tricyclic aziridine was prepared by nitrene interception [Nagata, 1968]. *N*-Acylation of the aziridine was accompanied by regioselective ring opening to provide the desired intermediate. The alternative mode of cleavage would have led to an azabicyclo[3.2.1]octane with two bulky axial substituents in a chair cyclohexane ring.

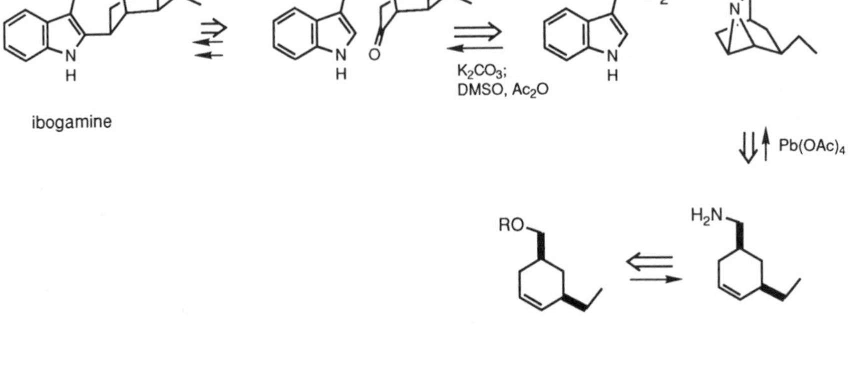

ibogamine

The less highly substituted benzylic carbon atoms of morphine and O-methylpallidinine are each under strong influence by the p-methoxy group. In principle, an intramolecular Friedel–Crafts alkylation is a useful way to bridge the morphinan skeleton. Since this acceptor site is also disjointly related to the nitrogen atom, a maneuver involving aziridinium ions would be profitable in the course of the synthesis [Evans, 1982b; McMurry, 1984].

There were some tactical adjustments in the morphine synthesis [Evans, 1982b], i.e., an α-amino aldehyde was created prior to the cyclization.

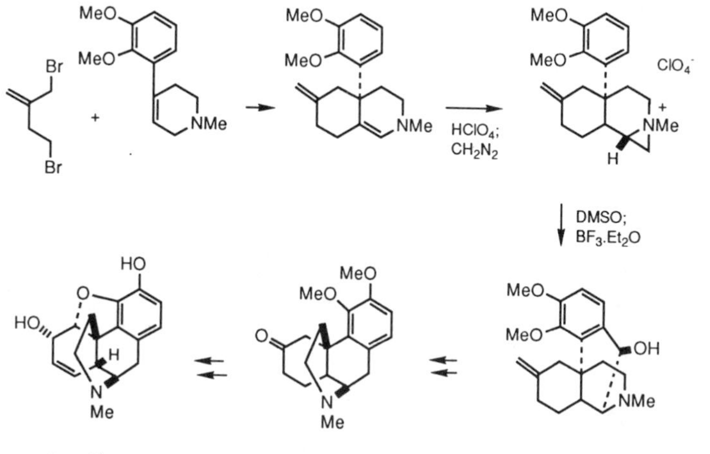

morphine

A route to the morphinane system via an aziridinium intermediate [Broka, 1988] not only resulted in a heterocycle expansion and carbocyclization, it also altered the through-bond relationship of the carbonyl and the amino group.

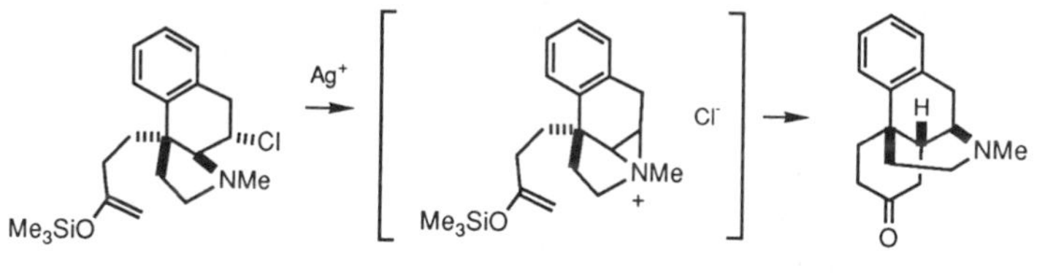

The focal point for synthesis of serratinine [Harayama, 1975] must be the fused/spiro array of the indolizidine portion of the molecule. The presence of an α-amino ketone system indicates an aziridine formation and opening as the most convenient steps to deliver the functional array.

It is interesting that while aziridination in this case gave rise to two diastereomers, only one of them can form the piperidine ring with a pendant emanating from the angular position of the hydrindane nucleus.

The 1,4-dioxygenation in the cyclohexane moiety, together with the requirement of a fused cyclopentene suggests a Diels–Alder approach.

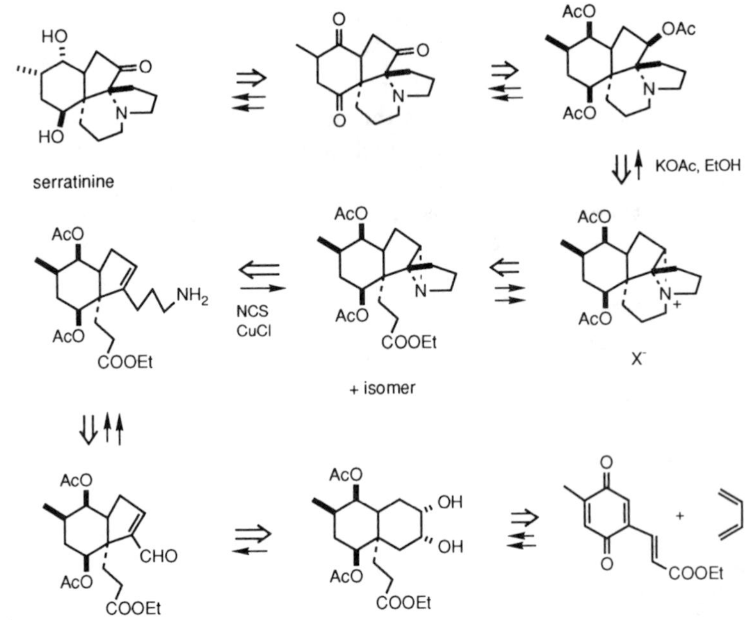

Many synthetic approaches to the pyrrolizidine alkaloids have been essayed. The one-carbon pendant asserts a donor role for its adjacent carbon in the ring system. While disconnection of this donor with the angular carbon yields a most favorable synthon, disconnection at the alternative C—C bond generates a species with an acceptor site disjoint to the nitrogen atom. This situation points again to the use of an aziridine intermediate for synthesis [Kametani, 1984].

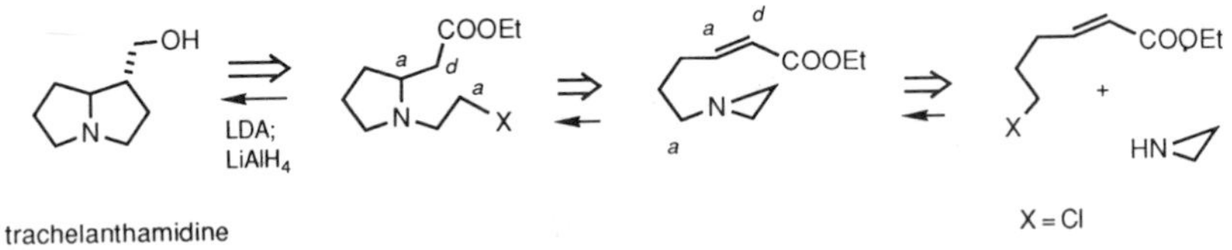

The synthetic applications of epoxides far antedate the use of aziridines because of their availability. The C—C bond formation with a carbon nucleophile represents a reliable method for extending a chain by two carbon units. Frequently the alkoxide ion generated by this ring opening would form an O—C bond with the donor position, as shown in a synthesis of physovenine [Onaka, 1971], and an intermediate en route to methyl nonactate [Lygo, 1987].

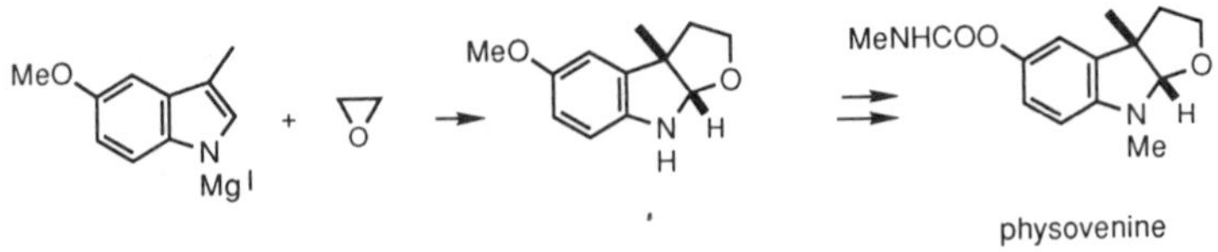

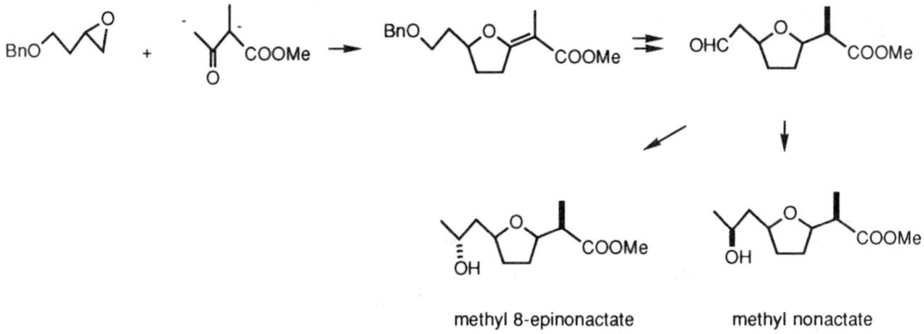

methyl 8-epinonactate   methyl nonactate

Aldols have been created by alkylation of 2-lithiofurans with epoxides and oxidation of the furan ring of the products [Ireland, 1990]. Interestingly, both furans and epoxides are disjoint molecules, their combination leads to conjoint products.

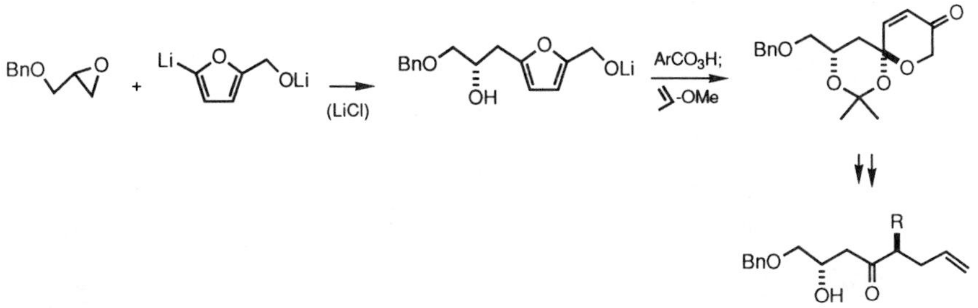

Reductive lithiation of epoxides gives rise to lithium β-lithioalkoxides [Cohen, 1990] which may be used to form 1,3-diols on reaction with carbonyl compounds. This procedure is much more convenient than those involving mercury/lithium or halogen/lithium exchange of β-hydroxy mercurials or halides.

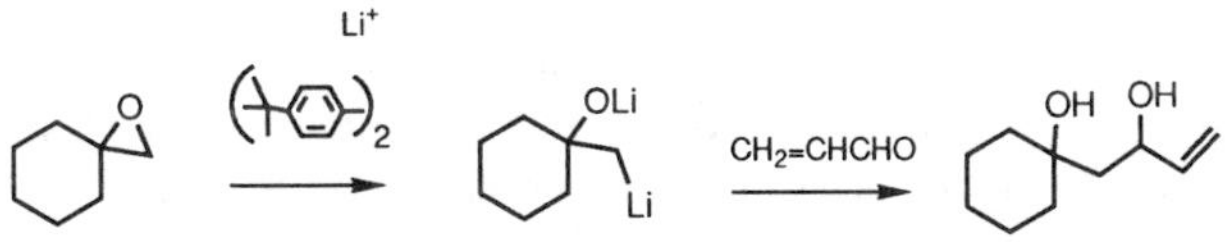

Again, it must be emphasized that reduction of a disjoint molecule results in a conjoint species.

It is of interest to note a chemoselective epoxide formation from a 1,6-anhydro-β-D-glucopyranoside derivative in the course of a synthesis of ($-$)-cis-rose oxide [T. Ogawa, 1978]. The internal displacement of a triol ditosylate led to the 3,4-epoxide.

The ditosylate is essentially symmetrical except for the two-atom bridge. Formation of the 3,4-epoxide is favored on the grounds that the acceptor character of C-2 is diminished by the extra oxygen (acetal group).

D-glucose

(-)-*cis*-rose oxide

The same epoxide was also generated *in situ* during an allylation step in an approach to thromboxane B$_2$ [A.G. Kelly, 1980].

Allylic epoxides and esters act as acceptors under palladium(0) catalysis. It is of special interest to discuss the S$_N$2' substitution of ω-vinyllactones with malonate anions [Trost, 1979b, 1981]. Depending on the size of the lactone ring, the displacement reaction may result in a product in which the carboxylic termini are either conjoint (from an even-membered lactone) or disjoint (from an odd-membered lactone).

Hexahydroindoles can be prepared via a Diels–Alder reaction of 1-acetoxy-1,3-butadiene and acrolein (and obviously their respective analogs), chain extension of the aldehyde to a homologous amine, and palladium(0)-catalyzed intramolecular *N*-alkylation [Trost, 1976a].

Consistent with our analysis of conjoint/disjoint ring system synthesis is the fact that the formation of a disjoint five-membered ring from a conjoint precursor such as a Diels–Alder adduct must involve an umpolung step. In the present case the step is the homologative amination, —CHO → —CH$_2$—CH$_2$—N.

13.3. CYCLOPENTANNULATION

On account of the disjoint cycle there are not as many methods for the synthesis of five-membered ring compounds as the six-membered homologs. Annulation that brings two components together to form the ring system, either in the [4 + 1] mode or the [3 + 2] mode, must involve contrapolarization of one of the components or use an umpoled (disjoint) species. Among recent developments two methods deserve special attention.

Annulation of enones with silylallenes in the presence of a Lewis acid [Danheiser, 1981a] is very efficient. The enhanced donor reactivity at the allenic terminus owes to the silyl substituent through a polarity alternation sequence.

Since the initial vinyl cationic species is reluctant to cyclize (to form a cyclobutane) migration of the silyl group ensues. Note that methyl migration is suppressed as it would result in an unfavorable α-silylvinyl cation in which two acceptors are directly connected. A formal contrapolarization occurs in the silyl migration step. In the original allene, the silylated trigonal carbon is a donor, and upon migration it becomes an acceptor, and the final product has a conjoint segment extending from the silicon atom to the carbonyl group.

The second significant annulation method is the palladium(0)-catalyzed cycloaddition of 2-(trimethylsilyl)methyl-2-propenyl acetate (and homologs) to alkenes. In view of the electron deficiency requirement on the part of the alkene addends, it is likely that the cycloaddition is initiated by attack of the donor end of the dipolar species. The reaction is then an asynchronous concerted reaction.

To illustrate its synthetic utility, a synthesis of brefeldin A [Trost, 1986a] was designed to incorporate this cycloaddition as the first step.

brefeldin-A

The trimethylenemethane-PdL$_2$ complex cycloadds to pyrones to give lactone-bridged cycloheptenes [Trost, 1989b]. This is a stepwise reaction involving nucleophilic attack by the complex at C-6 of the pyrone system, and ring closure with the α-carbon of the dienolate ion. Acceptor substituents can divert the reaction pathway toward a [3 + 2] cycloaddition.

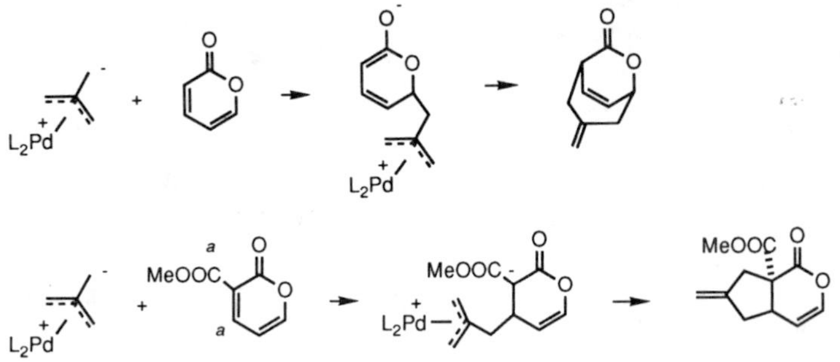

2-Oxyallyl cations, conveniently generated from α,α'-dibromoketones by treatment with an iron carbonyl reagent, are capable of undergoing [3 + 2]cycloadditions with alkenes. This methodology is the basis for a one-step preparation of α-cuparenone [Hayakawa, 1978], albeit in low yield.

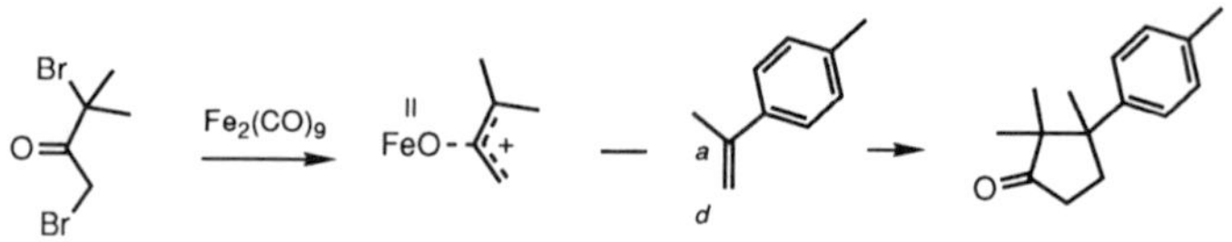

α-cuparenone

Note that the product is the more crowded isomer. Perhaps the higher stability of the intermediate with a tetrasubstituted iron enolate favored its formation and the eventual cyclization.

An intramolecular version of the cycloaddition led to camphor [Noyori, 1979].

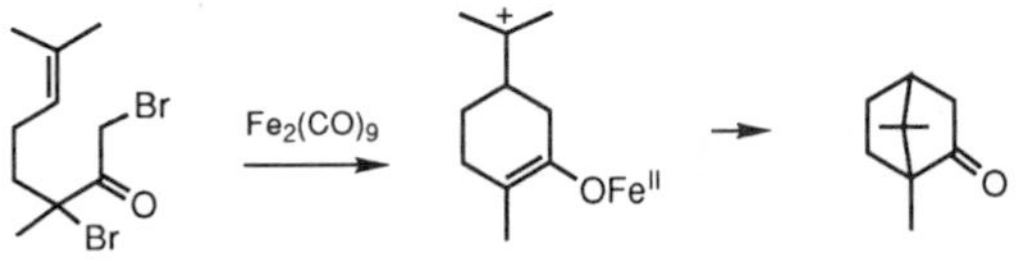

camphor

The $^+CH_2$—CO—$CH^=$ synthon which is valuable for cyclopentenone synthesis has a synthetic equivalent in $BrCH_2COCH$=PPh_3 [Altenbach, 1979]. A reaction of this conjunctive reagent with an aldehyde enolate has resulted in a cyclopentenone in low yield.

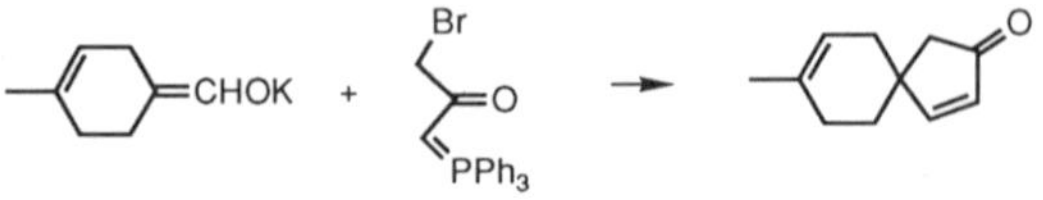

(It is of some interest to compare this reagent with the conjoint methyl α-bromomethylacrylate which reacts with enamines to give cyclohexanone derivatives [Dunham, 1971].)

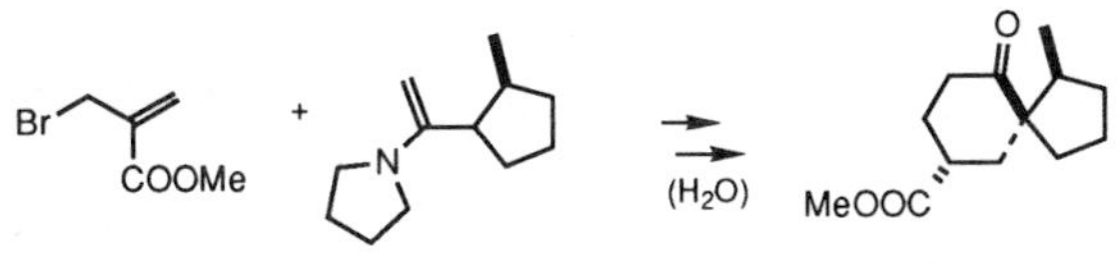

13.4. INTRAMOLECULAR ALDOL AND RELATED CONDENSATIONS

Self-condensation of 1,4- and 1,6-dicarbonyl compounds with at least one α-hydrogen leads to five-membered ring products. As mentioned previously (Sec. 4.1.) the direction of the condensation may be controlled by inherent structural features and/or reagents.

The synthesis of 1,4-dioxygenated substrates has been discussed in a separate chapter. A re-emphasis here is that their formation involving an internal C—C bond formation must employ some umpolung maneuver or umpoled (disjoint) precursor. With respect to the latter tactics it is perhaps not redundant to indicate a synthesis of methyl jasmonate [Torii, 1977]. The fumaraldehydic acid ester acetal is such an entity.

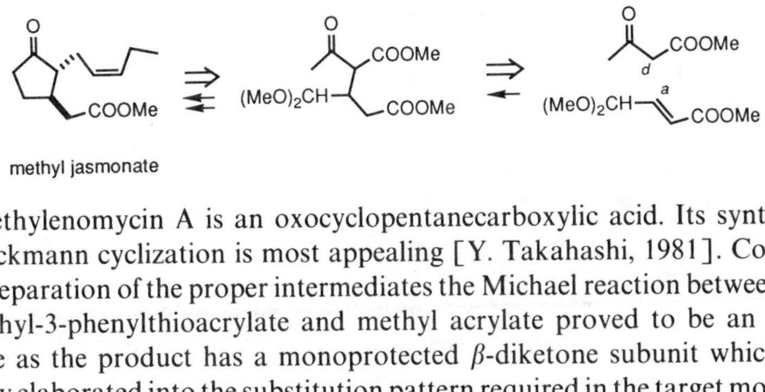

methyl jasmonate

Methylenomycin A is an oxocyclopentanecarboxylic acid. Its synthesis via a Dieckmann cyclization is most appealing [Y. Takahashi, 1981]. Concerning the preparation of the proper intermediates the Michael reaction between methyl 2-methyl-3-phenylthioacrylate and methyl acrylate proved to be an excellent choice as the product has a monoprotected β-diketone subunit which can be readily elaborated into the substitution pattern required in the target molecule.

The Michael donor is a disjoint species, perforce in the formation of a five-membered ring (each disjoint cycle must be derived from one disjoint component or by a contrapolarizing reaction).

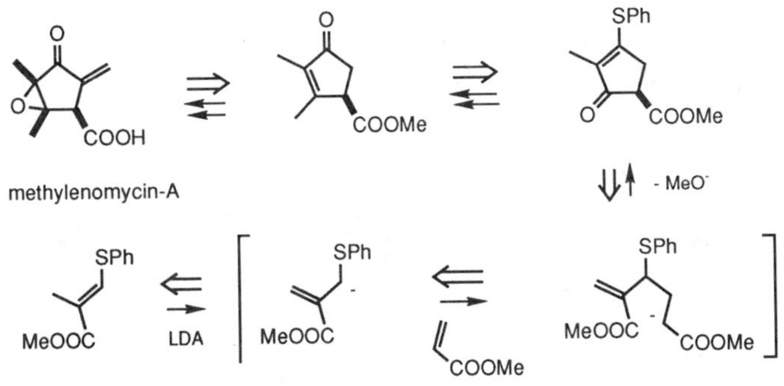

methylenomycin-A

The first prostaglandin A_2 synthesis based on carbohydrates is the elaboration of PGA_2 from L-rhamnose [Stork, 1976]. Two Claisen rearrangements were used to extend chains to the proper length and with an ester in each of them. The α,β-disubstituted adipic ester was the precursor of the five-membered ring (by Dieckmann cyclization).

The interesting aspect of the reaction sequence is that two transpositional carbomethoxymethylation steps performed on a *vic*-diol system gave rise to the disjoint diester. The second rearrangement involving transfer of chirality rendered the synthesis stereospecific.

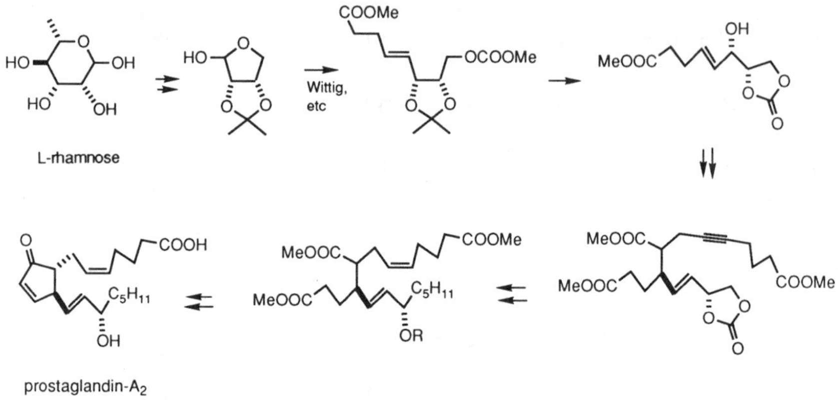

prostaglandin-A_2

1,6-Dicarbonyl compounds may be obtained conveniently by oxidative cleavage of cyclohexenes. The powerful Diels–Alder transform connected with the latter class of molecules renders the chemical operation encompassing the cycloaddition, ring cleavage, and intramolecular condensation most valuable for synthesis of cyclopentanoids with at least one carbon chain. It is impossible to mention the applications of the reaction sequence to even the most important syntheses. With this severe limitation one may single out the service in the synthesis of cholesterol [Woodward, 1952], prostaglandin E_1 [Corey, 1968a], gibberellic acid [Corey, 1978a,b], and retigeranic acid [Corey, 1985c].

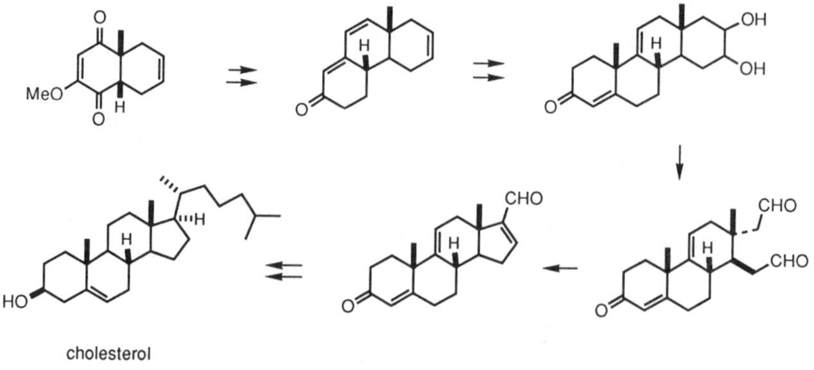

cholesterol

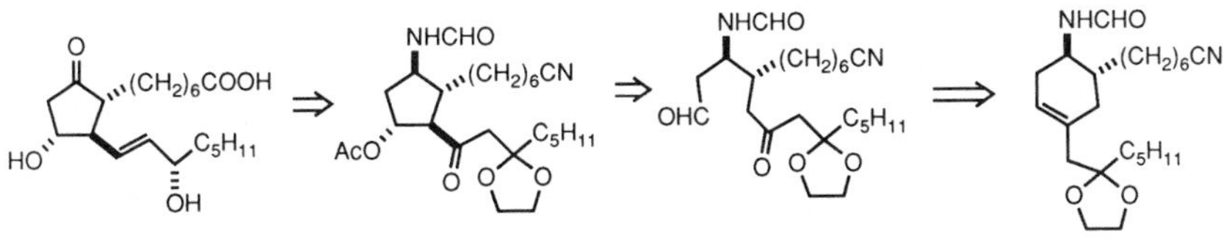

prostaglandin-E₁

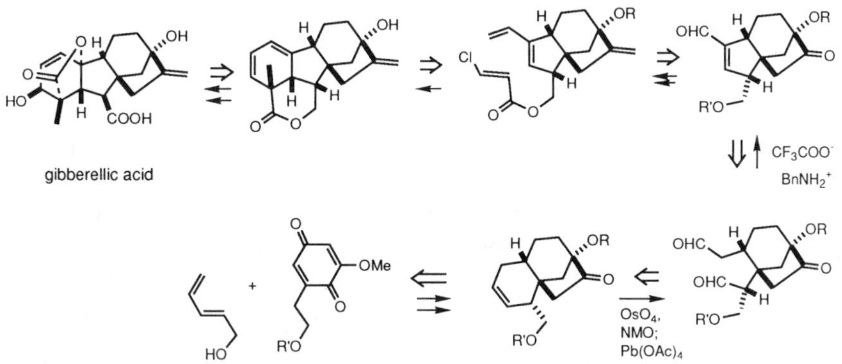

gibberellic acid

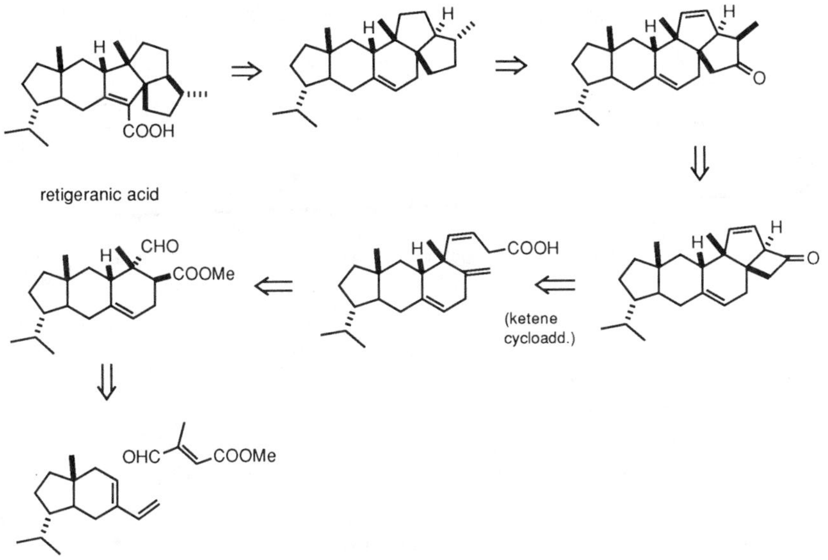

retigeranic acid

Condensation of cyclopentadienecarboxylic esters with aldehydes leads to fulvene-2-carboxylic esters. A synthesis of $\Delta^{9(12)}$-capnellene [Y. Wang, 1990] was accomplished by manipulating such a fulvene containing a terminal aldehyde group (released from a masked form) via a Michael–aldol type annulation.

$\Delta^{9(12)}$-capnellene

The unique electronic nature of cyclopentadiene enabled carbon atoms at both the β and γ positions with respect to the ester substituent to act as a donor. In the fulvene formation step and the Michael reaction the donor site is at C-β, whereas the aldol ring closure required the C-γ to attack the aldehyde acceptor. Note the substrate is a 1,6-dioxo species.

13.5. MODIFICATION OF EXISTING CYCLOPENTANOIDS AND SYNTHESIS FROM DISJOINT PRECURSORS

In the synthesis of molecules containing difficult structural elements the efficiency is greatly increased if construction of such elements can be avoided. The early synthesis of aromatic compounds by modification of other available aromatic substances was a matter of necessity as well as expediency. The necessity was due to the paucity of methodology for their construction.

Cyclopentanoids are relatively difficult to assemble in comparison with their homologs, the cyclohexanes. Consequently, chemists always look for ways to disconnect relevant synthetic targets to yield readily available precursors with a pre-existing cyclopentane subunit. Norbornane derivatives and dicyclopentadiene are distinctly popular starting materials. Dicyclopentadiene may be used to provide a diquinane or a 2,3-disubstituted norbornane framework, but most frequently it is thermally dissociated into the monomer to engage in the [2 + 2] and [4 + 2] cycloadditions.

A retrosynthetic analysis of verbenalol easily reaches a diquinane which contains an *exo* methyl group. This methyl group is accessible from the corresponding enone by a stereospecific conjugate addition (cuprate chemistry or Kharasch reaction). A diquinane precursor for the synthesis is meritorious because the *cis* ring juncture stereochemistry can be unequivocally maintained.

The 1,4-relationship between the ketone group and the ester is best elaborated from a precursor containing a latent one-carbon pendant, either oxygenated or as part of a double bond, so that oxidation generates a carboxyl derivative. This crucial requirement led to the consideration of using 1-oxodicyclopentadiene as the starting material for a synthesis of verbenalol [Sakan, 1968].

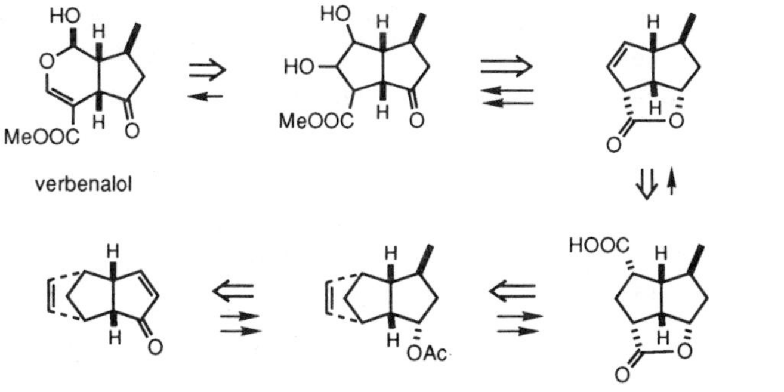

verbenalol

A bicyclic lactone (Corey lactone) is a versatile intermediate for all families of prostaglandins [Corey, 1969b]. It is remarkable that all the peripheral oxygen functions are pairwise conjoint, except those forming the lactone ring. Large-scale preparation of this lactone has been developed [Bindra, 1973; Peel, 1974] starting from a Prins reaction on norbornadiene.

Double-bond participation in the Prins reaction effectively contrapolarizes a terminus of the skip diene system. The resulting tricycle contains a disjoint 1,4-dioxy segment and a conjoint 1,5-dioxy segment, depending on how the bond paths are traced. Oxygen atom insertion by the Baeyer–Villiger reaction of a halonorbornanone carboxylic acid which was obtained in two steps from the Prins reaction product served to establish the oxygenation pattern except that of the γ-lactone. The latter was set by an intramolecular S_N2 displacement.

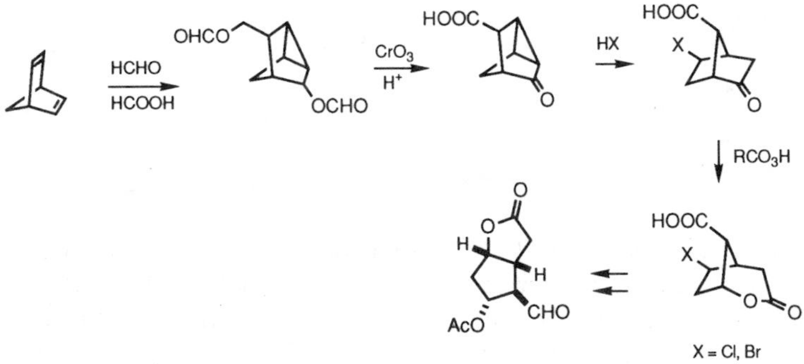

X = Cl, Br

Assembly of the Corey lactone by cyclization of a linear chain is possible only if the latter is equipped with a disjoint segment. The disjoint is between C-β and C-γ of the lactone ring. Disconnection of the potential linear precursor suggests (S)-malic acid to be an ideal point of departure [Paul, 1976]. Chain extension at both carboxyl termini can be performed in one operation.

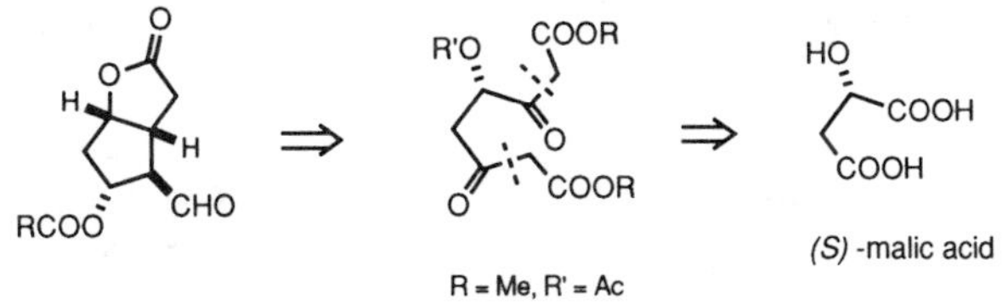

(S)-malic acid

R = Me, R' = Ac

For contemplating a synthesis of terrein it is advantageous to ignore temporarily the ethylidene group from the sidechain. Disconnection of the resulting cyclopentenone reveals a symmetric predecessor [Altenbach, 1990]. This compound can be correlated with tartaric acid.

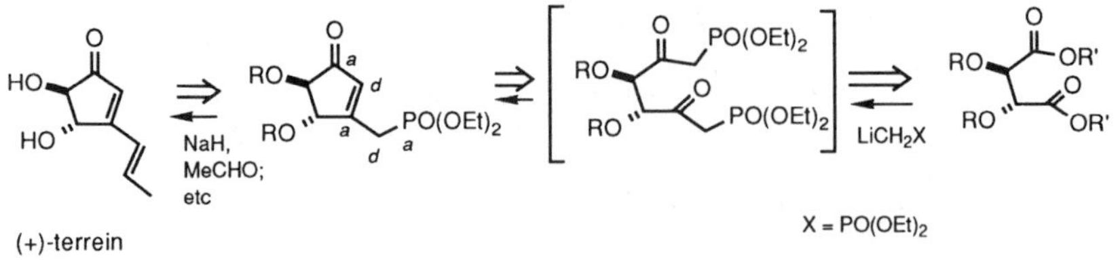

(+)-terrein

Neplanocin A may be considered similarly, the most conspicuous structural features of both molecules being the presence of a *vic*-diol subunit in the five-membered ring. In the case of (−)-neplanocin A [Bestmann, 1990] the retrosynthetic analysis must deviate slightly because the dihydroxyl system does not correspond to L-tartaric acid. However, protection of the diol in the form of an acetonide resolved the problem, as cyclization to the carbocycle also induced epimerization of the ring juncture to the thermodynamically more stable configuration.

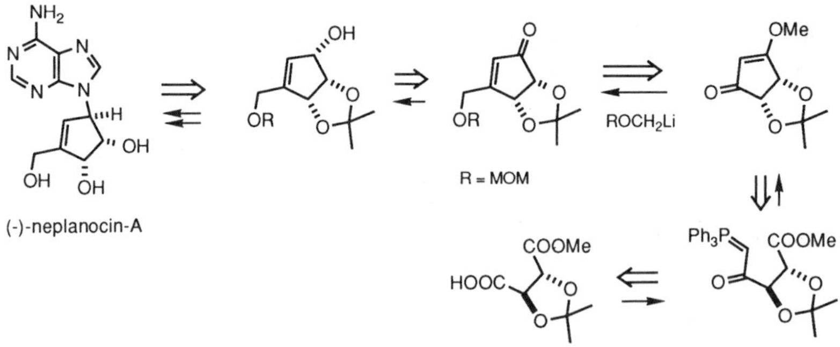

(-)-neplanocin-A

Retrosynthetic analysis of tetrahydroalstonine with attention to exploiting a hidden element of symmetry resulted in the formulation of a synthetic route [Hölscher, 1990] starting from a *cis*-bicyclo[3.3.0]octane-3,7-dione derivative which is readily available by a double Michael reaction of 4-acetoxy-2-cyclopentenone with an acetonedicarboxylic ester. The new ring inherits its disjointness from the "cyclopentadienone" template.

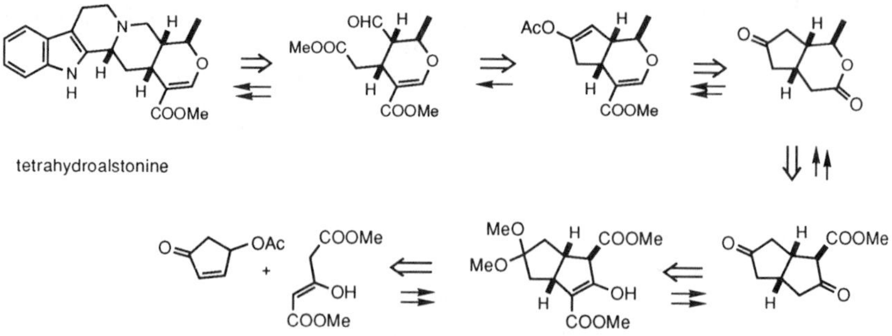

tetrahydroalstonine

In retronecine the two oxygen functions are disjoint while each of them is conjoint to the nitrogen atom along noncrossing circuits. The implication of this arrangement to synthetic strategy is that one can graft a disjoint O—C$_4$—O synthon to the remainder of the target molecule [Geissman, 1962].

The Michael addition of an amine to a fumaric ester produces the shorter disjoint chain. This also keys the closure of an oxygenated pyrrolidine by a Dieckmann condensation since a C$_3$ ester chain can be easily attached to the nitrogen atom (by Michael reaction). To proceed forward it needs to introduce another two-carbon pendant emanating from the amino function to the monocyclic intermediate already containing an acetic chain at an α-carbon of the pyrrolidine ring. In other words, a conjoint chain must meet with a disjoint segment in the formation of a five-membered ring.

The effectiveness of a Michael-initiated Dieckmann cyclization is shown in a synthesis of sarkomycin [Kodpinid, 1984]. This antibiotic is an oxocyclopentanecarboxylic acid, and the applicability of the smooth condensation reaction to close the ring depends on a disjoint precursor, as in the present case, dimethyl itaconate (in a protected form).

(S)-4-Hydroxy-2-hydroxymethyl-2-cyclopentenone has been obtained from (−)-quinic acid by a sequence of reactions including protection of the three functional groups except the vic-cis diol system [Elliott, 1980]. The free diol was cleaved (NaIO$_4$) and the resulting dialdehyde subjected to aldolization with pyrrolidinium acetate to afford a five-membered ring product.

13.6. HETEROCYCLIC FIVE-MEMBERED RINGS

Five-membered heterocycles do not fare better electronically than cyclopentane as two of the nuclear atoms still must pair in a disjoint fashion. In fact, certain arrangements are conducive to ring fission. For example, isoxazoles in which C-3 is unsubstituted or is acylated–carboxylated are very sensitive to bases. Their fragmentation to give the conjoint β-keto nitriles occurs readily. The nitrogen atom contrapolarizes ($d \rightarrow a$) as the reaction proceeds.

Despite the intrinsic electronic malaise in odd-membered rings, and its magnification on substitution with heteroatom(s), there are still many methods for the construction of heterocycles. We can only select a very few to discuss. Because of the commonality of pyrrolidine and tetrahydrofuran subunits in many natural products, synthesis of these ring systems deserves particular attention.

The most logical disconnection of a pyrrolidine ring is an internal C—N bond. An N-Alkylation transform or a Schiff-reduction transform is conceivable. From the latter transform 4-amino carbonyl compounds are required to reconstitute the ring. These disjoint compounds necessitate an umpolung technique to assemble them.

The utility of nitro compounds in these circumstances is evident. Reaction of conjugate nitroalkenes with aldehydes and ketones in the presence of a base is one option employed in a synthesis of polyzonimine [Smolanoff, 1975].

polyzonimine

The more popular version is the Michael addition of nitroalkanes to enones. Nitroalkanes are more readily available and much more stable than nitroalkenes. A simple example describing the use of this method is a synthesis of myosmine [Stein, 1957].

myosmine

A synthesis of monomorine I [Stevens, 1982b] started from the preparation of a nitro ketone and its reduction to afford a *cis*-2,5-dialkylpyrrolidine. Since catalytic hydrogenation of 1-pyrrolines is expected to be stereoselective, construction of the indolizidine nucleus from a pyrrolidine template seems better assured than the alternative assembly from a piperidine intermediate. Admittedly it requires another C=N bond reduction at the end of the synthesis, however, reduction of a fused ring system is subject to stereoelectronic control which is easier to manage. Actually, the task was accomplished by using a borohydride reagent.

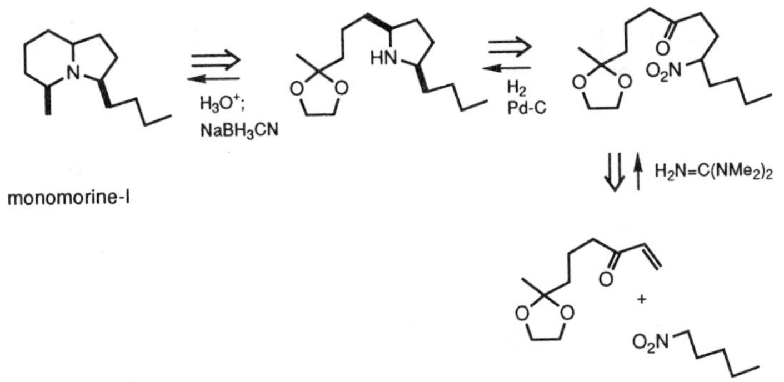

monomorine-I

γ-Nitro esters are similarly accessible by the Michael reaction. On reduction of the nitro group, cyclization ensues. Thus, γ-lactams such as anantine [Tchissambou, 1982] may be synthesized by this route.

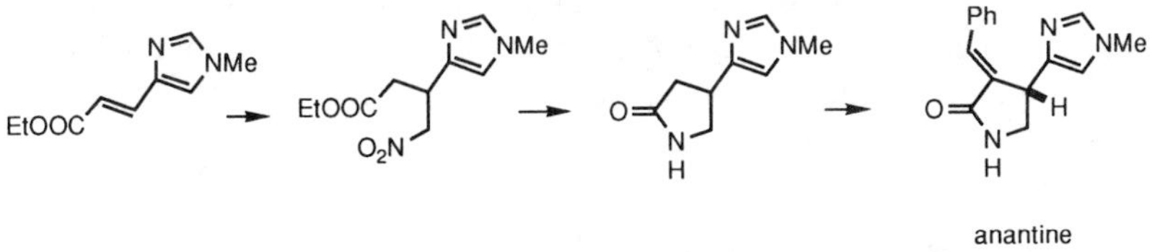

anantine

There are many ways to disconnect mesembrine. For its synthesis the formation of the nonaromatic portion by an intramolecular Michael reaction represents a most direct method. Schiff condensation followed by reduction is a tactical alternative.

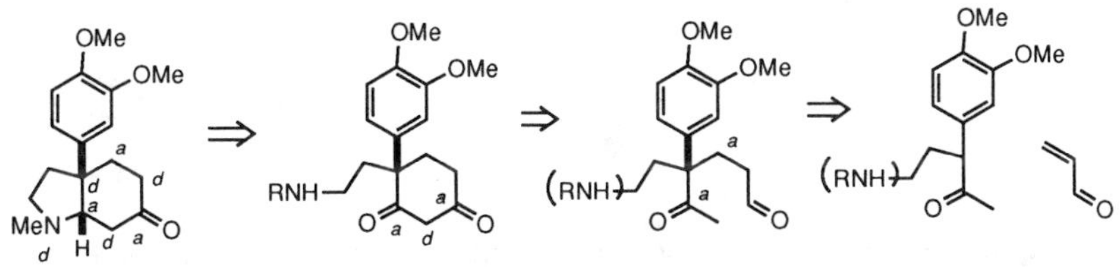

This latter approach can be further extrapolated, for the sake of operational facility, to a bisannulation [Oh-ishi, 1968].

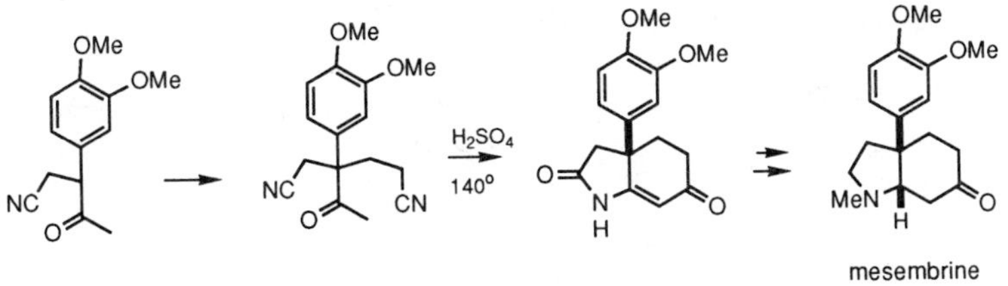

mesembrine

A physostigmine approach [Takano, 1982b] based on reductive formation of the aminal system benefited from the chiral synthesis of γ-lactones. The asymmetric centers governed the steric course of the methylation at the benzylic site.

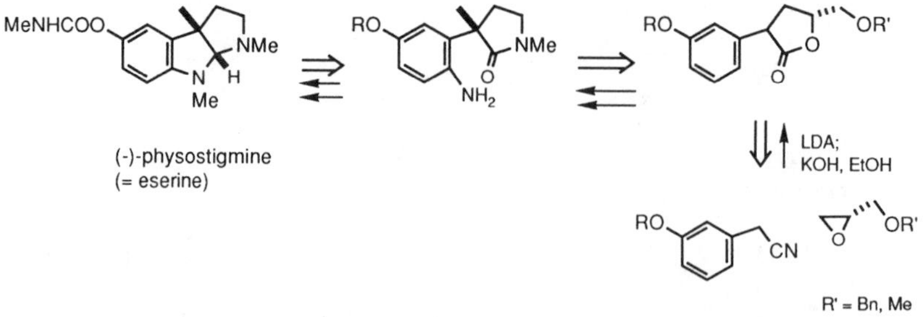

(-)-physostigmine
(= eserine)

R' = Bn, Me

A less straightforward plan for synthesis of the flavopereirine ring system [M. Takahashi, 1971] was apparently inspired by the quinoline to indole transformation (cf. reversed biosynthesis relationship of cinchonamine with the quinine bases).

Disconnection of the N—C linkages of flavopereirine generates a 2-acylpyridine. Reconnection of the terminal carbonyl group with the anilino nitrogen gives rise to a 4-acyltetrahydroquinolinone. Synthesis of the latter type of compounds requires a contrapolarizing step (e.g., oxidation) as the 1,4-dicarbonyl segment is disjoint.

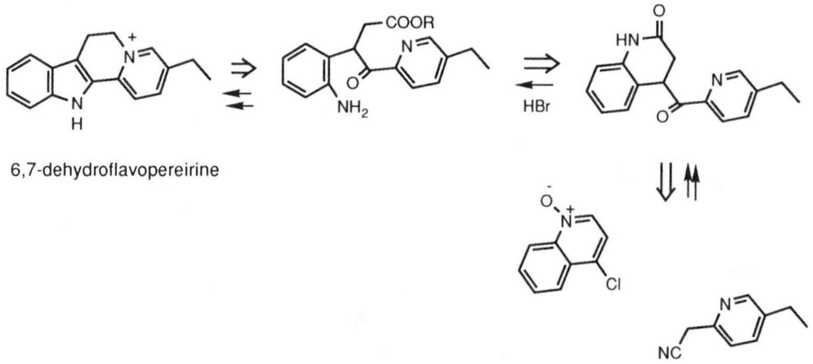

6,7-dehydroflavopereirine

5-Oxyindole derivatives have been recognized as tautomers of the Schiff bases derived from 2-(aminoethyl)-1,4-benzoquinones. Accordingly, a convenient approach to these substances involves redox manipulations of *o,m*-dioxygenated phenylacetonitriles. Bufotenine and physostigmine have been acquired [Harley-Mason, 1954a,b] by such a method.

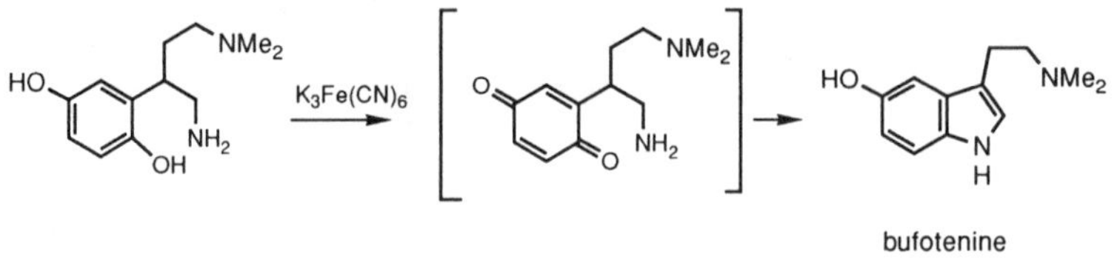

bufotenine

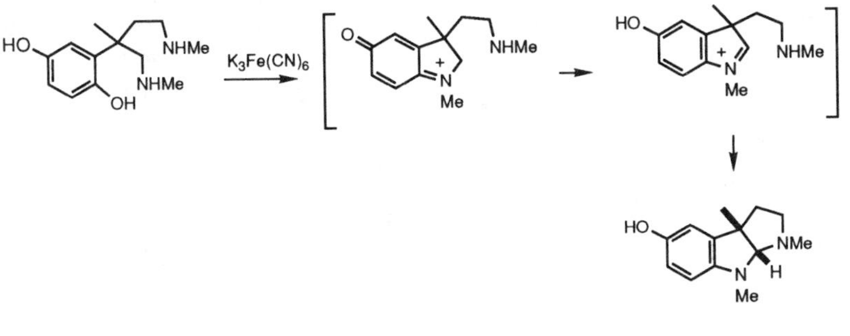

eseroline

Condensation of 1,4-dicarbonyl compounds with ammonia gives pyrroles. Essentially, a donor atom is used twice, with the second step effecting the cyclization. The synthesis of metacycloprodigiosin [Wasserman, 1969] incorporated such a process.

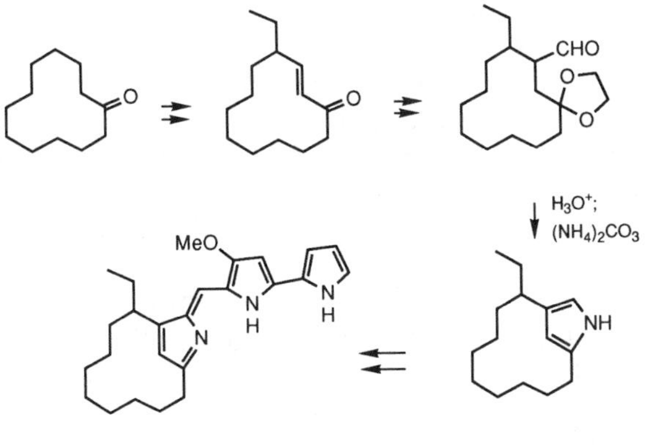

metacycloprodigiosin

Transannular Schiff reaction and *in situ* reduction form the basis of a δ-coniceine synthesis [Garst, 1982]. Preparation of the required azacyclo-nonanone in an *N*-protected form relied on a reductive carbonylation using a cyclic borane as template.

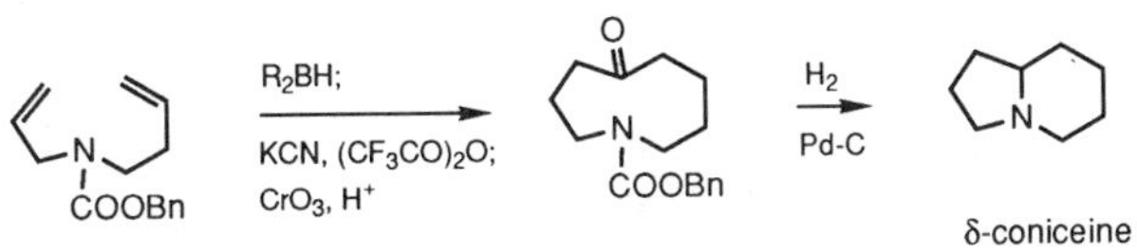

δ-coniceine

An enlightening approach to chanoclavine I [Plieninger, 1976] was based on C—N and C=C bond disconnection and reconnection of the two aldehyde groups. This analysis results in a benzobicyclo[2.2.2]octadiene precursor. The secondary amine is *trans-vic* to the aldehyde which is distal to the anilino nitrogen, and it must be introduced in a masked form before dissolution of the bridged ring system. In principle, the relative stereochemistry of the vicinal substituents is of no consequence because the aldehyde pendant provides a handle for equilibration. Considering the incursion of a Wittig reaction to be at a later stage, the required configuration is certain to be attained.

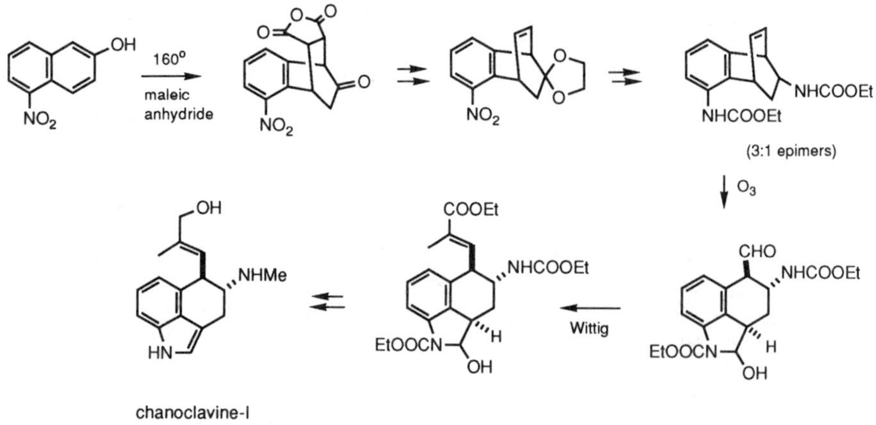

chanoclavine-I

The benzobicyclic intermediate was assembled from a Diels–Alder reaction of 5-nitro-2-naphthol with maleic anhydride. The nitro group was destined to be the key element of the indole nucleus. The carboxyl functions of the anhydride were eliminated to leave a double bond.

The Diels–Alder adduct contains a ketone group and this served as the amination handle.

The critical step of a conessine synthesis must be the attachment of the pyrrolidine moiety. As an extension of the hydrochrysene approach to steroid synthesis this problem was resolved via conjugate hydrocyanation of an 18-nor-$\Delta^{13(17)}$-pregnen-20-one and subsequent reductive cyclization [Johnson, 1966].

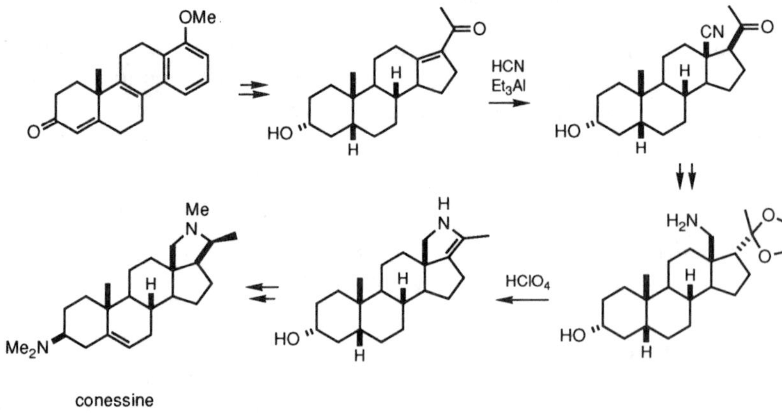

conessine

(An alternative method for conessine synthesis [Stork, 1962a] embodied a displacement of an 18-tosyloxy group with hydroxylamine which immediately was followed by nitrone cyclization with the 20-keto group.)

The Diels–Alder reaction of dienes with *N*-acyl nitroso compounds affords an entry to 1,4-N,O difunctional compounds and hence pyrrolidines. Intramolecular versions of this hetero Diels–Alder reaction are the crucial steps in the synthesis of monomorine I [Iida, 1986], heliotridine and retronecine [Keck, 1980], and cephalotaxine [Burkholder, 1988].

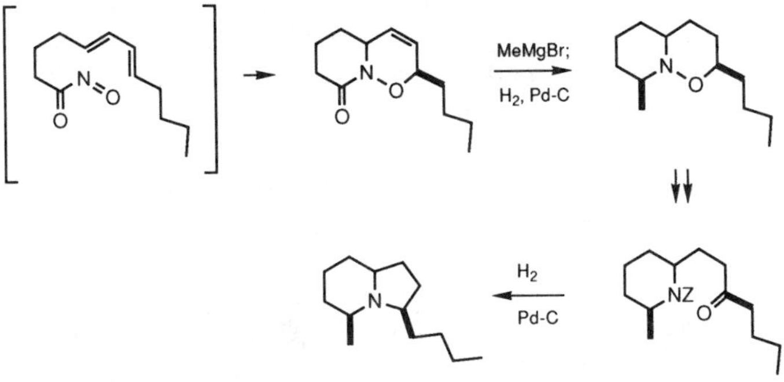

monomorine-I

Z = COOBn

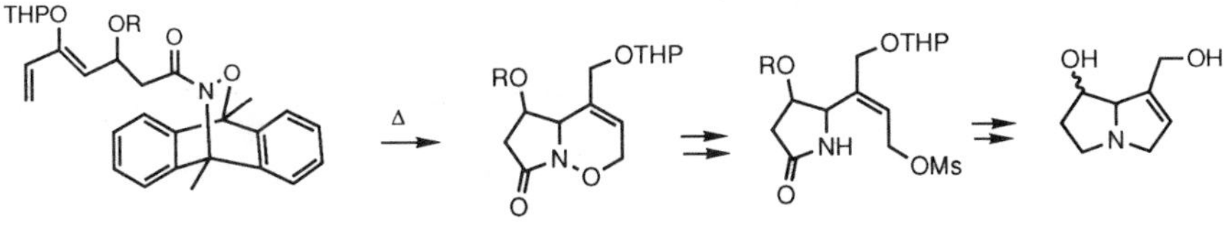

α-OH   heliotridine
β-OH   retronecine

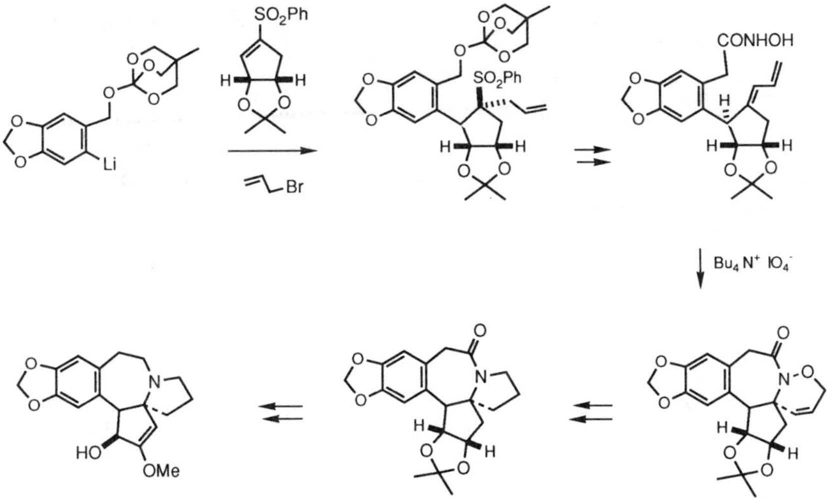

cephalotaxine

(major)

An interesting aspect of the cephalotaxine synthesis is the generation of both *anti-cis* and *anti-trans* 1,2-oxazines, the latter (minor) adduct resulting from reaction with an "anti" tether. However, this compound also proved to be useful for the synthesis as it is possible to achieve equilibration of the α-diketone intermediate which already possesses the complete skeleton of cephalotaxine via a retro-Michael–Michael reaction sequence.

1-Chloro-1-nitrosocyclohexane behaves as an avid dienophile, resulting in a net transfer of the nitroso group the conjugate diene. Reductive cleavage of the N—O bond of the adduct requires less stringent conditions to furnish a 1,4-amino alcohol. As the nitrogen atom is not acylated, it is even easier to form pyrrolidines from these cycloadducts, and an intramolecular N-alkylation may be used to achieve the cyclization.

A synthesis of pseudotropine [Iida, 1984] from 6-benzyloxy-1,3-cycloheptadiene revealed the occurrence of the Diels–Alder reaction from the *syn* face of the benzyloxy group. This is due to a sofa conformation of the diene molecule.

pseudotropine

Intramolecular amidomercuration is useful for azacycle synthesis. The regioselectivity for the cyclization of N-allyloxymethyl carbamates has shed light on the polar effects [Harding, 1984b, 1988a]. Substrates with no other substituents on the double bond invariably give oxazolidines, but the presence of an alkyl group at the terminus favors the mode of cyclization leading to six-membered ring compounds as the major products, at least in one case. Even more intriguing is that a benzyloxymethyl substituent on the terminus reverses the trend completely. (This is analogous to the effect of an oxygen substituent on iodocyclization of allylic imidates, which has also been observed [Bongini, 1986].)

In terms of polar effects, mercuronium ions derived from the vinyl substrates are reluctant to collapse into incipient primary carbenium species, even though the ring opening may be more or less in concert with the attack by the carbamate function. The situation is different for the substrates containing a 1,2-dialkylated double bond; while the two carbon ends of the corresponding mercuronium ions are hardly different electronically, the effect of the ethereal oxygen is that the proximal sp^2-carbon acquires a donor character and the distal carbon, that of an acceptor. Such an effect is canceled by another allyl ether, and the 5-*exo* closure predominates.

These cyclization reactions permit stereoselective synthesis of the 1,2-amino alcohol subunit which is present in many biologically active substances.

It is not redundant to stress the usefulness of the strategy for regioselective and stereoselective functionalization of alkenes which relies on intramolecular delivery of a donor in a cyclization pathway. The attacking donor is usually anchored onto an existing donor group, and depending on the number of bonds separating the acceptor site (alkene terminus) and the participating donor atom, a conjoint or disjoint polyfunctional system may be created.

The preparation of 2-amino-2-deoxytetritol derivatives from 2-butene-1,4-diol monobenzyl ether [Bongini, 1985] via iodination of the trichloroacetamidate demonstrates the introduction of an amino group vicinal to the anchoring oxygen function. Upon facile internal displacement of the iodide by the oxygen donor, the process is completed. Since both reactions involve formation of five-membered heterocycles, all the polar substituents in the product(s) are disjoint.

Homoallylic carbamates give rise to oxazinones, as witnessed in an amination reaction en route to N-benzoylristosamine [Hirama, 1985]. In this case the amino group and the oxygen anchor are conjoint.

N-benzoylristosamine

In the final section of this chapter we focus our attention on a synthetic approach to *Erythrina* alkaloids. From the retrosynthetic perspective the hydrindolone portion of 3-demethoxyerythratidinone is conjoint except for the spirocyclic center. Disconnection of the cyclohexenone results in a 3-oxopyrrolidine in which C-2 (spiro center) is an acceptor site, therefore a precursor with an *a–a* array would be required for assemblage of the pyrrolidine ring. Furthermore, it is advantageous to employ a longer functionalized segment than the four-carbon minimum to react with the dimethoxyphenethylamine so that a more convenient linkage point to the remaining structural elements is provided, and an accentuated donor site is prepared for the polarity inversion, i.e., conversion of a methylene group into a ketone.

Logically, the acceptor species is the oxo derivative of a Nazarov keto ester [Wasserman, 1989]. And according to this scheme the condensation with the phenethylamine was effected via a Michael addition, a Schiff reaction, and a Pictet–Spengler cyclization. The resulting tricyclic product is equipped with desirable functional groups to achieve the final annulation.

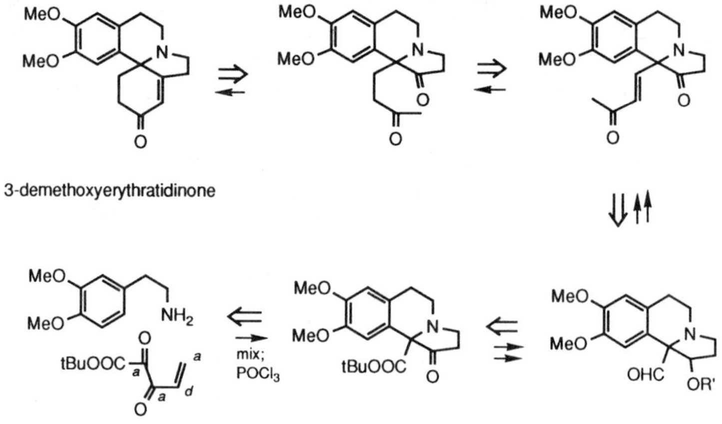

3-demethoxyerythratidinone

14

PHOTOCHEMICAL REACTIONS

14.1. CYCLOADDITION OF α,β-UNSATURATED CARBONYL COMPOUNDS

[2 + 2] Photocycloadditions that are of synthetic significance usually involve α,β-unsaturated carbonyl compounds in the $\pi-\pi^*$ excited states. The singlet excited state may then undergo cycloaddition with a ground state addend, or it may collapse to the lower-energy triplet before the cycloaddition. The reaction in the excited singlet state is stereoselective, while cycloadditions in both singlet and triplet states are usually regiospecific.

An excited state chromophore reverses its ground state polarity. This phenomenon is reflected in the frequently preferred formation of head-to-head dimers of enones and related substances. The umpolung action also simplifies many synthetic operations.

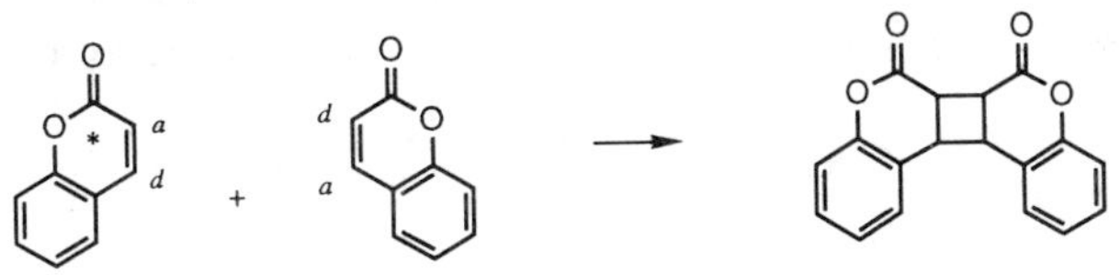

The intriguing observation of sensitized photodissociation of the head-to-head dimer of dimethylthymine but not the head-to-tail dimer [Majima, 1980] may be explained here. In the polarity-inverted excited state cleavage of the C—C bond joining the α-positions of the carbonyl groups would afford a 1,4-zwitterion. Fragmentation either with or without participation of one of the nitrogen atoms is most propitious.

A word of caution about the photochemical reactions described in this section is that they are not ionic, and exciplexes are probably involved in many of them. Nevertheless, the polar character of various reacting centers defines the mode of bond formation and breakage.

Of particular relevance is that even when the polarization effects of the substituents are overruled owing to the absence of an electronic contribution to the activation energy, marked differences in quantum yields have been observed [E. Fischer, 1989]. In other words, the quantum yields for reactions leading to the disjoint products are higher.

A most felicitous application of the [2 + 2] photocycloaddition is found in the preparation of the bicyclic ketone(s) for a synthesis of caryophyllene and isocaryophyllene [Corey, 1964b]. The report demonstrated clearly the unusual regioselectivity of enone/alkene cycloaddition [Corey, 1964a] and it harbingered the extensive exploitation of such reactions in the synthesis of numerous complex molecules in which the four-membered ring is preserved or serves as a latent structural element in the target compounds.

Although two epimers were obtained in the cycloaddition of 2-cyclohexenone and isobutene, alkali treatment isomerized the major *trans*-fused isomer to the *cis* isomer to maximize the yield of a useful synthetic intermediate for the synthesis of caryophyllene and isocaryophyllene. As the relative stability of the ring juncture changes on enlargement to the larger (9-membered) ring, stereochemical problems never surfaced. As demanded by the constitution of the caryophyllenes, the *gem*-dimethyl group must be attached to the γ-carbon of the ketone group, and this requirement was fulfilled by the pairwise donor–acceptor matching of the isobutene with the excited cyclohexenone.

caryophyllene   isocaryophyllene

The same bicyclic ketone has been used to start a synthesis of quadrone [Takeda, 1983].

Piperitone cycloadds to methyl cyclobutenecarboxylate in a head-to-head fashion, as predicted by the polar transition state. The adduct has been converted into 10-epijunenol [Wender, 1978] in three steps.

10-epijunenol

The polar transition state for this class of cycloaddition has been questioned [Lange, 1990] on the basis of the exclusive formation of a head-to-tail co-dimer of a cyclohexenone and a cyclohexenecarboxylic ester. However, these authors did not consider the possibility of the product being arisen from union of an excited enone and an excited ester. Both addends have similar chromophores and are expected to undergo excitation with approximately equal efficiency. Alternatively, the excited ester could be derived from energy transfer from the excited enone. On the other hand, the population of the excited cyclobutenecarboxylic ester may be lower as a result of the unfavorable eclipsed conformation of the tetrahedral carbanion-like β-carbon, as opposed to a staggered conformation of the excited cyclohexenecarboxylic ester.

The bicyclo[4.2.0]octane skeleton of a head-to-head cycloadduct was preserved in the preparation of a penitrem D intermediate [A.B. Smith, 1989].

The most interesting part of a synthesis of the *Ormosia* alkaloids [Liu, 1970] was the lateral attachment of a piperidine ring to the bridged system. This was achieved via a head-to-tail photocycloaddition. The process, combined with the subsequent fragmentation of an α-bromo ketone as triggered by a cyclobutanoxide ion served to introduce an acetaldehyde chain to the α-position of the enone. It was followed by chain elongation to an amide, Michael addition, and reduction.

ormosanine

Perhaps the most extensively employed [2 + 2] photocycloaddition reaction in synthesis is that of the enone + allene process. The orientation is almost exclusively head-to-head, with only a few exceptions. The stereochemistry of the cycloaddition is also interesting, and it seems that the adducts are derived from a thermodynamically controlled pathway. It has been proposed that in the $\pi-\pi^*$ excited state of the ene carbonyl chromophore the β-carbon has much sp^3 character [Wiesner, 1975].

For illustration of the synthetic utility of this reaction a synthesis of atisine [Guthrie, 1966] is outlined.

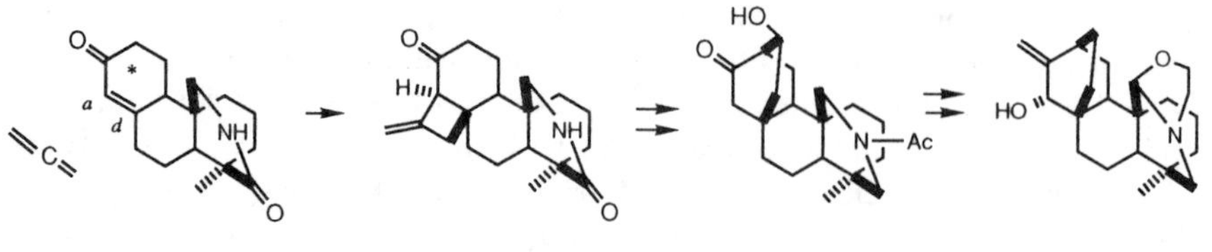

atisine

An exquisite synthesis of (−)-perhydrohistrionicotoxin [Winkler, 1989] embodies an intramolecular head-to-head dioxenone photocyloaddition. The substrate was prepared from (+)-glutamic acid. Formation of four asymmetric centers in one step is the salient feature of this route. Two of the centers are retained in the target molecule.

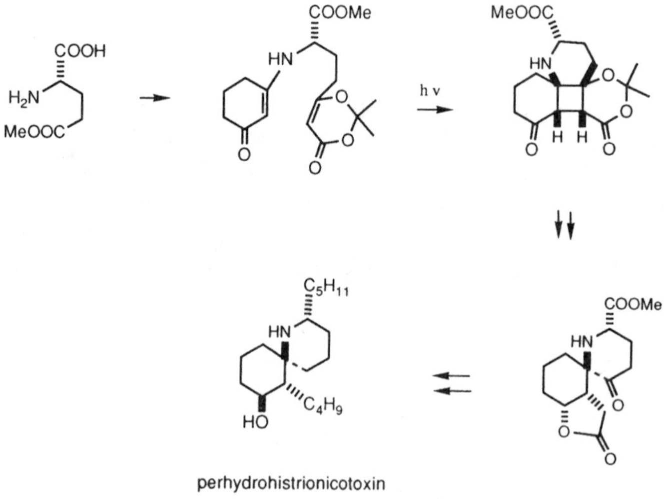

perhydrohistrionicotoxin

It is significant that the thermally induced oxidopyrylium cycloaddition which provided a key intermediate for the synthesis of phorbol [Wender, 1990] may be effected at room temperature under irradiation (15 min., 350 nm), albeit in low yield. The thermal reaction proceeded at 200°C (48 h).

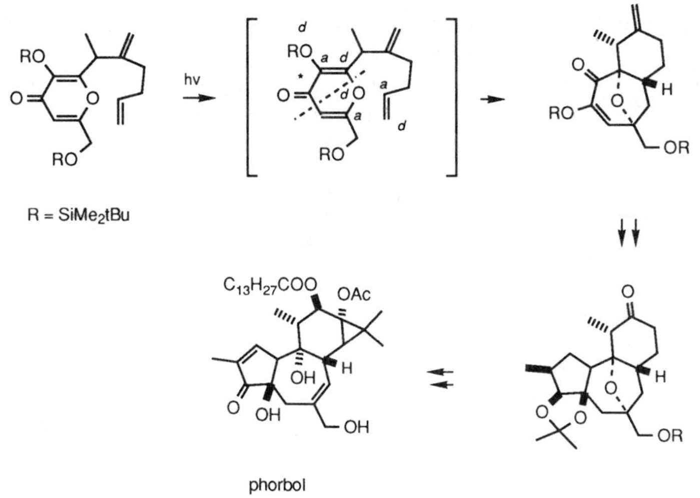

R = SiMe₂tBu

phorbol

The efficiency of the photoreaction may be explained as the following. Excitation of the enone system inverted the polarity of the sp^2-atoms such that the β-carbon became donor-like, the α-carbon acceptor was then stabilized by the alkoxy substituent instead of being destabilized as in the ground state. Assuredly this temporary electronic reorganization is conducive to the cycloaddition. (Note the pyrone may be considered as an ensemble of two chromophores, i.e., an excited enone and a ground-state enol ether.)

14.2. PATERNO–BÜCHI REACTION

The Paterno–Büchi reaction [Jones, 1981] consists of oxetane formation from a carbonyl compound and an alkene(arene)/alkyne moiety. Products with disjoint array of functional groups generally can be accounted for by the fact that the $n-\pi^*$ excited carbonyl is umpoled (a shift of electron density from oxygen to carbon).

The regioselectivity of this reaction may be exploited in synthesis. For example, esters of 3′,5′-dimethoxybenzoin esters release carboxylic acids on irradiation [Sheehan, 1971] because the intramolecular Paterno–Büchi products undergo fragmentation readily. Note that the oxetane formation proceeds from polarity-matched components. The new C—C bond bridges the methoxy group(s) to the carboxyl into a conjoint segment. (Note also the contrapolarizing circuitry due to cyclopropanation.)

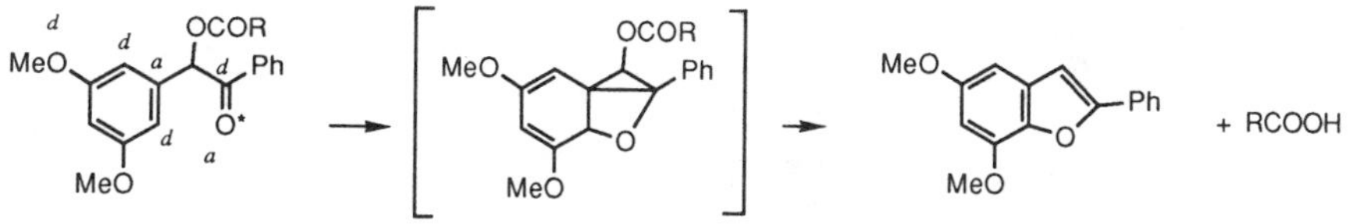

An application of the Paterno–Büchi reaction is found in a synthesis of asteltoxin [S.L. Schreiber, 1984b] and that of perillaketone [Zamojski, 1984]. The remarkable aspect is that while an enol ether and a carbonyl compound cycloadd to give a 3-alkoxyoxetane, furan is an exception.

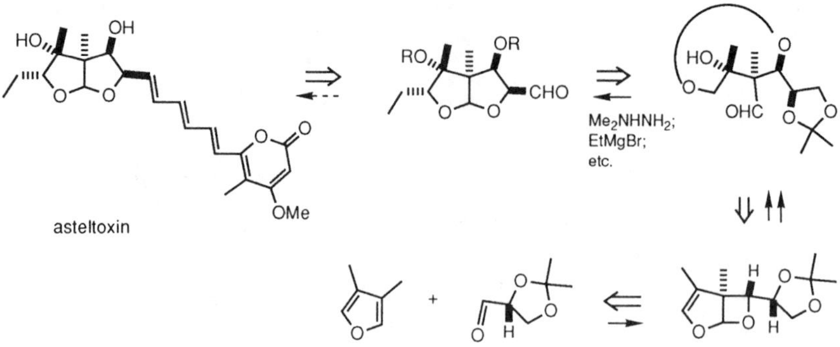

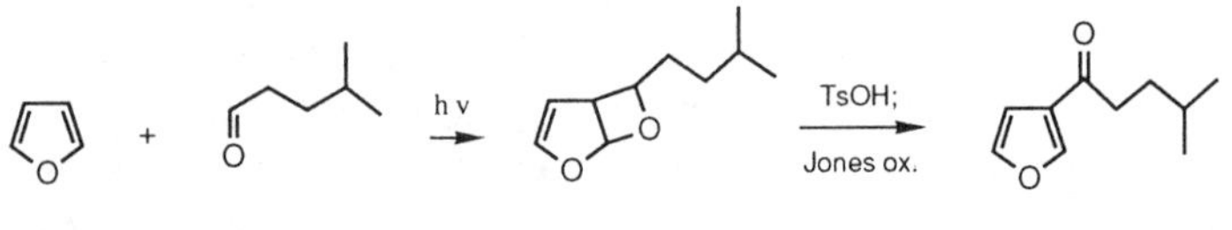

asteltoxin

perillaketone

We must remember that odd-membered heterocycles are disjoint molecules and as such they often behave in ways opposite to the expected. Electrophilic substitution at C-2 of a furan ring is well known. This site is adjacent to the donor oxygen and should be an acceptor. However, the same carbon atom is also a donor if the complete conjugate system is taken into consideration (i.e., counting through O, C-5, C-4...).

With a digression to 5-trimethylsilylcyclopentadiene, it is noted that a similarity exists in certain of its reactivities attributable to the parallelism of Me_3SiCH to O of the furan ring. For example, cycloaddition of the hydrocarbon with dichloroketene showed a regioselectivity with which a bicyclic ketone having proximal silyl and carbonyl groups was obtained. This reaction has been exploited in a synthesis of loganin aglucone acetate [Au-Yeung, 1977].

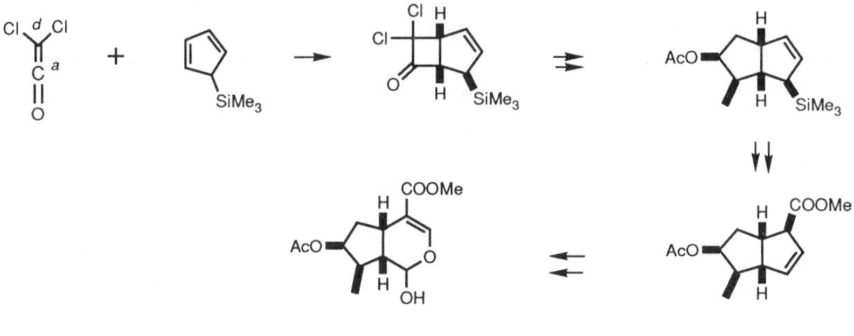

loganin aglucone
acetate

It may be worth mentioning an interesting observation of a photochemically induced reaction of imines with the Fischer carbene complexes. In these complexes the α-methylene or methyl group is strongly acidified by the metal atom. Alkylidenation at this site has been observed when the complexes are reacted with imines. Interestingly, under irradiation, totally different results emerge [Hegedus, 1984]. The reason for the diverse behavior is that the electrophilic carbenes acquire significant nucleophilic characters in the excited states (contrapolarization).

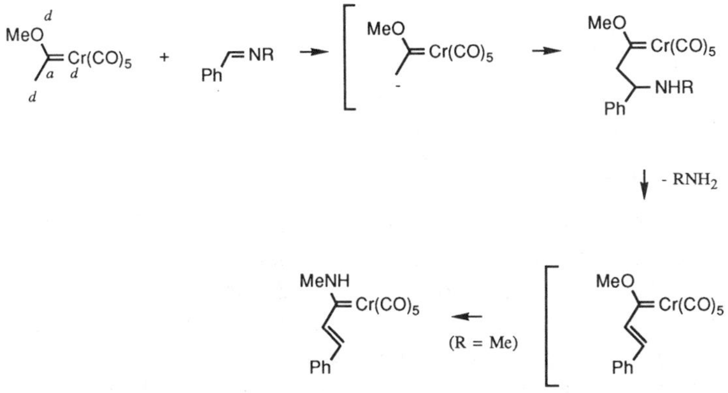

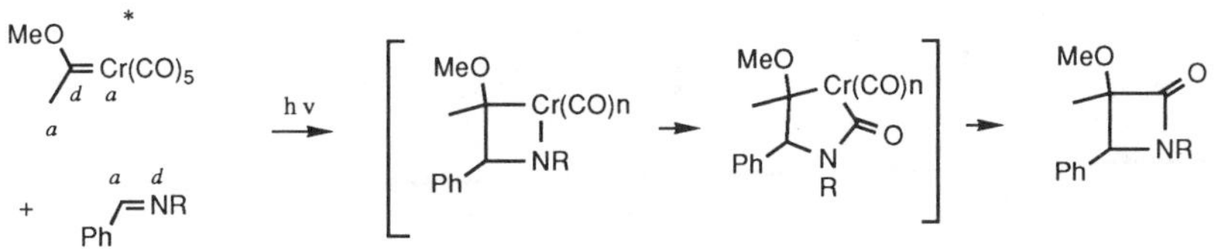

14.3. ELECTROCYCLIZATIONS AND REARRANGEMENTS

The behavior of 1-substituted cycloheptatrienes in the photoexcited states has been rationalized by a sudden-polarization model [Tezuka,1981]. This could also be easily predicted on the basis of polar-group stabilization or destabilization of zwitterions.

In terms of polarity alternation, the stabilization of a positive charge by an *ipso* donor is commonsense. In the acceptor-substituted cycloheptatrienes the initial zwitterions are expected to be less stable than those in which the positive charge is adjacent to the *ipso* carbon (which is linked to the acceptor group, and hence completing a short polarity alternating sequence).

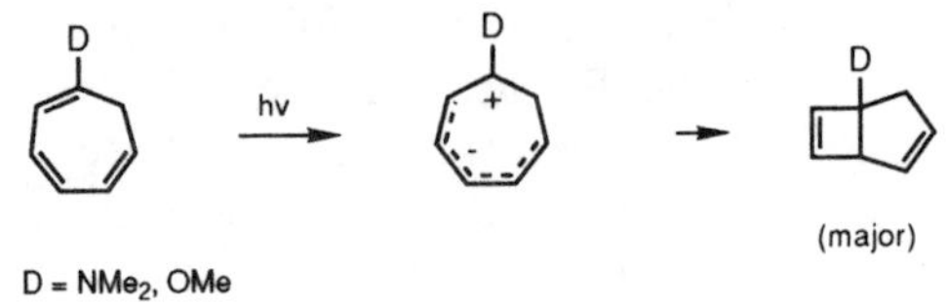

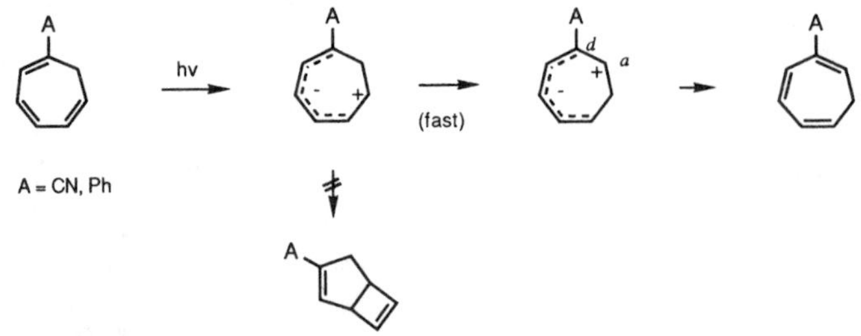

D = NMe₂, OMe

(major)

A = CN, Ph

In a synthesis of brefeldin A [Baudouy, 1977] a key feature is the use as starting material of a bicyclic enone obtained by photolysis of tropolone α-methyl ether [Crabbé, 1976]. The primary product underwent a rearrangement which apparently involved a tricyclic zwitterion. The intervention of a cyclopropane (formation and rupture) was essential to the transformation of a disjoint species to a conjoint molecule. Such transformation is also crucial to the generation of a cyclopentanone with a two-carbon sidechain.

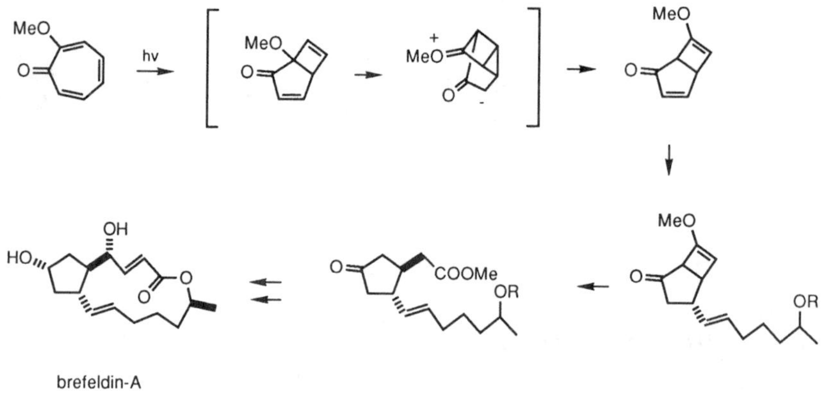

brefeldin-A

Many fascinating photochemical rearrangements have been elucidated. The formation of isophotosantonic acid lactone from α-santonin [Barton, 1962] is also of synthetic significance because many guaianolide sesquiterpenes may be elaborated from the photo product.

The transformation is thought to involve bridging between C-1 and C-5 of the excited molecule and collapse of the bridged diradical to an oxyallyl zwitterion ensues. The cationic site is conjugated to the cyclopropane and naturally it invokes opening of the strained ring.

As expected, there is umpolung in this transformation. In α-santonin the angular carbon C-10 is a donor site as it is γ to the carbonyl group. The same atom in the product is an acceptor (carbinolic).

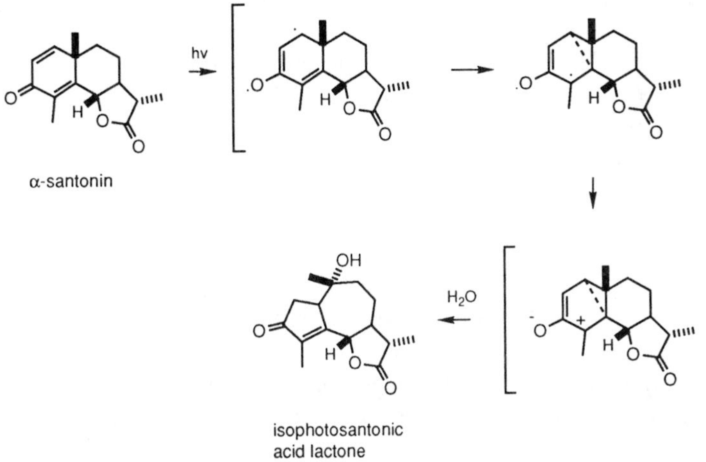

α-santonin

isophotosantonic
acid lactone

Functional-group protection by photosensitive residues [Pillai, 1987] has benefited many synthetic operations. The protected molecules are usually stable to many reagents under thermal conditions. After transformations elsewhere in a molecule are completed, the interfering functional group may be recovered from its masked form by irradiation in a proper medium.

A carbonyl protection as the ketal of o-nitrophenylethylene glycol [Hebert, 1974] is one example. The benzylic position which is a donor (conjoint to the nitro substituent) is converted into an acceptor after irradiation. Collapse of the hemiacetal intermediate to liberate the carbonyl compound is expected.

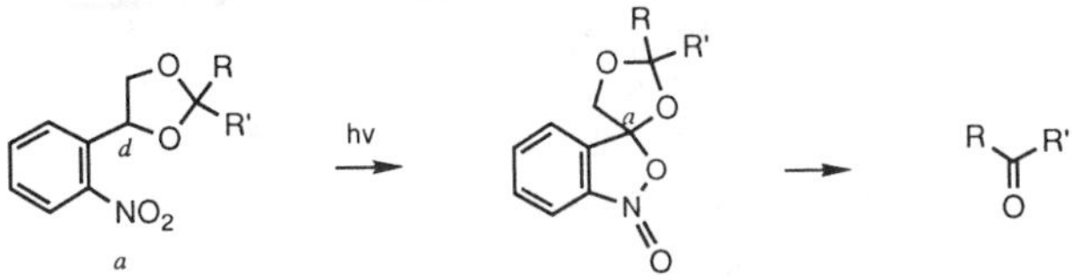

The two substituents on the aromatic ring of an m-nitrophenyl carboxylate are disjoint. However, photosensitivity of such esters have been shown [Wieland, 1966]. The reason for this reactivity is the formation of bicyclic zwitterionic species which activate the carboxyl group toward nucleophiles such as water.

3,5-Dimethoxybenzyl esters are also photosensitive [Chamberlain, 1966]. Note the homologation and change of the nuclear substituent from an acceptor to a donor.

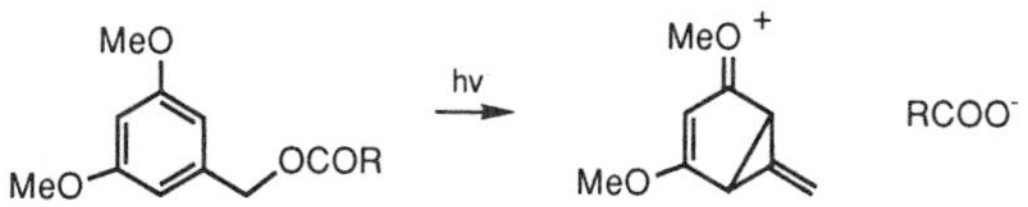

ω-Aminoalkyl *m*-nitrophenyl ethers undergo photo-Smiles rearrangement [Wubbels, 1989]. The aromatic carbon atom directly attached to the ethereal oxygen atom behaves as a donor during its bonding with the amino cation radical. This contrapolarization is effected in the excited state.

Photoisomerization of many α,β-epoxy ketones has been observed, they include acyclic, semicyclic and cyclic varieties. A simple case is the conversion of 2-benzoyl-3-phenyloxirane to 1,3-diphenyl-1,3-propanedione [Bodforss, 1918]. Related to this reaction is the isomerization of oxaziranes to amides [Kaminsky, 1966].

The above disjoint to conjoint transformation contrasts with the conjoint to disjoint rearrangement of triphenyl-1,3-diazabicyclo[3.1.0]hex-3-ene to the dihydropyrazine [Padwa, 1970]. The common feature is the photoinduced change of the electronic relationship between the heteroatoms.

Finally, a synthesis of biotol (α- and β-) [Grewel, 1987] may be taken as an example to show the current popularity of the photochemically induced oxa-di-π-methane rearrangement to access diquinanes. A tentative mechanistic description with a contrapolarized carbonyl group (cf. that which intervenes in the Paterno–Büchi reaction) attacking the double bond would be favored by the presence of an ester at the far terminus, although such a substitution pattern is not a requirement for the rearrangement.

E = COOMe   III

α-biotol

15

FRAGMENTATION REACTIONS

Surprisingly, fragmentation reactions permeate the literature of synthesis. The purpose of their use is to convert a ring structure into chains or a polycyclic framework into a simpler one. The fundamental concept underlying such synthetic strategies is that stereocontrol for introducing substituents in nonaromatic ring systems, especially five- and six-membered rings, is much easier to accomplish, and methods abound for the construction of a molecular skeleton containing rings of fewer than seven members.

The stereocontrol must be coupled with control of the fragmentation process in order to be able to dictate the geometry of the alkenes or cycloalkenes.

Generally, a molecule fragments only when a conjoint electronic circuit is present. For example, the Grob-type fragmentation [Grob, 1969] occurs readily when an activated $d3d$ array is set up, i.e., two donor groups being separated by three atoms.

15.1. GROB FRAGMENTATIONS

Despite the development of many stereoselective olefination methods in recent years, the synthesis of 1,2-di- and trisubstituted alkenes is by no means perfected. An excellent method for the assembly of certain alkenes might not be applicable to others. Consequently, the fragmentation strategy may be employed to great advantage. However, it must be emphasized that this type of fragmentation reaction requires an antiperiplanar arrangement of the single bonds to be broken.

The most prominent aspect of a synthesis of the cecropia juvenile hormone [Zurflüh, 1968] is the unfolding of the two (Z)-trisubstituted double bonds of a C_{15} intermediate from a bicyclic precursor by two fragmentation reactions involving a 1,3-diol monotosylate subunit.

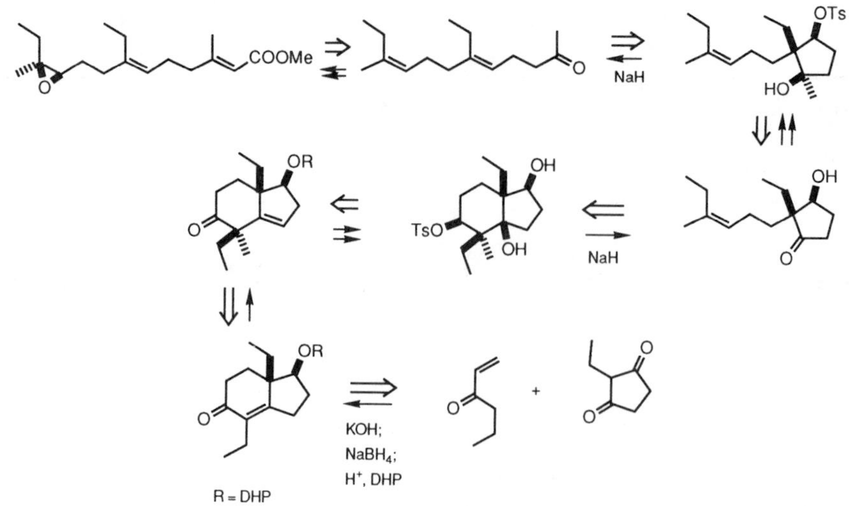

R = DHP

To achieve this result, in each instance the leaving group (tosyloxy) must be placed *cis* to the ethyl chain at the adjacent carbon atoms. This was easily done by complex metal hydride reduction of the corresponding ketones in the hydrindene skeleton, a reaction course which has many analogies in the steroid field. The triggering hydroxyl for the first fragmentation was introduced via an epoxidation–reduction sequence, the double bond that underwent epoxidation having been set during the methylation step. The methylation was directed to the α-face by virtue of steric shielding of the alkoxy and the angular methyl substituents. On the other hand, epoxidation was likely assisted by the homoallylic hydroxyl by hydrogen bonding.

Conversion of the eudesmane to the elemane skeleton requires scission of the C-2/C-3 bond. While many elemene syntheses relied on oxidative cleavage of a Δ²-derivative, a route featuring peripheral fragmentation is most interesting [Ando, 1978]. The alkoxide trigger was generated *in situ* by isomerization of an epoxide under the Meerwein–Pondorff reduction conditions. The product was then elaborated into saussurea lactone.

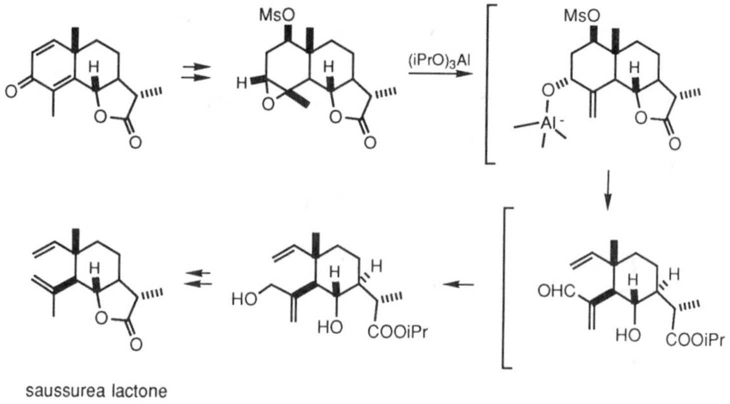

saussurea lactone

The *cis* 1,3-relationship of the isopropanol and the double bond in the spiroannulated six-membered ring of hinesol was noted and an outstanding synthetic design formulated [Marshall, 1969] based on the simultaneous unveiling of a vinyl chain and a ketone by a Grob fragmentation.

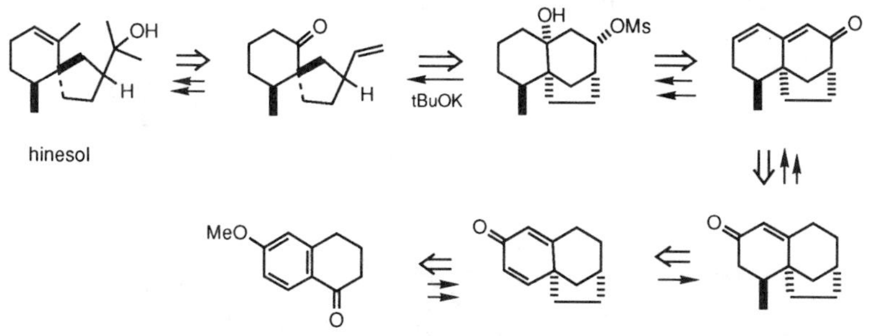

When β-elimination and retroaldol reaction are suppressed in a β-oxy carbonyl system, fragmentation may be induced on base treatment, to produce, accordingly, an unsaturated carboxylic acid derivative. The tetrahedral adduct is a special 1,3-diol.

Molecular strain always accelerates fragmentation when the donor–acceptor distribution is conducive to such a process. A synthesis of the dimethyl ester of a monarch butterfly pheromone [Trost, 1978a] exploited this factor with great success.

An approach to α-pinene [Larsen, 1977] called for the formation and then the fragmentation of a tricyclic ketone. The latter step also served to establish the double bond.

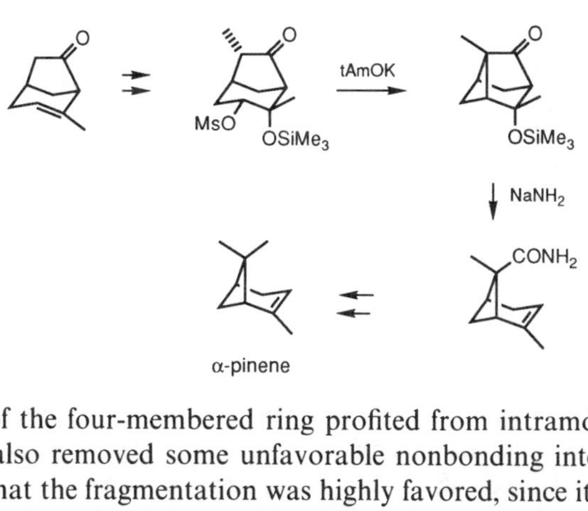

Formation of the four-membered ring profited from intramolecularity in a process which also removed some unfavorable nonbonding interactions. Also noteworthy is that the fragmentation was highly favored, since it occurred with an unactivated departing group. The favorable factors are good orbital alignment and the relief of angular strain.

Stereoselective creation of two adjacent quaternary carbon centers through direct C—C bond formation is rather difficult. When at least one of the carbon atoms is also part of a six-membered ring, the task is simplified by employing the Diels–Alder reaction and proper manipulation of the adduct, as illustrated in a fine synthesis of trichodermol [Still, 1980].

trichodermol

III

R = SiMe₃

The approach features ring contraction of the enedione portion of the adduct that was derived from p-benzoquinone, and a fragmentation step that gave rise to the 6-link-5 ring system.

In a route to longifolene [McMurry, 1972] there was a sterically enforced intramolecular aldolization during delivery of a methyl group to an enone. Fortunately, the 1,3-dioxygenated pattern is amenable to recovering the original ring system by a fragmentation maneuver.

longifolene

The difficulties in the direct and controlled construction of medium-sized and large rings in the past have elicited fragmentation strategies to circumvent them. Thus, the problem confronting a total synthesis of caryophyllene/isocaryophyllene was particularly severe in the early 1960s when methodologies were much more limited. The first solution [Corey, 1964b] epitomizes the ingenuity and logic of creating the cyclononene in either (E)- or (Z)-form from a hydrindane precursor. Cleavage of the intercyclic bond of the isomeric 1,3-diol monotosylates generated the norketone epimers of the sesquiterpenes. Wittig reaction completed the synthesis.

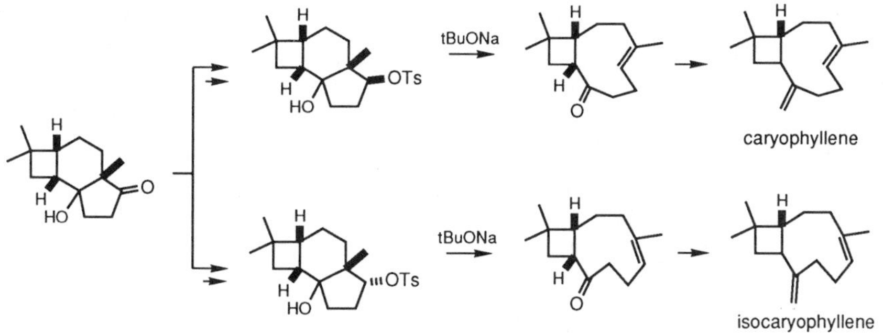

caryophyllene

isocaryophyllene

The geometry of the cyclononene unsaturation correlates with the relative configuration of the angular methyl group and the tosylate.

Germacrenes are vexatious targets for synthesis because of their great tendency to undergo transannular reactions (e.g., Cope rearrangement, cyclization). Such susceptibility to dismutation undermines the synthetic pursuit and demands nonacidic conditions at relatively low temperatures for the installation of the diene chromophore.

The first synthesis of hedycaryol [Wharton, 1972] made use of a fragmentation pathway. Accordingly, an octalin tosylate was assembled and submitted to hydroboration. Alkaline treatment of the organoborane (one of the isomers containing an *a4d* circuit) caused the cleavage of the intercyclic bond.

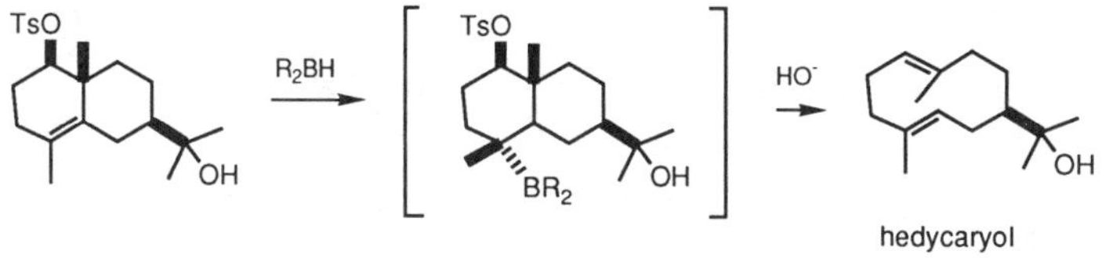

hedycaryol

Recalling the great virtue of the caryophyllene synthesis involving fragmentation unfolding of the medium-sized ring, a route to ceroplastol I [Boeckman, 1989a] was also designed with a variation on the same theme. The fragmentation was set up from a bridged ring system and with a hemiketal (ketone + MeONa) as the trigger. The ester thus generated was to participate in a Dieckmann cyclization to form a cyclopentanone unit.

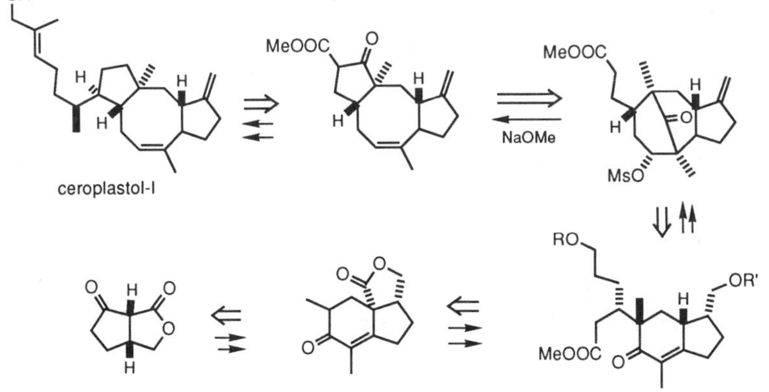

ceroplastol-I

The fascinating *meta* photocycloaddition of an arene to an alkene has been developed into an extremely efficient methodology for synthesis. In an ingenious approach to rudmollin [Wender, 1986] a photocycloadduct was converted into a mesylate which fragmented on reduction with lithium aluminum hydride, simultaneously creating the hydroxymethyl substituent at the ring juncture and a double bond which served as the anchoring site for the γ-lactone.

The bridged ring system of the photoadduct rendered stereocontrol extremely facile in the introduction of the two-carbon subunit for eventual elaboration into the lactone, as well as the establishment of the *endo* mesylate.

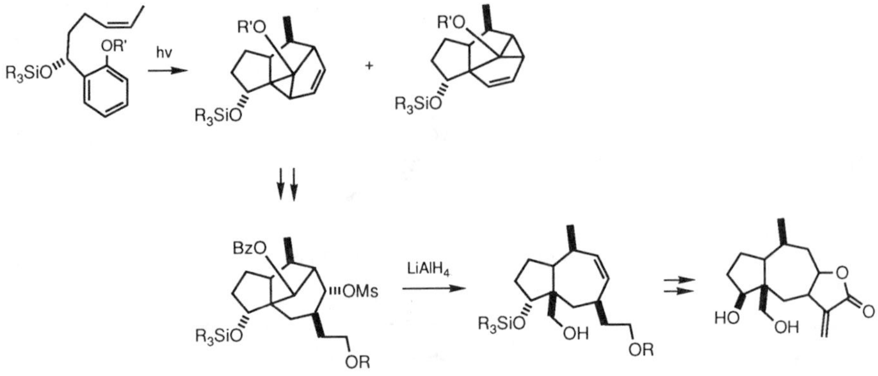

rudmollin

Pleuromutilin demands imaginative and meticulous planning for its synthesis, particularly in the creation of the eight-membered ring which is adorned with seven asymmetric centers. As delineated in the following scheme [Gibbons, 1982], this last problem was vigorously tackled by a brominative fragmentation.

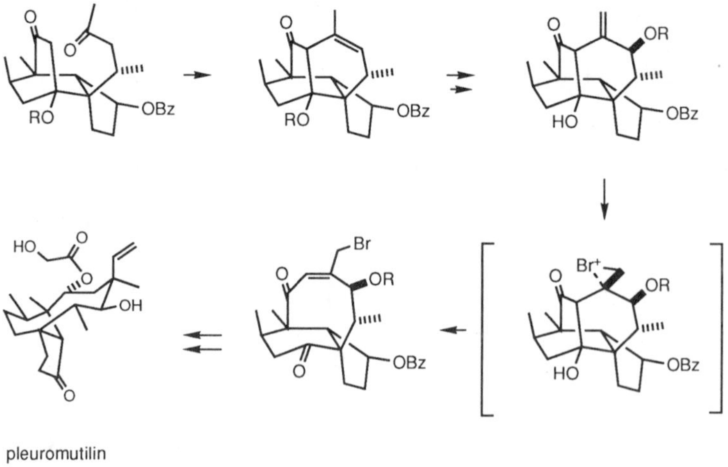

pleuromutilin

It should be noted that retroaldol fission of the tetracyclic system could not be achieved because the C-α/C-β bond is restricted to an orthogonal conformation with respect to the π-bond of the carbonyl group.

Fragmentation of 1,3-diols in which the substitution pattern at one of the carbon termini favors cation formation (e.g., tertiary) may be promoted by an acid. Of course, such fragmentation is likely to be stepwise.

A highly efficient synthesis of nootkatone [Dastur, 1974] involved not only the fragmentative transformation of a bicyclo[2.2.2]octene substrate into a cyclohexenone. Under the solvolytic conditions, polyene cyclization to afford an octalone system also took place.

nootkatone

15.2. RETRO-CLAISEN REACTION AND RELATED PROCESSES

The Claisen reaction leading to 1,3-dioxo compounds is a reversible process, the products are stabilized as the enolates as they are formed. The cleavage of 1,3-dioxo compounds, acquired by other means, into two fragments upon base treatment may then be considered as a legitimate transform in retrosynthetic analysis.

A synthesis of brefeldin A [LeDrian, 1982] from a photochemical product of α-tropolone methyl ether adequately illustrates the use of monoprotected 1,3-diketones whose functionalities would be separated and transformed into others. Here, the cyclobutene ring which is fused to the cyclopentenone provided steric shielding to direct the attachment of an alkenyl chain from the *exo* face of the molecule, effectively setting a *trans*-3,4-disubstituted cyclopentanone when the retro-Claisen fission was to be performed. The latter reaction is chemoselective, as a much greater relief of strain attends the scission of the four-membered ring.

brefeldin-A

A most popular and useful synthetic sequence of indirect, reductive oxoethylation of an α,β-enone is via photocycloaddition with allene, oxidative cleavage of the double bond, and hydrolytic fragmentation. One of the early examples is found in a synthesis of atisine [Guthrie, 1966].

A functionality reorganization involving a retro-Claisen fission is shown in the following approach to pyrrolizidines [Ohnuma, 1983]. The eight-membered ketolactam intermediate underwent transannular bond formation readily.

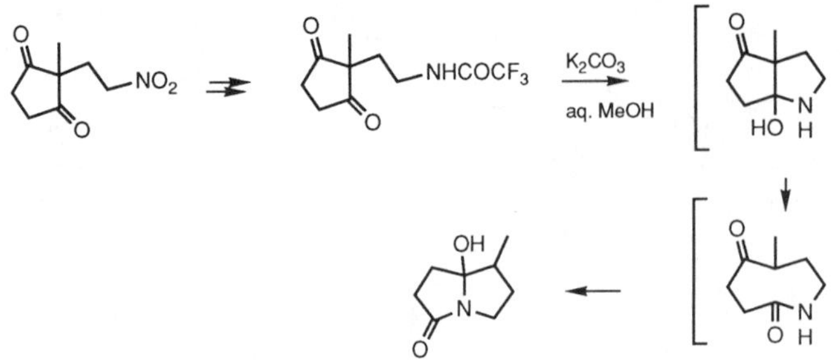

A Claisen condensation/retro-Claisen fragmentation sequence linked up ethyl nicotinate and N-methylpyrrolidone while retaining a 1,4-N,O separation in the sidechain that was to be fashioned into the pyrrolidine portion of nicotine [Späth, 1928].

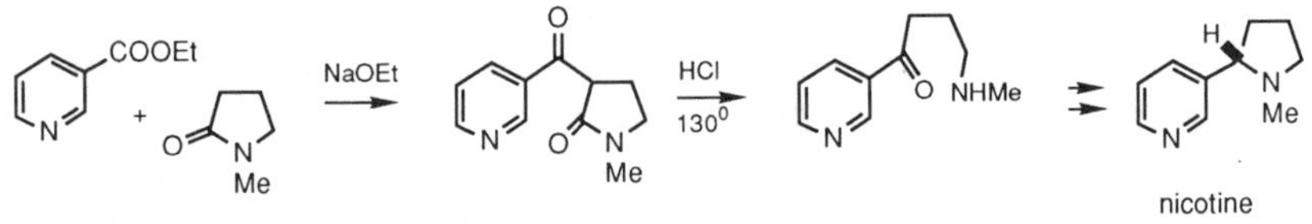

A retro-aldol fragmentation constitutes an important step of a new method for converting 2,2-disubstituted 1,3-cyclohexanediones into 3-substituted 2-cyclohexenones by reaction with dimethyl lithiomethylphosphonate [Yamamoto, 1990]. The fragmentation generated β,ζ-diketo phosphonate intermediates which underwent a Wittig–Horner reaction. A formal synthesis of α-acoradiene illustrates its utility, although in this case the spirocyclization by the Sakurai–Hosomi reaction was stereorandom and only the product arising from one of the synclinal transition states proved to be useful for the synthesis.

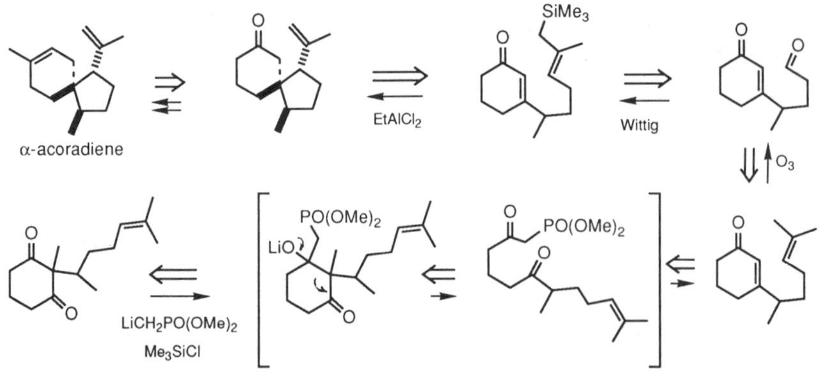

15.3. α-CLEAVAGE REACTIONS

The classic synthesis of quinine [Woodward, 1945] teaches stereocontrol during establishment of two vicinal carbon chains on a cyclic structure by degradation of a bicyclic precursor. The subtarget of the synthesis, quinotoxine, was disconnected into a homomeroquinene derivative, and further into a *cis*-perhydroisoquinolinone.

It must be remembered that in the early part of the 1940s the repertoire of olefin-forming methods was small, and the Hofmann elimination of a quaternary ammonium hydroxide was perhaps one of the most reliable by which a vinyl chain instead of the ethylidene could be created.

The reason for pursuing the synthesis from an isoquinoline derivative may be twofold. First of all, isoquinolines were readily available, their catalytic hydrogenation leading to the requisite *cis* ring juncture that corresponds to the arrangement of the homomeroquinene sidechains well precedented; secondly, even if the oxocycloheptanopyridine could be obtained, a serious regiochemical issue pertaining to the cleavage (e.g., Beckmann rearrangement, etc.) would have arisen. Thus, in the historical context, the choice is hard to criticize.

The chain-releasing process was also distinguished by an elegance in that an ester group was generated together with an oxime in the other chain via *C*-nitrosation and *in-situ* fragmentation. The oxime was converted into the quaternary ammonium group.

When the nitrosation site is a methylene group, tautomerization of the α-nitroso ketone to the α-oximino ketone usually takes precedence to fragmentation. However, these substances are susceptible to Beckmann fragmentation. The ring cleavage en route to showdomycin [Inoue, 1980a] is an example that is also favored by the strain release.

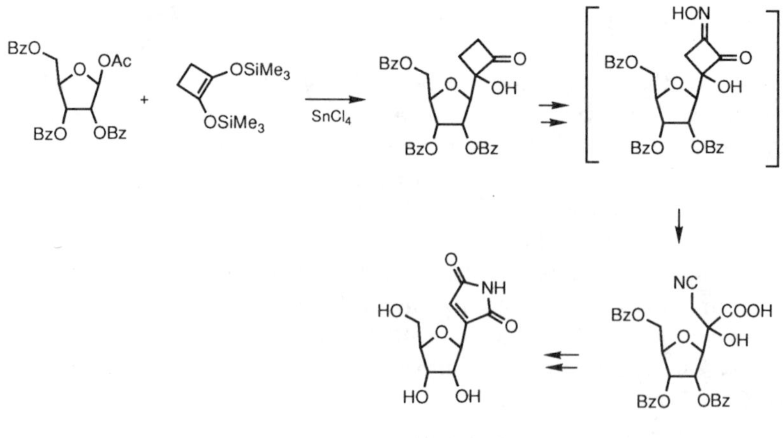

showdomycin

The Japp-Klingemann arylhydrazone formation [Phillips, 1959] at an activated methine group, involving coupling with an arenediazonium ion and cleavage of the acceptor group (e.g. C=O) from the products, is analogous to the ketone nitrosation–fragmentation. This reaction is useful for the *de novo* preparation of 3-substituted indoles, e.g., serotonin [Merlini, 1975].

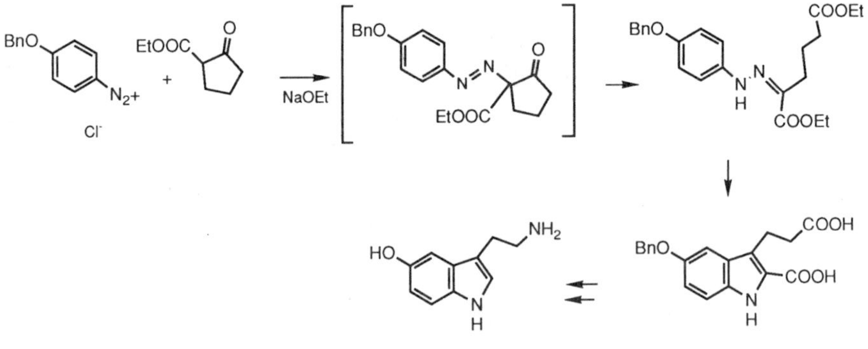

serotonin

Other strong acceptor substituents such as NO_2, SO_2R, which are capable of supporting a carbanion at the *ipso* position can induce fragmentation analogously to the retro-Claisen fission.

A currently favorite method for the devolution of carbon chains from cyclic ketones is that consisting of α-bisalkylthiolation and fragmentation. The advantage of the process is its generality and the two chains of the products being functionally differentiated. The application of this reaction sequence to synthesis is extensive, only one example is presented here.

The significance of norcamphor as a building block for natural product synthesis may be appreciated by an analysis of its role in an approach to protoemetine [Takano, 1982a], with pertinence to the fragmentation.

protoemetine

X = OH

KOH, tBuOH;
MeI, aq.MeCN

CrO₃;
TsS(CH₂)₃STs,
Et₃N

The carbon atoms adjoining the nitrogen atom of protoemetine are, by nature, acceptors. FGI introduces carbonyl groups at these sites in differentiable oxidation states. Both carbonyls are related to the formyl group of protoemetine by the same number of intervening C—C bonds, and association of each pair of the chains results in cyclopentane derivatives. Further association of these structures leads to, in one case, norcamphor which stands out as the choiced starting material, considering its availability and stereocontrol in C-ethylation (actually done at the lactone stage to avoid regiorandomization during the Baeyer–Villiger reaction).

After amidation the cyclopentane ring was cleaved accordingly. Sulfenylation was regioselective for steric reasons (at methylene next to a secondary instead of a tertiary carbon atom) and from this point the synthesis became straightforward.

15.4. RETRO-MANNICH FISSION

β-Amino ketones, β-aminomethylindoles and related compounds are readily formed by the Mannich reaction. The conjoint array of the functionalities renders the reaction reversible, therefore the components may be regenerated from the Mannich bases. To be synthetically useful, the retro-Mannich reaction (and all other retrograde reactions) has to be effected after modification of another portion of the molecule, or it is performed on Mannich bases prepared by other methods. In the following examples, the retro-Mannich reaction constitutes an important step toward achieving the synthetic goal.

A synthetic approach to olivacine [Besselièvre, 1976; Naito, 1981] based on biogenetic considerations is interesting. Intermediates were assembled by facile C—C bond-forming reactions (Mannich-type and Friedel–Crafts reactions). Fragmentation of the C—N bond resulted in a carbazole bearing an ideal sidechain for a subsequent cyclization.

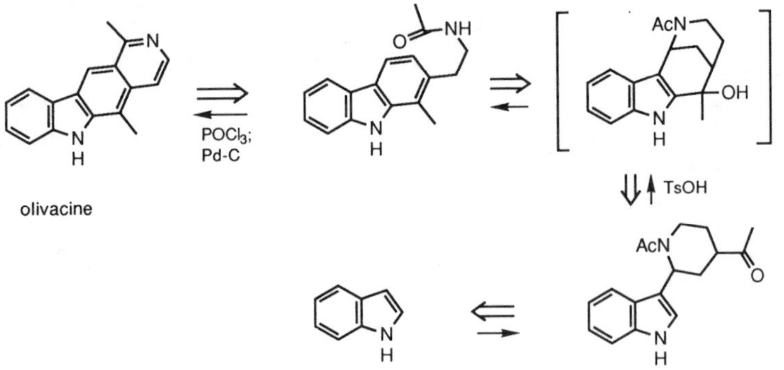

olivacine

Aspidosperma alkaloids, when accessible as the indolenine forms, can undergo isomerization via a retro-Mannich–Mannich reaction manifold. Thus, *trans → cis* equilibration at the ring juncture of the perhydroquinoline portion apparently occurred during the Fischer indole synthesis step at the late stage of a synthesis of aspidospermine [Stork, 1963b].

Advantage has been taken of this behavior in many other synthetic endeavors. For example, the conversion of vincadifformine into vincamine [Hugel, 1981] can be achieved by solvolysis of its hydroxylated derivative. Recyclization of the chanoiminium intermediate, itself the result of a retro-Mannich process, at the α-position of the indole nucleus, was followed by a second fragmentation to give the tetracyclic precursor of vincamine.

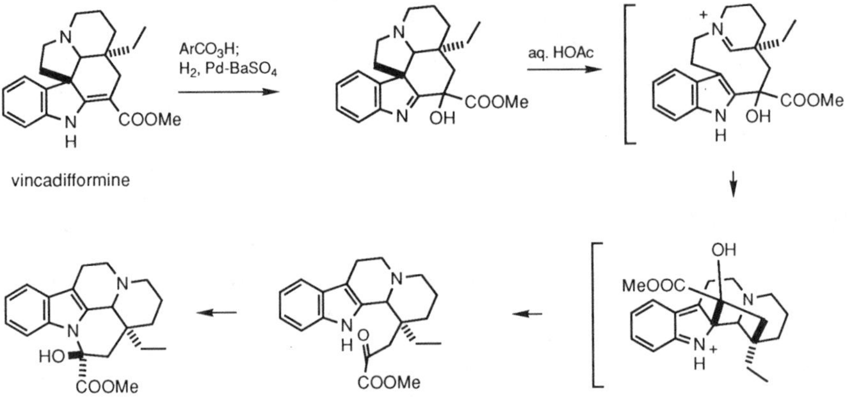

vincadifformine

vincamine

Secodine is a plausible biosynthetic intermediate of *Aspidosperma* alkaloids. Many approaches to secodine-type compounds have been investigated, among them the indoloazepine route is most versatile and elegant. Indeed, a secodine derivative, arisen from a sequence of Mannich reaction, *N*-alkylation, and two retro-Mannich fragmentations, cyclized immediately to provide a direct precursor of tabersonine [Kuehne, 1985].

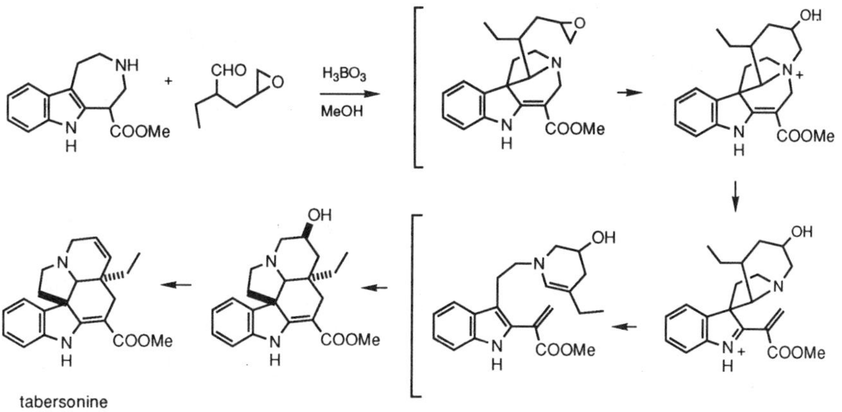

tabersonine

Of enormous synthetic potential and apparent generality is the very concise reaction sequence of intramolecular photocycloaddition of β-amino α,β-enone and alkene which is followed by a retro-Mannich fragmentation *in situ*. The facility of the latter reaction is due to angular strain of the four-membered ring. The resulting imino ketone can be caused to undergo a Mannich reaction at the alternative donor site flanking the carbonyl group, after N-alkylation. A synthesis of mesembrine [Winkler, 1988] exemplifies the method.

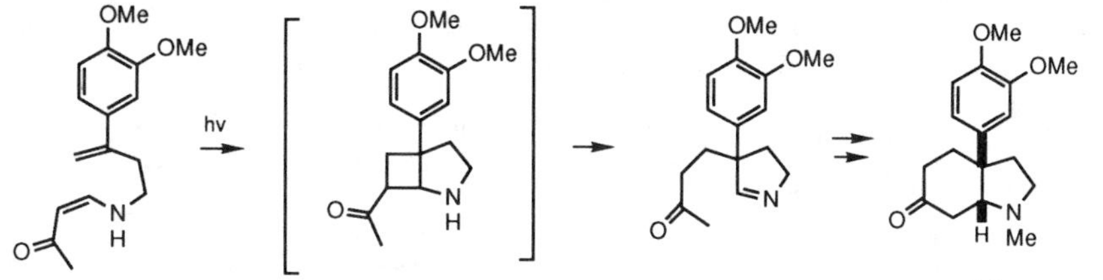

mesembrine

Related to the retro-Mannich reaction is the reductive fragmentation of an N—C—C—C—O system which highlighted a synthesis of daphnilactone A [Ruggeri, 1989]. Preparation of the hexacyclic precursor was very efficient in that the participation of two double bonds and two aldehyde groups created four rings in one step.

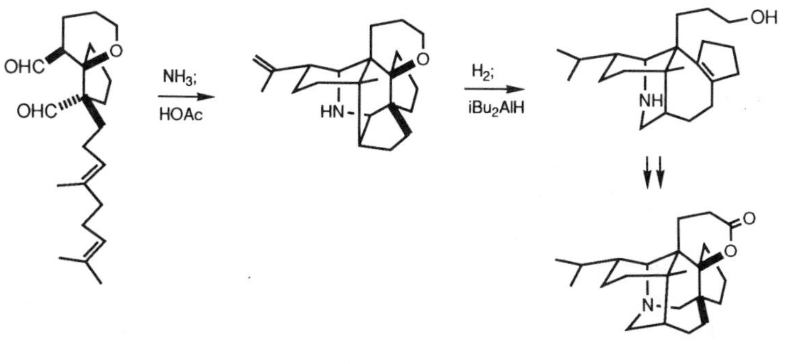

daphnilactone-A

A partial aza-Cope rearrangement retron is present in homoallylic amines. Consequently, it is advantageous to analyze seco-ergot alkaloids retrosynthetically on this basis to derive and evaluate synthetic routes.

Application of an aza-Cope rearrangement transform to the iminium salt of isochanoclavine I yields an azacyclodecadiene which may be correlated with a perhydroquinoline system by a fragmentation reaction [Kiguchi, 1989].

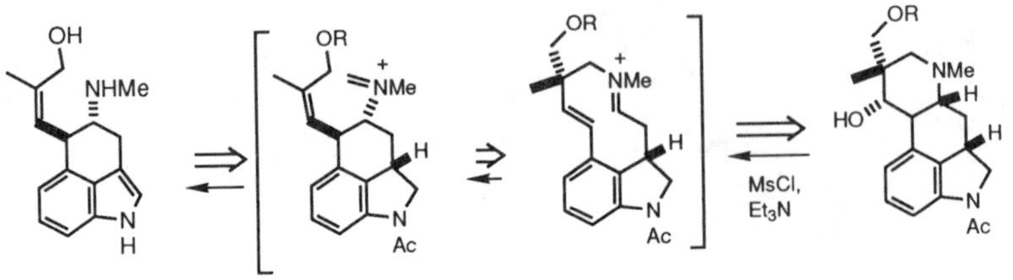

isochanoclavine-I

A priori, the retrosynthetic analysis could indicate a direct fragmentative transform. However, fragmentation of simpler hydroquinolines prefers an internal pathway over the peripheral counterpart. The (Z)-configuration of the double bond arises from a *cis* relationship of the oxygenated functionalities in the piperidine ring.

It may be mentioned that β-amino carbonyl compounds are also subject to retro-Michael reactions. These competing reactions may be favored by activating the amino residue, e.g., by N-acylation. A synthesis of laurifonine [Bremner, 1985] typifies an excellent approach to dibenzo[*d, f*]azonines.

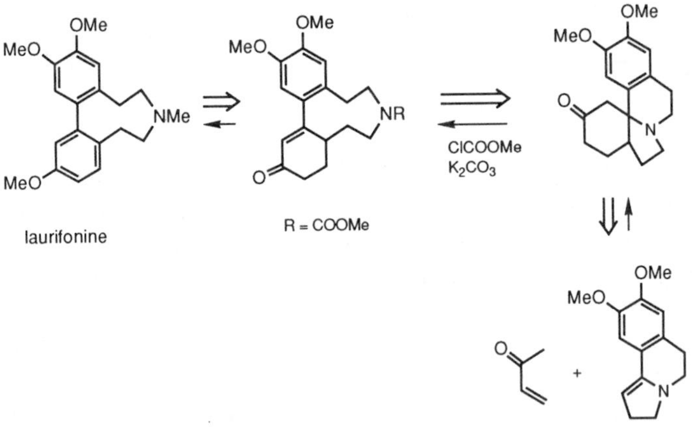

laurifonine

R = COOMe

15.5. FRAGMENTATION OF EXTENDED SEGMENTS

Carbenoid addition to double bonds, including those within aromatic rings, leads to cyclopropanes. While norcaradiene derivatives tend to tautomerize to afford cycloheptatrienes, this propensity defers to the fragmentation pathway when the addend is an acylcarbenoid, the substrate is a phenol, and the cycloaddition is directed toward the $C_{(meta)}$—$C_{(para)}$ bond. In fact, the initial adducts cannot be isolated. Fragmentation is favored both electronically (conjoint systems) and energetically (strain factor).

To great advantage, this procedure has been employed in a synthesis of α-chamigrene [Iwata, 1979].

α-chamigrene

It is not coincidental that a synthesis of byssochlamic acid [Stork, 1972b], which contains a 9-membered ring like caryophyllene, should rely on a fragmentative pathway. In this instance a bicyclo[4.3.1]decan-9-one core was degraded by a Beckmann fragmentation, leaving a secondary cyano group to be converted into an ethyl chain. The fragmentation was favored by the formation of a tertiary benzylic cation which was further stabilized, and perhaps instigated by a methoxy group in the peri position (conjoint).

byssochlamic acid

Two other points that deserve further note are (1) the employment of 1,4-dimethoxybenzene residues as latent maleic anhydride subunits, to facilitate construction of the central framework, and (2) the dissolved metal reduction to furnish a *cis* stereochemistry of the two alkyl substituents. The stilbene chromophore is susceptible to the thermodynamically controlled reduction, with the result of the propyl group in an equatorial-like conformation which would also avoid eclipsing the peri-methoxy group.

An unerringly important fragmentation reaction developed in recent times is the Eschenmoser–Tanabe fragmentation of α,β-epoxy carbonyl compounds via substituted hydrazones. The substituent on the hydrazine must play the role of a donor during fragmentation. Many variants of this fragmentation are now known.

This process forms the basis of a synthesis of muscone [Eschenmoser, 1967] via transitory annulation to insert a four-carbon subunit to cyclododecanone.

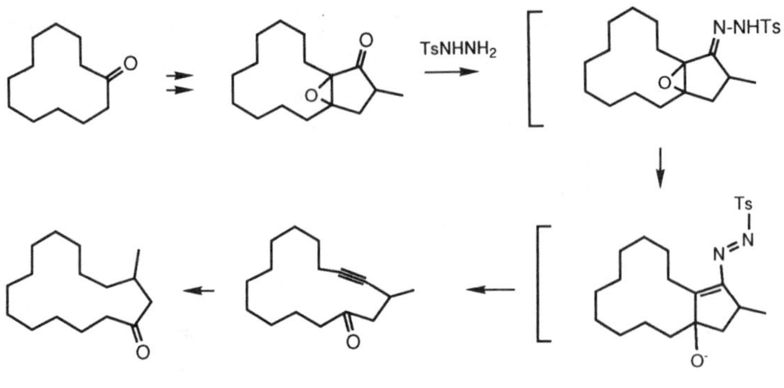

muscone

A synthesis of brefeldin A [Nakatani, 1985] was hinged on a fragmentation reaction and avoided a great many of the stereochemical problems. Using a hydropentalenone template the configuration of the cyclopentanol is easily controlled. After fragmentation the sidechain containing the ketone group can be epimerized. The isolated (*E*)-double bond corresponds to a *trans-vic* substitution pattern, which is not difficult to fashion since equilibration of the ketone precursor may be implemented.

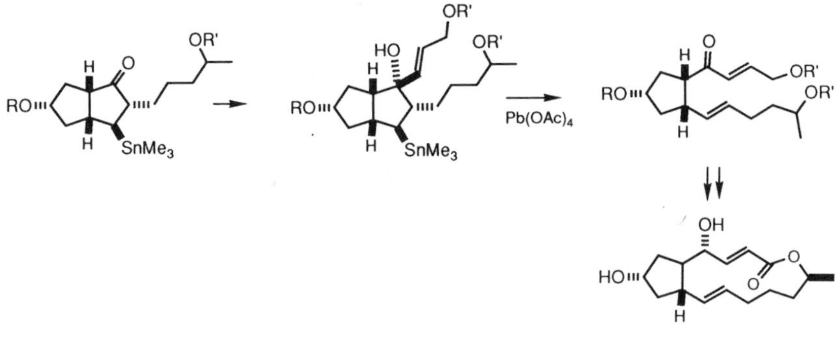

brefeldin-A

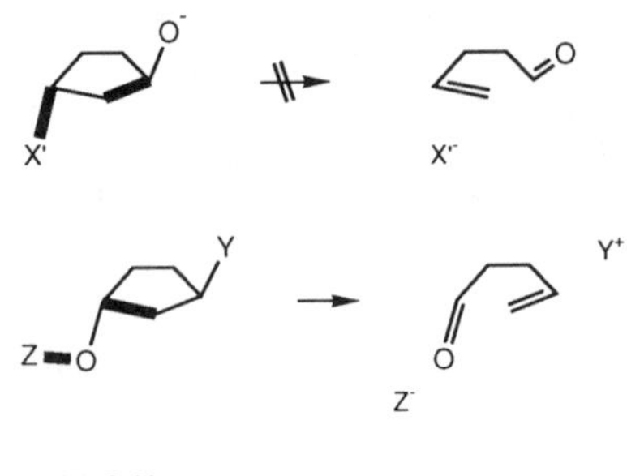

Y = SnMe$_3$
Z = Pb(OAc)$_3$

Selection of the "leaving group" is of critical importance. A 1,3-diol monosulfonate system would not be serviceable in this case since the C—C and C—X' bonds that would undergo cleavage cannot assume an antiperiplanar orientation. On the other hand, an oxidative fragmentation of a γ-hydroxy stannane may accommodate such stereoelectronic requirements.

A unique feature of an aphidicolin synthesis [Ireland, 1984] based on spiroannulation via a Claisen rearrangement is the fragmentation of a pentacyclic intermediate which was made possible by the presence of both the silyl and the carbonyl groups. The intricate ring system of the precursor was formed by an intramolecular ketene/alkene cycloaddition.

Like the tin atom in the oxidative fragmentation, the excellent acceptor (electrofugal) characteristics of silicon, combined with the strain factor, makes the fragmentation pathway extremely smooth.

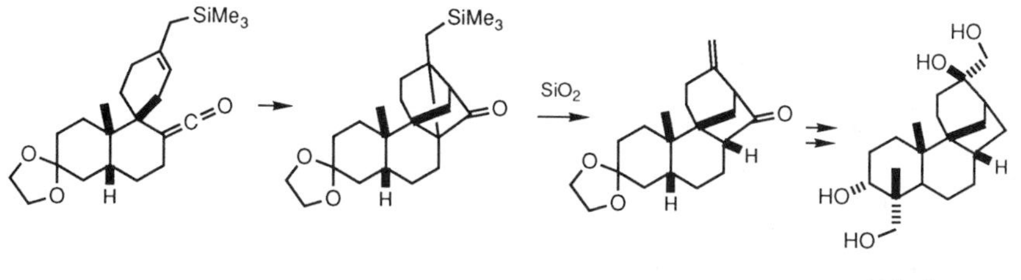

aphidicolin

There is an impressive unfolding of unsaturated macrolide rings based on a decarboxylative double fragmentation [Shibuya, 1979, Sternbach, 1979]. A ketal group relayed the electron flow which resulted in the expulsion of a tosylate. Favorable stereoelectronic alignment of the relevant bonds is noted.

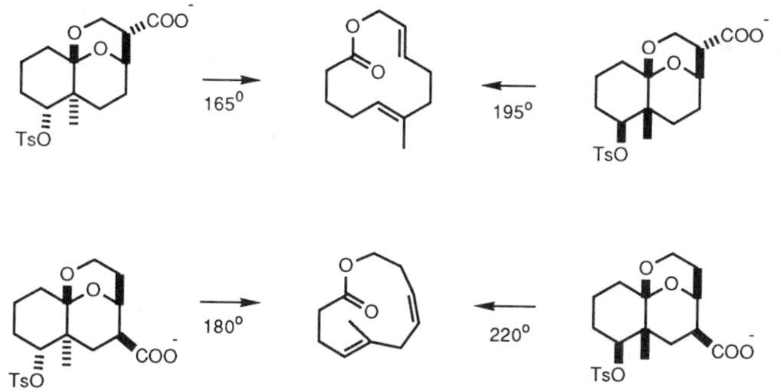

16

OXIDOREDUCTIONS AND SOME
OTHER UMPOLUNG PROCESSES

16.1. OXIDATIONS

Oxidoreduction always involves changes of polarity of the participating atoms. This phenomenon has been named contrapolarization, in contradistinction to the manipulated umpolung processes.

We cannot deal with so many redox reactions. A very few are selected to illustrate their utility in synthesis. To begin, let us examine two general types of polar redox processes which are exemplified by the chromic oxidation of alcohols and hydride abstraction of silyl ethers.

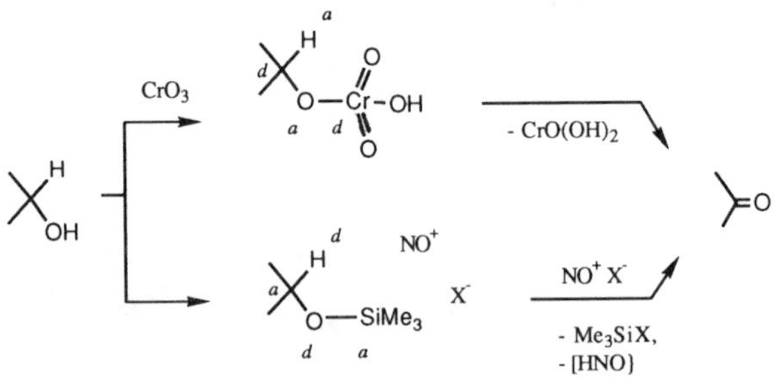

In chromic oxidation the decomposition of the chromate esters involves pairwise contrapolarization of the ethereal oxygen atom and chromium. On the other hand, the hydride transfer reaction is instigated by the accentuation of polarity alternation by the silicon atom, which bestows the α-hydrogen with much donor character.

While it is not commonly considered an oxidation, α-halogenation or selenylation of carbonyl compounds does belong to this class of reactions. In the products the α-carbon behaves as an acceptor, as reflected in displacement reactions, and in eliminations that specify the leaving group as a donor and the atom it departs from, an acceptor.

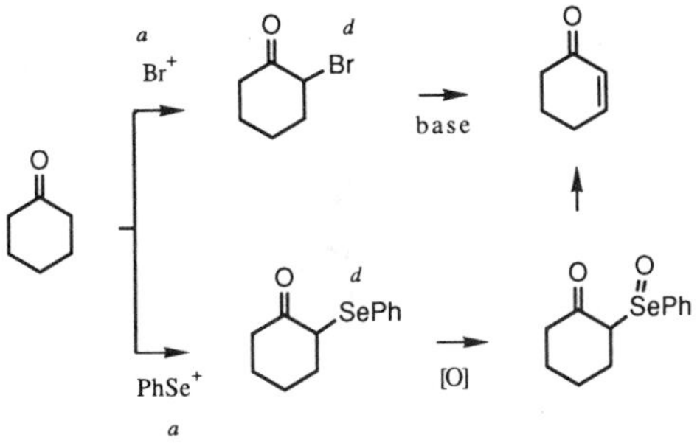

A synthesis of acoragermacrone [Still, 1977] featured an anionic oxy-Cope rearrangement to unveil the cyclodecadienone system. The product is a (2Z, 6E)-dienone isomer, and the configuration of the conjugated double bond must be inverted.

This inversion was achieved by reaction with trimethylstannyllithium and trapping the enolate as a trimethylsilyl ether. Oxidation of the latter gave acoragermacrone.

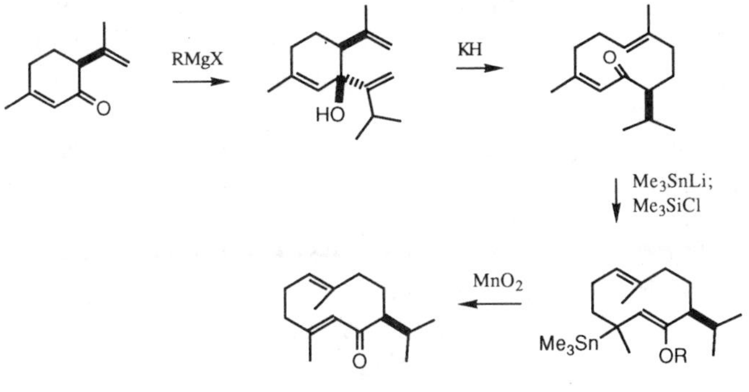

acoragermacrone   R = SiMe₃

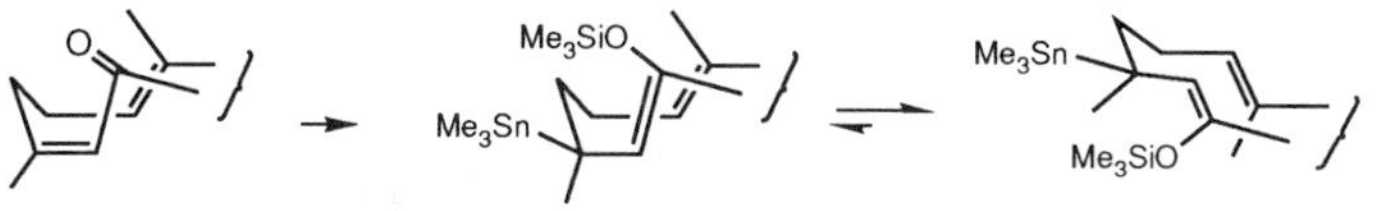

The oxidation at the C—Sn bond changed the carbon to an acceptor and desilylation of the enol silyl ether regenerated the enone chromophore. Thus, the stannyl group was a necessary conformation regulator. Oxidation of simple enol silyl ethers also leads to enones, formally a hydride ion is transferred to the oxidizing agent in such cases.

The incompatibility of an oxygen function at the 6a-position (tazettine numbering) to the tandem oxy-Cope rearrangement/Mannich cyclization precluded the exploitation of this efficient methodology to the synthesis of tazettine. However, replacing the oxygen function in the precursor with an acceptor, specifically, a silyl group, removed the obstacle [Overman, 1989]. This requires oxidative maneuver at the silyl group after establishing the hydroindolone system. This is a case in which change of a polar substituent is mandatory.

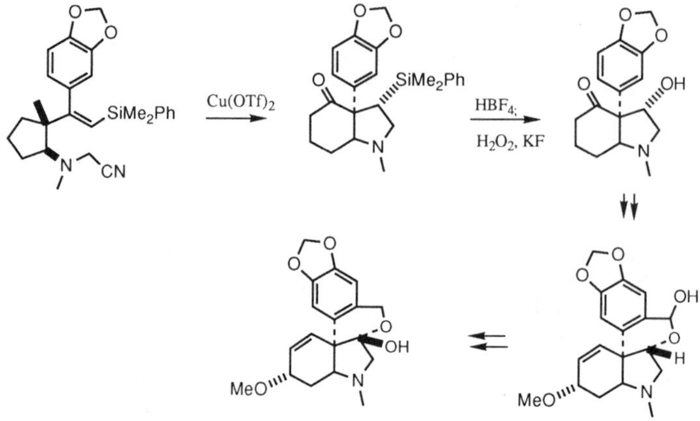

tazettine

Siloxolanes formed by intramolecular hydrosilylation of 3-alkynyl silyl ethers have been converted into β-hydroxy ketones on oxidation [Tamao, 1988]. The salient feature of this operation is the disjoint cycle to conjoint chain inversion as effected by oxidation.

Traditionally, β-hydroxy esters cannot be prepared by hydration of the corresponding conjugated esters, as the equilibrium favors the unsaturated compounds. Since the Michael reaction using silylcuprate reagents is a good reaction, and peracid oxidation of the C—Si bond in the presence of fluoride ion proceeds with retention of configuration at the carbon center, the indirect method represents a useful stereoselective and regiospecific pathway to 1,3-dioxygenated substances. This method is applicable to a synthesis of the Prelog–Djerassi lactone [Chow, 1985].

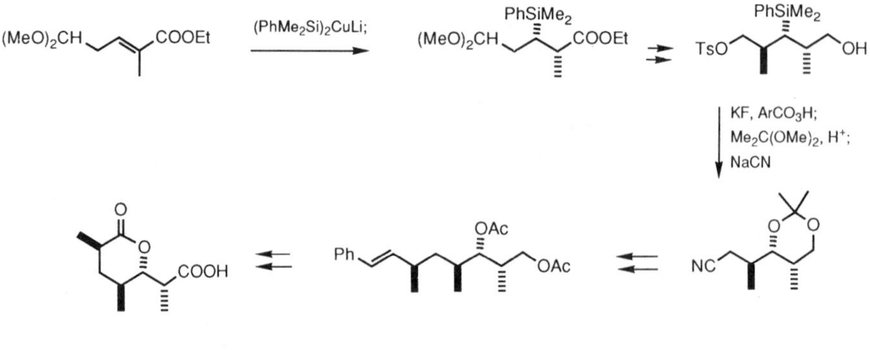

Prelog-Djerassi lactone

Protoberberine synthesis based on a strategy involving B-ring formation by a Michael reaction must equip the precursor with a terminal functionality that can contrapolarize in order to enable a Friedel–Crafts alkylation to complete the C-ring. A vinyl sulfoxide fulfills this requirement on account of the ability of the sulfinyl group to undergo a Pummerer rearrangement [Pyne, 1987]. This rearrangement is a redox reaction, i.e., reduction at sulfur and oxidation at its adjacent carbon atom.

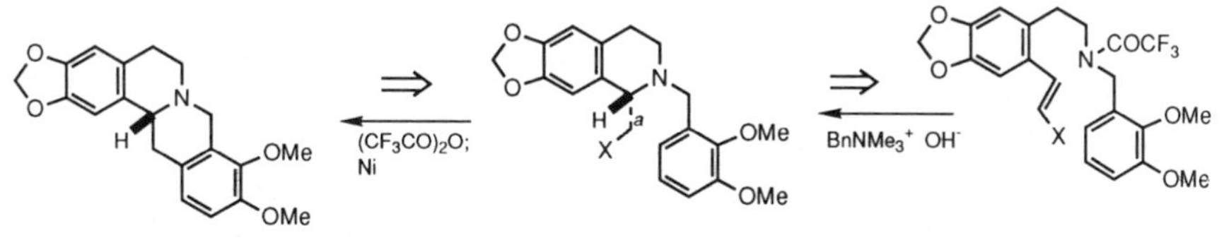

(+)-canadine

X = SOTol

Contrapolarization at the α- and β-positions of the indole nucleus occurred in a synthetic approach to (–)-vindoline [P.L. Feldman, 1987]. Implementation of a pertinent reaction sequence was necessary to avoid destruction of the chirality center which was painstakingly constructed from L-aspartic acid.

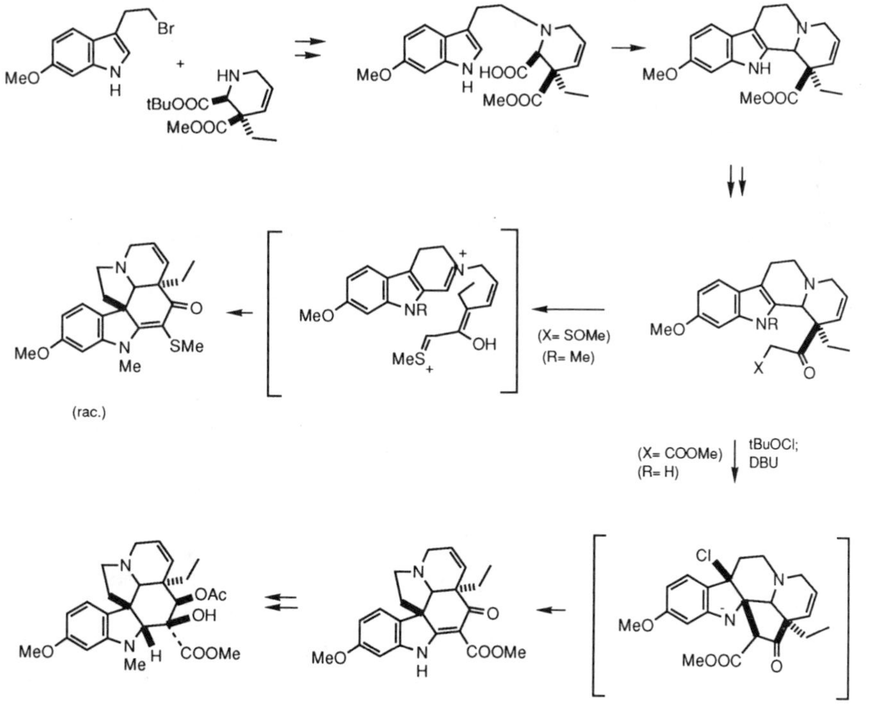

vindoline

Accordingly, the tetracyclic intermediate containing a β-keto ester chain was chlorinated (*t*-butyl hypochlorite) and treated with a base (DBU). These conditions induced an anionic Mannich reaction which was followed by rearrangement. Cyclization of a structurally akin α-sulfinyl ketone via the Pummerer rearrangement intermediate led to a racemic pentacycle owing to the intervention of a retro-Mannich reaction. The Pummerer rearrangement led to an acylthionium ion in which two acceptor carbons were linked to one another directly. Such inherently unstable species tend to seek out reaction courses to relieve the unfavorable electronic interactions, and retro-Mannich fragmentation provides a temporary attenuation of the acceptor properties of the carbonyl group.

An excellent C—C bond-formation reaction originally devised for the assembly of the BC-ring system of vitamin B_{12} [Woodward, 1968, Eschenmoser, 1969] consists of coupling a thiolactam with an acyl enamine. The sulfur atom is then extracted by triethyl phosphite.

The coupling step was actually performed in the presence of an oxidant to give a disulfide which acts as an acceptor. The episulfide tautomer of the coupled product was subject to attack by the phosphite reagent. The salient features are the formation of a disjoint system in the coupling, and reversion to a conjoint state upon desulfurization via a three-membered ring intermediate.

The nonoxidative, alkylation route based on thiocarboxylic acids and α-haloketones [Roth, 1971] works equally well to afford 1,3-diketones. A further modification has permitted synthesis of macrolides which contain a β-keto lactone system, e.g., diplodialide A [Ireland, 1980a].

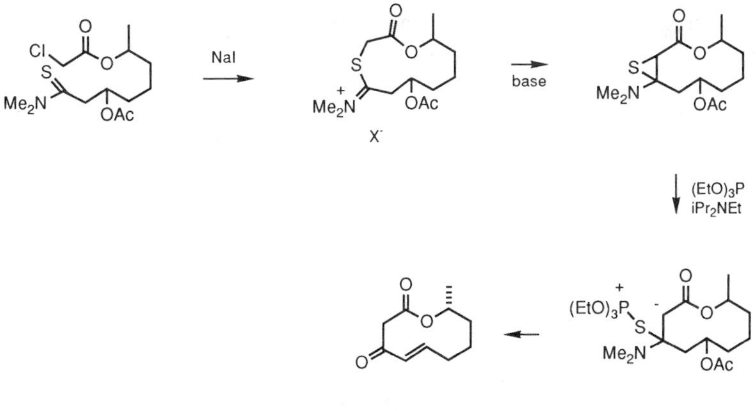

diplodialide-A

Let us examine some oxidative reactions. Ozonization is a popular and widely applicable method for scission of C=C linkages. It involves a cycloaddition–cycloreversion sequence to give a carbonyl fragment and a zwitterionic hydroperoxide which would recombine by the 1,3-dipolar cycloaddition. The zwitterionic species may be trapped by reactive solvent molecules.

It has been shown that functionality differentiation of the product is possible [S.L. Schreiber, 1982]. Of particular significance is the nature of the dialdehyde monoacetal products deriving from allylic and homoallylic acetates. In short, the donor group has a decisive effect on the collapse of the primary ozonides. The original investigators noted that the preferred mode of cleavage for 5-*endo*-acetoxynorbornene is opposite to what is expected by considering the inductive effect. However, the results are in perfect agreement with that predicted on the basis of polarity alternation.

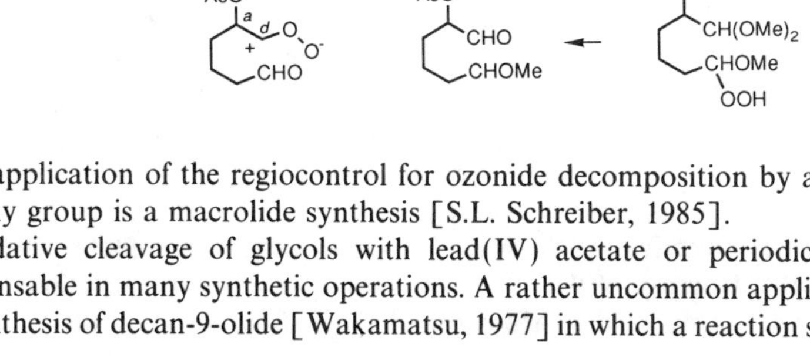

An application of the regiocontrol for ozonide decomposition by an allylic methoxy group is a macrolide synthesis [S.L. Schreiber, 1985].

Oxidative cleavage of glycols with lead(IV) acetate or periodic acid is indispensable in many synthetic operations. A rather uncommon application is in a synthesis of decan-9-olide [Wakamatsu, 1977] in which a reaction sequence incorporating the oxidation provided a solution to the notoriously difficult problem of medium-sized-ring formation.

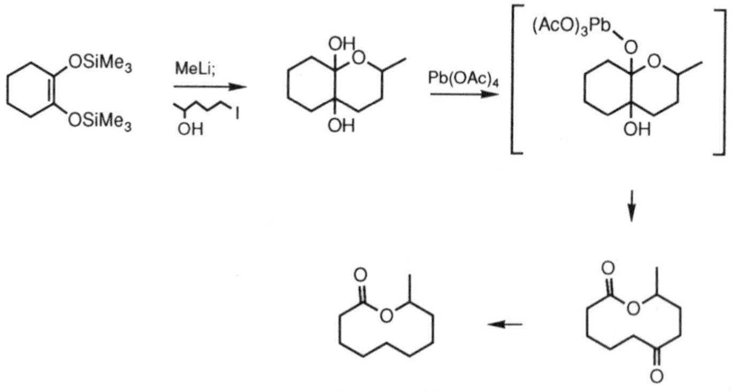

decan-9-olide

The facile access to functionalized tricyclic systems by a tandem intra-molecular Michael and aldol reactions of cyclohexenones containing an aldehyde sidechain was instrumental to the development of several syntheses of complex natural products. An isotwistane derivative was identified as a suitable precursor for coriamyrtin [Niwa, 1984b].

Key steps of the synthesis were bridgehead hydroxylation and subsequent cleavage of a glycol. By the latter reaction a *cis*-perhydroindanone was unveiled.

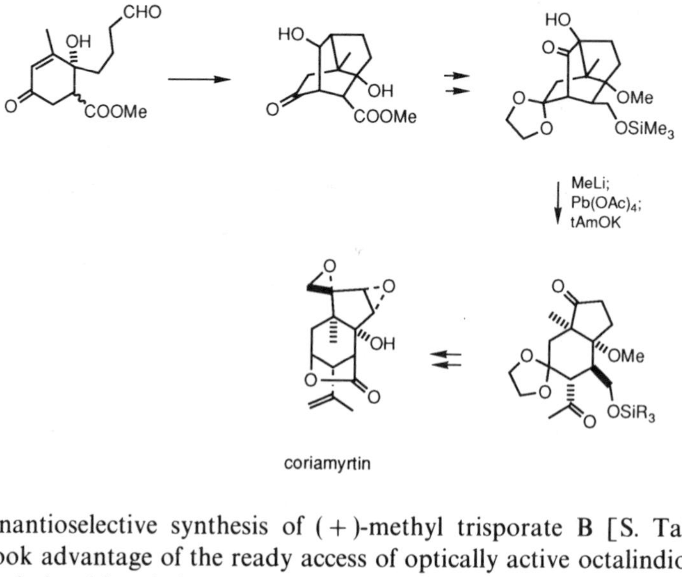

coriamyrtin

An enantioselective synthesis of (+)-methyl trisporate B [S. Takahashi, 1988] took advantage of the ready access of optically active octalindiones and the 1,6-relationship of the carboxyl and the fully substituted olefinic carbon atom of the long sidechain. The oxidative cleavage of a properly modified α-ketol liberated the desired intermediate.

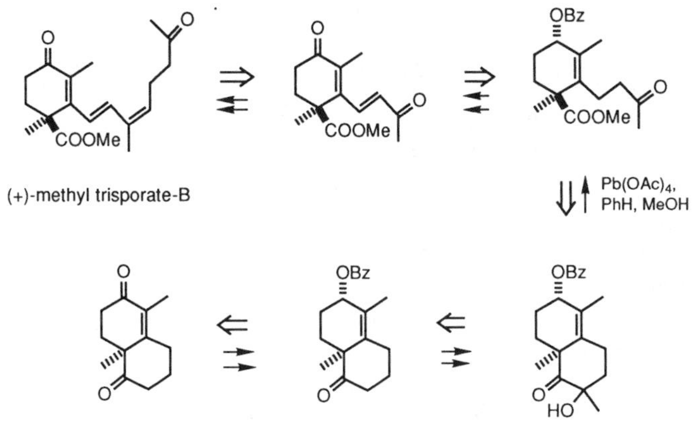

(+)-methyl trisporate-B

Pb(OAc)₄,
PhH, MeOH

The susceptibility to aromatization of betalamic acid argues for a structural containment to be put into its synthetic intermediates. Outstandingly designed in this context is a synthesis [Büchi, 1977] in which the potential piperideinedicarboxylic acid chromophore is locked up in a tropane skeleton until the very last step.

N-Benzylnorteloidinone, obtained from a Robinson–Schöpf reaction, proved to be an ideal intermediate for the synthetic pursuit. With the glycol bridge serving as a latent dicarboxylic acid and the ketone group ready to be extended to the unsaturated aldehyde, the molecule is neatly packaged and protected for every need in its journey to the target. The final step of lead(IV) acetate cleavage of the α-diketone bridge also induced β-elimination and movement of the double bond into conjugation.

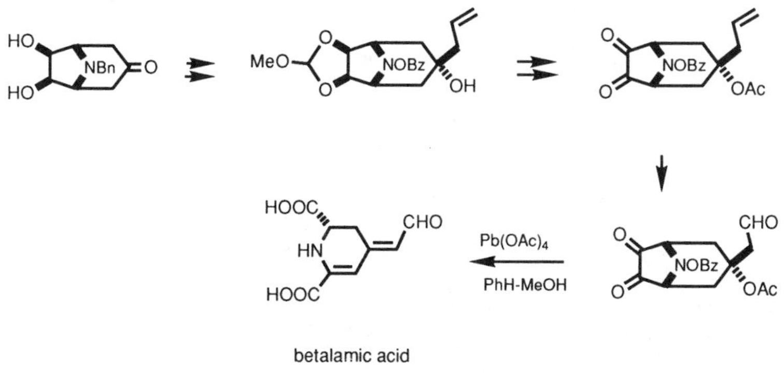

betalamic acid

Oxidative cleavage of the C—C bond between the functionalized atoms in 1,2-bis(trimethylsiloxy)cyclopropanes generates 1,3-diketones. This disjoint-to-conjoint transformation has been gainly applied in a ring expansion route to muscone [Ito, 1977].

muscone

In brefeldin A there is a concentration of functionalities in the upper chain of the macrolactone, and it continues along the edge of the fused cyclopentane down to the isolated double bond. It would be an extremely valuable consideration if in its synthesis there was a correlation of all these functional groups so that they may be created simultaneously. The peri hydroxyl is 1,4-related to the nonconjugated double bond, or 1,5-related to the distal sp^2-terminus, reconnection of these carbon atoms into a cyclopentane subunit that permits stereoselective elaboration of the other substituents deserves particular attention. The key operation is then a fragmentation process to generate the double bond and the oxygenated site [Nakatani, 1985].

The fragmentation circuit extending three carbon atoms offers many opportunities to succeed at the desired transformation. The actual fragmentable precursor that was selected depended on its availability in stereoselective and economic terms.

Oxidative cleavage of γ-hydroxyalkyl stannanes appeared to fulfill most of the requirements, and the proper derivative could be obtained readily from a diquinane derivative. The oxidation was promoted by lead(IV) acetate, and although the two carbon chains of the product were *cis*, equilibration at a later stage was assured by the presence of a ketone group.

brefeldin-A

The dense functionality and multiple asymmetric centers present in picrotoxinin are tantalizing features that attract synthetic efforts. Oxidation reactions figured prominently in a successful execution of a magnificient plan [Corey, 1979c].

The plan recognized the *vic*-dioxygenation on the cyclohexane template as clearable, i.e., regio- and stereospecific reconstitution of the two lactone rings from the dicarboxylic acid is feasible, and that the alkylation pattern of this cyclohexane corresponds to a *p*-menthane skeleton. Furthermore, the relative positioning of the tertiary hydroxyl group and the required double bond suggests carvone as the starting material. α-Alkylation of the enone system shifts the unsaturation into the location as desired.

picrotoxinin

(-)-carvone

The elegance of this synthesis lies in how the two carboxyl groups were created stereospecifically. With respect to the cyclohexane nucleus the two carboxyl groups are *cis* and 1,6-related, therefore they can be conveniently reconnected into another six-membered ring. Such a projected precursor is apparently accessible from carvone via an alkylation–aldolization sequence.

Missing from the bicyclo[3.3.1]nonane skeleton is a two-carbon subunit for the cyclopentane component. The two-carbon chain is conjoint, the donor end would link to the original carbonyl of carvone and the acceptor end to the α-carbon of the future fused lactone. The acceptor site must be in a higher oxidation state in order that the epoxide may be introduced more conveniently.

In the above are the essential considerations for synthetic operations besides protection of the isopropenyl sidechain. Also important is the oxidation of the bridged ring intermediate into an α-diketone, not only because it is required in a subsequent oxidative cleavage to liberate the two carboxylic acids, the cyclization to give the cyclopentane also demands activation of the donor site in an aldol condensation step.

The simultaneous formation of the two lactone rings was achieved oxidatively. The ethereal oxygen atom of a lead(IV) carboxylate underwent contrapolarization to induce the C—O bond formation with its proximal trigonal carbon of the double bond. The incipient carbocation was immediately trapped by the other carboxyl group which hovered in the vicinity.

Triquinacene is a unique hydrocarbon which was thought to be an excellent precursor of dodecahedrane. The first synthesis of triquinacene [Woodward, 1964] exploited the special chemistry made possible by the spatial proximity of the functional groups.

The first double bond was introduced at the expense of two extraneous C—C bonds by lead(IV) acetate oxidation of a 1,4-diketone hydrate. The anhydride which was formed simultaneously was well suited for degradation.

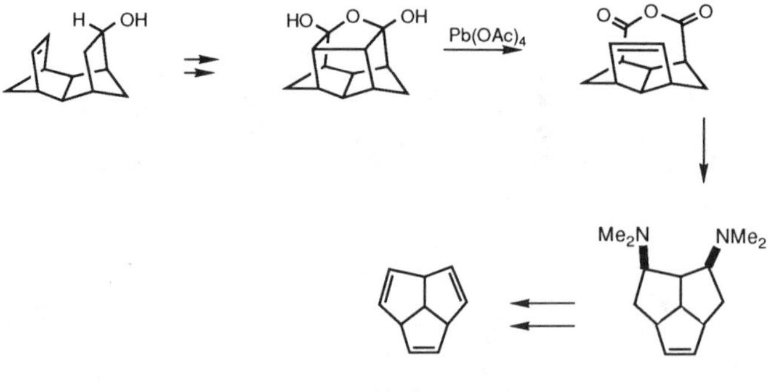

triquinacene

γ-Keto carboxylic acids undergo oxidative decarboxylation on treatment with lead(IV) acetate in the presence of a cupric salt [McMurry, 1974a]. The incipient carbocation is conjoint to the ketone and its generation is favored. Deprotonation of this species follows logically.

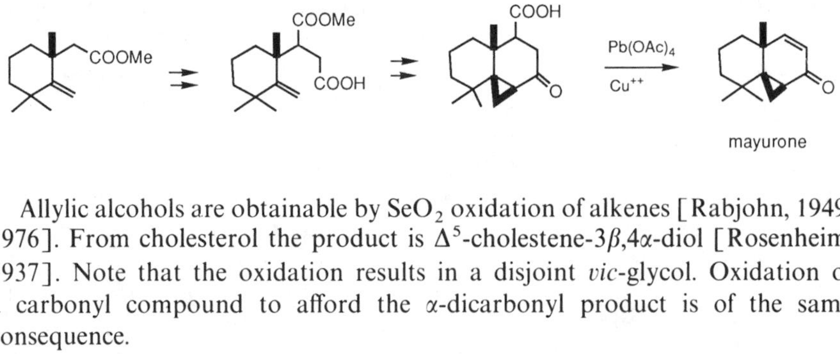

mayurone

Allylic alcohols are obtainable by SeO_2 oxidation of alkenes [Rabjohn, 1949, 1976]. From cholesterol the product is Δ^5-cholestene-3β,4α-diol [Rosenheim, 1937]. Note that the oxidation results in a disjoint *vic*-glycol. Oxidation of a carbonyl compound to afford the α-dicarbonyl product is of the same consequence.

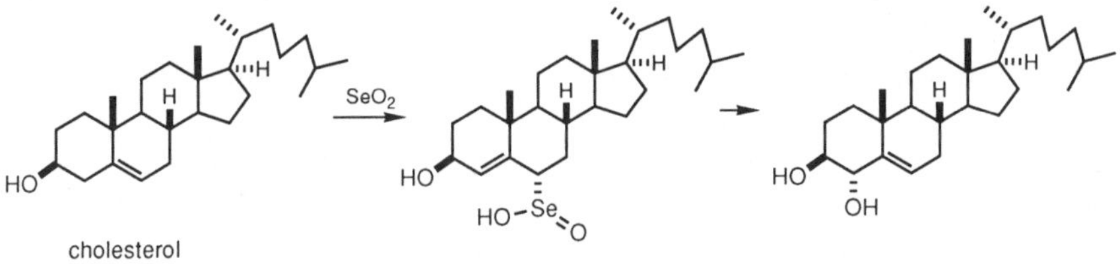

cholesterol

Polarity inversion accompanies every redox process, whether one- or two-electron transfer is involved. Oxidative phenolic coupling definitely proceeds via free radicals, and the products are invariably disjoint.

The intramolecular *o,p*-coupling product of an *N*-arylphenethylamine usually undergoes further cyclization to give a hydrodibenzofuran. Naturally the umpolung bequeaths a facility in the formation of the five-membered ring (cf. a synthesis of galanthamine [Kametani, 1969]).

+ p,p-product

galanthamine

The application of organotransition metal complexes to synthesis is on the rise. Accordingly, mild procedures for liberating the organic ligands are needed. For metal carbonyl complexes, oxidative degradation by treatment with trimethylamine oxide represents a useful method. It is thought that this decomplexation proceeds via attack on the coordinated carbonyl [Blumen, 1979]. The change in the electronic state of the metal should be noted.

16.2. REDUCTIONS

The umpolung action of halogen/metal exchange of a halocarbon is the basis for the extensive organometal-mediated C—C bond-forming reactions. These reactions include the Grignard, Reformatskii reactions and those involving organolithiums. Strictly speaking, the *ipso* atom of the halocarbon molecule undergoes reduction by the metallic reagent during formation of the organometallic species. In the formation of Reformatskii reagents the somewhat disjoint α-bromoalkanoic esters are converted into conjoint and less reactive organozinc compounds.

Among modern synthetic methods related to these classical reactions, that using π-allylnickel reagents represents a significant development. One example of its utility is a synthesis of frullanolide [Semmelhack, 1981].

frullanolide

The first step of forming the π-allylnickel complex is similar to any other C—X bond insertion by metallic reagents, except that the π-bonding between carbon and nickel is driven by thermodynamics. Upon achievement of umpolung and completion of the intramolecular reaction with the aldehyde, a $[B—CO—Ni(CO)_2]^-$ mediated oxidative-addition of the vinylic bromide follows.

It must be emphasized that the stoichiometric reactions of π-allylnickel complexes as shown above differ fundamentally from reactions of π-allylpalladium complexes derived from palladium(0) and allyl esters. The palladium(0) species is catalytic in metal which serves to activate the allylic acceptor and to provide stereocontrol for the substitution reactions. There is no change in polarity of the reagents.

Discussion of several reduction processes in terms of polarity control is now in order. It is possible to rationalize the remarkable regioselectivity attending the borohydride reducton of one of the enedione carbonyls in each of the early intermediates for synthesis of reserpine [Woodward, 1958] and tetrodotoxin [Kishi, 1972]. Rationalization can be achieved by the notion that an infused donor character of the carbonyl group proximal to the polar substituent via the polarity alternation mechanism results in decreased reactivity. However, steric factors should not be ignored.

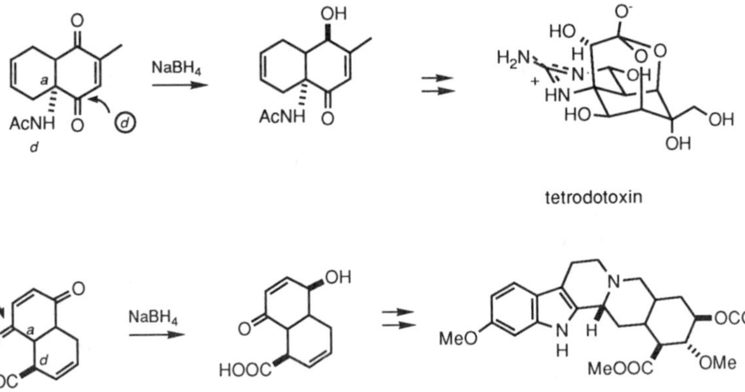

tetrodotoxin

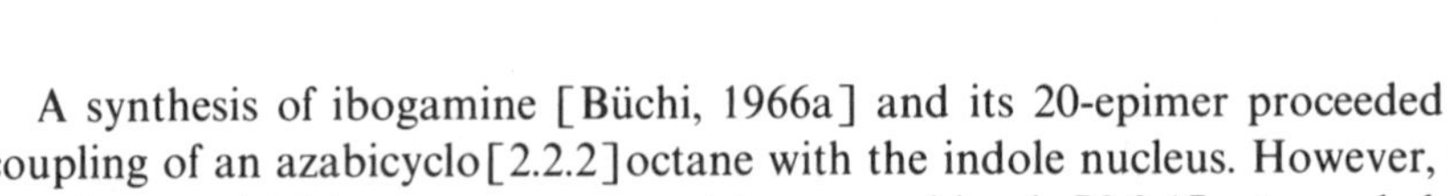

reserpine   [Ar = 3,4,5-(MeO)$_3$C$_6$H$_2$]

A synthesis of ibogamine [Büchi, 1966a] and its 20-epimer proceeded via coupling of an azabicyclo[2.2.2]octane with the indole nucleus. However, this reaction resulted in a rearrangement to an azabicyclo[3.2.1]octane skeleton due to intervention of an aziridinium species. This rearrangement also disturbed the conjoint array spanning the nitrogen atom and the oxygen function of the ethyl chain. Re-entry to the more symmetrical network was achieved by a reduction–Michael sequence.

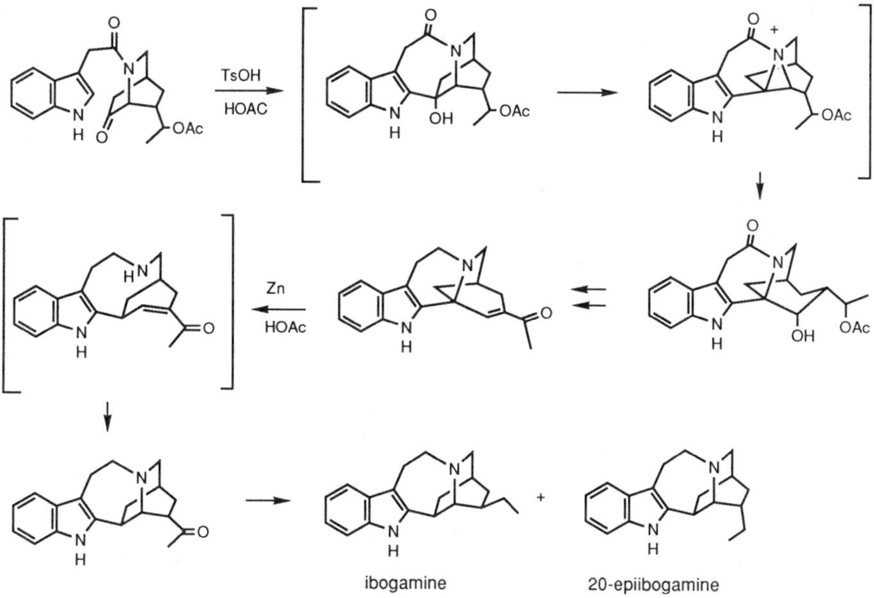

ibogamine   20-epiibogamine

Regiospecific generation of cyclic silyl enol ethers through a Brook rearrangement of α-silylated allylic alcohols is in effect a reductive *O*-silylation of enones [Koreeda, 1990]. During the rearrangement a transient siloxy anion emerges, hence contrapolarization intervenes.

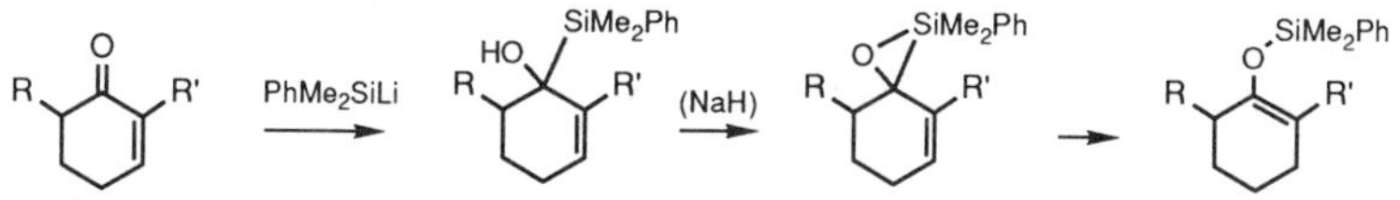

It is significant that in the exposure of (trimethylsilyl)methyl-1,4-benzo-quinones to donor reagents the silyl group is displaced and the quinone moiety is reduced [Karabelas, 1990]. The reduction is coupled to the polarity inversion at the benzylic site.

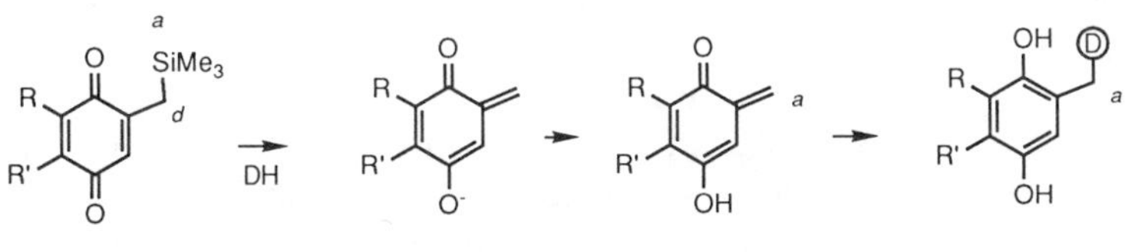

D = OAc, OR...

A quadrone precursor lacking the lactone ring is a 1,4-diketone. Formation of this segment necessarily involves an umpolung or contrapolarizing event. In a rather unusual synthesis [Sowell, 1990] in which reductive and oxidative C—C bond formation processes are most prominent, this segment was established (actually before the closure of the five-membered ring) by electroreduction. In fact, the *gem*-dimethylated cyclopentane ring (disjoint) was also similarly created by an electrochemically induced process (aldehyde + unsaturated ester).

The cyclopentanone ring was the last to assemble. A less well-known intramolecular free-radical cyclization involving a nitrile as the acceptor proceeded particularly well. The radical center was generated from oxidative decarboxylation of an acid.

16.3. SOME UMPOLUNG PROCESSES

In formal valence terms the nitrogen atom of an amine and that of a nitro group represent two oxidation states. Umpolung action may be designed to assemble a molecular network containing one of the functional groups by using the other. Because of the general difficulties in directly functionalizing the α-position of an amine (especially that containing an active hydrogen), the umpolung technique is frequently applied to amine synthesis starting from nitro compounds. Reduction of the nitro group is facile.

Tricyclic ergot alkaloids such as chanoclavine I are conceivably accessible from 3,4-disubstituted indoles. Formation of the six-membered ring must also permit the establishment of a *trans* stereochemistry of the two sidechains.

Based on this consideration, FGI of the amino group with the nitro group would be most expedient because the allylic chain easily accommodates an acceptor center. Moreover, the nitroethyl pendant is conjoint with the indole nucleus and therefore readily assembled (Vilsmeier + Henry reactions followed by double-bond reduction). In fact, reduction of the nitroalkene moiety was accompanied by the C—C bond formation (S_N2' displacement). At this juncture nitro group modification was the only remaining task of the synthesis [Somei, 1986].

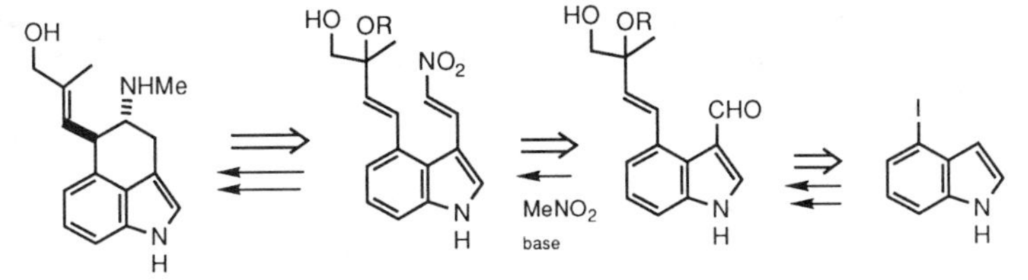

chanoclavine-I

Lycoramine contains a 1,3-dioxycyclohexane subunit which can be constructed easily from the appropriate arylacetaldehyde because of the conjoint functionalities. However, both oxygen atoms form a disjoint relationship with the amino groups, therefore it is necessary to use umpoled species to create chains between the different heteroatoms, should a synthesis of lycoramine be pursued with a Robinson annulation to assemble the cyclohexane ring.

In a synthesis [Sanchez, 1984] according to this strategy the assembly of the shortest link between oxygen and nitrogen was accomplished by a Michael addition of nitromethane to a substituted cinnamonitrile. The nitromethyl group is a formyl equivalent.

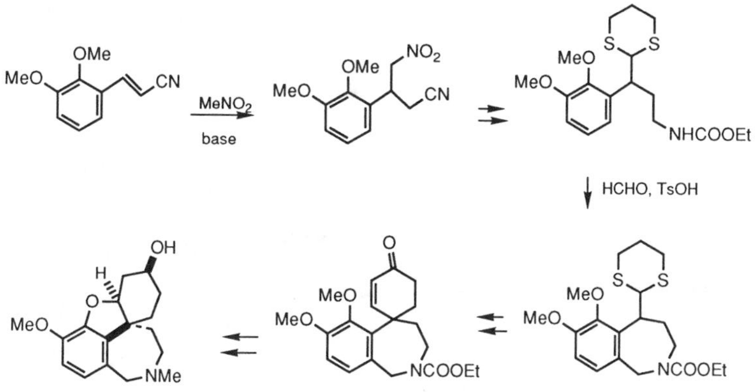

lycoramine

α-Halocarboxylic esters are disjoint species and naturally they have found numerous uses in circumstances when umpolung manipulation is needed. It has been stated several times that odd-membered-ring formation must rely on such a strategy and a simple approach to the pyrrolizidine alkaloids such as isoretronecanol and trachelanthamidine [Pinnick, 1978] from 2-carbethoxy-methylenepyrrolidine indeed proceeded via an alkylation step which employed ethyl bromoacetate.

isoretronecanol

Like α-halocarboxylic acid derivatives α-aminonitriles are also captodative (disjoint). These nitrogenated compounds are very versatile precursors of iminium ions as well as α-amino carbanions. The synthesis of both optical isomers of coniine [Guerrier, 1983] from a common intermediate obtained from glutaraldehyde, (+)-norephedrine, and KCN is a proof of the maneuverability of the α-aminonitriles.

Accordingly, in later steps the donor property of the cyano group was accentuated by coordination with silver ion, and using a strong base, deprotonation at the α-carbon was accomplished.

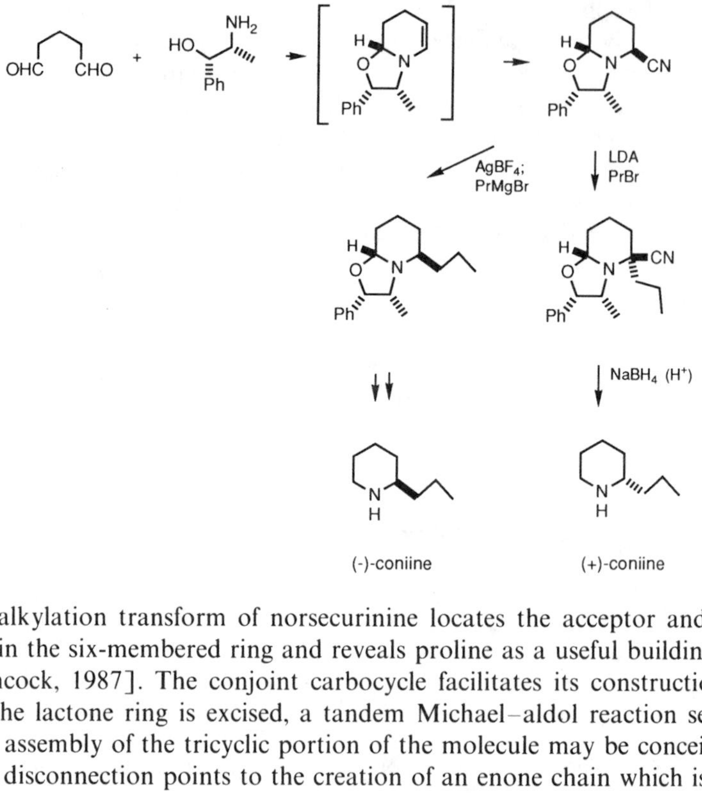

An alkylation transform of norsecurinine locates the acceptor and donor atoms in the six-membered ring and reveals proline as a useful building block [Heathcock, 1987]. The conjoint carbocycle facilitates its construction, and when the lactone ring is excised, a tandem Michael–aldol reaction sequence for the assembly of the tricyclic portion of the molecule may be conceived.

The disconnection points to the creation of an enone chain which is linked to the proline carbonyl. In terms of synthetic expediency, a precursor which already contains the carbon chain elements for eventual closure of the lactone ring would be ideal. Prefabrication of the disjoint segment (from ketone to ester) would forestall any unexpected difficulties in its elaboration.

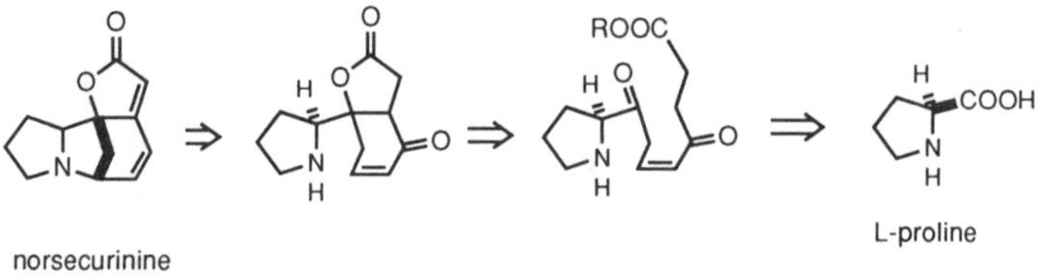

Interestingly, a six-carbon chain properly punctuated with oxygen substituents was utilized to couple with the proline-derived β-keto phosphonate. This chain was fashioned from a 1,4-cyclohexanedione monoketal.

Two significant observations during the synthesis were made. The keto-phosphonate prepared from methyl prolinate emerged as a racemate. Secondly, a skeletal rearrangement intervened at the tricyclic level during the conversion of the ketone into a double bond using sulfoxide pyrolysis. An azetidinium salt was formed in the thiolysis of the mesylate intermediate. Fortunately, the sulfoxide pyrolysis also proceeded via such an ion pair to regenerate a portion of the olefin product with the desired skeleton.

norsecurinine

Umpolung of allylic acetates, especially tertiary acetates, is very difficult. It requires the formal generation of the allyl anions and trapping them with the desired acceptor reagents at the tertiary site. It has now been possible to achieve this transformation by means of silicon chemistry. Thus, a silyl cuprate reagent displaces the acetoxy group via an S_N2' mechanism [Fleming, 1984], and the ensuing allylsilane can then react with the acceptor under the influence of a Lewis acid, again transpositionally [Fleming, 1979]. The net result is a polarity inversion at the acetoxylated carbon of the original allyl acetate.

Much simpler cases involving umpolung include metallation of organohalides, and displacement of an organohalide with a phosphine, which is the routine step preceding a Wittig reaction. Hydration of unsaturated compounds (alkenes, alkynes) via hydroboration must be followed by an oxidation step, with typical umpolung.

17

REARRANGEMENTS

17.1. 1,2-REARRANGEMENTS

A 1,2-rearrangement is commonly induced by electron deficiency and most effectively when an *a–a* array is present or created, as evident by examination of the following reactions.

Pinacol rearrangement:

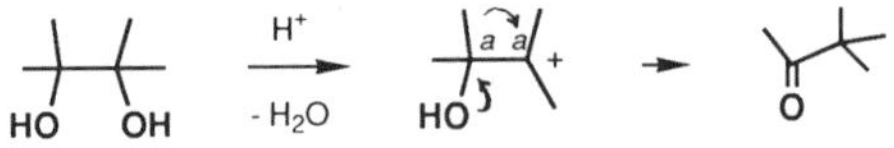

Wolff rearrangement:

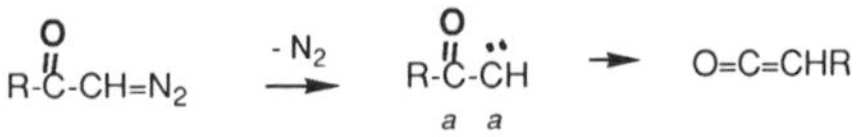

Hofmann rearrangement:

Baeyer–Villiger reaction:

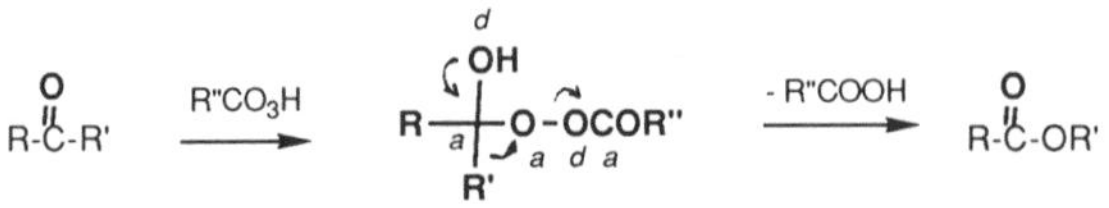

The Lewis-acid-catalyzed rearrangement of epoxides is actually a version of the pinacol rearrangement. Of special interest is the ring expansion of oxaspiro[2.2]pentanes [Trost, 1975a].

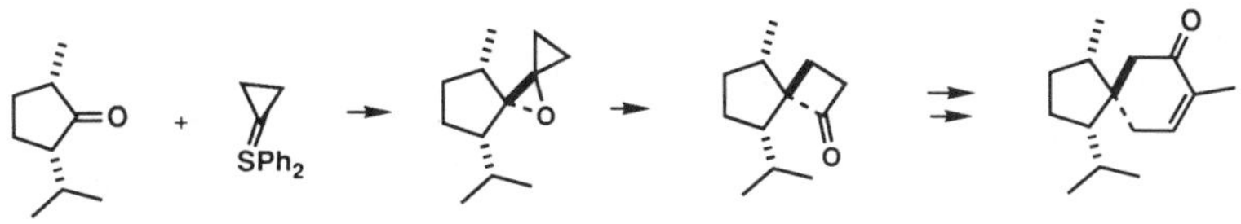

acorenone-B

Rearrangement reactions are accelerated when they are accompanied by strain relief. Expansion of small rings by rearrangement enjoys such effects.

Alkylation of 1,2-bis(trimethylsiloxyl)cyclobutene with acetals sets up a favorable system for ring expansion by a pinacol-type rearrangement [Shimada, 1984]. There are many synthetic applications of this method, but a convenient synthesis of cuparene suffices to demonstrate its potential.

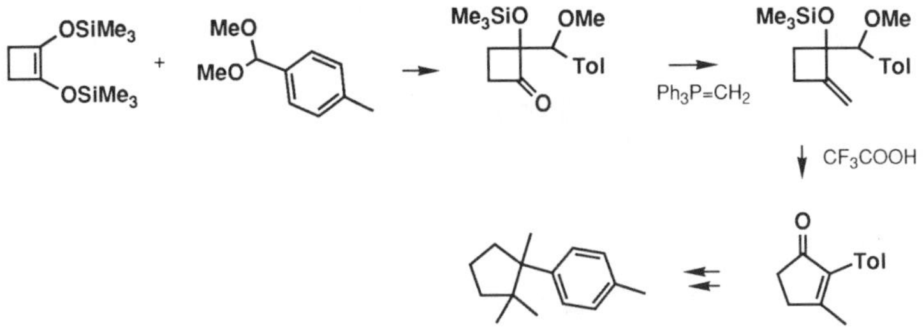

cuparene

A synthesis of α-cuparenone [Halazy,1982] via two ring-expansion steps has been achieved.

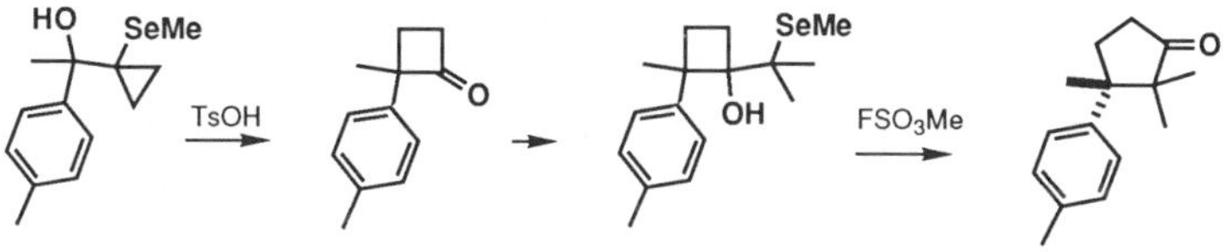

α-cuparenone

A most elegant, and perhaps extreme exploitation of the 1,2-rearrangement to synthesis is the cascade migration, leading to, for example, [6.5]coronane [Wehle, 1987].

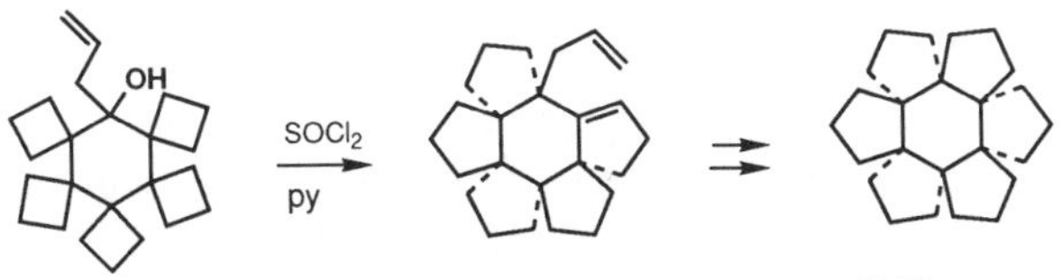

[6.5]coronene

This efficient and novel reaction has great potential for a total synthesis of modhephene [Fitjer, 1988]. A trisnormodhephene has been obtained without difficulty.

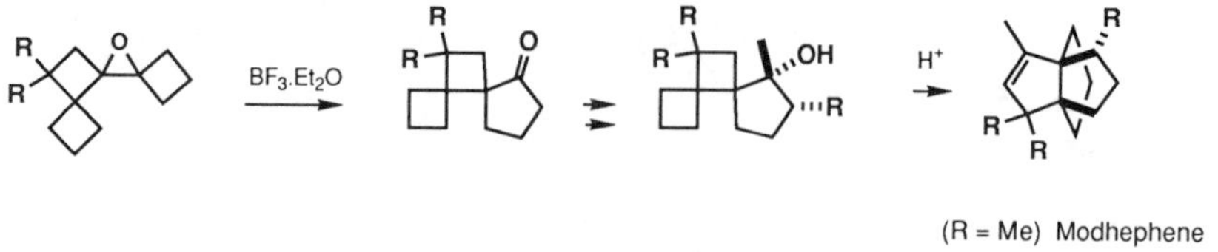

(R = Me) Modhephene

A salient feature in a synthesis of verrucarol [Roush, 1983] is that although two Diels–Alder adducts were obtained from the cycloaddition of *x*-methyl-5-trimethylsilylcyclopentadiene with methyl acrylate, the epoxides of both adducts underwent desilylative rearrangement to afford the same bicyclic alcohol.

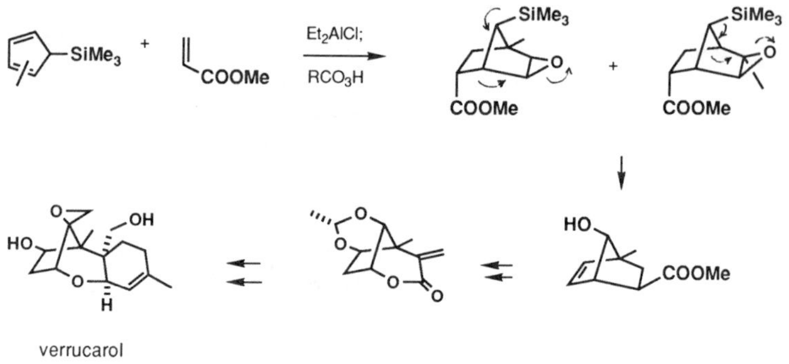

verrucarol

The direction of epoxide ring opening, of one isomer, was governed by the relative stability of the two incipient carbocation (tertiary vs secondary). The more symmetrical epoxide may relieve a certain amount of strain due to a 2:6 *endo* interaction between the ester group and the methine hydrogen by the observed rearrangement mode. Interestingly, ring opening in the other direction would have generated a tertiary carbocation.

There is no question that polarity control is of utmost importance in the Lewis-acid-catalyzed isomerization of α,β-epoxy ketones [House, 1956]. The generation of α-diketones is expected as this pathway avoids incipient α-oxocarbenium ions (disjoint species).

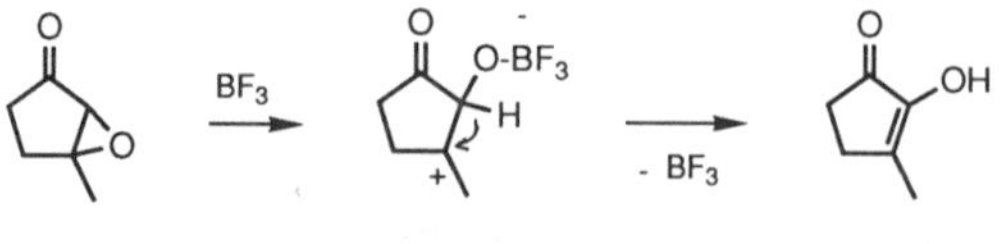

By contrast, exposure of the same compounds to palladium(0) reagents leads to 1,3-diketones [Suzuki, 1980]. Necessarily, the mechanisms are totally different in the two reactions.

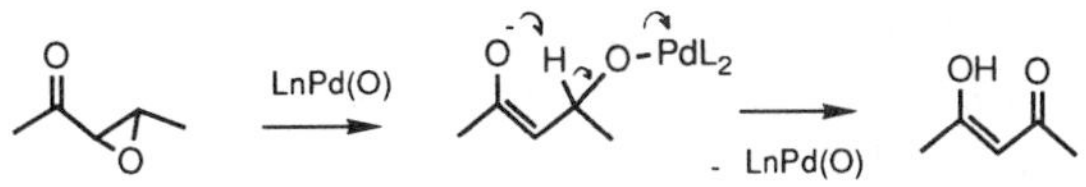

No matter what method is used to generate α-alkoxy carbocations the result (rearrangement) is the same. Thus, 3-methoxycyclopentyl ketones have been prepared by double-bond interception of a remote methoxycarbenium ion [Hirst, 1989] as shown below.

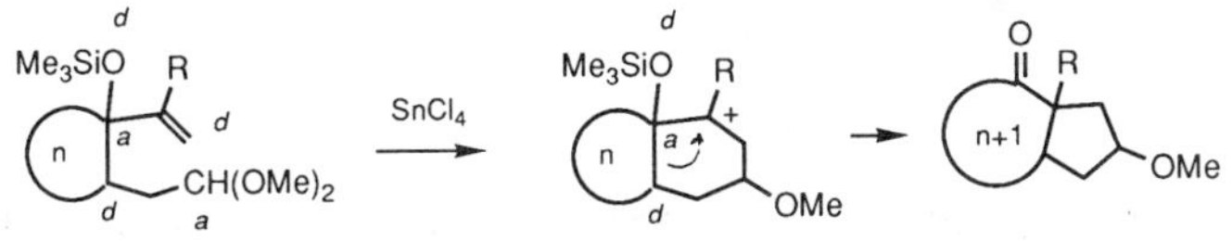

Protonation of γ-alkoxy-α,β-enones may induce rearrangement to afford 1,4-diketones. This reaction is extremely facile in the following case which culminated in a synthesis of grandisol [Wenkert, 1978a]. The rearrangement was facilitated by the transformation into a less strained bicyclic framework.

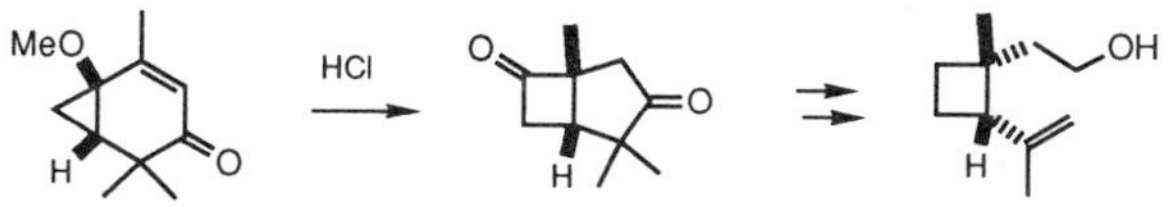

grandisol

Extensive use of the pinacol rearrangement, especially the version of solvolytic transformation of a *vic*-diol monosulfonate, has been recorded in the chemical literature. Many of these examples deal with a fused ring system, e.g., $m, n \rightarrow (m + 1), (n - 1)$ in terms of ring size readjustment (cf. synthesis of (−)-aromadendrene [Büchi, 1966b]). It is also possible to convert one bridged ring skeleton into another, as shown in an approach to stachenone [Monti, 1979].

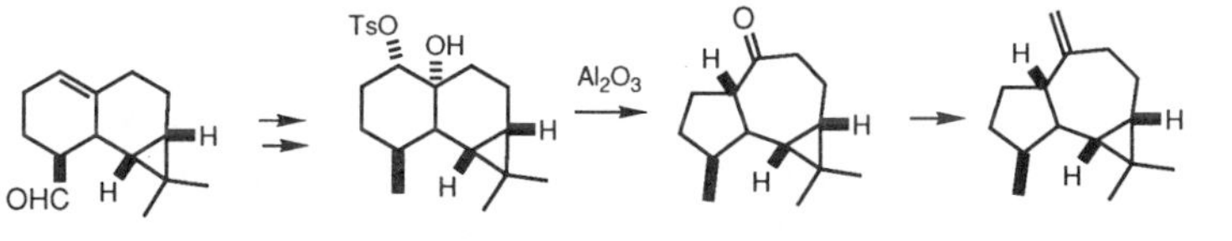

(-)-aromadendrene

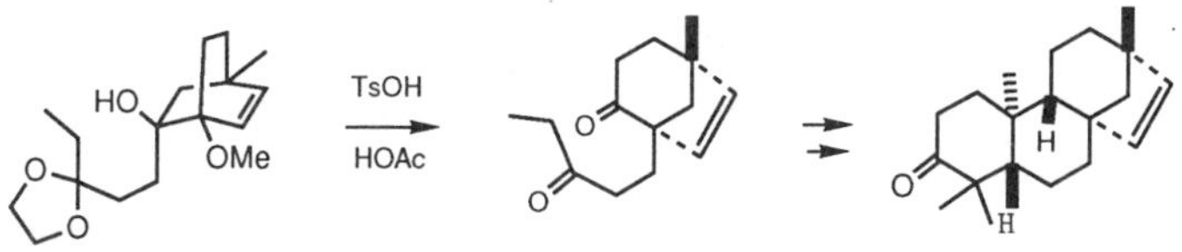

stachenone

An interesting observation pertains to the desulfonylative ring expansion of β-hydroxy sulfones in which substitutents are also present at the α-position. A phenylthio group apparently does not affect the rearrangement (whether it is ameliorating cannot be said) whereas a methoxy group has an inhibitory effect except for reactions leading to less strained products, α-Phenylthiocycloalkanones are electronically more favorable to expand than the corresponding α-methoxycycloalkanones because the sulfur atom has substantial acceptor character.

X = SPh   all n
X = OMe   n = 4, 5 only

The Nazarov cyclization of a cross-conjugated trienone in which the β-carbon of the vinyl group is trimethylsilylated, furnished a cyclopentenone with a migrated silylvinyl residue [Denmark, 1988]. Rearrangement occurred immediately after the ring closure. It should be noted that in the trienone the carbonyl group and the silyl substituent are disjoint, and the 1,2-rearrangement alters the relationship to a conjoint segment.

Fragmentation of certain halocyclopropanes may also be considered as a 1,2-rearrangement. A synthesis of nezukone [Birch, 1968] incorporated this protocol.

nezukone

Electron-deficient carbenoid centers promote 1,2-alkyl migrations. With further inducement by a donor substituent in an adjacent carbon atom, expansion of cyclic ketones may be effected [Hiyama, 1979].

nootkatone

An interesting example in which a 1,2-rearrangement is followed by fragmentation [Woodward, 1968] shows a change of a disjoint $a-a$ system to a conjoint $a1a$ intermediate. The latter is liable to fragment because of ring strain.

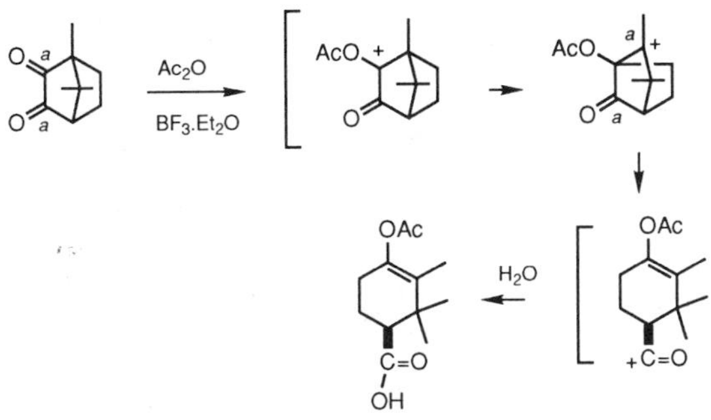

A skeletal rearrangement accompanied the thermal elimination of methanol from the photocycloadduct of piperitone with 1,1-dimethoxyethylene, changing the fused ring framework into a bicyclo[3.2.1]octenone [Yanagiya, 1979]. The transformation was perhaps unexpected but the bridged system is easily recognized as a convenient precursor of helminthosporal and many other sesquiterpenes.

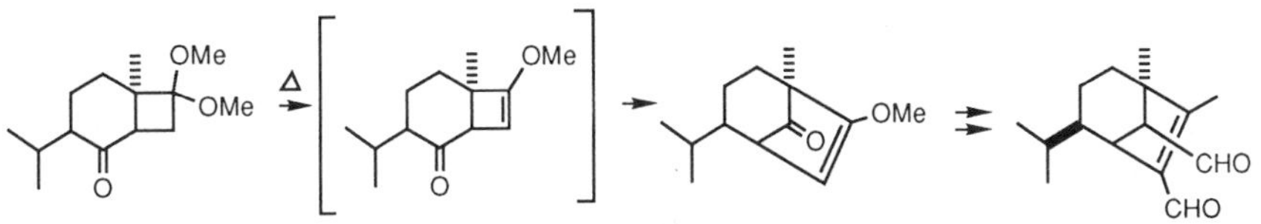

helminthosporal

The first step of the rearrangement must be the ring expansion to relieve the strain of the four-membered ring. This bond migration must also be favored by the formation of a transient zwitterion in which the cationic portion is stabilized (methoxyallyl).

σ-Bond insertion by diazoalkanes is a well-known reaction. It is particularly facile for homologation of ketones. In a synthesis of zizaene [Coates, 1972] the complete skeleton developed from a norcamphor derivative by an intramolecular C—C bond insertion.

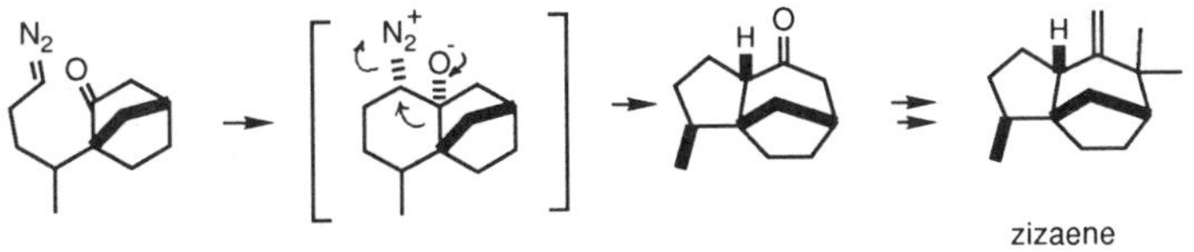

zizaene

The Wolff rearrangement has been frequently employed to create a cycloalkanecarboxylic acid subunit from the homologous cycloalkanone. It also occurs during the Arndt–Eistert synthesis. The mildness of the reaction conditions for the conversion of a cycloalkanone into the α-diazoketone and the rearrangement which can be effected photochemically, offers a great advantage in situations dealing with complex molecules. In an alcohol solvent the corresponding ester is produced.

Skeletal modification by this rearrangement may become a necessity in many syntheses. For example, isocomene has been elaborated from a tricyclic ketone via this ring contraction [Oppolzer, 1979]. The ketone was obtained from an ene cyclization, and it proved impossible to prepare the corresponding angular triquinane directly by the same method because of excessive ring strain.

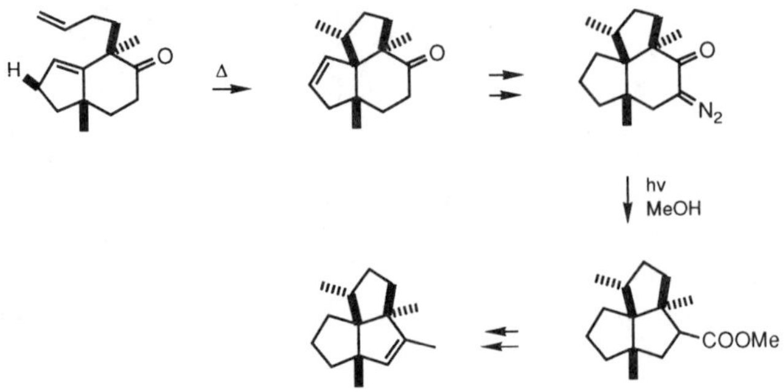

isocomene

An approach to pentalenic acid [Ihara, 1987] was keyed on a reflexive Michael reaction and also required the ring contraction. It is interesting that this transformation did not alter the $O—C_5—O$ relationship, because the rearrangement step involved movement of an out-of-the-chain atom only.

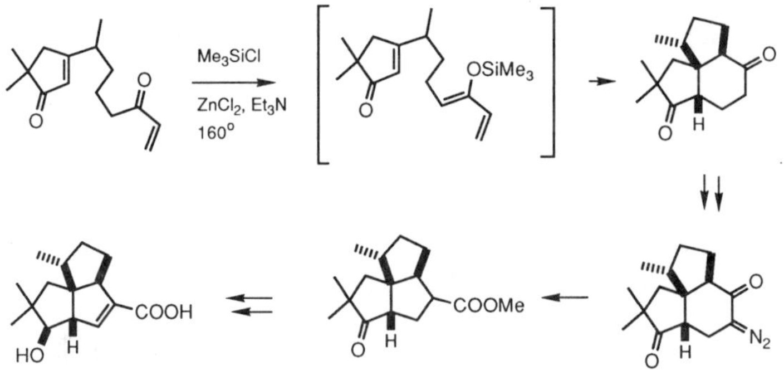

pentalenic acid

The Favorskii rearrangement also converts a cyclic ketone into a cyclo-alkanecarboxylic acid derivative with one less nuclear atom. It embodies α-halogenation and treatment of the product with a base, and in rare circumstances by silver-ion-assisted solvolysis.

The general mechanism prescribes α,α'-dehydrohalogenation to give a cyclopropanone intermediate, but for small-ring ketones (e.g., cyclobutanone), the semibenzilic-type rearrangement prevails. A synthesis of sirenin [Harding, 1988b] employed this ring contraction as a key step.

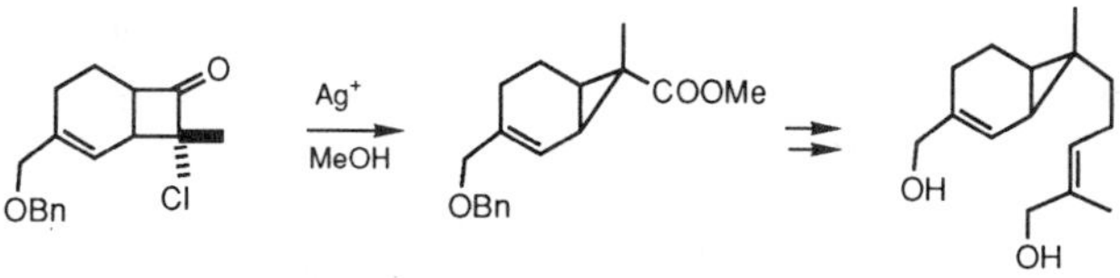

sirenin

The salient feature of the Favorskii rearrangement [Kende, 1960] and the Ramberg–Bäcklund rearrangement [Paquette, 1977] is that deprotonation occurs at the nonhalogenated α'-carbon atom. In view of the captodative situation (acceptor C=O, donor halogen), deprotonation at the α-carbon is definitely less favorable.

Functionalization of a double bond with rearrangement may be achieved on metallosolvation with oxidizing metal salts. Thallium(III) salts are very effective [McKillop, 1973]. (A synthesis of sativene [Oppolzer, 1984] using a longifolene intermediate hinged on a ring contraction to generate an acetylcyclohexane subunit by thallium(III) oxidation.)

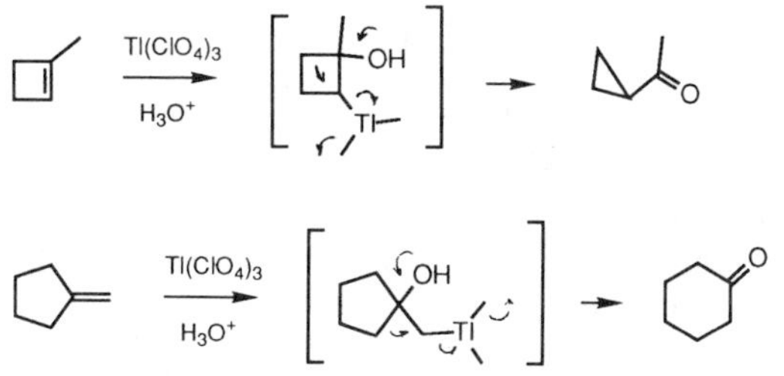

Lead(IV) acetate is somewhat less reactive than thallium(III) species, and acetoxylation seems to be the major reaction it brings about. However, a synthesis of cephalotaxine [Kuehne, 1988] involved a ring contraction of an acyl enamine by the action of lead(IV) acetate. The migrating group was the nitrogen atom.

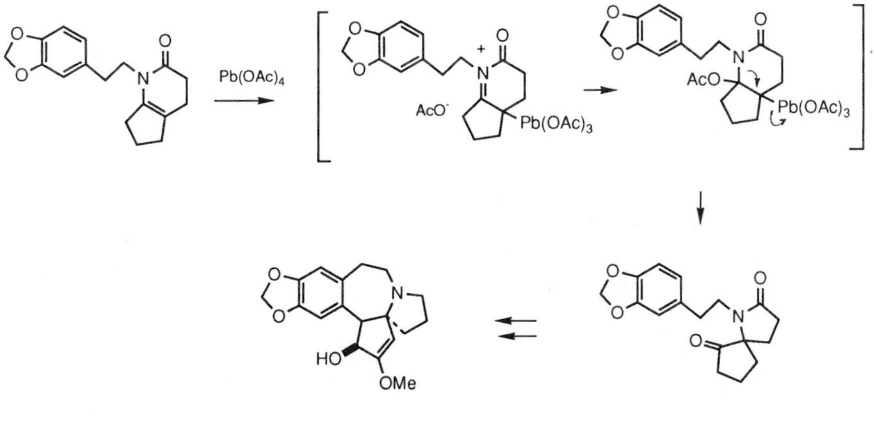

cephalotaxine

It has been known that an azaspirocyclic enone is a synthetic precursor of perhydrohistrionicotoxin. An alternative route to this enone, based on a Robinson annulation transform, must address the potential reactivity problem on the part of the piperidinecarboxaldehyde intermediate which is captodative. Furthermore, its preparation would not be straightforward.

A rearrangement pathway to gain access to the aldehyde [Duhamel, 1986] resolved this synthetic ambiguity.

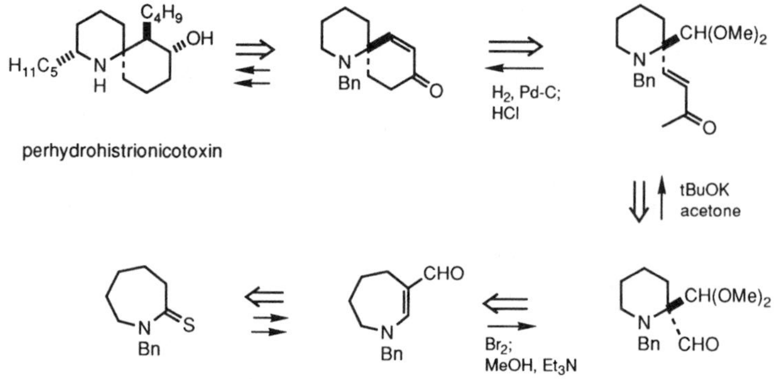

perhydrohistrionicotoxin

The Beckmann rearrangement gives amides/lactams [Donaruma, 1960] via imino carbenium ion intermediates. It has been successful to intercept these intermediates with other donors present in the reaction media. An alkene is effective as an internal nucleophile, as demonstrated in a synthesis of muscopyridine [Sakane, 1983].

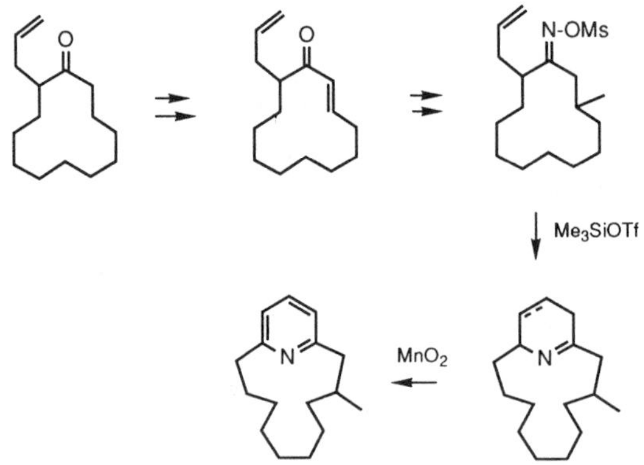

muscopyridine

Of many disconnections of ibogamine a Fischer-indole transform simplifies many of the projected operations in synthesis. The tricyclic β-amino ketone is not particularly amenable to a Michael or a Mannich transform because of stereochemical issues and the unavailability of its precursors. On the other hand, disconnection of the N—CH$_2$ bond and a Beckmann rearrangement transform reveals a cis-decalindione system. At this point the Diels–Alder approach is evident [Salley, 1967].

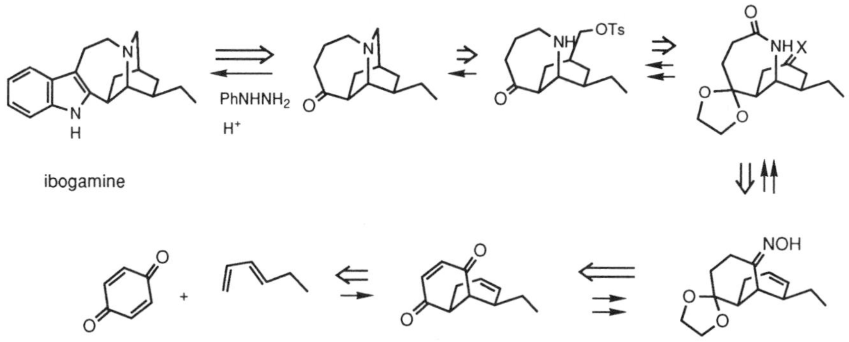

It is noted that the disjoint functionality of benzoquinone was converted into the conjoint circuit of a β-amino ketone via the 1,2-rearrangement step.

It is instructive to compare the effect of a Lewis acid on norbornenone oxime with the corresponding diol acetonide [VerHaeghe, 1989]. Beckmann fragmentation resulted in the unsaturated oxime, whereas the acetonide underwent rearrangement. The rearrangement propensity of the latter was enhanced by the suppression of fragmentation by the proximal oxygen atom of the heterocycle, as the incipient carbocation arising from fragmentation would be disjoint to the oxygen.

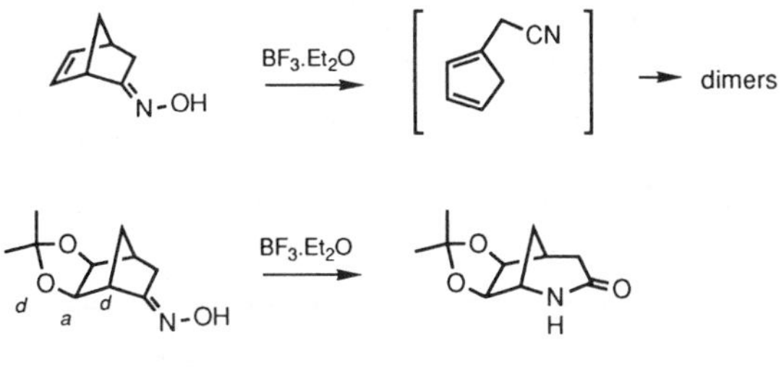

Formation of a 1,3-diamine by reductive Beckmann rearrangement of a symmetrical 1,4-dioxime [Papageorgiou, 1987] must be a stepwise process. The regiochemistry of the second stage of the reaction was apparently affected by the electronic set-up of the semi-rearranged intermediate. It is interesting that the product is conjoint while the dioxime is not.

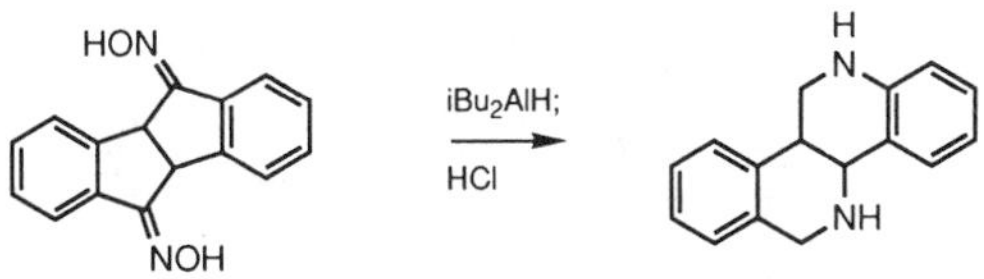

Although the Hofmann rearrangement is infrequently employed in complex synthesis, one recent application is noteworthy in that a γ-lactone was specifically transformed into a 1,3-amino alcohol subunit in an approach to (−)-histrionicotoxin and (−)-histrionicotoxin 235A [Stork, 1990b] which also featured an internal N-alkylation to create the azaspirocyclic system.

(-)-histrionicotoxin-235A

The synthesis started from allylation of methyl 6-oxohexanoate with an optically active *B*-allyldiisopinocampheylborane. After protection of the hydroxyl group, dialkylation of the ester with a bromide containing an allylic epoxide led to the bridged lactone. The second stage of the alkylation, i.e., epoxide ring opening by the ester enolate, is regio- and stereoselective [Stork, 1990a]. Thus with all the stereogenic centers established a Hofmann-like rearrangement was conducted. The remaining steps are rather straightforward.

The bridged lactone intermediate served as a branching point toward elaboration of (−)-histrionicotoxin. The double bonds were subjected to ozonolysis, and the resulting dialdehyde was converted by a Wittig reaction to two (*Z*)-iodoalkene subunits, each of them was then coupled to an acetylene.

The significant retrosynthetic consideration of picrotoxinin is the association of the two lactone carbonyls. With this thought the subtarget becomes a bicyclo[3.2.1]octane derivative with an additional bridge, the bridging element being part of a *p*-menthane. The small bridge may be disconnected (alkylation transform) and in so doing an aboriginal role of (−)-carvone for a synthesis is evident [Corey, 1979c].

picrotoxinin

(-)-carvone

A tricarbocyclic intermediate was fashioned accordingly by two aldol condensations. Cleavage of the α-diketone with NaOCl was followed by an oxidative bislactonization. The epoxide ring was introduced stereospecifically to a sterically biased α,β-unsaturated lactone intermediate.

A nonenolizable α-diketone with its unfavorable electronic and dipole–dipole interactions is susceptible to attack by various donors. Thus, in the present example, it seems likely that an adduct with ClO⁻ was formed, in whatever small amounts, and this adduct underwent 1,2-acyl migration to the oxygen atom while expelling the chloride ion.

Bridgehead migration during the Baeyer–Villiger reaction of N-carbobenzoxy-2-azabicyclo[2.2.2]octan-5-one [Krow, 1981] is expected, as its donor trait is reinforced by the nitrogen atom. Remarkably, an ester group at C-3 completely reverses the regioselectivity [Baxter, 1977]. No steric factor could be involved.

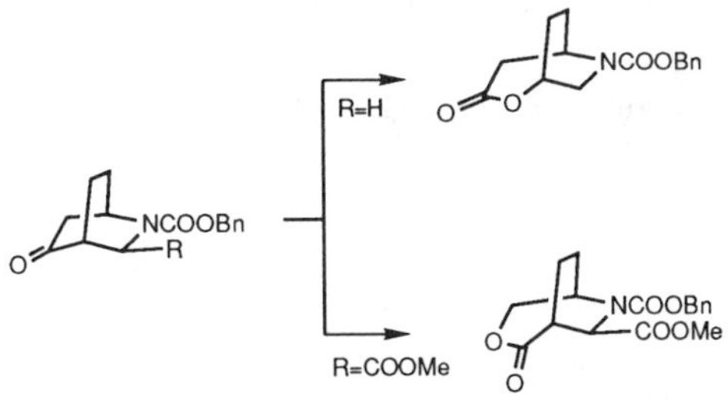

Also evident is the electronic effect of a γ-substituent outside the bridgehead of an oxabicyclic ketone [Noyori, 1980]. It is noted that a better donor α-methylene has a higher migratory aptitude.

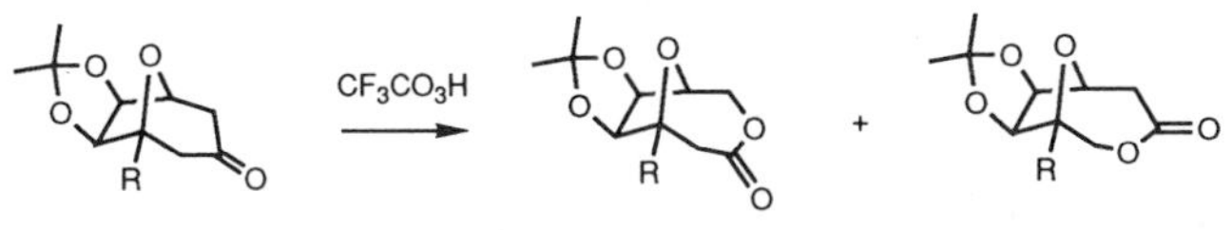

R = Ph	88	12
CH₂OTf	86	14
CH₃	47	53

Baeyer–Villiger reaction of locally symmetrical ketones gives rise to mixtures of esters or lactones. Variation of substitution patterns at the β-carbon and beyond is not expected to have a profound influence on the regioselectivity of the reaction. However, a dramatic effect of a β-silyl group has been discovered [Hudrlik, 1980]. This distinct regioselectivity has now been and continues to be exploited in the synthesis of many natural products, including (+)-α-cuparenone [Asaoka, 1988].

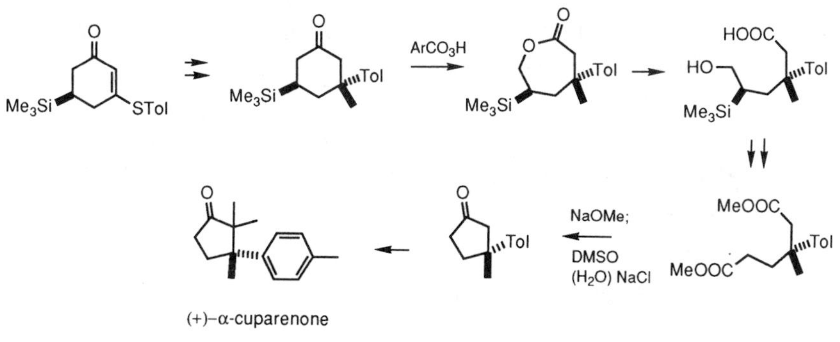

(+)–α-cuparenone

The origin of the silicon effect is not clear. It is known that a silicon atom favors the concentration of a positive charge at a β-atom, and the electron deficiency of such an atom would decrease its migratory attitude toward an electron-deficient center. On the other hand, migration of this atom which corresponds to the α-carbon of 3-trimethylsilylcyclohexanone would result in a conjoint molecule with an alternating polarity sequence of atoms extending from the carbonyl group to the silicon. The rearrangement rectifies the unfavorable electronic interactions along the functionalized chain of atoms in the ketone.

The oxidative degradation of 1,3-diketones with hydrogen peroxide to generate two carboxylic acids (Payne rearrangement) is a special kind of 1,2-rearrangement in which the leaving group is a carbon fragment. This reaction has been employed in a synthesis of cis-dihydrochrysanthemolactone from 3-carene [Ho, 1983].

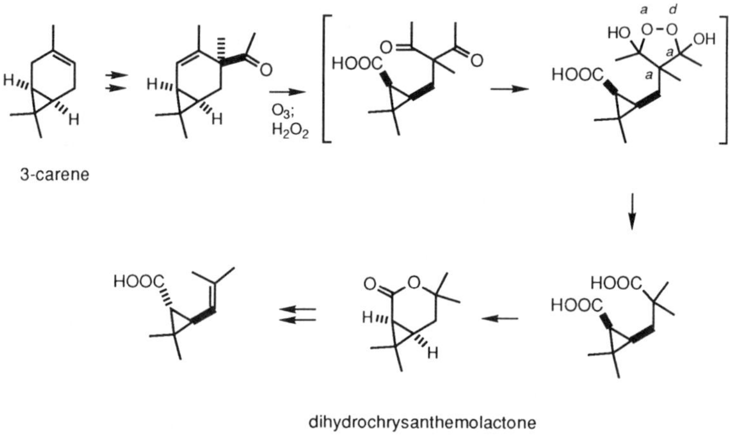

3-carene

dihydrochrysanthemolactone

17.2. SIGMATROPIC REARRANGEMENTS

Allylic sulfoxides undergo [2.3]sigmatropic rearrangement to afford transposed allylic alcohols. This is a useful contrapolarizing functionalization which has found numerous applications in synthesis, e.g., (E)-nuciferol [Evans, 1973], and a potential intermediate of verrucarol [Trost, 1978b].

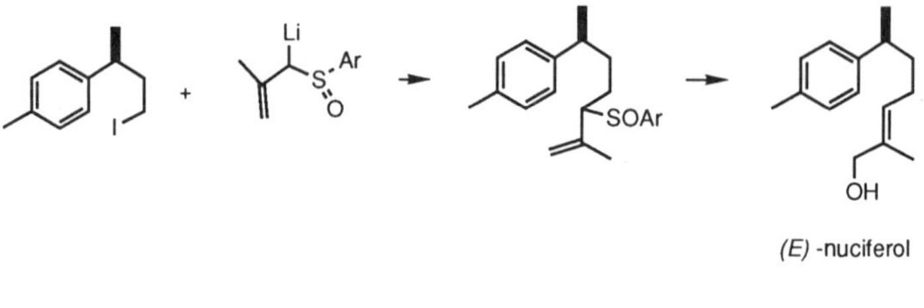

(E)-nuciferol

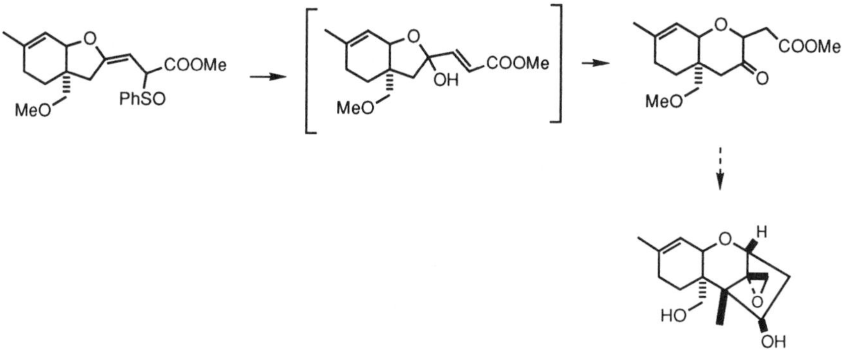

verrucarol

Stereochemical rectification of the $(13Z, 15R)$-prostaglandin E$_1$ derivative which was obtained from conjugate addition of a (Z)-vinylcuprate reagent to a cyclopentenone has been achieved via the sulfenate ester to sulfoxide rearrangement [J.G. Miller, 1974]. The 14-ene $(13R)$-sulfoxide was treated with trimethyl phosphite. Any $(15S)$-sulfenate ester in equilibrium was destroyed by the phosphite.

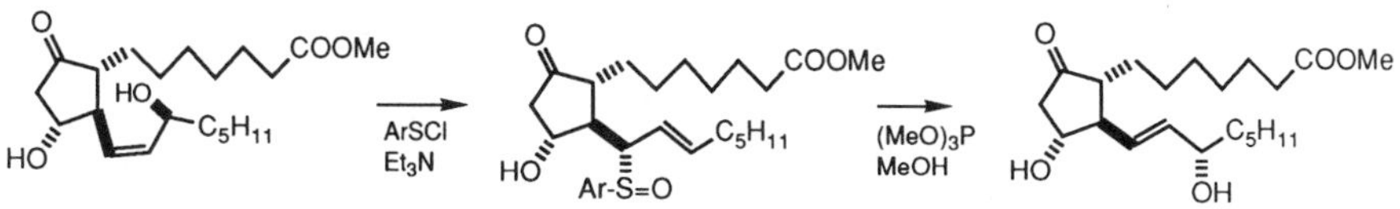

Allylic sulfoxides embedded in a six-membered carbocycle may be procured via the Diels–Alder reaction of inverse electron demand. It has a significant implication in the assembly of the hasubanan alkaloid skeleton which contains disjoint functionalities (O,N). While installation of these functional groups accompanying a direct annulation is impossible without umpolung maneuver, sigmatropic rearrangement of an allylic sulfoxide can be exploited to great advantage [Evans, 1972].

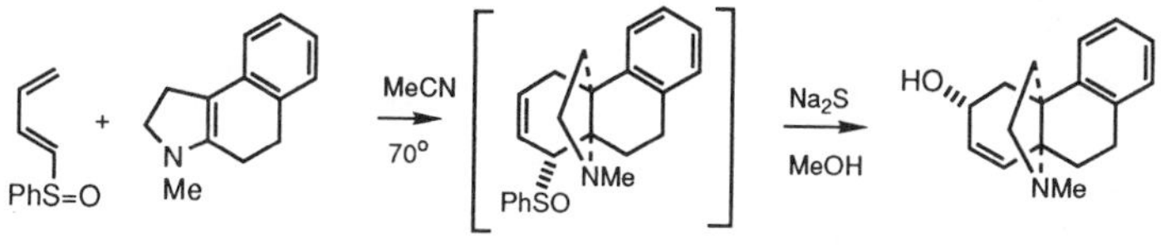

Configuration inversion of a tertiary alcohol is not easy. For allylic alcohols there is an excellent solution, i.e., by consecutive [2.3]sigmatropic rearrangements of the sulfenate and then the sulfoxide. The second rearrangement can be rendered irreversible by desulfurization with an organic phosphite [Morera, 1983].

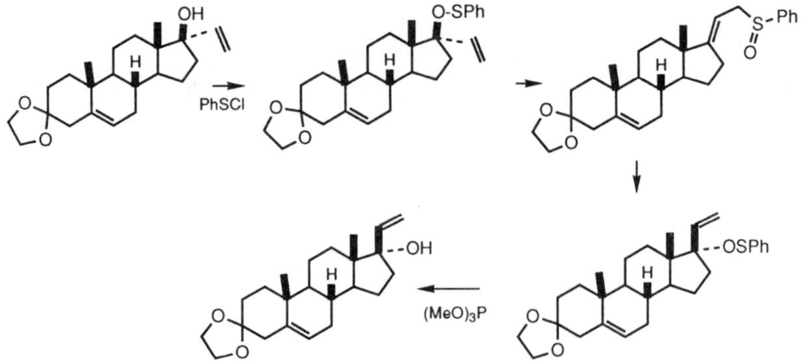

Stereoselective synthesis of (Z)-trisubstituted homoallylic alcohols by a [2,3]sigmatropic Wittig rearrangement has been observed [Still, 1978]. The stereoselectivity results from the preference of a transition state in which the alkyl substituent geminal to the oxygen atom occupies a pseudoaxial conformation in order to avoid $A^{(1,2)}$ strain, due to its spatial proximity to the vinylic substituent.

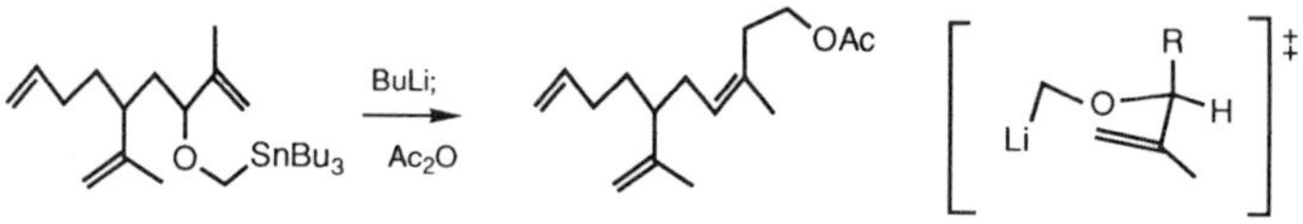

red scale triene

The Wittig rearrangement of allyl propargyl ethers under strongly basic conditions has found use in a synthesis of (−)-talaromycin A [Midland, 1985]. This approach was based on the ready generation of a 1,5-enyne-4-ol subunit which maps the string of atoms from the primary alcohol to the spirocyclic center. Thus, hydration of the triple bond with regiocontrol by the hydroxyl group originally unveiled from the rearrangement, and cleavage of the double bond. The five-center transition state necessarily involves changes of electronic relationships, and this is evident by comparing the ether (1,4-dioxy) and the alcohol product (1,3-dioxy).

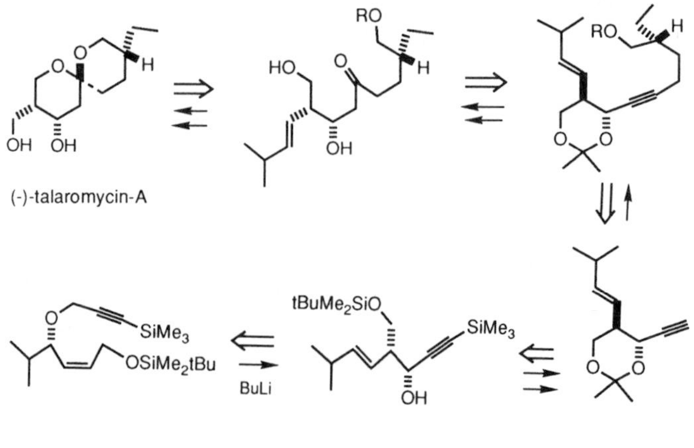

(-)-talaromycin-A

1,1-Dipoles are, by definition, disjoint entities, capture of α-acylcarbenoids with a donor species necessarily leads to disjoint products. Since ylides would be formed from nonionized *n*-donors, further transformations of synthetic significance may be contemplated. With the use of an allyl ether as an internal donor to trap an acylcarbenoid, an oxonium ylide may be formed. It undergoes [2,3]sigmatropic rearrangement [Pirrung, 1990]. A novel synthesis of griseofulvin based on this process is shown below. Note the O → C chirality transfer during the rearrangement step.

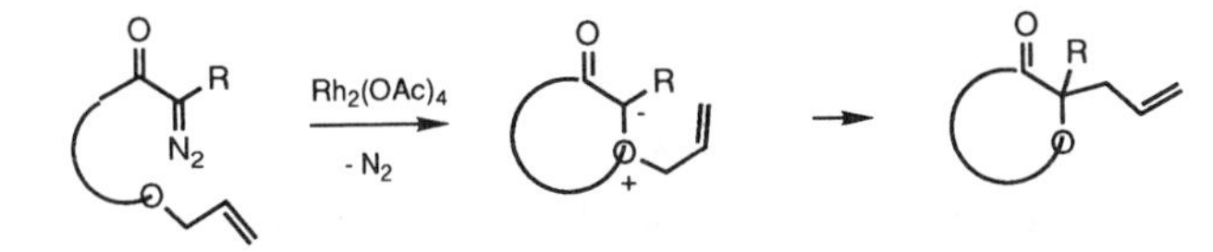

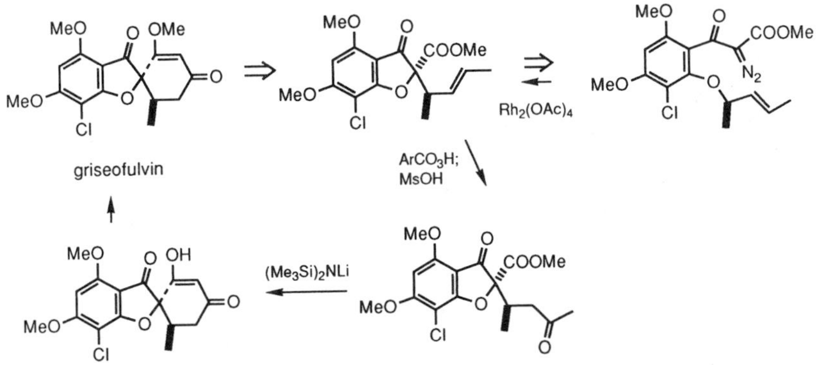

griseofulvin

An application of the [2.3]sigmatropic rearrangement of the enamonium ylide to establish the spirocyclic center of (+)-polyzonimine has been reported (Sugahara, 1984]. The intramolecular rearrangement obviated difficulties in C—C bond formation at a neopentylic carbon. More interestingly, the *N*-cyanomethylamino component of the product is easily degraded (retro-Strecker reaction) to the aldehyde. The contrapolarizability of the cyano group makes it a very versatile activating device.

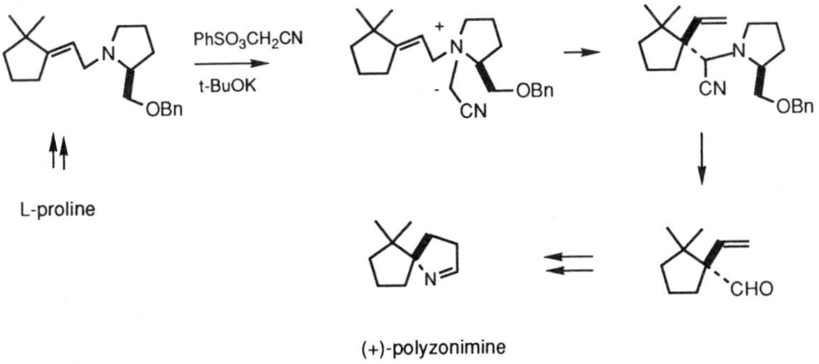

L-proline

(+)-polyzonimine

A Claisen rearrangement proceeds via a conjoint six-center transition state and it preserves the C*—C—C—O chain as the terminal carbon (C*) of the allylic system is concerned. Only the oxidation states of several atoms are changed.

There has been some evidence for the effect of a polar group on the rates of Claisen rearrangement [Curran, 1984]. Electronic complementarity at the terminal atoms leads to acceleration.

$$k_O / k_{CH_2} = 10\text{-}25$$

The Claisen rearrangement provides a general method of chain homologation with redox functionalization. It plays a pivotal role in a synthesis of eremophilone [Ziegler, 1977] and an access of the Inhoffen–Lythgoe diol precursor of vitamin D_3 [Trost, 1979a].

eremophilone

A more profound consequence of functionalization is witnessed in the Claisen rearrangement of the vinyl ether of N-carbethoxy-3-ethyl-5-hydroxy-Δ^3-piperideine, leading to an intermediate of tabersonine [Ziegler, 1973a].

tabersonine

In the allylic alcohol the nitrogen and the oxygen atoms are separated by two carbons in the shortest path, but in the product they are four atoms apart.

Many bislactone metabolites contain a Δ^4-unsaturated carbonyl segment which includes the ring juncture atoms. Such a segment is an ester Claisen rearrangement retron. Furthermore, the stereogenic center that is not at the ring juncture could serve as a diastereocontrol element in the rearrangement step such that the new C—C bond formation occurs antarafacial to the alkyl substituent after the lactone system is reorganized. A stereorational synthesis of ethisolide according to this scheme has been realized [Burke, 1986].

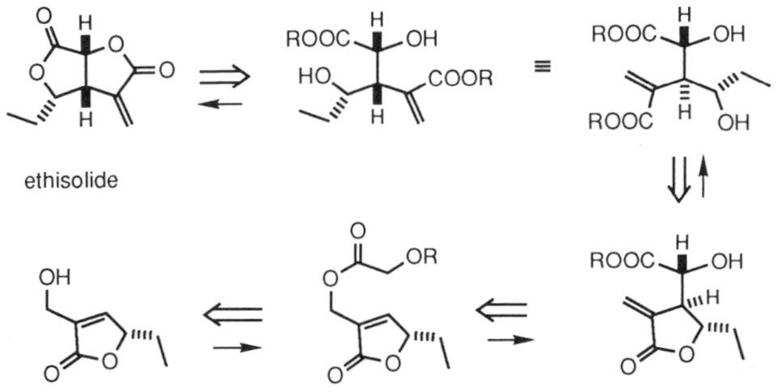
ROOC
OH
HO
COOR
ROOC
OH
OR

ethisolide

There have been many other neat applications of the intramolecular lactone Claisen rearrangement to generate functionalized carbocycles. Essentially a ring contraction attends the process. An outstanding example of the structural metamorphosis is found in a synthesis of quadrone [Funk, 1986].

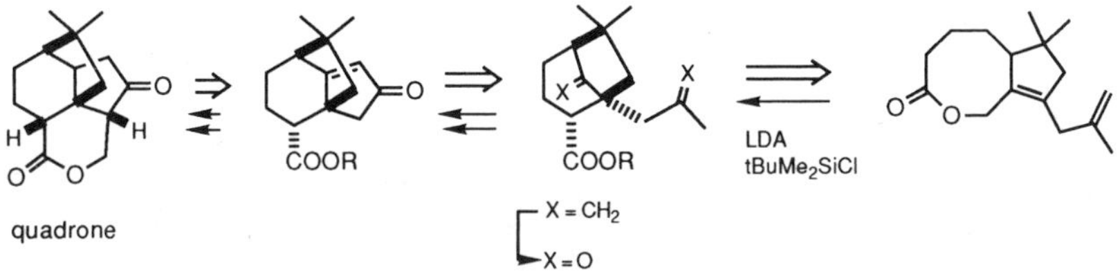
COOR
COOR
LDA
tBuMe2SiCl
X = CH2
X = O

quadrone

It has been demonstrated that the FGA (conjugated double bond) of a tricarbocyclic ketone precursor is most profitable because an aldol transform may be performed on it. A Wittig transform then reveals a γ,δ-unsaturated acid/ester system which permits the ester (lactone) Claisen rearrangement approach to be considered.

The smooth methylenation of a lactone carbonyl with the Tebbe reagent has enabled the formulation of intriguing synthetic routes based on the Claisen rearrangement. As an illustration a synthesis of precapnelladiene [Kinney, 1985] is shown below. The 1,5-cyclooctadiene skeleton may be derived from a 4-cyclooctenone, and whence a δ-valerolactone.

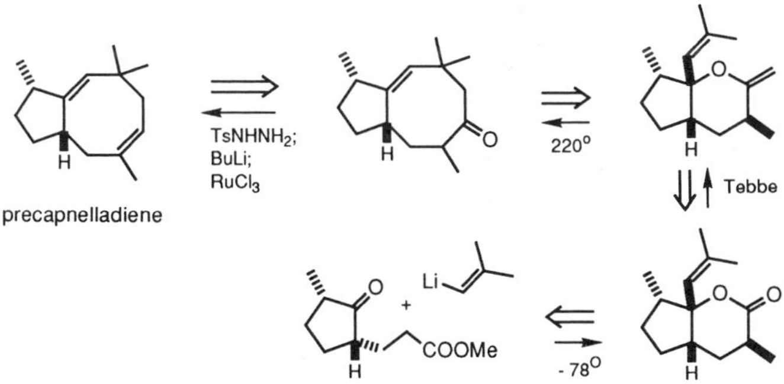
TsNHNH2;
BuLi;
RuCl3
220°
Tebbe
Li
COOMe
- 78°

precapnelladiene

The Cope rearrangement (including oxy-Cope and variants) is polarity conservative as an even-numbered pericycle is involved. Accordingly, a disjoint substrate gives rise to a disjoint product, as in the case of 3-substituted cyclohexanone synthesis en route to *erythro*-juvabione [Evans, 1980].

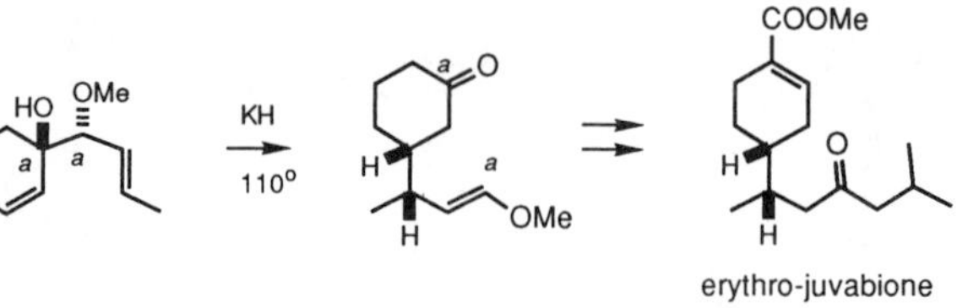

erythro-juvabione

Stripping the oxygen functionality of periplanone B and retaining only a ketone group at C-10 simplifies the retrosynthetic analysis [S.L. Schreiber, 1984a].The synthetic equivalency of the diene subunit to a cyclobutene via a thermal process enables the operation of an oxy-Cope transform. In turn, a photocycloadduct of cryptone and allene is reached.

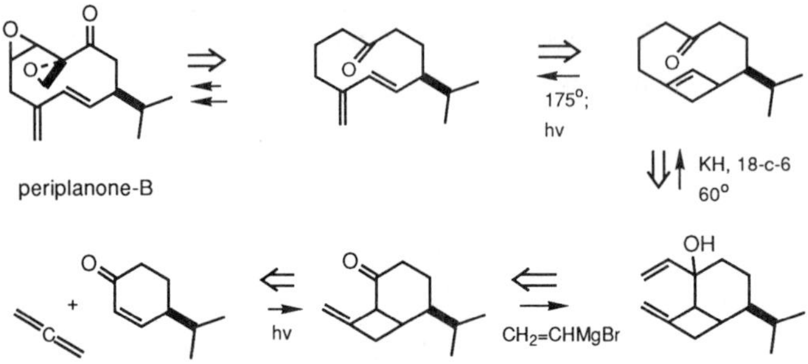

periplanone-B

In this approach the stereoselectivity of the photocycloaddition and the Grignard reaction is of no consequence because after the rearrangement and the electrocycloreversion only the original asymmetric center of cryptone is maintained.

As a result of polarity matching in the Diels–Alder reaction of 1-azadienes with dienophiles, an adduct from a dihydropyridine and methyl α-methoxyacrylate is a valuable building block for the DE-ring segment of reserpine [Wender, 1980]. Conversion of the bridged system into a *cis*-hydroisoquinoline skeleton needs chain extension and modification of the *endo* ester and a Cope rearrangement. Chain elongation to a β-methoxy α,β-unsaturated ester subunit met the requisite for the generation of the condensed ring system and it also set the carbomethoxy group in place stereochemically correctly.

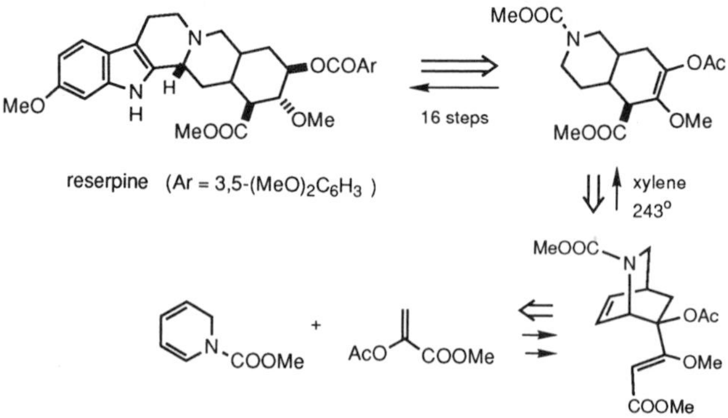

reserpine   (Ar = 3,5-(MeO)$_2$C$_6$H$_3$)

A tactical variation of the above approach involved an ionic aza-Cope rearrangement [Kunng, 1983]. Here, a C—N bond cleavage occurred instead of a C—C bond scission in the skeletal transformation. Admittedly, the E-ring of the product is less densely functionalized.

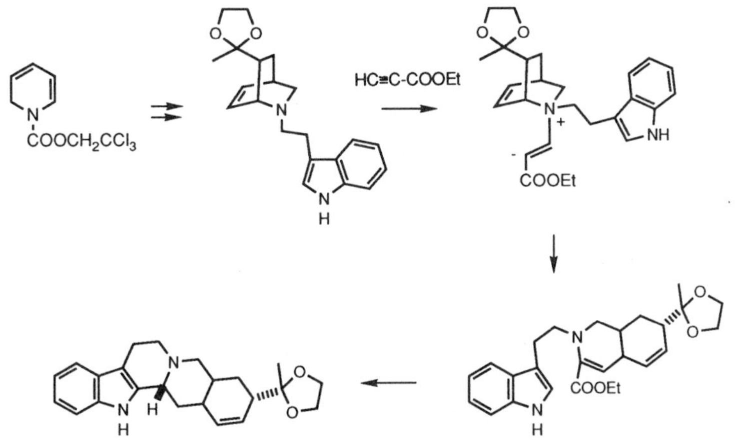

The Fischer indole synthesis [B. Robinson, 1963, 1969], exemplified by the generation of 1,2,3,4-tetrahydrocarbazole from cyclohexanone phenylhydrazone, is initiated by a [3.3]sigmatropic rearrangement. As arylhydrazones (and their ene hydrazine tautomers) are disjoint molecules with two directly linked nitrogen atoms, their rearrangement involves contrapolarization at the azaallyl moiety is a necessity. Once the bond exchange is made, a normal cyclization of the 1,4-difunctional intermediate leads to the disjoint product.

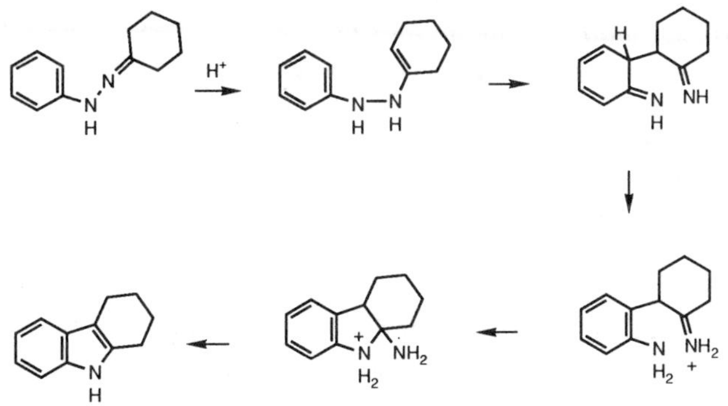

18

EPILOG

Chemical synthesis planning by disconnection (retrosynthetic analysis) should benefit from a survey of functional-group distribution in the target molecule and structures after appropriate FGI/FGA operations. Since C—C bond formation is central to carbogenic synthesis let us have a final look at various possibilities in making a short chain anchored by two functional groups.

(1) For molecules with substituents of the same polarity (A or D) which are separated by odd-numbered bonds (1,2-dioxy compounds, hydroxy amines, 1,4-diketones, etc.) the inherent disruption of polarity alternation demands a contrapolarizing action or umpolung techniques for their assembly. Redox processes are commonly used. The McMurry reaction, acyloin/benzoin condensations, and the Stetter reaction all belong to the unpolung category.

It should be noted that compounds with such substitution patterns are prone to rearrangement, e.g., pinacol rearrangement, Tiffeneau–Demyanov ring expansion, Wolff rearrangement.

(2) Molecules with substituents of the same polarity but separated by even-numbered bonds are most readily constructed. A great selection of condensation reactions are available for the formation of these conjoint molecules. Among these reactions are the aldol condensation and variants such as the Claisen–Schmidt, Knoevenagel, and Perkin reactions, the Mannich reaction, Claisen and Dieckmann condensations, the Michael reaction and effective combinations of two or more of these fundamental processes including the Robinson annulation, the Stork enamine reaction, and the Friedländer reaction.

The key step of all the facile reactions leading to conjoint products is reversible. The products may be stabilized by dehydration, enolization, or other means. Related to the reversibility due to conjoint functionalities is the Grob fragmentation which surprisingly has found many applications in synthesis.

(3) Molecules with substituents of opposite polarity which are separated by odd-numbered bonds are also amenable to facile construction. The alkene linkage can be considered as a special case and in this context the effectiveness of the Wittig, Peterson, and Henry reactions is understandable.

(4) For molecules with substituents of opposite polarity which are separated by even-numbered bonds, contrapolarization or umpolung maneuver for their assembly is indicated. The use of disjoint compounds, notably cyclopropanes may facilitate solution of many synthetic problems.

In this concluding section a brief analysis of a most recently reported synthesis of (+)-norpatchoulenol [Liu, 1990] should remind the reader how the consequence of one type of reaction affects the next and the whole sequence.

(+)-norpatchoulenol

If the target molecule is disconnected at the tertiary alcohol the precursor is a bicyclic ketone with a sidechain bearing an allylic anion which would ordinarily be generated by reductive methods (organometallic reagent formation is a reduction). The functionalized sidechain is earmarked for assembly by vinylation of an aldehyde group. This aldehyde is disjoint to the ketone in the other branch of the bridged ring skeleton. In order to utilize the more facile C—C bond-formation methods it might be advantageous to consider the corresponding norketone, i.e., a bicyclo[2.2.2]octanedione as a useful relay, with the necessary one-carbon chain extension to follow. The diverse steric environments of the two ketone groups would validate the strategy. Obviously, application of a Claisen condensation transform to the diketone is feasible and an alkyl 2,2,4-trimethyl-3-oxocyclohexaneacetate becomes an early intermediate.

Many routes to the keto ester are conceivable. The one involving elaboration of the desmethyl cyclohexenone has the advantage of its correlation with methyl campholenate by an aldol–oxidative cleavage transform sequence, and is accordingly favored.

The synthesis started from (+)-camphor and fragmentation of camphor-10-sulfonic acid. This was then converted into a cyclohexenone via oxidative cleavage of a five-membered ring and aldol condensation. The bridged ring system was created by an intramolecular Claisen condensation and the nonenolizable diketone was homologated *in situ* at the less hindered carbonyl group by a Wittig olefination to give the keto aldehyde. Vinyllithium reaction was followed by etherification and finally a reductive cyclization.

Regarding electronic events accompanying the various steps, there was preservation of the status quo after the fragmentation and the Claisen condensation, and inversion resulted in the Wittig homologation and the final reductive cyclization. Since the two inversions (conjoint to disjoint and vice versa) canceled out during the course of building the final ring, it is natural to expect the conjoint 1,3-diketone would lead to a conjoint product (such as a six-membered ring).

REFERENCES

Abe, Y., Harukawa, T., Ishikawa, H., Miki, T., Sumi, M., and Toga, T. (1956) *J. Am. Chem. Soc.* **78**: 1422.

Adam, G., Zibuck, R. and Seebach, D. (1987) *J. Am. Chem. Soc.* **109**: 6176.

Adam, W. and Rucktaschel, R. (1972) *J. Org. Chem.* **37**: 4128.

Adam, W., Liu, J. C., and Rodriguez, O. (1973) *J. Org. Chem.* **38**: 2269.

Agami, C., Kazakos, A., Levisalles, J., and Sevin, A. (1980) *Tetrahedron* **36**: 2977.

Ager, D.J. (1984) *Synthesis* 384.

Agranat, I., Bentor, Y., and Shih, Y.S. (1977) *J. Am. Chem. Soc.* **99**: 7068.

Ahlbrecht, H. and Pfaff, K. (1980) *Synthesis* 413.

Ahond, A., Cave, A., Kan-Fan, C., Husson, H.-P., deRostolan, J., and Potier, P. (1968) *J. Am. Chem. Soc.* **90**: 5622.

Akers, J.A. and Bryson, T.A. (1989) *Tetrahedron Lett.* **30**: 2187.

Akiba, K., Nakatani, M., Wada, M., and Yamamoto, Y. (1985) *J. Org. Chem.* **50**: 63.

Alcazar, V., Tapia, I., and Moran, J.R. (1990) *Tetrahedron* **46**: 1057.

Alder, A. and Bellus, D. (1983) *J. Am. Chem. Soc.* **105**: 6712.

Alexakis, A., Chapdelaine, M.J., and Posner, G.H. (1978) *Tetrahedron Lett.* 4209.

Allway, P.A., Sutherland, J.K., and Joule, J.A. (1990) *Tetrahedron Lett.* **31**: 4781.

Altenbach, H.-J. (1979) *Angew. Chem.* **91**: 1005.

Altenbach, H.-J. and Holzapfel, W. (1990) *Angew. Chem. Int. Ed. Engl.* **29**: 67.

Aly, M.F. and Grigg, R. (1985) *Chem. Commun.* 1523.

Amri, H., Rambaud, M., and Villieras, J. (1989) *Tetrahedron Lett.* **30**: 7381.

Andersen, S.H., Das, N.B., Jorgensen, R.D., Kjeldsen, G., Knudsen, J.S., Sharma, S.C., and Torsell, K.B.G. (1982) *Acta Chem. Scand.* **B36**: 1.

Anderson, P.C., Clive, D.L.J., and Evans, C.F. (1983) *Tetrahedron Lett.* **24**: 1373.

Ando, M., Büchi, G., and Ohnuma, T. (1975) *J. Am. Chem. Soc.* **97**: 6880.

Ando, M., Tajima, K., and Takase, K. (1978) *Chem. Lett.* 617.

Andriamialisoa, R.Z., Diatta, L., Rasoanaivo, P., Langlois, N., and Potier, P. (1975) *Tetrahedron* **31**: 2347.

Andriamialisoa, R.Z., Langlois, N., and Langlois, Y. (1985) *J. Org. Chem.* **50**: 961.

Angell, E.C., Fringuelli, F., Pizzo, F., Minuti, L., Taticchi, A., and Wenkert, E. (1989) *J. Org. Chem.* **54**: 1217.

Aoki, S., Fujimura, T., Nakamura, E., and Kuwajima, I. (1989) *Tetrahedron Lett.* **30**: 6541.

Arai, Y., Kamikawa, T., and Kubota, T. (1972) *Tetrahedron Lett.* 1615.

Asaoka, M., Yanagida, N., Sugimura, N., and Takei, H. (1980) *Bull. Chem. Soc. Jpn.* **53**: 1061.

Asaoka, M., Mukuta, T., and Takei, H. (1981) *Tetrahedron Lett.* 735.

Asaoka, M., Takenouchi, K., and Takei, H. (1988) *Tetrahedron Lett.* **29**: 325.

Ashe, A.J., III (1985) *Top. Curr. Chem.* **105**: 125.

Attur-ur-Rahman, Beisler, J.A., and harley-Mason, J. (1980) *Tetrahedron* **36**: 1063.

Auerbach, J. and Weinreb, S.M. (1972) *J. Am. Chem. Soc.* **94**: 7172.

Auricchio, S., Morrocchi, S., and Rica, A. (1974) *Tetrahedron Lett.* 2793.

Au-Yeung, B.W. and Fleming, I. (1977) *Chem. Commun.* 81.

Bachmann, W.E., Cole, W., and Wilds, A.L. (1940) *J. Am. Chem. Soc.* **62**: 824.

Baciocchi, E., Ruzziconi, R., and Sebastiani, G.V. (1983) *J. Am. Chem. Soc.* **105**: 6114.

Bäckvall, J.E. (1978) *Tetrahedron Lett.* 163.

Bäckvall, J.E. and Juntunen, S.K. (1987) *J. Am. Chem. Soc.* **109**: 6396.

Bäckvall, J.E. and Plobeck, N.A. (1990) *J. Org. Chem.* **55**: 4528.

Bakuzis, P., Bakuzis, M.L.F., and Weingarten, T.F. (1978) *Tetrahedron Lett.* 2371.

Baldwin, J.E. (1976) *Chem. Commun.* 734, 738.

Baldwin, J.E., Bailey, P.D., Gallacher, G., Singleton, K.A., and Wallace, P.M. (1983) *Chem. Commun.* 1490.

Baldwin, J.E., Adlington, R.M., Bottaro, J.C., Jain, A.U., Kolhe, J.N., Perry, M.W.D., and Newington, I.M. (1984) *Chem. Commun.* 1095.

Baldwin, S.W. and Doll, R.J. (1979) *Tetrahedron Lett.* 3275.

Baldwin, S.W. and Landmesser, N.G. (1982) *Tetrahedron Lett.* **23**: 4443.

Baldwin, S.W., Martin, G.F., and Nunn, D.S. (1985) *J. Org. Chem.* **50**: 5720.

Ban, Y., Seto, M., and Oishi, T. (1972) *Tetrahedron Lett.* 2113.

Baraldi, P.G., Barco, A., Benetti, S., Pollini, G.P., and Simoni, D. (1983) *Tetrahedron Lett.* **24**: 5669.

Baraldi, P.G., Barco, A., Benetti, S., Moroder, F., Pollini, C.P., and Simoni, D. (1985) *J. Org. Chem.* **48**: 1297.

Barco, A., Benetti, S., Casolari, A., Pollini, G.P., and Spalluto, G. (1990) *Tetrahedron Lett.* **31**: 4917.

Barluenga, J., Jimenez, C., Najera, C., and Yus, M. (1982) *Synthesis* 414.

Barros, M.T., Geraldes, C.F.G.C., Maycock, C.D., and Silva, M.I. (1968) *Tetrahedron* **44**: 2283.

Bartlett, P.A., Johnson, W.C., and Elliott, J.D. (1983) *J. Am. Chem. Soc.* **105**: 2088.

Barrett, A.G.M., Cheng, M.-C., Spilling, C.D., and Taylor, S.J. (1989) *J. Org. Chem.* **54**: 992.

Barton, D.H.R., Levisalles, J.E.D., and Pinhey, J.T. (1962) *J. Chem. Soc.* 3472.

Barton, D.H.R., Motherwell, W.B., and Zard, S.Z. (1983) *Bull. Soc. Chim. Fr. II*: 61.

Batt, D.G., Ganem, B. (1978) *Tetrahedron Lett.* 3323.

Battersby, A.R. and Turner, J.C. (1960) *J. Chem. Soc.* 717.

Baudouy, R., Crabbé, P., Greene, A.E., Le Drian, C., and Orr, A.F. (1977) *Tetrahedron Lett.* 2973.

Baxter, A.J.G. and Holmes, A.B. (1977) *J. Chem. Soc. Perkin Trans. I*: 2343.

Beak, P. and Snieckus, V. (1982) *Acc. Chem. Res.* **15**: 306.

Beames, D.J., Halleday, J.A., and Mander, L.N. (1974) *Aust. J. Chem.* **27**: 2645.

Bee, L.K., Garratt, P.J., and Mansuri, M.M. (1980) *J. Am. Chem. Soc.* **102**: 7076.

Belavadi, V.K. and Kulkarni, S.N. (1976) *Indian J. Chem.* **14B**: 901.

Beltrame, P., Beltrame, P.L., Cattania, M.G., and Simonetta, M. (1973) *J. Chem. Soc. Perkin Trans. II*: 63.

Bergman, J. and Pelcman, B. (1989) *J. Org. Chem.* **54**: 824.

Bergmann, E.D., Ginsburg, D., and Pappo, R. (1959) *Org. React.* **10**: 179.

Bertele, E., Boos, H., Dunitz, J.D., Elsinger, F., Eschenmoser, A., Felner, I., Gribi, H.P., Gschwend, H., Meyer, E.F., Pesaro, M., and Scheffold, R. (1964) *Angew. Chem.* **76**: 393.

Berti, G., Canedoli, S., Crotti, P., and Macchia, F. (1984) *J. Chem. Soc. Perkin Trans. I*: 1183.

Bertrand, M. and Le Gras, J. (1967) *Bull. Soc. Chim. Fr.* 4336.

Bertrand, M., Dulcere, J.P., and Gil, G. (1980) *Tetrahedron Lett.* **21**: 1945.

Besselièvre, R. and Husson, H.-P. (1976) *Tetrahedron Lett.* 1873.

Bestmann, H.J. and Schade, G. (1982) *Tetrahedron Lett.* **23**: 3543.

Bestmann, H.J. and Roth, D. (1990) *Angew. Chem. Int. Ed. Engl.* **29**: 99.

Biehl, E.R., Patrizi, R., and Reeves, P.C. (1971) *J. Org. Chem.* **36**: 3252.

Bindra, J.S., Grodski, A., Schaaf, T.K., and Corey, E.J. (1973) *J. Am. Chem. Soc.* **95**: 7522.

Birch, A.J., Brown, J.M., and Subba Rao, G.S.R. (1964) *J. Chem. Soc.* 3309.

Birch, A.J. and Keeton, R. (1968) *J. Chem. Soc. (C)* 109.

Birch, A.J., MacDonald, P.L., and Powell, V.M. (1970) *J. Chem. Soc. (C)* 1469.

Birch, A.J., Chamberlain, K.B., and Oloyeda, S.S. (1971) *Aust. J. Chem.* **24**: 2179.

Bishop, P.M., Pearson, J.R., and Sutherland, J.K. (1983) *Chem. Commun.* 123.

Black, K.A. and Vogel, P. (1986) *J. Org. Chem.* **51**: 5341.

Blagg, J., Davies, S.G., and Mobbs, B.E. (1985) *Chem. Commun.* 619.

Bloomfield, J.J., Owsley, D.C., and Nelke, J.M. (1976) *Org. React.* **23**: 259.

Blumen, D.J., Barnett, K.W., and Brown, T.L. (1979) *J. Organomet. Chem.* **173**: 71.

Bodforss, S. (1918) *Chem. Ber.* **51**: 214.

Boeckman, R.K. (1973) *J. Am. Chem. Soc.* **95**: 6867.

Boeckman, R.K., Blum, D.M., and Arthur, S.D. (1979) *J. Am. Chem. Soc.* **101**: 5060.

Boeckman, R.K. and Ko, S.S. (1980) *J. Am. Chem. Soc.* **102**: 7146.

Boeckman, R.K. and Ko, S.S. (1982) *J. Am. Chem. Soc.* **104**: 1033.

Boeckman, R.K., Goldstein, S.W., and Walters, M.A. (1988) *J. Am. Chem. Soc.* **110**: 8250.

Boeckman, R.K., Arvanitis, A., and Voss, M.E. (1989a) *J. Am. Chem. Soc.* **111**: 2737.

Boeckman, R.K., Weidner, C.H., Perni, R.B., and Napier, J.J. (1989b) *J. Am. Chem. Soc.* **111**: 8036.

Boger, D.L. and Mullican, M.D. (1982) *Tetrahedron Lett.* **23**: 4555.

Boger, D.L. (1989) *Strat. Tactics Org. Synth.* **2**: 1.

Bohlmann, F., Winterfeldt, E., Laurent, H., and Ude, W. (1963) *Tetrahedron* **19**: 195.

Bohlmann, F., Habek, D., Poetsch, E., and Schumann, D. (1967) *Chem. Ber.* **100**: 2742.

Bohlmann, F., Forster, J.-J., and Fischer, C.H. (1976) *Liebigs Ann. Chem.* 1487.

Bongini, A., Cardillo, G., Orena, M., Sandri, S., and Tomasini, C. (1985) *J. Chem. Soc. Perkin Trans. I*, 935.; (1986) *J. Org. Chem.* **51**: 4905.

Bornack, W.K., Bhagwat, S.S., Ponton, J., and Helquist, P. (1981) *J. Am. Chem. Soc.* **103**: 4467.

Bosch, J., Bennasar, M.-L., Zulaica, E., and Feliz, M. (1984) *Tetrahedron Lett.* **25**: 3119.

Bosch, J. and Bonjoch, J., (1988) in *Studies in natural products chemistry* (Attar-ur-Rahman, Ed.) Vol. 1, Elsevier, Amsterdam, p. 81.

Bottini, A.T. and Gal, J. (1971) *J. Org. Chem.* **36**: 1718.

Brade, W. and Vasella, A. (1989) *Helv. Chim. Acta* **72**: 1649.

Brady, W.T. and Cheng, T.C. (1977) *J. Org. Chem.* **42**: 732.

Brady, W.T. and Gu, Y.-Q. (1988) *J. Org. Chem.* **53**: 1353.

Brandange, S. and Lindblom, L. (1976) *Acta Chem. Scand.* (1976) **B30**: 93.

Brandsma, L., Bos, H.J.T., Arens, J.F. (1968) in *The chemistry of acetylene* (Viehe, H.G, Ed.) Dekker, New York, Chap. 11.

Bremner, J.B. and Dragar, C. (1985) *Heterocycles* **23**: 1451.

Brendel, J. and Weyerstahl, P. (1989) *Tetrahedron Lett.* **30**: 2371.

Breuer, E. and Zbaida, S. (1975) *Tetrahedron* **31**: 499.

Brieger, G. and Bennett, J.N. (1980) *Chem. Rev.* **80**: 63.

Bringmann, G. (1985) *Liebigs Ann. Chem.* 2105.

Broka, C.A. and Gerlits, J.F. (1988) *J. Org. Chem.* **53**: 2144.

Brown, C.A. and Yamaichi, A. (1979) *Chem. Commun.* 100.

Brown, E., Lavoue, J., and Dhal, R. (1973) *Tetrahedron* **29**: 455.

Brownbridge, P. (1983) *Synthesis* 2, 85.

Bryson, T.A. and Reichel, C.J. (1980) *Tetrahedron Lett.* **21**: 2381.

Buchanan, G.L. and Young, G.A.R. (1971) *Chem. Commun.* 643.

Büchi, G., Coffen, D.L., Kocsis, K., Sonnet, P.E., and Ziegler, F.E. (1966a) *J. Am. Chem. Soc.* **88**: 3099.

Büchi, G., Hofheinz, W., and Paukstelis, J.V. (1966b) *J. Am. Chem. Soc.* **88**: 4113.

Büchi, G. and Wüest, H. (1966c) *J. Org. Chem.* **31**: 977.

Büchi, G., Foulkes, D.M., Kurono, M., Mitchell, G.F., and Schneider, R.S. (1967) *J. Am. Chem. Soc.* **89**: 6745.

Büchi, G. and Wüest, H. (1968) *Tetrahedron* **24**: 2049.

Büchi, G., Gould, S.J., and Naf, F. (1971) *J. Am. Chem. Soc.* **93**: 2492.

Büchi, G., Carlson, J.A., Powell, J.E., and Tietze, L.-F. (1973) *J. Am. Chem. Soc.* **95**: 540.

Büchi, G., Fliri, H., and Shapiro, R. (1977) *J. Org. Chem.* **42**: 2192.

Burgstahler, A.W. and Bithos, Z.J. (1959) *J. Am. Chem. Soc.* **81**: 503.

Burke, S.D. and Grieco, P.A. (1979) *Org. React.* **26**: 361.

Burke, S.D., Murtiashaw, C.W., Saunders, J.O., and Dike, M.S. (1982) *J. Am. Chem. Soc.* **104**: 872.

Burke, S.D., Murtiashaw, C.W., Saunders, J.O., Oplinger, J.A., and Dike, M.S. (1984) *J. Am. Chem. Soc.* **106**: 4558.

Burke, S.D., Cobb, J.E., and Takeuchi, T. (1985) *J. Org. Chem.* **50**: 3420.

Burke, S.D. and Pacofsky, G.J. (1986) *Tetrahedron Lett.* **27**: 445.

Burke, S.D., Cobb, J.E., and Takeuchi, K. (1990) *J. Org. Chem.* **55**: 2138.

Burkholder, T.P. and Fuchs, P.L. (1988) *J. Am. Chem. Soc.* **110**: 2341.

Burnell, R.H., Jean, M., and Poirier, D. (1987) *Can. J. Chem.* **65**: 775.

Butenandt, A., Schiedt, U., Biekert, E., and Cromartie, R.J.T. (1954) *Liebigs Ann. Chem.* **590**: 75.

Buttery, C.D., Cameron, A.G., Dell, C.P., and Knight, D.W. (1990) *J. Chem. Soc. Perkin Trans. 1*: 1601.

Cabal, M.P., Coleman, R.S., and Danishefsky, S.J. (1990) *J. Am. Chem. Soc.* **112**: 3253.

Cacchi, S., Ciattini, P.G., Morera, E., and Ortar, G. (1988) *Tetrahedron Lett.* **29**: 3117.

Cane, D.E. and Thomas, P.J. (1984) *J. Am. Chem. Soc.* **106**: 5295.

Carey, J.T., Knors, C., and Helquist, P. (1986) *J. Am. Chem. Soc.* **108**: 8313.

Carrupt, P.A. and Vogel, P. (1982) *Tetrahedron Lett.* **23**: 2563.

Carruthers, W. and Moses, R.C. (1987) *Chem. Commun.* 509.

Cartier, D., Ouahrani, M., and Levy, J. (1989) *Tetrahedron Lett.* **30**: 1951.

Castells, J., Lopez-Calahorra, F., Bassedas, M., and Urrios, P. (1988) *Synthesis* 314.

Celerier, J.P., Haddad, M., Jacoby, D., and Lhommet, G. (1987) *Tetrahedron Lett.* **28**: 6597.

Chalmers, J.R., Dickson, G.T., Elks, J., and Hems, B.A. (1949) *J. Chem. Soc.* 3424.

Chamberlain, J.W. (1966) *J. Org. Chem.* **31**: 1658.

Chamberlin, A.R. and Chung, J.Y.L. (1982) *Tetrahedron Lett.* **23**: 2619.

Chamberlin, A.R., Nguyen, H.D., and Chung, J.Y.L. (1984) *J. Org. Chem.* **49**: 1682.

Chambers, M.S. and Thomas, E.J. (1989) *Chem. Commun.* 23.

Chapman, O.L., Engel, M.R., Springer, J.P., and Clardy, J.C. (1971) *J. Am. Chem. Soc.* **93**: 6696.

Chen, J., Browne, L.J., and Gonnela, N.C. (1986) *Chem. Commun.* 905.

Chen, R. and Rowand, D.A. (1980) *J. Am. Chem. Soc.* **102**: 6609.

Chong, J.M. and Sharpless, K.B. (1985) *J. Org. Chem.* **50**: 1560.

Chow, H.-F. and Fleming, I. (1985) *Tetrahedron Lett.* **26**: 397.

Ciganek, E. (1984) *Org. React.* **32**: 1.

Coates, R.M. and Sowerby, R.L. (1972) *J. Am. Chem. Soc.* **94**: 5386.

Cohen, T., Jeong, I.-H., Mudryk, B., Bhupathy, M., and Awad, M.M.A. (1990) *J. Org. Chem.* **55**: 1528.

Coleman, R.S. and Grant, E.B. (1990) *Tetrahedron Lett.* **31**: 3677.

Collum, D.B., McDonald, J.H., III, and Still, W.C. (1980) *J. Am. Chem. Soc.* **102**: 2120.

Colvin, E.W., Malchenko, S., Raphael, R.A., and Roberts, J.S. (1973) *J. Chem. Soc. Perkin Trans. I*, 1989.

Colvin, E.W. and Monteith, M. (1990) *Chem. Commun.* 1230.

Comins, D.L., Dehghani, A., Killpack, M.O., LaMunyon, D.H., and Morgan, L.A. (1990) 199th ACS Nat. Meet. ORGN 239.

Confalone, P.N., Pizzolato, G., Lollar-Confalone, D.L., and Uskokovic, M.R. (1980) *J. Am. Chem. Soc.* **102**: 1954.

Cooke, M.P. and Parlman, R.M. (1977) *J. Am. Chem. Soc.* **99**: 5222.

Cookson, R.C. and Smith, S.A. (1979) *J. Chem. Soc. Perkin Trans. I*, 2447.

Corey, E.J., Ohno, M., Vatakencherry, P.A., and Mitra, R.B. (1961) *J. Am. Chem. Soc.* **83**: 1251.

Corey, E.J. and Nozoe, S. (1963) *J. Am. Chem. Soc.* **85**: 3527.

Corey, E.J., Bass, J.D., LeMahieu, R., and Mitra, R.B. (1964a) *J. Am. Chem. Soc.* **86**: 5570.

Corey, E.J., Mitra, R.B., and Uda, H. (1964b) *J. Am. Chem. Soc.* **86**: 485.

Corey, E.J., Ohno, M., Vatakencherry, P.A., and Mitra, R.B. (1964c) *J. Am. Chem. Soc.* **86**: 478.

Corey, E.J., Andersen, N.H., Carlson, R.M., Paust, J., Vedejs, E., Vlattas, I., and Winter, R.E.K. (1968a) *J. Am. Chem. Soc.* **90**: 3245.

Corey, E.J., Katzenellenbogen, J.A., Gilman, N.W., Roman, S.A., and Erickson, B.W. (1968b) *J. Am. Chem. Soc.* **90**: 5618.

Corey, E.J. and Hegedus, L.S. (1969a) *J. Am. Chem. Soc.* **91**: 4926.

Corey, E.J., Weinshenker, N.M., Schaaf, T.K., and Huber, W. (1969b) *J. Am. Chem. Soc.* **91**: 5675.

Corey, E.J. and Wipke, W.T. (1969c) *Science* **166**: 178.

Corey, E.J. and Balanson, R.D. (1973) *Tetrahedron Lett.* 3153.

Corey, E.J. and Balanson, R.D. (1974) *J. Am. Chem. Soc.* **96**: 6516.

Corey, E.J., Arnett, J.F., and Widiger, G.N. (1975a) *J. Am. Chem. Soc.* **97**: 430.

Corey, E.J., Crouse, D.N., and Anderson, J.E. (1975b) *J. Org. Chem.* **40**: 2140.

Corey, E.J., Nicolaou, K.C., and Toru, T. (1975c) *J. Am. Chem. Soc.* **97**: 2287.

Corey, E.J. and Balanson, R.D. (1976a) *Heterocycles* **5**: 445.

Corey, E.J. and Enders, D. (1976b) *Tetrahedron Lett.* 11.

Corey, E.J. and Wollenberg, R.H. (1976c) *Tetrahedron Lett.* 4705.

Corey, E.J., Danheiser, R.L., Chandrasekaran, S., Keck, G.E., Gopalan, B., Larsen, S.D., Siret, P., and Gras, J.-L. (1978a) *J. Am. Chem. Soc.* **100**: 8034.

Corey, E.J., Danheiser, R.L., Chandrasekaran, S., Siret, P., Keck, G.E., and Gras, J.-L. (1978b) *J. Am. Chem. Soc.* **100**: 8031.

Corey, E.J. and Estreicher, H. (1978c) *J. Am. Chem. Soc.* **100**: 6294.

Corey, E.J., Behforouz, M., and Ishiguro, M. (1979a) *J. Am. Chem. Soc.* **101**: 1608.

Corey, E.J. and Ishiguro, M. (1979b) *Tetrahedron Lett.* 2745.

Corey, E.J. and Pearce, H.L. (1979c) *J. Am. Chem. Soc.* **101**: 5841.

Corey, E.J., Tius, M.A., and Das, J. (1980) *J. Am. Chem. Soc.* **102**: 1742.

Corey, E.J. and Pyne, S.G. (1983) *Tetrahedron Lett.* **24**: 2821.

Corey, E.J. and Boaz, N.W. (1985a) *Tetrahedron Lett.* **26**: 6019.

Corey, E.J. and Desai, M.C. (1985b) *Tetrahedron Lett.* **26**: 3535.

Corey, E.J., Desai, M.C., and Engler, T.A. (1985c) *J. Am. Chem. Soc.* **107**: 4339.

Corey, E.J., Long, A.K., and Rubenstein, S.D. (1985d) *Science* **228**: 408.

Corey, E.J. and Magriotis, P.A. (1987a) *J. Am. Chem. Soc.* **109**: 287.

Corey, E.J. and Su, W.-g. (1987b) *J. Am. Chem. Soc.* **109**: 7534.

Corey, E.J., Wess, G., Xiang, Y.B., and Singh, A.K. (1987c) *J. Am. Chem. Soc.* **109**: 4717.

Corey, E.J., Da Silva Jardine, P., and Rohloff, J.C. (1988a) *J. Am. Chem. Soc.* **110**: 3672.

Corey, E.J. and Gavai, A.V. (1988b) *Tetrahedron Lett.* **29**: 3201.

Corey, E.J., Kang, M.-c., Desai, M.C., Ghosh, A.K., and Houpis, I.N. (1988c) *J. Am. Chem. Soc.* **110**: 649.

Corey, E.J. and Su, W.-g. (1988d) *Tetrahedron Lett.* **29**: 3423.

Corey, E.J. and Carpino, P. (1989a) *J. Am. Chem. Soc.* **111**: 5472.

Corey, E.J. and Cheng, X.-M. (1989b) *The logic of chemical synthesis*, Wiley, New York.

Corey, E.J., Imwinkelried, R., Pikul, S., and Xiang, Y.B. (1989c) *J. Am. Chem. Soc.* **111**: 5493.

Coté, R., Bouchard, P., Couture, Y., Furstoss, R., Waegell, B., and Lessard, J. (1989) *Heterocycles* **28**: 987.

Crabbé, P., Barreiro, E., Cruz, A., Depres, J.-P., Meana, M.d.C., and Greene, A.E. (1976) *Heterocycles* **5**: 725.

Cremins, P.J., Saengchantara, S.T., and Wallace, T.W. (1987) *Tetrahedron* **43**: 3075.

Crenshaw, L., Khanapure, S.P., Siriwadane, U., and Biehl,, E.R. (1988) *Tetrahedron Lett.* **29**: 3777.

Crimmins, M.T. and DeLoach, J.A. (1984) *J. Org. Chem.* **49**: 2076.

Crimmins, M.T. and O'Mahony, R. (1989) *J. Org. Chem.* **54**: 1157.

Crimmins, M.T. and Nantermet, P.G. (1990) *J. Org. Chem.* **55**: 4235.

Curran, D.P. and Suh, Y.G. (1984) *J. Am. Chem. Soc.* **106**: 5002.

Curran, D.P., Choi, S.-M., Gothe, S.A., and Lin, F.-t. (1990) *J. Org. Chem.* **55**: 3710.

Dainter, R.S., Suschitzky, H., Wakefield, B.J., Hughes, N., and Nelson, A.J. (1988) *J. Chem. Soc. Perkin Trans. I* 227.

Danheiser, R.L., Carini, and D.J., Basak, A. (1981a) *J. Am. Chem. Soc.* **103**: 1604.

Danheiser, R.L., Martinez-Davila, C., Auchus, R.J., and Kadonaga, J.T. (1981b) *J. Am. Chem. Soc.* **103**: 2443.

Danheiser, R.L. and Fink, D.M. (1985) *Tetrahedron Lett.* **26**: 2509.

Danheiser, R.L. and Cha, D.D. (1990) *Tetrahedron Lett.* **31**: 1527.

Danikiewicz, W., Jaworski, T., and Kwiatkowski, S. (1983) *Synth. Commun.* **13**: 255.

Danishefsky, S., Hatch, W.E., Sax, M., Abola, E., and Pletcher, J. (1973) *J. Am. Chem. Soc.* **95**: 2410.

Danishefsky, S. and Cain, P. (1975) *J. Am. Chem. Soc.* **97**: 5282.

Danishefsky, S. and Hirama, M. (1977a) *J. Am. Chem. Soc.* **99**: 7740.

Danishefsky, S., McKee, R., and Singh, R.K. (1977b) *J. Am. Chem. Soc.* **99**: 4783.

Danishefsky, S., Vaughn, K., Gadwood, R., and Tsuzuki, K. (1980) *J. Am. Chem. Soc.* **102**: 4262.

Danishefsky, S. (1981) *Acc. Chem. Res.* **14**: 400.

Danishefsky, S., Chackalamannil, S., Silvestri, M., and Springer, J. (1983) *J. Org. Chem.* **48**: 3615.

Danishefsky, S., Larson, E., and Springer, J. (1985) *J. Am. Chem. Soc.* **107**: 1274.

Danishefsky, S.J., Selnick, H.G., DeNinno, M.P., and Selle, R.E. (1987) *J. Am. Chem. Soc.* **109**: 1572.

Danishefsky, S.J. and Simoneau, B. (1988) *Pure Appl. Chem.* **60**: 1555.

Danishefsky, S. (1989a) *Chemtracts-Org. Chem.* **2**: 273.

Danishefsky, S., Cabal, M.P., and Chow, K. (1989b) *J. Am. Chem. Soc.* **111**: 3456.

Danishefsky, S.J., Armistead, D.M., Wincott, F.E., Selnick, H.G., and Hungate, R. (1989c) *J. Am. Chem. Soc.* **111**: 2967.

Dastur, K.P. (1974) *J. Am. Chem. Soc.* **96**: 2605.

Dauben, W.G. and Krabbenhoft, H.O. (1977) *J. Org. Chem.* **42**: 282.

Davis, D.A. and Gribble, G.W. (1990) *Tetrahedron Lett.* **31**: 1081.

Deighton, M., Hughes, C.R., and Ramage, R. (1975) *Chem. Commun.* 662.

Demuth, M. (1984) *Chimia* **38**: 257.

Denmark, S.E. and Jones, T.K. (1982) *J. Am. Chem. Soc.* **104**: 2642.

Denmark, S.E., Cramer, C.J., and Sternberg, J.A. (1986a) *Tetrahedron Lett.* **27**: 3693.

Denmark, S.E. and Sternberg, J.A. (1986b) *J. Am. Chem. Soc.* **108**: 8277.

Denmark, S.E. and Hite, G.A. (1988) *Helv. Chim. Acta* **71**: 195.

DeShong, P. and Kell, D.A. (1986) *Tetrahedron Lett.* **27**: 3979.

DeShong, P. and Sidler, D.R. (1988) *J. Org. Chem.* **53**: 4892.

Dickinson, R.A., Kubela, R., MacAlpine, G.A., Stojanac, Z., and Valenta, Z. (1972) *Can. J. Chem.* **50**: 2377.

Dickman, D.A. and Heathcock, C.H. (1989) *J. Am. Chem. Soc.* **111**: 1528.

Disanayaka, B.W. and Weedon, A.C. (1985) *Chem. Commun.* 1282.

Dolby, L.J., Nelson, S.J., and Senkovich, D. (1972) *J. Org. Chem.* **37**: 3691.

Donaruma, L.G. and Heldt, W.Z. (1960) *Org. React.* **11**: 1.

Dötz, K.H. (1984) *Angew. Chem. Int. Ed. Engl.* **23**: 587.

Duhamel, P., Kotera, M., and Monteil, T. (1986) *Bull. Chem. Soc. Jpn.* **59**: 2353.

Dunham, D.J. and Lawton, G.R. (1971) *J. Am. Chem. Soc.* **93**: 2074.

Earley, W.G., Jacobsen, E.J., Meier, G.P., Oh, T., and Overman, L.E. (1988a) *Tetrahedron Lett.* **29**: 3781.

Earley, W.G., Oh, T., and Overman, L.E. (1988b) *Tetrahedron Lett.* **29**: 3785.

Elad, D. and Ginsburg, D. (1954) *J. Chem. Soc.* 3052.

Elliott, J.D., Hetmanski, M., Stoodley, R.J., and Palfreyman, M.N. (1980) *Chem. Commun.* 924.

Escalone, H. and Maldonado, L.A. (1980) *Synth. Commun.* **10**: 857.

Eschenmoser, A., Felix, D., and Ohloff, G. (1967) *Helv. Chim. Acta* **50**: 708.

Eschenmoser, A. (1969) *Pure Appl. Chem.* **20**: 1.

Evans, D.A., Bryan, C.A., and Sims, C.L. (1972) *J. Am. Chem. Soc.* **94**: 2891.

Evans, D.A., Andrews, G.C., Fujimoto, T.T., and Wells, D. (1973) *Tetrahedron Lett.* 1389.

Evans, D.A. and Nelson, J.V. (1980) *J. Am. Chem. Soc.* **102**: 774.

Evans, D.A. (1982a) *Aldrichimica Acta* **15**: 23.

Evans, D.A. and Mitch, C.H. (1982b) *Tetrahedron Lett.* **23**: 285.

Evans, D.A., Thomas, E.W., and Cherpeck, R.E. (1982c) *J. Am. Chem. Soc.* **104**: 3695.

Evans, D.A. and Biller, S.A. (1985) *Tetrahedron Lett.* **26**: 1907.

Feldman, P.L. and Rapoport, H. (1987) *J. Am. Chem. Soc.* **109**: 1603.

Feldman, K.S. and Simpson, R.E. (1989) *Tetrahedron Lett.* **30**: 6985.

Ficini, J., d'Angelo, J., Genet, J.P., and Noire, J. (1971a) *Tetrahedron Lett.* 1569.

Ficini, J. and Genet, J.P. (1971b) *Tetrahedron Lett.* 1565.

Ficini, J. (1976) *Tetrahedron* **32**: 1449.

Ficini, J. and Touzin, A.M. (1977) *Tetrahedron Lett.* 1081.

Ficini, J., Guigant, A., and D'Angelo, J. (1979) *J. Am. Chem. Soc.* **101**: 1318.

Ficini, J., Revial, G., and Genet, J.P. (1981) *Tetrahedron Lett.* **22**: 629, 633.

Fink, M., Gaier, H., and Gerlach, H. (1982) *Helv. Chim. Acta* **65**: 2563.

Fischer, E. and Gleiter, R. (1989) *Angew. Chem. Int. Ed. Engl.* **28**: 925.

Fischer, M.J. and Overman, L.E. (1990) *J. Org. Chem.* **55**: 1447.

Fitjer, L., Majewski, M., and Kanschik, A. (1988) *Tetrahedron Lett.* **29**: 1263.

Fleming, I. (1976) *Frontier orbitals and organic chemical reactions*, Wiley, Chichester.

Fleming, I. and Paterson, I. (1979) *Synthesis* 446.

Fleming, I. (1981) *Chem. Soc. Rev.* **10**: 83.

Fleming, I. and Waterson, D. (1984) *J. Chem. Soc. Perkin Trans. I* 1809.

Fleming, I. and Lawrence, N.J. (1988) *Tetrahedron Lett.* **29**: 2073.

Fliri, A. and Hohenlohe-Oehringer, K. (1980) *Chem. Ber.* **113**: 607.

Floyd, M.B. (1978) *J. Org. Chem.* **43**: 1641.

Franck-Neumann, M., Sedrati, M., and Mokhi, M. (1987) *J. Organomet. Chem.* **326**: 389.

Fraser-Reid, B., Anderson, R.C., Hicks, D.R., and Walker, D.L. (1977) *Can. J. Chem.* **55**: 3986.

Fráter, G. (1974) *Helv. Chim. Acta* **57**: 172.

Freear, J. and Tipping, A.E. (1969) *J. Chem. Soc.* (C) 411.

Froborg, J. and Magnusson, G. (1978) *J. Am. Chem. Soc.* **100**: 6728.

Fuji, K., Usami, Y., Sumi, K., Ueda, M., and Kajiwara, K. (1986) *Chem. Lett.* 1655.

Fujimoto, R., Kishi, Y., and Blount, J.F. (1980) *J. Am. Chem. Soc.* **102**: 7154.

Fukuyama, T., Dunkerton, L.V., Aratani, M., and Kishi, Y. (1975) *J. Org. Chem.* **40**: 2011.

Fukuyama, Y., Kirkemo, C.L., and White, J.D. (1977) *J. Am. Chem. Soc.* **99**: 646.

Funk, R.L. and Abelman, M.M. (1986) *J. Org. Chem.* **51**: 3247.

Fürstner, A. (1989) *Synthesis* 571.

Furuta, K., Miwa, Y., Iwanaga, K., and Yamamoto, H. (1988) *J. Am. Chem. Soc.* **110**: 6254.

Furuya, S. and Okamoto, T. (1988) *Heterocycles* **27**: 2609.

Garner, P. and Park, J.M. (1990) *J. Org. Chem.* **55**: 3772.

Garst, M.E., Bonfiglio, J.N., and Marks, J. (1982) *J. Org. Chem.* **47**: 1494.

Gasa, S., Hamanaka, N., Matsunaga, S., Okuno, T., Takeda, N., and Matsumoto, T. (1976) *Tetrahedron Lett.* 553.

Gassman, P.G. and Talley, J.J. (1980) *J. Am. Chem. Soc.* **102**: 1214.

Gassman, P.G., Singleton, D.A., Wilwerding, J.J., and Chavan, S.P. (1987) *J. Am. Chem. Soc.* **109**: 2182.

Gates, M. and Tschudi, G. (1956) *J. Am. Chem. Soc.* **78**: 1380.

Gawley, R.E. (1976) *Synthesis* 777.

Geissman, T.A. and Waiss, A.C. (1962) *J. Org. Chem.* **27**: 139.

Gerlach, H. and Thalmann, A. (1977) *Helv. Chim. Acta* **60**: 2866.

Gibbons, E.G. (1982) *J. Am. Chem. Soc.* **104**: 1767.

Gilbertson, S.R. and Wulff, W.D. (1989) *Synlett.* 47.

Gleiter, R., Heilbronner, E., and Hornung, V. (1970) *Angew. Chem. Int. Ed. Engl.* **9**: 901.

Goldberg, S.I. and Lipkin, A.H. (1972) *J. Org. Chem.* **37**: 1823.

Godleski, S.A., Heacock, D.J., Meinhart, J.D., and Van Wallendael, S. (1983) *J. Org. Chem.* **48**: 2101.

Govindachari, T.R., Sidhaye, A.R., and Viswanathan, N. (1970) *Tetrahedron* **26**: 3829.

Greenlee, M.L. (1981) *J. Am. Chem. Soc.* **103**: 2425.

Greenlee, W.J. and Woodward, R.B. (1976) *J. Am. Chem. Soc.* **98**: 6075.

Grethe, G., Lee, H.L., and Uskokovic, M.R. (1976) *Helv. Chim. Acta* **59**: 2268.

Greuter, H., Schmid, H., and Frater, G. (1977) *Helv. Chim. Acta* **60**: 1701.

Grewel, R.S., Hayes, P.C., Sawyer, J.F., and Yates, P. (1987) *Chem. Commun.* 1290.

Grieco, P.A., Oguri, T., Gilman, S., and De Titta, G.T. (1978) *J. Am. Chem. Soc.* **100**: 1616.

Grob, C.A. (1969) *Angew. Chem. Int. Ed. Engl.* **8**: 535.

Gröbel, B.-T. and Seebach, D. (1977) *Synthesis* 357.

Guerrier, L., Royer, J., Grierson, D.S., and Husson, H.-P. (1983) *J. Am. Chem. Soc.* **105**: 7754.

Gupta, R.B. and Franck, R.W. (1989) *J. Am. Chem. Soc.* **111**: 7668.

Guthrie, R.W., Vatenta, Z., and Wiesner, K. (1966) *Tetrahedron Lett.* 4645.

Hagiwara, H., Uda, H., and Kodama, T. (1979) *Koen Yoshishu-Koryo, Terupen oyobi seiyu Kagaku ni Kansuru Toron Kai.* **23**: 118.

Hagiwara, H., Okano, A., and Uda, H. (1985) *Chem. Commun.* 1047.

Hagiwara, H., Akama, T., and Uda, H. (1989) *Chem. Lett.* 2067.

Halazy, S., Zutterman, F., and Krief, A. (1982) *Tetrahedron Lett.* **23**: 4385.

Hamanaka, H. and Matsumoto, T. (1972) *Tetrahedron Lett.* 3087.

Hanessian, S. (1983) *Total synthesis of natural products: the 'chiron' approach*, Peramon, Oxford.

Harayama, T., Ohtani, M., Oki, M., and Inubushi, Y. (1975) *Chem. Pharm. Bull.* **23**: 1511.

Harding, K.E. and Burks, S.R. (1984a) *J. Org. Chem.* **49**: 40.

Harding, K.E., Stephens, R., and Hollingsworth, D.R. (1984b) *Tetrahedron Lett.* **25**: 4631.

Harding, K.E. and Hollingsworth, D.R. (1988a) *Tetrahedron Lett.* **29**: 3789.

Harding, K.E., Strickland, J.B., and Pommerville, J. (1988b) *J. Org. Chem.* **53**: 4877.

Harley-Mason, J. and Jackson, A.H. (1954a) *J. Chem. Soc.* 1165; (1954b) *J. Chem. Soc.* 3651.

Harris, T.M., Harris, C.M. (1986) *Pure Appl. Chem.* **58**: 283.

Hart, D.J., Cain, P.A., Evans, D.A. (1978) *J. Am. Chem. Soc.* **100**: 1548.

Hase, T.A. (1987) *Umpoled synthons*, Wiley, New York.

Hashimoto, S., Sakata, S., Sonegawa, M., and Ikegami, S. (1988) *J. Am. Chem. Soc.* **110**: 3670.

Hassner, A. and Murthy, K.S.K. (1986) *Tetrahedron Lett.* **27**: 1407.

Hauser, F.M., Hewawasam, P., and Rho, Y.S. (1989) *J. Org. Chem.* **54**: 5110.

Hayakawa, Y., Shimizu, F., and Noyori, R. (1978) *Tetrahedron Lett.* 993.

Hayashi, Y., Nishizawa, M., and Sakan, T. (1975) *Chem. Lett.* 387.

Heathcock, C.H., Ellis, J.E., McMurry, J.E., and Coppolino, A. (1971) *Tetrahedron Lett.* 4995.

Heathcock, C.H., White, C.T., Morrison, J.J., and VanDerveer, D. (1981) *J. Org. Chem.* **46**: 1296.

Heathcock, C.H., Kleinman, E.F., and Binkley, E.S. (1982a) *J. Am. Chem. Soc.* **104**: 1054.

Heathcock, C.H., Taschner, M.J., Rosen, T., Thomas, J.A., Hadley, C.R., and Popjak, G. (1982b) *Tetrahedron Lett.* **23**: 4747.

Heathcock, C.H., Tice, C.M., and Germroth, T.C. (1982c) *J. Am. Chem. Soc.* **104**: 6082.

Heathcock, C.H. (1984) in *Asymmetric Synthesis*, (Morrison, J.D., Ed.) Vol. 3, Chap. 2, Academic Press, New York.

Heathcock, C.H., Davidsen, S.K., Mills, S., and Sanner, M.A. (1986) *J. Am. Chem. Soc.* **108**: 5650.

Heathcock, C.H. and von Geldern, T.W. (1987) *Heterocycles* **25**: 75.

Heathcock, C.H., Blumenkopf, T.A., and Smith, K.M. (1989) *J. Org. Chem.* **54**: 1548.

Hebert, J. and Gravel, D. (1974) *Can. J. Chem.* **52**: 187.

Heck, R.F. (1979) *Acc. Chem. Res.* **12**: 146.

Hegedus, L.S., McGuire, M.A., Schultze, L.M., Yijun, C., and Anderson, O.P. (1984) *J. Am. Chem. Soc.* **106**: 2680.

Hegedus, L.S. and Perry, R.J. (1985) *J. Org. Chem.* **50**: 4955.

Heitz, M.P. and Overman, L.E. (1989) *J. Org. Chem.* **54**: 2591.

Helquist, P. (1990) priv. commun.

Henne, A.L. and Kaye, S. (1950) *J. Am. Chem. Soc.* **72**: 3369.

Hess, J. (1972) *Chem. Ber.* **105**: 441.

Hiemstra, H. and Speckamp, W.N. (1983) *Tetrahedron Lett.* **24**: 1407.

Hikino, H., Suzuki, N., and Takemoto, T. (1966) *Chem. Pharm. Bull.* **14**: 1441.

Hine, J. and Ehrenson, S.J. (1958) *J. Am. Chem. Soc.* **80**: 824.

Hirai, Y., Hagiwara, A., and Yamazaki, T. (1986) *Heterocycles* **24**: 571.

Hirai, Y., Terada, T., Okaji, Y., Yamazaki, T., and Momose, T. (1990) *Tetrahedron Lett.* **31**: 4755.

Hirama, M., Shigemoto, T., Yamazaki, Y., and Ito, S. (1985) *Tetrahedron Lett.* **26**: 4133.

Hirama, M., Noda, T., Ito, S., and Kabuto, C. (1988) *J. Org. Chem.* **53**: 706.

Hirst, G.C., Howard, P.N., and overman, L.E. (1989) *J. Am. Chem. Soc.* **111**: 1514.

Hiyama, T., Shinoda, M., and Nozaki, H. (1979) *Tetrahedron Lett.* 3529.

Ho, T.-L. (1968) Unpubl. results, Univ. Alberta.

Ho, T.-L. (1971) *Chem. Ind.* 487.

Ho, T.-L. and Wong, C.M. (1974) *Synth. Commun.* **4**: 135.

Ho, T.-L. (1977) *Hard and soft acids and bases principle in organic chemistry* Academic Press, New York, New York.

Ho, T.-L. (1979) *Synth. Commun.* **9**: 37.

Ho, T.-L. (1983) *Synth. Commun.* **13**: 761.

Ho, T.-L. (1988) *Rev. Chem. Interm.* **9**: 117.

Ho, T.-L. (1989a) *Res. Chem. Interm.* **11**: 157.

Ho, T.-L. (1989b) *J. Chem. Educ.* **66**: 785.

Hoffmann, R. (1971) *Acc. Chem. Res.* **4**: 1.

Holmquist, C.R. and Roskamp, E.J. (1989) *J. Org. Chem.* **54**: 3258.

Holmwood, G.M. and Roberts, J.C. (1971) *Tetrahedron Lett.* 833.

Hölscher, P., Knölker, H.-J., and Winterfeldt, E. (1990) *Tetrahedron Lett.* **31**: 2705.

Holton, R.A., Kennedy, R.M., Kim, H.-B., and Krafft, M.E. (1987) *J. Am. Chem. Soc.* **109**: 1597.

Horton, M. and Pattenden, G. (1983) *Tetrahedron Lett.* **24**: 2125.

Hosomi, A., Saito, M., and Sakurai, H. (1980) *Tetrahedron Lett.* **21**: 355.

Hosomi, A., Hayashida, H., and Tominaga, Y. (1989) *J. Org. Chem.* **54**: 3254.

House, H.O. and Wasson, R.L. (1956) *J. Am. Chem. Soc.* **78**: 4394.

House, H.O., Crumrine, D.S., Teranishi, A.Y., and Olmstead, H.D. (1973) *J. Am. Chem. Soc.* **95**: 3310.

Hudlicky, T., Fleming, A., and Radesca, L. (1989) *J. Am. Chem. Soc.* **111**: 6691.

Hudrlik, P.F., Peterson, D., and Rona, R.J. (1975) *J. Org. Chem.* **40**: 2263.

Hudrlik, P.F., Hudrlik, A.M., Nagendrapper, G., Yimenu, T., Zellers, E.T., and Chin, E. (1980) *J. Am. Chem. Soc.* **102**: 6894.

Hugel, G., Massiot, G., Lévy, J., and LeMen, J. (1981) *Tetrahedron* **37**: 1369.

Hutchinson, C.R., Ikeda, T., and Yue, S. (1985) *J. Org. Chem.* **50**: 5193.

Ibuka, T., Tanaka, K., and Inubushi, Y. (1970) *Tetrahedron Lett.* 4811.

Ihara, M., Kirihara, T., Fukumoto, K., and Kametani, T. (1985) *Heterocycles* **23**: 1097.

Ihara, M., Katogi, M., Fukumoto, K., and Kametani, T. (1987) *Chem. Commun.* 721.

Ihara, M., Suzuki, M., Fukumoto, K., and Kabuto, C. (1990) *J. Am. Chem. Soc.* **112**: 1164.

Iida, H., Watanabe, Y., and Kibayashi, C. (1984) *Tetrahedron Lett.* **25**: 5094; (1986) *Tetrahedron Lett.* **27**: 5513.

Iio, H., Isobe, M., Kawai, T., and Goto, T. (1979) *Tetrahedron* **35**: 941.

Imanishi, T., Shin, H., Yagi, N., and Hanaoka, M. (1980) *Tetrahedron Lett.* **21**: 3285.

Imanishi, T., Yagi, N., Shin, H., and Hanaoka, M. (1981) *Tetrahedron Lett.* **22**: 4001.

Inoue, T. and Kuwajima, I. (1980a) *Chem. Commun.* 251.

Inoue, T. and Mukaiyama, T. (1980b) *Bull. Chem. Soc. Jpn.* **53**: 174.

Inouye, Y., Niwayama, S., Okada, T., and Kakisawa, H. (1987) *Heterocycles* **25**: 109.

Inubushi, Y., Kikuchi, T., Ibuka, T., Tanaka, K., Saji, I., and Tokane, K. (1972) *Chem. Commun.* 1252.

Ireland, R.E. and Brown, F.R. (1980a) *J. Org. Chem.* **45**: 1868.

Ireland, R.E. and Häbich, V.D. (1980b) *Tetrahedron Lett.* 1389.

Ireland, R.E., McGarvey, G.J., Anderson, R.C., Badoud, R., Fitzsimmons, B., and Thaisrivongs, S. (1980c) *J. Am. Chem. Soc.* **102**: 6178.

Ireland, R.E., Daub, J.P., Mandel, G.S., and Mandel, N.S. (1983) *J. Org. Chem.* **48**: 1312.

Ireland, R.E., Dow, W.C., Godfrey, J.D., and Thaisrivongs, S. (1984) *J. Org. Chem.* **49**: 1001.

Ireland, R.E., Wipf, P., Miltz, W., and Vanasse, B. (1990) *J. Org. Chem.* **55**: 1423.

Irie, H., Kishimoto, T., and Uyeo, S. (1968) *J. Chem. Soc.* (*C*) 3051.

Irie, H., Katakawa, J., Mizuno, Y., Udaka, S., Taga, T., and Osaki, K. (1978) *Chem. Commun.* 717.

Ishibashi, H., Sato, T., Takahashi, M., Hayashi, M., and Ikeda, M. (1988) *Heterocycles* **27**: 2787.

Isobe, M., Iio, H., Kawai, T., and Goto, T. (1978) *J. Am. Chem. Soc.* **100**: 1940.

Ito, Y. and Saegusa, T. (1977) *J. Org. Chem.* **42**: 2326.

Ito, Y., Kato, H., Imai, H., and Saegusa, T. (1982) *J. Am. Chem. Soc.* **104**: 6449.

Ito, Y., Nakajo, E., Nakatsuka, M., and Saegusa, T. (1983) *Tetrahedron Lett.* **24**: 2881.

Iwashita, T., Kusumi, T., and Kakisawa, H. (1979) *Chem. Lett.* 1337.

Iwata, C., Yamada, M., and Shinoo, Y. (1979) *Chem. Pharm. Bull.* **27**: 274.

Jacobi, P.A., Blum, C.A., DeSimone, R.W., and Udodong, U.E.S. (1989) *Tetrahedron Lett.* **30**: 7173.

Jacquesy, J.-C., Jouannetaud, M.-P., and Makani, S. (1980) *Chem. Commun.* 110.

Jahangir, MacLean, D.B., Brook, M.A., and Holland, H.L. (1986) *Chem. Commun.* 1608.

Jasor, Y., Luche, M.J., Gaudry, M., and Marquet, A. (1974) *Chem. Commun.* 253.

Jernow, J., Tautz, W., Rosen, P., and Blount, J.F. (1979) *J. Org. Chem.* **44**: 4210.

Johnson, W.S., Petersen, J.W., and Gutsche, C.D. (1947) *J. Am. Chem. Soc.* **69**: 2942.

Johnson, W.S. and Daub, G.H. (1951) *Org. React.* **6**: 1.

Johnson, W.S., Marshall, J.A., Keana, J.F.W., Franck, R.W., Martin, D.G., and Bauer, V.J. (1966) *Trahedron* (Suppl. 8: Part II) 541.

Johnson, W.S. (1981) *Stud. Org. Chem.* (*Amsterdam*) **6**: 1.

Jones, G., II. (1981) in *Org. Photochem.* (A. Padwa, Ed.) **5**: 1.

Julia, M., LeGoffic, F., Igolen, J., and Baillarge, M. (1969) *Tetrahedron Lett.* 1569.

Jung, M.E. (1976) *Tetrahedron* **32**: 3.

Jung, M.E. and Lowe, J.A. (1978) *Chem. Commun.* 95.

Jung, M.E., McCombs, C.A., Takeda, Y., and Pan, Y.-G. (1981) *J. Am. Chem. Soc.* **103**: 6677.

Jung, M.E., Usui, Y., and Vu, C.T. (1987) *Tetrahedron Lett.* **28**: 5977.

Just, G. and Simonovitch, C. (1967) *Tetrahedron Lett.* 2093.

Kahn, S.D., Pau, C.F., Overman, L.E., and Hehre, W.J. (1986) *J. Am. Chem. Soc.* **108**: 7381.

Kalaus, G., Kiss, M., Kajtar-Peredy, M., Brlik, J., Szabo, L., and Szantay, C. (1985) *Heterocycles* **23**: 2783.

Kametani, T., Shibuya, S., Seino, S., and Fukumoto, K. (1964) *J. Chem. Soc.* 4146.

Kametani, T., Yamaki, K., Yagi, H., and Fukumoto, K. (1969) *J. Chem. Soc.* (*C*) 2602.

Kametani, T. and Ihara, M. (1971) *J. Chem. Soc.* (*C*) 999.

Kametani, T., Kajiwara, M., Takahashi, T., and Fukumoto, K. (1975) *J. Chem. Soc. Perkin Trans. 1*: 737.

Kametani, T., Higa, T., Loc, C.V., Ihara, M., Koizumi, M., and Fukumoto, K. (1976) *J. Am. Chem. Soc.* **98**: 6186.

Kametani, T., Suzuki, Y., Terasawa, H., and Ihara, M. (1979) *J. Chem. Soc. Perkin Trans. I*: 1211.

Kametani, T., Higashiyama, K., Honda, T., and Otomasu, H. (1982) *J. Chem. Soc. Perkin Trans. I*: 2935.

Kametani, T., Higashiyama, K., Otomasu, H., and Honda, T. (1984) *Heterocycles* **22**: 729.

Kametani, T., Takeda, H., Suzuki, Y., Kasai, H., and Honda, T. (1986) *Heterocycles* **24**: 3385.

Kaminsky, L.S. and Lamchen, M. (1966) *J. Chem. Soc. (C)* 2295.

Kang, S.H., Kim, W.J., and Chae, Y.B. (1988) *Tetrahedron Lett.* **29**: 5169.

Karabelas, K. and Moore, H.W. (1990) *J. Am. Chem. Soc.* **112**: 5372.

Kashima, C., Mukai, N., Yamamoto, Y., Tsuda, Y., and Omote, Y. (1977) *Heterocycles* **7**: 241.

Katritzky, A.R., Fan, W.-Q., and Li, Q.-L. (1987) *Tetrahedron Lett.* **28**: 1195.

Katsumura, S. and Isoe, S. (1982) *Chem. Lett.* 1689.

Kawamata, T., Harimaya, K., and Inayama, S. (1988) *Bull. Chem. Soc. Jpn.* **61**: 3770.

Keana, J.F.W. and Eckler, P.E. (1976) *J. Org. Chem.* **41**: 2850.

Keck, G.E. and Nickell, D.G. (1980) *J. Am. Chem. Soc.* **102**: 3632.

Keely, S.L. and Tahk, P.C. (1968) *J. Am. Chem. Soc.* **90**: 5584.

Keely, S.L., Martinez, A.J., and Tahk, F.C. (1970) *Tetrahedron* **26**: 4729.

Kelly, A.G. and Roberts, J.S. (1980) *Chem. Commun.* 228.

Kelly, R.B., Harley, M.L., and Alward, S.J. (1980) *Can. J. Chem.* **58**: 755.

Kelly, T.R., Gillard, J.W., Goerner, R.N., and Lyding, J.M. (1977) *J. Am. Chem. Soc.* **99**: 5513.

Kelly, T.R. (1978a) *Tetrahedron Lett.* 1387.

Kelly, T.R. and Montury, M. (1978b) *Tetrahedron Lett.* 4309.

Kelly, T.R., Magee, J.A., and Weibel, F.R. (1980) *J. Am. Chem. Soc.* **102**: 798.

Kelly, T.R., Bell, S.H., Ohashi, N., and Armstrong-Chong, R.J. (1988) *J. Am. Chem. Soc.* **110**: 6471.

Kende, A.S. (1960) *Org. React.* **11**: 261.

Kende, A.S., Liebeskind, L.S., Mills, J.E., Rutledge, P.S., and Curran, D.P. (1977) *J. Am. Chem. Soc.* **99**: 7082.

Kende, A.S., Luzzio, M.J., and Mendoza, J.S. (1990) *J. Org. Chem.* **55**: 918.

Kido, F., Tsutsumi, K., Maruta, R., and Yoshikoshi, A. (1979) *J. Am. Chem. Soc.* **101**: 6420.

Kieczykowski, G.R., Pogonowski, C.S., Richman, J.E., and Schlessinger, R.H. (1977) *J. Org. Chem.* **42**: 175.

Kieczykowski, G.R. and Schlessinger, R.H. (1978) *J. Am. Chem. Soc.* **100**: 1938.

Kigoshi, H., Imamura, Y., Niwa, H., and Yamada, K. (1989) *J. Am. Chem. Soc.* **111**: 2302.

Kiguchi, T., Kuninobu, N., Naito, T., and Ninomiya, I. (1989) *Heterocycles* **28**: 19.

Kinney, W.A., Coghlan, M.J., and Paquette, L.A. (1985) *J. Am. Chem. Soc.* **107**: 7352.

Kirmse, W. and Goer, B. (1990) *J. Am. Chem. Soc.* **112**: 4556.

Kishi, Y., Aratani, M., Fukuyama, T., Nakatsubo, F., Goto, T., Inoue, S., Tanino, H., Sugiura, S., and Kakoi, H. (1972) *J. Am. Chem. Soc.* **94**: 9217, 9219.

Kitahara, H., Tozawa, Y., Fujita, S., Tajiri, A., Morita, N., and Asao, T. (1988) *Bull. Chem. Soc. Jpn.* **61**: 3362.

Klatte, F., Rosentretter, U., and Winterfeldt, E. (1977) *Angew. Chem. Int. Ed. Engl.* **16**: 878.

Klein, J. (1988) *Tetrahedron* **44**: 503.

Kodpinid, M., Siwapinyoyos, T., and Thebtaranonth, Y. (1984). *J. Am. Chem. Soc.* **106**: 4862.

Koft, E.R. and Smith, A.B., III (1984a) *J. Org. Chem.* **49**: 832; (1984b) *J. Am. Chem. Soc.* **106**: 2115.

Kolaczkowski, L. and Reusch, W. (1985) *J. Org. Chem.* **50**: 4766.

Kondo, K., Umemoto, T., Takahatake, Y., and Tunemoto, D. (1977) *Tetrahedron Lett.* 113.

Koreeda, M. and Koo, S. (1990) *Tetrahedron Lett.* **31**: 831.

Kozar, L.G., Clark, R.D., and Heathcock, C.H. (1977) *J. Org. Chem.* **42**: 1386.

Kozikowski, A.P. and Chen, Y.-Y. (1981) *J. Org. Chem.* **46**: 5248.

Kozikowski, A.P. and Stein, P.D. (1982) *J. Am. Chem. Soc.* **104**: 4023.

Kozikowski, A.P., Chen, Y.-Y., Wang, B.C., and Xu, Z.-B. (1984a) *Tetrahedron* **40**: 2345.

Kozikowski, A.P., Greco, M.N., and Springer, J.P. (1984b) *J. Am. Chem. Soc.* **106**: 6873.

Kozikowski, A.P., Konoike, T., and Nieduzak, T.R. (1986) *Chem. Commun.* 1350.

Krafft, M.E. (1988) *J. Am. Chem. Soc.* **110**: 968.

Krapcho, A.P. and Vivelo, J.A. (1985) *Chem. Commun.* 233.

Kraus, G.A., Roth, B., Frazier, K., and Shimagaki, M. (1982) *J. Am. Chem. Soc.* **104**: 1114.

Kraus, G.A. and Hon. Y.-S. (1985a) *J. Am. Chem. Soc.* **107**: 4341.

Kraus, G.A. and Nagy, J.O. (1985b) *Tetrahedron* **41**: 3537.

Kraus, G.A., Molina, M.T., and Walling, J.A. (1987) *J. Org. Chem.* **52**: 1273.

Kraus, G.A. and Sy, J.O. (1989) *J. Org. Chem.* **54**: 77.

Kretchmer, R.A. and Thompson, W.J. (1976) *J. Am. Chem. Soc.* **98**: 3379.

Krohn, K. (1980) *Tetrahedron Lett.* **21**: 3557.

Krohn, K. and Baltus, W. (1986) *Synthesis* 942.

Krow, G.R. and Fan, D.M. (1974) *J. Org. Chem.* **39**: 2674.

Krow, G.R. (1981) *Tetrahedron* **37**: 2697.

Krow, G.R., Shaw, D.A., Lynch, B., Lester, W., Szczepanski, S.W., Raghavachari, R., and Derome, A.E. (1988) *J. Org. Chem.* **53**: 2258.

Kuchkova, K.I., Semenov, A.A., and Terent'eva, I.V. (1971) *Acta Chim. (Budapest)* **69**: 367.

Kuehne, M.E. (1964) *J. Am. Chem. Soc.* **86**: 2946.

Kuehne, M.E. and Podhorez, D.E. (1985) *J. Org. Chem.* **50**: 924.

Kuehne, M.E., Bornmann, W.G., Parsons, W.H., Spitzer, D.T., Blount, J.F., and Zubieta, J. (1988) *J. Org. Chem.* **53**: 3439.

Kuehne, M.E. and Pitner, J.B. (1989) *J. Org. Chem.* **54**: 4553.

Kuhnke, J. and Bohlmann, F. (1988) *Liebigs Ann. Chem.* 743.

Kulkarni, Y.S. and Snider, B.B. (1985) *J. Org. Chem.* **50**: 2809.

Kündig, E.P., Desobry, V., Simmons, D.P., and Wenger, E. (1989) *J. Am. Chem. Soc.* **111**: 1804.

Kunng, F.A., Gu, J.-M., Chao, S., Chen, Y., and Mariano, P.S. (1983) *J. Org. Chem.* **48**: 4262.

Kurihara, T., Terada, T., and Yoneda, R. (1986) *Chem. Pharm. Bull.* **34**: 442.

Kuwajima, I., Kato, M., and Mori, A. (1980) *Tetrahedron Lett.* **21**: 4291.

Kuwajima, I. and Urabe, H. (1981) *Tetrahedron Lett.* **22**: 5191.

Lange, G.L., Organ, M.G., and Lee, M. (1990) *Tetrahedron Lett.* **31**: 4689.

Langlois, Y., Langlois, N., and Potier, P. (1975) *Tetrahedron Lett.* 955.

Langlois, Y., Pouilhes, A., Genin, D., Andriamialisoa, R.Z., and Langlois, N. (1983) *Tetrahedron* **39**: 3755.

Lansbury, P.T., Spagnuolo, C.J., Zhi, B., and Grimm, E.L. (1990) *Tetrahedron Lett.* **31**: 3965.

Lapworth, A. (1898) *J. Chem. Soc.* **73**: 495.

Larsen, S.D. and Monti, S.A. (1977) *J. Am. Chem. Soc.* **99**: 8015.

LeDrian, C. and Greene, A.E. (1982) *J. Am. Chem. Soc.* **104**: 5473.

Lee, J.G. and Bartsch, R.A. (1979) *J. Am. Chem. Soc.* **101**: 228.

Lee, S.Y., Niwa, M., and Snider, B.B. (1988) *Tetrahedron Lett.* **53**: 2356.

Leete, E. (1979) *Chem. Commun.* 821.

Le Gall, T., Lellouche, J.P., and Beaucourt, J.P. (1989) *Tetrahedron Lett.* **30**: 6521.

Lenoir, D. (1989) *Synthesis* 883.

Leonard, N.J. and Blum, S.W. (1960) *J. Am. Chem. Soc.* **82**: 503.

Leonard, N.J. and Sato, T. (1969) *J. Org. Chem.* **34**: 1066.

Levin, J.I. and Weinreb, S.M. (1983) *J. Am. Chem. Soc.* **105**: 1397.

Ley, S.V., Somovilla, A.A., Broughton, H.B., Craig, D., Slawin, A.M.Z., Toogood, P.L., and Williams, D.J. (1989) *Tetrahedron* **45**: 2143.

Li, T.-t. and Wu, Y.L. (1981) *J. Am. Chem. Soc.* **103**: 7007.

Liotta, D., Zima, G., and Saindane, M. (1982) *J. Org. Chem.* **47**: 1258.

Lipshutz, B.H. and Elworthy, T.R. (1990) *Tetrahedron Lett.* **31**: 477.

Liu, H.-J., Valenta, Z., and Yu, T.T.J. (1970) *Chem. Commun.* 1116.

Liu, H.-J. and Ogino, T. (1973) *Tetrahedron Lett.* 4937.

Liu, H.-J. and Majumdar, S.P. (1975) *Synth. Commun.* **5**: 125.

Liu, H.-J. and Chan, W.H. (1979) *Can. J. Chem.* **57**: 708.

Liu, H.-J. Browne, E.N.C., and Chew, S.Y. (1988) *Can. J. Chem.* **66**: 2345.

Liu, H.-J. and Ralitsch, M. (1990) *Chem. Commun.* 997.

Lombardo, L., Mander, L.N., and Turner, J.V. (1980) *J. Am. Chem. Soc.* **102**: 6626.

Lounasmaa, M. and Jokela, R. (1986) *Heterocycles* **24**: 1663.

Lygo, B. and O'Connor, N. (1987) *Tetrahedron Lett.* **28**: 3597.

MacAlpine, G.A., Raphael, R.A., Shaw, A., Taylor, A.W., and Wild, H.-J. (1976) *J. Chem. Soc. Perkin Trans. 1*: 410.

Maercker, A. (1965) *Org. React.* **14**: 270.

Magnus, P., Cairns, P.M., and Moursounidis, J. (1987) *J. Am. Chem. Soc.* **109**: 2469.

Magnus, P., Mugrage, B., DeLuca, M.R., and Cain, G.A. (1990) *J. Am. Chem. Soc.* **112**: 5220.

Majetich, G., Defauw, J., and Ringold, C. (1988) *J. Org. Chem.* **53**: 50.

Majima, T., Pac, C., Kubo, J., and Sakurai, H. (1980) *Tetrahedron Lett.* **21**: 377.

Manas, A.-R.B. and Smith, R.A.J. (1975) *Chem. Commun.* 216.

Mandell, L., Singh, K.-P., Gresham, J.T., and Freeman, W. (1963) *J. Am. Chem. Soc.* **85**: 2682.

Mander, L.N. and Sethi, S.P. (1983) *Tetrahedron Lett.* **24**: 5425.

Marazano, C., LeGoff, M.-T., Fourrey, J.-L., and Das, B.C. (1981) *Chem. Commun.* 389.

Marini-Bettolo, R., Tagliatesta, P., Lupi, A., and Bravetti, D. (1983) *Helv. Chim. Acta* **66**: 760, 1922.

Marino, J.P., Silveira, C., Comasseto, J., and Petragnani, N. (1987) *J. Org. Chem.* **52**: 4139.

Marshall, J.A. and Fanta, W.I. (1964) *J. Org. Chem.* **29**: 2501.

Marshall, J.A., Cohen, N., and Hochstetler, A.R. (1966) *J. Am. Chem. Soc.* **88**: 3408.

Marshall, J.A. and Brady, S.F. (1969) *Tetrahedron Lett.* 1387.

Marshall, J.A. and Ruden, R.A. (1970) *Tetrahedron Lett.* 1239.

Marshall, J.A. and Ellison, R.H. (1976) *J. Am. Chem. Soc.* **98**: 4312.

Martel, J. and Huynh, C. (1967) *Bull. Soc. Chim. Fr.* 985.

Martin, S.F., Dappen, M.S., Dupre, B., and Murphy, C.J. (1987) *J. Org. Chem.* **52**: 3706.

Martin, S.F. (1989a) in *Strategies and tactics in organic synthesis* (Lindberg, T., Ed.), Vol. 2, Chap. 9, Academic Press, San Diego.

Martin, S.F., Pacofsky, G.J., Gist, R.P., and Lee, W.-C. (1989b) *J. Am. Chem. Soc.* **111**: 7634.

Maruoka, K., Hashimoto, S., Kitagawa, Y., Yamamoto, H., and Nozaki, H. (1977) *J. Am. Chem. Soc.* **99**: 7705.

Masaki, Y., Nagata, K., Serizawa, Y., and Kaji, K. (1982) *Tetrahedron Lett.* **23**: 5553.

Masaki, Y., Hashimoto, K., Serizawa, Y., and Kaji, K. (1984) *Bull. Chem. Soc. Jpn.* **57**: 3476.

Matsumoto, T., Shirahama, H., Ichihara, A., Shin, H., Kagawa, S., Sakan, F., and Miyano, K. (1971) *Tetrahedron Lett.* 2049.

McCourbery, A. and Mathieson, D.W. (1949) *J. Chem. Soc.* 696.

McKillop, A., Hunt, J.D., Kienzle, F., Bigham, E., and Taylor, E.C. (1973) *J. Am. Chem. Soc.* **95**: 3635.

McMurry, J.E. and Glass, T.E. (1971) *Tetrahedron Lett.* 2575.

McMurry, J.E. and Isser, S.J. (1972) *J. Am. Chem. Soc.* **94**: 7132.

McMurry, J.E. and Blaszczak, L.C. (1974a) *J. Org. Chem.* **39**: 2217.

McMurry, J.E., Melton, J., and Padgett, H. (1974b) *J. Org. Chem.* **39**: 259.

McMurry, J.E. and Matz, J.R. (1982) *Tetrahedron Lett.* **23**: 2723.

McMurry, J.E. and Miller, D.D. (1983) *Tetrahedron Lett.* **24**: 1885.

McMurry, J.E., Farina, V., Scott, W.J., Davidsen, A.H., Summers, D.R., and Shenvi, A. (1984) *J. Org. Chem.* **49**: 3803.

McNamara, J.M. and Kishi, Y. (1984) *Tetrahedron* **40**: 4685.

Mehta, G., Krishnamurthy, N., and Karra, S.R. (1989) *Chem. Commun.* 1299.

Melching, K.H., Hiemstra, K., Klaver, W.J., and Speckamp, W.N. (1986) *Tetrahedron Lett.* **27**: 4799.

Merlini, L., Nasini, G., and Palamareva, M. (1975) *Gazz. Chim. Ital.* **105**: 339.

Meyers, A.I. and Nazarenko, N. (1973a) *J. Org. Chem.* **38**: 175.

Meyers, A.I., Nolen, R.L., Collington, E.W., Narwid, T.A., and Strickland, R.C. (1973b) *J. Org. Chem.* **38**: 1974.

Meyers, A.I. and Sturgess, M.A. (1989) *Tetrahedron Lett.* **30**: 1741.

Midland, M.M. and Gabriel, J. (1985) *J. Org. Chem.* **50**: 1143.

Miller, J.G., Kurz, W., Untch, K.G., and Stork, G. (1974) *J. Am. Chem. Soc.* **96**: 6774.

Miller, R.B., Tsang, T. (1988) 3rd N. Am. Chem. Congr. ORGN p. 184.

Minot, C., Laporterie, A., and Dubac, J. (1976) *Tetrahedron* **32**: 1523.

Misaka, Y., Mizutani, T., Sekido, M., and Uyeo, S. (1968) *J. Chem. Soc. (C)* 2954.

Mitschka, R., Cook, J.M., and Weiss, U. (1978) *J. Am. Chem. Soc.* **100**: 3973.

Miyano, M. (1981) *J. Org. Chem.* **46**: 1846.

Miyashita, M., Yanami, T., and Yoshikoshi, A. (1976) *J. Am. Chem. Soc.* **98**: 4679.

Miyashita, M., Kumazawa, T., and Yoshikoshi, A. (1981) *Chem. Lett.* 593.

Miyashita, M., Hoshino, M., and Yoshikoshi, A. (1988) *Tetrahedron Lett.* **29**: 347.

Mizuno, Y., Tomita, M., and Irie, H. (1980) *Chem. Lett.* 107.

Monti, S.A. and Yang, Y.-L. (1979) *J. Org. Chem.* **44**: 897.

Monti, S.A. and Dean, T.R. (1982) *J. Org. Chem.* **47**: 2679.

Moody, C.J. and Warrellow, G.J. (1987) *Tetrahedron Lett.* **28**: 6089.

Morera, E. and Ortar, G. (1983) *J. Org. Chem.* **48**: 119.

Moss, W.O., Bradbury, R.H., Hales, N.H., and Gallagher, T. (1990) *Chem. Commun.* 51.

Mukaiyama, T. (1976) *Angew. Chem.* **88**: 111.

Mukaiyama, T. (1977) *Angew. Chem. Int. Ed. Engl.* **16**: 817.

Mukhopadhyay, T. and Seebach, D. (1982) *Helv. Chim. Acta* **65**: 385.

Müller, W.H., Preuss, R., and Winterfeldt, E. (1977) *Chem. Ber.* **110**: 2424.

Muratake, H., Takahashi, T., Natsume, M. (1983) *Heterocycles* **20**: 1963.

Muxfeldt, H., Hardtmann, G., Kathawala, F., Vedejs, E., and Moobery, J.B. (1968) *J. Am. Chem. Soc.* **90**: 6534.

Myers, M.R. and Cohen, T. (1989) *J. Org. Chem.* **54**: 1290.

Nagata, W., Hirai, S., Okumura, T., and Kawata, K. (1968) *J. Am. Chem. Soc.* **90**: 1650.

Nagato, N., Ogawa, M., and Naito, T. (1973) Japan Kokai 7386816.

Naito, T., Iida, N., and Ninomiya, I. (1981) *Chem. Commun.* 44.

Naito, T., Kojima, N., Miyata, O., and Ninomiya, I. (1986) *Chem. Pharm. Bull.* **34**: 3530.

Najera, C. and Yus, M. (1987) *Tetrahedron Lett.* **28**: 6709.

Nakatani, K. and Isoe, S. (1985) *Tetrahedron Lett.* **26**: 2209.

Narasimhan, N.S. and Gokhale, S.M. (1985) *Chem. Commun.* 85.

Narasimhan, N.S. and Patil, P.A. (1986) *Tetrahedron Lett.* **27**: 5133.

Nemoto, H., Kurobe, H., Fukumoto, K., and Kametani, T. (1985a) *Chem. Lett.* 259.

Nemoto, H., Nagai, M., Fukumoto, K., and Kametani, T. (1985b) *Tetrahedron* **41**: 2361.

Newton, R.F. and Roberts, S.M. (1980) *Tetrahedron* **36**: 2163.

Nicolaou, K.C., Daines, R.A., Chakraborty, T.K., and Ogawa, Y. (1988) *J. Am. Chem. Soc.* **110**: 4685.

Nielsen, A.T. and Houlihan, W.J. (1968) *Org. React.* **16**: 1.

Niwa, H., Hasegawa, T., Ban, N., and Yamada, K. (1984a) *Tetrahedron Lett.* **25**: 2797.

Niwa, H., Wakamatsu, K., Hida, T., Niiyama, K., Kigoshi, H., Yamada, M., Nagase, H., Suzuki, M., and Yamada, K. (1984b) *J. Am. Chem. Soc.* **106**: 4547.

Nokami, J., Mandai, T., Watanabe, H., Ohyama, H., and Tsuji, J. (1989) *J. Am. Chem. Soc.* **111**: 4126.

Nomoto, T., Nasui, N., and Takayama, H. (1984) *Chem. Commun.* 1646.

Nossin, P.M.M. and Speckamp, W.N. (1979) *Tetrahedron Lett.* 4411.

Noyori, R., Nishizawa, M., Shimizu, F., Hayakawa, Y., Maruoka, K., Hashimoto, S., Yamamoto, H., and Nozaki, H. (1979) *J. Am. Chem. Soc.* **101**: 220.

Noyori, R., Sato, T., and Kobayashi, H. (1980) *Tetrahedron Lett.* 2569.

Noyori, R. and Suzuki, M. (1990) *Chemtracts-Org. Chem.* **3**: 173.

Ogawa, M., Kitagawa, Y., and Natsume, M. (1987) *Tetrahedron Lett.* **28**: 3985.

Ogawa, T., Takasaka, N., and Matsui, M. (1978) *Carbohydr. Res.* **60**: C4.

Ohashi, M., Maruishi, T., and Kakisawa, H. (1968) *Tetrahedron Lett.* 719.

Oh-ishi, T. and Kugita, H. (1968) *Tetrahedron Lett.* 5445.

Ohnuma, T., Tabe, M., Shiiya, K., Ban, Y., and Date, T. (1983) *Tetrahedron Lett.* **24**:4249.

Oinuma, H., Dan, S., and Kakisawa, H. (1983) *Chem. Commun.* 654.

Oishi, T., Takechi, H., and Ban, Y. (1974) *Tetrahedron Lett.* 3757.

Onaka, T. (1971) *Tetrahedron Lett.* 4391.

Ono, N., Miyaki, H., and Kaji, A. (1982) *Chem. Commun.* 33.

Oppolzer, W. and Petrzilka, M. (1976) *J. Am. Chem. Soc.* **98**: 6722.

Oppolzer, W. and Godel, T. (1978) *J. Am. Chem. Soc.* **100**: 2583.

Oppolzer, W., Bättig, K., and Hudlicky, T. (1979) *Helv. Chim. Acta* **62**: 1493.

Oppolzer, W., Francotte, E., and Bättig, K. (1981) *Helv. Chim. Acta* **64**: 678.

Oppolzer, W., Grayson, J.I., Wegmann, H., and Urrea, M. (1983a) *Tetrahedron* **39**: 3695.

Oppolzer, W. and Robbiani, C. (1983b) *Helv. Chim. Acta* **66**: 1119.

Oppolzer, W. and Godel, T. (1984a) *Helv. Chim. Acta* **67**: 1154.

Oppolzer, W. and Mirza, S. (1984b) *Helv. Chim. Acta* **67**: 730.

Oppolzer, W., Robbiani, C., and Bättig, K. (1984c) *Tetrahedron* **40**: 1391.

Overman, L.E. and Campbell, C.B. (1974) *J. Org. Chem.* **39**: 1474.

Overman, L.E. and Jessup, P.J. (1978) *J. Am. Chem. Soc.* **100**: 5179.

Overman, L.E. and Fukaya, C.(1980) *J. Am. Chem. Soc.* **102**: 1454.

Overman, L.E. and Bell, K.L. (1981) *J. Am. Chem. Soc.* **103**: 1851.

Overman, L.E., Sworin, M., and Burk, R.M. (1983) *J. Org. Chem.* **48**: 2685.

Overman, L.E. and Burk, R.M. (1984) *Tetrahedron Lett.* **25**: 5739.

Overman, L.E., Castaneda, A., and Blumenkopf, T.A. (1986) *J. Am. Chem. Soc.* **108**:1303.

Overman, L.E. and Sharp, M.J. (1988) *J. Am. Chem. Soc.* **110**: 612.

Overman, L.E. and Wild, H. (1989) *Tetrahedron Lett.* **30**: 647.

Overman, L.E. and Ricca, D.J. (1991) in *Comprehensive Org. Synth.* (Trost, B.M., Ed.) Pergamon, London.

Padwa, A., Clough, S., and Glazer, E. (1970) *J. Am. Chem. Soc.* **92**: 1778.

Padwa, A. (1984) *1,3-Dipolar cycloaddition chemistry*, Wiley, New York.

Page, C.B. and Pinder, A.R. (1964) *J. Chem. Soc.* 4811.

Page, P.C.B., Rayner, C.M., and Sutherland, I.O. (1986) *Chem. Commun.* 1408.

Pansegrau, P.D., Rieker, W.F., and Meyers, A.I. (1988) *J. Am. Chem. Soc.* **110**: 7178.

Papageorgiou, C. and Borer, X. (1987) *J. Org. Chem.* **52**: 4403.

Paquette, L.A. (1977) *Org. React.* **25**: 1.

Paquette, L.A. and Han, Y.K. (1979) *J. Org. Chem.* **44**: 4014.

Paquette, L.A. and Kinney, W.A. (1982) *Tetrahedron Lett.* **23**: 5127.

Paquette, L.A., Macdonald, D., Anderson, L.G., and Wright, J. (1989)*J. Am. Chem. Soc.* **111**: 8037.

Parker, R.E. and Isaacs, N.S. (1959) *Chem. Rev.* **59**: 737.

Parker, W., Raphael, R.A., and Wilkinson, D.I. (1959) *J. Chem. Soc.* 2433.

Partridge, J.J., Chadha, N.K., and Uskokovic, M.R. (1973) *J. Am. Chem. Soc.* **95**: 532.

Patterson, I. (1988) *Chem. Ind.* 390.

Paul, K.G., Johnson, F., and Favara, D. (1976) *J. Am. Chem. Soc.* **98**: 1285.

Pearlman, B.A. (1979) *J. Am. Chem. Soc.* **101**: 6404.

Peel, R. and Sutherland, J.K. (1974) *Chem. Commun.* 151.

Pesaro, M., Bozzato, G., and Schudel, P. (1968) *Chem. Commun.* 1152.

Petrov, M.L. and Petrov, A.A. (1972) *Zh. Oshch. Khim.* **42**: 2345; (1973) **43**: 691.

Pflengle, W. and Kunz, H. (1989) *J. Org. Chem.* **54**: 4261.

Phillips, R.R. (1959) *Org. React.* **10**: 143.

Piers, E. and Llinas-Brunet, M. (1989) *J. Org. Chem.* **54**: 1483.

Pietrusiewicz, K.M. and Salamonczyk, I. (1988) *J. Org. Chem.* **53**: 2837.

Piettre, S. and Heathcock, C.H. (1990) *Science* **248**: 1532.

Pike, R.A., McMahon, J.E., Jex, V.B., Black, W.T., and Bailey, D.L. (1959) *J. Org. Chem.* **24**:1939.

Pillai, V.N.R. (1987) *Org. Photochem.* (Padwa, A., ed.) **9**: 225.

Pindur, U. and Abdoust-Houshang, E. (1989) *Liebigs Ann. Chem.* 277.

Pinnick, H.W. and Chang, Y.-H. (1978) *J. Org. Chem.* **43**: 4662; (1979) *Tetrahedron Lett.* 837.

Pirrung, M. (1990) 199th ACS Nat. Meet., ORGN 190.

Plattner, J.J., Gless, R.D., and Rapoport, H. (1972) *J. Am. Chem. Soc.* **94**: 8613.

Plieninger, H. and Schmalz, D. (1976) *Chem. Ber.* **109**: 2140.

Posner, G.H., Mallamo, J.F., and Black, A.Y. (1981) *Tetrahedron* **37**: 3921.

Posner, G.H. (1986) *Chem. Rev.* **86**: 831.

Posner, G.H., Haces, A., Harrison, W., and Kinter, C.M. (1987)*J. Org. Chem.* **52**:48036.

Posner, G.H., Canella, K.A., and Silversmith, E.F. (1988) *Proc. Indian Acad. Sci.* **100**: 81. [(1989) *Chem. Abstr.* **110**: 75015g].

Prinzbach, H., Fuchs, R., Kitzing, R., and Achenbach, H. (1968) *Angew. Chem. Int. Ed. Engl.* **7**: 727.

Prisbylla, M.P., Takabe, K., and White, J.D. (1979) *J. Am. Chem. Soc.* **101**: 762.

Pyne, S.G. (1987) *Tetrahedron Lett.* **28**: 4737.

Qian, C.-P. and Nakai, T. (1988) *Tetrahedron Lett.* **29**: 4119.

Qian, L. and Ji, R. (1989) *Tetrahedron Lett.* **30**: 2089.

Quesada, M.L., Kim, D., Ahn, S.K., Jeong, N.S., Hwang, Y., Kim, M.Y., and Kim, J.W. (1987) *Heterocycles* **25**: 283.

Quinkert, G., Schwartz, U., Stark, H., Weber, W.D., Baier, H., Adam, F., and Durner, G. (1982) *Liebigs Ann. Chem.* 1999.

Quirion, J.-C., Grierson, D.S., Royer, J., and Husson, H.-P. (1988) *Tetrahedron Lett.* **29**: 3311.

Rabjohn, N. (1949) *Org. React.* **5**: 331; (1976) *Org. React.* **24**: 261.

Rakshit, D. and Thomas, S.E. (1987) *J. Organomet. Chem.* **333**: C3.

Ramirez, F., Bhatia, S.B., and Smith, C.P. (1967) *Tetrahedron* **23**: 2067.

Rao, Y.K. and Nagarajan, M. (1988) *Tetrahedron Lett.* **29**: 107.

Rathke, M.W. (1975) *Org. React.* **22**: 423.

Raunio, E.K. and Frey, T.G. (1971) *J. Org. Chem.* **36**: 345.

Rebek, J., McCready, R.M. (1979) *Tetrahedron Lett.* 4337.

Rebek, J., Shaber, S.H., Shue, Y.-K., Gehret, J.-C., and Zimmerman, S. (1984) *J. Org. Chem.* **49**: 5164.

Reddy, C.L. and Nagarajan, M. (1988) *Tetrahedron Lett.* **29**: 4151.

Remiszewski,S.W., Stouch, T.R., and Weinreb, S.M. (1985) *Tetrahedron* **41**: 1173.

Reuter, J.M., Sinha, A., and Salomon, R.G. (1978) *J. Org. Chem.* **43**: 2438.

Ried, W. and Mangler, H. (1964) *Liebigs Ann. Chem.* **678**: 113.

Rigby, J.H. and Senanayake, C. (1987) *J. Am. Chem. Soc.* **109**: 3147.

Rigby, J.H. and Qabar, M. (1989) *J. Org. Chem.* **54**: 5852.

Riss, B.P. and Muckensturm, B. (1986) *Tetrahedron Lett.* **27**: 4979.

Roberts, B.W., Poonian, M.S., and Welch, S.C. (1969) *J. Am. Chem. Soc.* **91**: 3400.

Roberts, J.C., Sheppard, A.H., Knight, J.A., and Roffey, P. (1968) *J. Chem. Soc. (C)* 22.

Roberts, M.R. and Schlessinger, R.H. (1981) *J. Am. Chem. Soc.* **103**: 724.

Robinson, B. (1963) *Chem. Rev.* **63**: 373; (1969) *Chem Rev.* 69: 227.

Robinson, R. (1917) *J. Chem. Soc.* 762.

Rosenheim, O. and Starling, W.W. (1937) *J. Chem. Soc.* 377.

Rosenmund, P. and Hosseini-Merescht, M. (1990) *Tetrahedron Lett.* **31**: 647.

Rosini, G., Ballini, R., and Marotta, E. (1989) *Tetrahedron* **45**: 5935.

Roskamp, E.J. and Pedersen, S.F. (1987) *J. Am. Chem. Soc.* **109**: 6551.

Roth, M., Dubs, P., Gotschi, E., and Eschenmoser, A. (1971) *Helv. Chim. Acta* **54**: 710.

Roush, W.R. and D'Ambra, T.E. (1983) *J. Am. Chem. Soc.* **105**: 1059.

Roush, W.R., Michaelides, M.R., Tai, D.F., Chong, W.K.M. (1987) *J. Am. Chem. Soc.* **109**: 7575.

Rubottom, G.M., Vazquez, M.A., Pelegrina, D.R. (1974) *Tetrahedron Lett.* 4319.

Rubottom, G.M., Gruber, J.M., and Mong, G.M. (1976) *J. Org. Chem.* **41**: 1673.

Ruggeri, R.B., McClure, K.F., and Heathcock, C.H. (1989) *J. Am. Chem. Soc.* **111**: 1530.

Russell, G.A. and Kaupp, G. (1969) *J. Am. Chem. Soc.* **91**: 3851.

Ruzicka, L. (1920) *Helv. Chim. Acta* **3**: 752.

Ryckman, D.M. and Stevens, R.V. (1987) *J. Am. Chem. Soc.* **109**: 4940.

Saá, C., Guitian, E., Castedo, L., Suau, R., Saá, J.M. (1986) *J. Org. Chem.* **51**: 2781.

Saimoto, H., Hiyama, T., and Nozaki, H. (1983) *Bull. Chem. Soc. Jpn.* **56**: 3078.

Sainsbury, M. and Uttley, N.L. (1977) *Chem. Commun.* 319.

Sakan, T. and Abe, K. (1968) *Tetrahedron Lett.* 2471.

Sakane, S., Matsumura, Y., Yamamura, Y., Ishida, Y., Mauoka, K., and Yamamoto, H. (1983) *J. Am. Chem. Soc.* **105**: 672.

Salley, S.I. (1967) *J. Am. Chem. Soc.* **89**: 6762.

Sanchez, I.H., Soria, J.J., Lopez, F.J., Larraza, M.I., and Flores, H.J. (1984) *J. Org. Chem.* **49**: 157.

Sardina, F.J., Howard, M.H., Koskinen, A.M.P., and Rapoport, H. (1989) *J. Org. Chem.* **54**: 4654.

Sato, M., Sekiguchi, K., and Kaneko, C. (1985) *Chem. Lett.* 1057.

Sato, T., Watanabe, M., Onoda, Y., and Murayama, E. (1988) *J. Org. Chem.* **53**: 1894.

Satoh, T., Iwamoto, K., and Yamakawa, K. (1987) *Tetrahedron Lett.* **28**: 2603.

Sauer, J. and Sustmann, R. (1980) *Angew. Chem. Int. Ed. Engl.* **19**: 779.

Saulnier, M.G. and Gribble, G.W. (1982) *J. Org. Chem.* **47**: 757.

Schaefer, J.P. and Bloomfield, J.J. (1967) *Org. React.* **15**: 1.

Scheffold, R. and Orlinski, R. (1983) *J. Am. Chem. Soc.* **105**: 7200.

Schinzer, D. (1988) *Synthesis* 263.

Schlessinger, R.H. and Nugent, R.A. (1982) *J. Am. Chem. Soc.* **104**: 1116.

Schmidt, R.S. and Talbiersky, J. (1971) *Angew. Chem. Int. Ed. Engl.* **16**: 853.

Schneider, W.P., Axen, U., Lincoln, F.H., Pike, J.E., and Thompson, J.L. (1968) *J. Am. Chem. Soc.* **90**: 5895.

Schoemaker, H.E. and Speckamp, W.N. (1978) *Tetrahedron Lett.* 4841.

Schöpf, C., Lehmann, G., and Arnold, W. (1937) *Angew. Chem.* **50**: 783.

Schow, S.R., Bloom, J.D., Thompson, A.S., Winzenberg, K.N., and Smith, A.B., III. (1986) *J. Am. Chem. Soc.* **108**: 2662.

Schreiber, J., Leimgruber, W., Pesaro, M., Schudel, P., Threlfall, T., and Eschenmoser, A. (1961) *Helv. Chim. Acta* **44**: 540.

Schreiber, J., Maag, H., Hashimoto, N., and Eschenmoser, A. (1971) *Angew. Chem. Int. Ed. Engl.* **10**: 330.

Schreiber, S.L., Claus, R.E., and Reagan, J. (1982) *Tetrahedron Lett.* **23**: 3867.

Schreiber, S.L. and Santini, C. (1984a) *J. Am. Chem. Soc.* **106**: 4038.

Schreiber, S.L. and Satake, K. (1984b) *J. Am. Chem. Soc.* **106**: 4186.

Schreiber, S.L. and Liew, W.-F. (1985) *J. Am. Chem. Soc.* **107**: 2980.

Schreiber, S.L., Meyers, H.V., and Wiberg, K.B. (1986) *J. Am. Chem. Soc.* **108**: 8274.

Schreiber, S.L., Kelly, S.E., Porco, J.A., Sammakia, T., and Suh, E.M. (1988a) *J. Am. Chem. Soc.* **110**: 6210.

Schreiber, S.L. and Meyers, H.V. (1988b) *J. Am. Chem. Soc.* **110**: 5198.

Schreiber, S.L. and Smith, D.B. (1989) *J. Org. Chem.* **54**: 9.

Schroth, W., Peschel, J., and Zschunke, A. (1969) *Z. Chem.* **9**: 110.

Schultz, A.G. and Godfrey, J.D. (1976) *J. Org. Chem.* **41**: 3494.

Schumann, D., Müller, H.-J., and Naumann, A. (1982) *Liebigs Ann. Chem.* 1700.

Schumann, D. and Naumann, A. (1984) *Liebigs Ann. Chem.* 1519.

Schwartz, M.A. and Scott, S.W. (1971) *J. Org. Chem.* **36**: 1827.

Scott, W.L. and Evans, D.A. (1972) *J. Am. Chem. Soc.* **94**: 4779.

Seebach, D., Seuring, B., Kalinowski, H.-O., Lubosch, W., and Renger, B. (1977) *Angew. Chem.* **89**: 270.

Seebach, D. (1979) *Angew. Chem. Int. Ed. Engl.* **18**: 239.

Seitz, D.E., Milius, R.A., and Quick, J. (1982) *Tetrahedron Lett.* **23**: 1439.

Semmelhack, M.F., Chong, B.C., and Jones, L.D. (1972) *J. Am. Chem. Soc.* **94**: 8629.

Semmelhack, M.F., Clark, G.R., Farina, R., and Saeman, M. (1979) *J. Am. Chem. Soc.* **101**: 217.

Semmelhack, M.F., Tomoda, S., and Hurst, K.M. (1980) *J. Am. Chem. Soc.* **102**: 7567.

Semmelhack, M.F. and Brickner, S.J.(1981) *J. Org. Chem.* **46**: 1723.

Semmelhack, M.F., Keller, L., Sato, T., Spiess, E.J., and Wulff, W. (1985) *J. Org. Chem.* **50**: 5566.

Seto, H., Fujimoto, Y., Tatsuno, T., and Yoshioka, H. (1985) *Synth. Commun.* **15**: 1217.

Seyferth, D. and Hui, R.C. (1985) *J. Am. Chem. Soc.* **107**: 4551.

Shapiro, R.H. (1976) *Org. React.* **23**: 405.

Sharpless, K.B. and Verhoeven, T.R. (1979) *Aldrichimica Acta* **12**: 63.

Sharpless, K.B., Behrens, C.H., Katsuki, T., Lee, A.W.M., Martin, V.S., Takatani, M., Viti, S.M., Walker, F.J., and Woodard, S.S. (1983) *Pure Appl. Chem.* **55**: 589.

Sheehan, J.C., Wilson, R.M., and Oxford, A.W. (1971) *J. Am. Chem. Soc.* **93**: 7222.

Shibuya, M., Jaisli, F., and Eschenmoser, A.,(1979) *Angew, Chem. Int. Ed. Engl.* **18**: 636.

Shimada, J., Hashimoto, K., Kim, B.H., Nakamura, E., and Kuwajima, I. (1984) *J. Am. Chem. Soc.* **106**, 1759.

Shimizu, I., Matsuda, N., Noguchi, Y., Zako, Y., and Nagasawa, K. (1990) *Tetrahedron Lett.* **31**: 4899.

Shishido, K., Komatsu, H., Fukumoto, K., and Kametani, T.(1989) *Heterocycles* **28**: 43.

Shono, T., Matsumura, Y., and Kanazawa, T. (1983) *Tetrahedron Lett.* **24**: 4577.

Shono, T., Kashimura, S., Yamaguchi, Y., and Kuwata, F. (1987) *Tetrahedron Lett.* **28**: 4411.

Sih, C.J., Massuda, D., Corey, P., Gleim, R.D., and Suzuki, F. (1979) *Tetrahedron Lett.* 1285.

Smith, A.B., III (1988) 21st Nat. Medicin. Chem. Symp. 236.

Smith, A.B., III, Haseltine, J.N., and Visnick, M. (1989) *Tetrahedron* **45**: 2431.

Smith, L.R., Williams, H.J., and Silverstein, R.M. (1978) *Tetrahedron Lett.* 3231.

Smolanoff, J., Kluge, A.F., Meinwald, J., McPhail, A., Miller, R.W., Hicks, K., and Eisner, T. (1975) *Science* **188**: 734.

Snider, B.B. and Cartaya-Marin, C.P. (1984) *J. Org. Chem.* **49**: 1688.

Snowden, R.L. (1981) *Tetrahedron Lett.* **22**: 201.

Somei, M., Makita, Y., and Yamada, F. (1986) *Chem. Pharm. Bull.* **34**: 948.

Sommer, L.H. and Pioch, R.P. (1954) *J. Am. Chem. Soc.* **76**: 1606.

Somoza, C., Darias, J., and Ruveda, E.A. (1989) *J. Org. Chem.* **54**: 1539.

Souppe, J., Namy, J.-L., and Kagan, H.B. (1984) *Tetrahedron Lett.* **25**: 2869.

Sowell, C.G., Wolin, R., and Little, R.D. (1990) *Tetrahedron Lett.* **31**: 485.

Späth, E. and Bretschneider, H. (1928) *Ber.* **61B**: 327.

Speckamp, W.N. and Hiemstra, H. (1985) *Tetrahedron* **41**: 4367.

Spitzner, D. and Wenkert, E. (1984) *Angew. Chem. Int. Ed. Engl.* **23**: 984.

Spitzner, D. and Klein, I. (1990) *Liebigs Ann. Chem.* 63.

Stein, M.L. and Burger, A. (1957) *J. Am. Chem. Soc.* **79**: 154.

Sternbach, D., Shibuya, M., Jaisli, F., Bonetti, M., and Eschenmoser, A. (1979) *Angew. Chem. Int. Ed. Engl.* **18**: 634.

Stetter, H. (1976) *Angew. Chem. Inter. Ed. Engl.* **15**: 639.

Stetter, H. and Bender, H.-J. (1978) *Angew. Chem. Int. Ed. Engl.* **17**: 131.

Stevens, R.V. and Wentland, M.P. (1968) *J. Am. Chem. Soc.* **90**: 5580.

Stevens, R.V., Fitzpatrick, J.M., Kaplan, M., and Zimmerman, R.L. (1971) *Chem. Commun.* 857.

Stevens, R.V. (1977) *Acc. Chem. Res.* **10**: 193.

Stevens, R.V. and Lee, A.W.M. (1979) *J. Am. Chem. Soc.* **101**: 7032.

Stevens, R.V. and Bisacchi, G.S. (1982a) *J. Org. Chem.* **47**: 2396.

Stevens, R.V. and Lee, A.W.M. (1982b) *Chem. Commun.* 102.

Stevens, R.V. and Hrib, N. (1983a) *Chem. Commun.* 1422.

Stevens, R.V. and Pruitt, J.R. (1983b) *Chem. Commun.* 1425.

Stevens, R.V., Beaulieu, N., Chan, W.H., Daniewski, A.R., Takeda, T., Waldner, A., Willard, P.G., and Zutter, U. (1986) *J. Am. Chem. Soc.* **108**: 1039.

Still, W.C. (1977) *J. Am. Chem. Soc.* **99**: 4186.

Still, W.C. and Mitra, A. (1978) *J. Am. Chem. Soc.* **100**: 1927.

Still, W.C. and Tsai, M.-Y. (1980) *J. Am. Chem. Soc.* **102**: 3654.

Still, W.C. and Gennari, C. (1983) *Tetrahedron Lett.* **24**: 4405.

Stille, J.K. and Divakaruni, R. (1979) *J. Org. Chem.* **44**: 3474.

Stork, G. and Clarke, F.H. (1961) *J. Am. Chem. Soc.* **83**: 3114.

Stork, G., Darling, S.D., Harrison, I.T., and Wharton, P.S. (1962a) *J. Am. Chem. Soc.* **84**: 2018.

Stork, G. and Tomasz, M. (1962b) *J. Am. Chem. Soc.* **84**: 310.

Stork, G., Brizzolara, A., Landesman, H., Szmuszkovicz, J., and Terrell, R. (1963a) *J. Am. Chem. Soc.* **85**: 207.

Stork, G. and Dolfini, J.E. (1963b) *J. Am. Chem. Soc.* **85**: 2872.

Stork, G. and McMurry, J.E. (1967) *J. Am. Chem. Soc.* **89**: 5464.

Stork, G. (1968) *Pure Appl. Chem.* **17**: 383.

Stork, G. and Gregson, M. (1969) *J. Am. Chem. Soc.* **91**: 2373.

Stork, G. and Schultz, A.G. (1971) *J. Am. Chem. Soc.* **93**: 4074.

Stork, G. and Guthikonda, R.N. (1972a) *J. Am. Chem. Soc.* **94**: 5109.

Stork, G., Tabak, J.M., and Blount, J.F. (1972b) *J. Am. Chem. Soc.* **94**: 4735.

Stork, G. and Ganem, B. (1973) *J. Am. Chem. Soc.* **95**: 6152.

Stork, G. and Raucher, S. (1976) *J. A. Chem. Soc.* **98**: 1583.

Stork, G. and Hagedorn, A.A., III (1978) *J. Am. Chem. Soc.* **100**: 3609.

Stork, G. and Singh, J. (1979) *J. Am. Chem. Soc.* **101**: 7109.

Stork, G., Clark, G., and Shiner, C.S. (1981) *J. Am. Chem. Soc.* **103**: 4948.

Stork, G. and Sherman, D.H. (1982a) *J. Am. Chem. Soc.* **104**: 3758.

Stork, G., Winkler, J.D., and Shiner, C.S. (1982b) *J. Am. Chem. Soc.* **104**: 3767.

Stork, G. and Livingston, D.A. (1987) *Chem. Lett.* 105.

Stork, G. (1989) *Pure Appl. Chem.* **61**: 439.

Stork, G., Kobayashi, Y., Suzuki, T., and Zhao, K. (1990a) *J. Am. Chem. Soc.* **112**: 1661.

Stork, G. and Zhao, K. (1990b) *J. Am. Chem. Soc.* **112**: 5875.

Strekowski, L., Wydra, R.L., Cegla, M.T., Czarny, A., Harden, D.B., Patterson, S.E., Battiste, M.A., and Coxon, J.M. (1990) *J. Org. Chem.* **55**: 4777.

Strunz, G.M. (1983) *J. Agric. Food Chem.* **31**: 185.

Suda, M. (1982) *Tetrahedron Lett.* **23**: 427.

Sugahara, T., Komatsu, Y., and Takano, S. (1984) *Chem. Commun.* 214.

Sustmann, R. (1971) *Tetrahedron Lett.* 2721.

Suzuki, M., Watanabe, A., and Noyori, R. (1980) *J. Am. Chem. Soc.* **102**: 2095.

Swenton, J.S. and Jurcak, J.G. (1988) *J. Org. Chem.* **53**: 1530.

Szantay, C., Szabo, L., and Kalaus, G. (1977) *Tetrahedron* **33**: 1803.

Szychowski, J. and MacLean, D.B. (1979) *Can. J. Chem.* **57**: 1631.

Taber, D.F. (1977) *J. Am. Chem. Soc.* **99**: 3513.

Takagi, M., Goto, S., Ishihara, R., and Matsuda, T. (1976) *Chem. Commun.* 993.

Takahashi, M. (1971) *Itsuu Kenkyusho Nempo* 65 [(1972) *Chem. Abstr.* **77**: 61854t].

Takahashi, S., Oritani, T., and Yamashita, K. (1988) *Tetrahedron* **44**: 7081.

Takahashi, T., Kasuga, K., Takahashi, M., and Tsuji, J. (1979) *J. Am. Chem. Soc.* **101**: 4072.

Takahashi, T., Kanda, Y., Nemoto, H., Kitamura, K., Tsuji, J., and Fukazawa, Y. (1986) *J. Org. Chem.* **51**: 3393.

Takahashi, Y., Isobe, K., Hagiwara, H., Kosugi, H., and Uda, H. (1981) *Chem. Commun.* 714.

Takahata, H., Yamabe, K., Suzuki, T., and Yamazaki, T. (1986) *Chem. Pharm. Bull.* **34**: 4523.

Takanishi, K., Urabe, H., and Kuwajima, I. (1987) *Tetrahedron Lett.* **28**: 2281.

Takano, S., Ogawa, N., and Ogasawara, K. (1981) *Heterocycles* **16**: 915.

Takano, S., Hatakeyama, S., Takahashi, Y., and Ogasawara, K. (1982a) *Heterocycles* **17**: 263.

Takano, S., Goto, E., Hirama, M., and Ogasawara, K. (1982b) *Chem. Pharm. Bull.* **30**: 2641.

Takeda, K., Shimono, Y., and Yoshii, E. (1983) *J. Am. Chem. Soc.* **105**: 563.

Tamao, K., Maeda, K., Tanaka, T., and Ito, Y. (1988) *Tetrahedron Lett.* **29**: 6955.

Tamariz, J., Schwager, L., Stibbard, J.H.A., and Vogel, P. (1983) *Tetrahedron Lett.* **24**: 1497.

Tamura, Y., Mohri, S., Maeda, H., Tsugoshi, T., Sasho, M., and Kita, Y. (1984) *Tetrahedron Lett.* **25**: 309.

Tanaka, A., Uda, H., and Yoshikoshi, A. (1967) *Chem. Commun.* 188.

Tanaka, K., Uchiyama, F., Sakamoto, K., and Inubushi, Y. (1982) *J. Am. Chem. Soc.* **104**: 4965.

Tang, C. and Rapoport, H. (1972) *J. Am. Chem. Soc.* **94**: 8615.

Tanis, S.P., Chuang, Y.H., and Head, D.B. (1985) *Tetrahedron Lett.* **26**: 6147.

Tanis, S.P. and Dixon, L.A. (1987) *Tetrahedron Lett.* **28**: 2495.

Taploczay, D.J., Thomas, E.J., and Whitehead, J.W.F. (1985) *Chem. Commun.* 143.

Tarnchompoo, B., Thebtaranonth, C., and Thebtaranonth, Y. (1986) *Synthesis* 785.

Tatsuta, K., Tanaka, A., Fujimoto, K., Kinoshita, M., and Umezawa, S. (1977) *J. Am. Chem. Soc.* **99**: 5826.

Tatsuta, K., Amemiya, Y., Maniwa, S., and Kinoshita, M. (1980) *Tetrahedron Lett.* 2837.

Taylor, E.C. and Martin, S.F. (1974) *J. Am. Chem. Soc.* **96**: 8095.

Tchissambou, L., Benechie, M., and Khuong-Huu, F. (1982) *Tetrahedron* **38**: 2687.

Tebbe, F.N., Parshall, G.W., and Reddy, G.S. (1978) *J. Am. Chem. Soc.* **100**: 3611.

Ternansky, R.J., Balogh, D.W., and Paquette, L.A. (1982) *J. Am. Chem. Soc.* **104**: 4503.

Terao, Y., Imai, N., Achiwa, K., and Sekiya, M. (1982) *Chem. Pharm. Bull.* **30**: 3167.

Tezuka, T., Kikuchi, O., Houk, K.N., Paddon-Row, M.N., Santiago, C.M., Rondan, N.G., Williams, J.C., and Gandour, R.W. (1981) *J. Am. Chem. Soc.* **103**: 1367.

Thomsen, I. and Torssell, K.B.G. (1988) *Acta Chem. Scand.* **42B**: 303.

Timms, G.H., Tupper, D.E., and Morgan, S.E. (1989) *J. Chem. Soc. Perkin Trans. I*: 817.

Tobe, Y., Yamashita, D., Takahashi, T., Inata, M., Sato, J., Kakiuchi, K., Kobiro, K., and Odaira, Y. (1990) *J. Am. Chem. Soc.* **112**: 775.

Torii, S., Tanaka, H., and Kobayashi, Y. (1977) *J. Org. Chem.* **42**: 3473.

Torssell, K.B.G. (1988) *Nitrile oxides, nitrones, and nitronates in organic synthesis*, VCH, New York.

Toth, J.E. and Fuchs, P.L. (1987) *J. Org. Chem.* **52**: 473.

Tou, J.S. and Reusch, W. (1980) *J. Org. Chem.* **45**: 5012.

Tramontini, M. (1973) *Synthesis* 703.

Trost, B.M., Hiroi, K., and Holy, N. (1975a) *J. Am. Chem. Soc.* **97**: 5873.

Trost, B.M. and Keely, D.E. (1975b) *J. Org. Chem.* **40**: 2013.

Trost, B.M. and Genet, J.P. (1976a) *J. Am. Chem. Soc.* **98**: 8516.

Trost, B.M., Taber, D.F., and Alper, J.B. (1976b) *Tetrahedron Lett.* 3857.

Trost, B.M., Bogdanowicz, M.J., Frazee, W.J., and Salzmann, T.N. (1978a) *J. Am. Chem. Soc.* **100**: 5512.

Trost, B.M. and Rigby, J.H. (1978b) *J. Org. Chem.* **43**: 2938.

Trost, B.M., Bernstein, P.R., and Funfschilling, P.C. (1979a) *J. Am. Chem. Soc.* **101**: 4378.

Trost, B.M. and Klun, T.P. (1979b) *J. Am. Chem. Soc.* **101**: 6756; (1981) **103**: 1864.

Trost, B.M., Shuey, C.D., and DiNinno, F. (1979c) *J. Am. Chem. Soc.* **101**: 1284.

Trost, B.M., Vladuchick, W.C., and Bridges, A.J. (1980) *J. Am. Chem. Soc.* **102**: 3554.

Trost, B.M. and McDougal, P.G. (1982a) *J. Am. Chem. Soc.* **104**: 6110.

Trost, B.M., Verhoeven, T.R. (1982b) in *Comprehensive organometallic chemistry*, (Wilkinson, G. and Stone, F.G.A., Eds) Vol 8, Pergamon, Oxford.

Trost, B.M. and Pearson, W.H. (1983) *J. Am. Chem. Soc.* **105**: 1054.

Trost, B.M., Lynch, J., Renaut, P., and Steinman, D.H. (1986a) *J. Am. Chem. Soc.* **108**: 284.

Trost, B.M. and Mignani, S. (1986b) *J. Org. Chem.* **51**: 3435.

Trost, B.M. and Mikhail, G.K. (1987) *J. Am. Chem. Soc.* **109**: 4124.

Trost, B.M. (1989a) *Angew. Chem. Int. Ed. Engl.* **28**: 1173.

Trost, B.M. and Schneider, S. (1989b) *Angew. Chem. Int. Ed. Engl.* **28**: 213.

Truce, W.E. and Tichenor, G.J.W. (1972) *J. Org. Chem.* **37**: 2391.

Tufariello, J.J. and Tette, J.P. (1971) *Chem. Commun.* 469.

Tufariello, J.J. and Ali, S.A. (1978a) *Tetrahedron Lett.* 4647.

Tufariello, J.J. and Gastrone, R.G. (1978b) *Tetrahedron Lett.* 2753.

Tufariello, J.J. and Ali, S.K. (1979a) *Tetrahedron Lett.* 4445.

Tufariello, J.J., Mullen, G.B., Tegeler, J.J., Trybulski, E.J., Wong, S.C., and Ali, S.A. (1979b) *J. Am. Chem. Soc.* **101**: 2435.

Tull, R., O'Neill, R.C., McCarthy, E.P., Pappas, J.J., and Chemerda, J.M. (1964) *J. Org. Chem.* **29**: 2425.

Uno, H., Naruta, Y., and Maruyama, K. (1984) *Tetrahedron* **40**: 4741.

Utimoto, K., Kato, S., Tanaka, M., Hoshino, Y., Fujikura, S., and Nozaki, H. (1982) *Heterocycles* **18**: 149.

Uyehara, T., Shida, N., and Yamamoto, Y. (1989) *Chem. Commun.* 113.

van Tamelen, E.E. and Baran, J.S. (1958a) *J. Am. Chem. Soc.* **80**: 4659.

van Tamelen, E.E., Shamma, M., Burgstahler, A.W., Wolinsky, J., Tamm, R., and Aldrich, P.E. (1958b) *J. Am. Chem. Soc.* **80**: 5006.

van Tamelen, E.E., Spencer, T.A., Allen, D.S., and Orvis, R.L. (1959) *J. Am. Chem. Soc.* **81**: 6341; (1961) *Tetrahedron* **14**: 8.

van Tamelen, E.E. and Foltz, R.L. (1960) *J. Am. Chem. Soc.* **82**: 2400.

van Tamelen, E.E., Placeway, C., Schiemenz, G.P., and Wright, I.G. (1969) *J. Am. Chem. Soc.* **91**: 7359.

van Tamelen, E.E. and Oliver, L.K. (1970) *J. Am. Chem. Soc.* **92**: 2136.

Vedejs, E. (1974) *J. Am. Chem. Soc.* **96**: 5944.

Vedejs, E. and Martinez, G.R. (1979) *J. Am. Chem. Soc.* **101**: 6452.

Vedejs, E. and Martinez, G.R. (1980) *J. Am. Chem. Soc.* **102**: 7993.

Vedejs, E., Campbell, J.B., Gadwood, R.C., Rodgers, J.D., Spear, K.L., and Watanabe, Y. (1982) *J. Org. Chem.* **47**: 1534.

Vedejs, E., Miller, W.H., and Pribish, J.R. (1983) *J. Org. Chem.* **48**: 3611.

Vedejs, E. and Marth, C.F. (1990) *J. Am. Chem. Soc.* **112**: 3905.

Veenstra, S.J. and Speckamp, W.N. (1981) *J. Am. Chem. Soc.* **103**: 4645.

VerHaeghe, D.G., Weber, G.S., and Pappalardo, P.A. (1989) *Tetrahedron Lett.* **30**: 4041.

Viehe, H.G., Janousek, Z., Merenyi, R., and Stella, L. (1985) *Acc. Chem. Res.* **18**: 148.

Vlattas, I., Harrison, I.T., Tokes, L., Fried, J., and Cross, A.D. (1968) *J. Org. Chem.* **33**: 4176.

Wada, M., Nishihara, Y., and Akiba, K. (1985) *Tetrahedron Lett.* **26**: 3267.

Wadsworth, W.S. (1977) *Org. React.* **25**: 73.

Wakamatsu, T., Akasaka, K., and Ban, Y. (1977) *Tetrahedron Lett.* 2755.

Wakamatsu, T., Miyachi, N., Ozaki, F., Shibasaki, M., and Ban, Y. (1988) *Tetrahedron Lett.* **29**: 3829.

Walts, A.E. and Roush, W.R. (1985) *Tetrahedron* **41**: 3463.

Wang, S.L.B. and Wulff, W.D. (1990) *J. Am. Chem. Soc.* **112**: 4550.

Wang, Y. (1990) *J. Org. Chem.* **55**: 4504.

Wanner, M.J., Koomen, G.J., and Pandit, U.K. (1981) *Heterocycles* **15**: 377.

Wanner, M.J. and Koomen, G.J. (1989) *Tetrahedron Lett.* **30**: 2301.

Wanzlick, H.-W., Gritzky, R., and Heidepriem, H. (1963) *Chem. Ber.* **96**: 305.

Wasserman, H.H., Keith, D.D., and Nadelson, J. (1969) *J. Am. Chem. Soc.* **91**: 1264.

Wasserman, H.H. and Amici, R.M. (1989) *J. Org. Chem.* **54**: 5843.

Watt, D.S. and Corey, E.J. (1972) *Tetrahedron Lett.* 4651.

Wehle, D. and Fitjer, L. (1987) *Angew. Chem. Int. Ed. Engl.* **26**: 130.

Weinreb, S.M. and Cvetovich, R.J. (1972) *Tetrahedron Lett.* 1233.

Welch, J.T., Plummer, J.S., and Herbert, R.W. (1988) 3rd N. Am. Chem. Congr. ORGN 356.

Welch, J.T., DeCorte, B., and DeKimpe, N. (1990) *J. Org. Chem.* **55**: 4981.

Welch, S.C. and Chayabunjonglerd, S. (1979) *J. Am. Chem. Soc.* **101**: 6768.

Wender, P.A. and Lechleiter, J.C. (1978) *J. Am. Chem. Soc.* **100**: 4321.

Wender, P.A., Schaus, J.M., and White, A.W. (1980) *J. Am. Chem. Soc.* **102**: 6157.

Wender, P.A. and Fisher, K. (1986) *Tetrahedron Lett.* **27**: 1857.

Wender, P.A. and McDonald, F.E. (1990) *J. Am. Chem. Soc.* **112**: 4956.

Wenkert, E. (1968a) *Acc. Chem. Res.* **1**: 78.

Wenkert, E., Dave, K.G., and Stevens, R.V. (1968b) *J. Am. Chem. Soc.* **90**: 6177.

Wenkert, E., Chang, H., Chawla, H., Cochran, D., Hagaman, E., King, J., and Orito, K. (1976) *J. Am. Chem. Soc.* **98**: 3645.

Wenkert, E., Berges, D.A., and Golob, N.F. (1978a) *J. Am. Chem. Soc.* **100**: 1263.

Wenkert, E., Buckwalter, B.L., Craveiro, A.A., Sanchez, E.L., and Sathe, S.S. (1978b) *J. Am. Chem. Soc.* **100**: 1267.

Wenkert, E., Hudlicky, T., and Showalter, H.D.H. (1978c) *J. Am. Chem. Soc.* **100**: 4893.

Wenkert, E. (1981a) *Pure Appl. Chem.* **53**: 1271.

Wenkert, E., Halls, T.D.J., Kwart, L.D., Magnusson, G., and Showalter, H.D.H. (1981b) *Tetrahedron* **37**: 4017.

Wenkert, E., St. Pyrek, J., Uesato, S., and Vankar, Y.D. (1982) *J. Am. Chem. Soc.* **104**: 2244.

Wenkert, E. (1984) *Heterocycles* **21**: 325.

Wenkert, E., Decorzant, R., and Naf, F. (1989) *Helv. Chim. Acta* **72**: 756.

Wharton, P.S., Sundin, C.E., Johnson, D.W., and Kluender, H.C. (1972) *J. Org. Chem.* **37**: 34.

White, D.R. (1976) *Tetrahedron Lett.* 1753.

Whitlock, H.W. and Smith, G.L. (1967) *J. Am. Chem. Soc.* **89**: 3600.

Wieland, T. and Lamperstorfer, C. (1966) *Makromol. Chem.* **31**: 1658.

Wieland, T., Birr, C., and Wiesenbach, H. (1969) *Angew. Chem.* **81**: 782.

Wiesner, K., Ho, P., Chang, D., Lam, Y.K., and Ren, W.Y. (1973) *Can. J. Chem.* **51**: 3978.

Wiesner, K. (1975) *Tetrahedron* **31**: 1655.

Wiesner, K. (1977) *Chem. Soc. Rev.* **6**: 413.

Wijnberg, B.P. and Speckamp, W.N. (1981) *Tetrahedron Lett.* **22**: 5079.

Wijnberg, B.P., Speckamp, W.N., and Oostveen, A.R.C. (1982) *Tetrahedron* **38**: 209.

Wilkens, J., Kühling, A., and Blechert, S. (1987) *Tetrahedron* **43**: 3237.

Williams, D.R. and Sit, S.-Y. (1982) *J. Org. Chem.* **47**: 2846.

Williams, R.M., Dung, J.S., Josey, J., Armstrong, R., and Meyers, H. (1983) *J. Am. Chem. Soc.* **105**: 3214.

Willstätter, R. and Bommer, M. (1921a) *Liebigs Ann. Chem.* **422**: 15.

Willstätter, R. and Pfannenstiel, M. (1921b) *Liebigs Ann. Chem.* **422**: 1.

Willstätter, R., Wolfes, O., and Maeder, H. (1923) *Liebigs Ann. Chem.* **434**: 111.

Wilson, S.R., Phillips, L.R., and Natalie, K.J. (1979) *J. Am. Chem. Soc.* **101**: 3340.

Winkler, J.D., Muller, C.L., and Scott, R.D. (1988) *J. Am. Chem. Soc.* **110**: 4831.

Winkler, J.D. and Hershberger, P.M. (1989) *J. Am. Chem. Soc.* **111**: 4852.

Winterfeldt, E., Radunz, H., and Korth, T. (1968) *Chem. Ber.* **101**: 3172.

Winterfeldt, E., Korth, T., Pike, D., and Boch, M. (1972) *Angew. Chem. Int. Ed. Engl.* **11**: 289.

Wiseman, J.R. and Lee, S.Y. (1986) *J. Org. Chem.* **51**: 2485.

Wittig, G. (1962) *Angew. Chem. Int. Ed. Engl.* **1**: 415.

Wittig, G., Frommeld, H.D., and Suchanek, P. (1963) *Angew. Chem. Int. Ed. Engl.* **2**: 683.

Wolinsky, J. and Barker, W. (1960) *J. Am. Chem. Soc.* **82**: 636.

Wolinsky, J., Slabaugh, M.R., and Gibson, T. (1964) *J. Org. Chem.* **29**: 3740.

Woodward, R.B. and Doering, W.v.E. (1945) *J. Am. Chem. Soc.* **67**: 860.

Woodward, R.B. and McLamore, W.M. (1949) *J. Am. Chem. Soc.* **71**: 379.

Woodward, R.B., Sondheimer, F., Taub, D., Heusler, K., and McLamore, W.M. (1952) *J. Am. Chem. Soc.* **74**: 4223.

Woodward, R.B., Bader, F.E., Bickel, H., Frey, A.J., and Kierstead, R.W. (1958) *Tetrahedron* **2**: 1.

Woodward, R.B. (1963a) *Pure Appl. Chem.* **6**: 651.

Woodward, R.B., Cava, M.P., Ollis, W.D., Hunger, A., Daeniker, H.U., and Schenker, K. (1963b) *Tetrahedron* **19**: 247.

Woodward, R.B. (1963-4) *The Harvey Lectures* **59**: 31.

Woodward, R.B., Fukunaga, T., and Kelly, R.C. (1964) *J. Am. Chem. Soc.* **86**: 3162.

Woodward, R.B. (1968) *Pure Appl. Chem.* **17**: 519.

Woodward, R.B. and Hoffmann, R. (1970) *The conservation of orbital symmetry*, Academic Press, New York.

Woodward, R.B., et al. (1981) *J. Am. Chem. Soc.* **103**: 3210, 3213, 3215.

Wovkulich, P.M., Tang, P.C., Chadha, N.K., Batcho, A.D., Barrish, J.C., and Uskokovic, M.R. (1989) *J. Am. Chem. Soc.* **111**: 2596.

Wu, Y.-J. and Burnell, D.J. (1988) *Tetrahedron Lett.* **29**: 4369.

Wubbels, G.G. and Cotter, W.D. (1989) *Tetrahedron Lett.* **30**: 6477.

Wulff, W.D. and Yang, D.C. (1983) *J. Am. Chem. Soc.* **105**: 6726.

Yamago, S. and Nakamura, E. (1989) *J. Am. Chem. Soc.* **111**: 7285.

Yamaguchi, M., Hasebe, K., Tanaka, S., and Minami, T. (1986) *Tetrahedron Lett.* **27**: 959.

Yamakawa, K. and Satoh, T. (1977) *Chem. Pharm. Bull.* **25**: 2535; *Heterocycles* **8**: 221.

Yamamoto, Y. and Furuta, T. (1990) *J. Org. Chem.* **55**: 3971.

Yanagiya, M., Kaneko, K., Kaji, T., and Matsumoto, T. (1979) *Tetrahedron Lett.* 1761.

Yeh, M.C.P., Knochel, P., Butler, W.M., and Berk, S.C. (1988) *Tetrahedron Lett.* **29**: 6693.

Zamojski, A. and Kozluk, T. (1984) *J. Org. Chem.* **42**: 1089.

Zhao, S.-K. and Helquist, P. (1990) *J. Org. Chem.* **55**: 5820.

Zheng, G.-C. and Kakisawa, H. (1989) *Bull. Chem. Soc. Jpn.* **62**: 602.

Ziegler, F.E. and Bennett, G.B. (1973a) *J. Am. Chem. Soc.* **95**: 7458.

Ziegler, F.E. and Spitzner, E.B. (1973b) *J. Am. Chem. Soc.* **95**: 7146.

Ziegler, F.E. and Wender, P.A. (1974) *Tetrahedron Lett.* 449.

Ziegler, F.E., Reid, G.R., Studt, W.L., and Wender, P.A. (1977) *J. Org. Chem.* **42**: 1991.

Ziegler, F.E., Fang, J.-M., and Tam, C.C. (1982) *J. Am. Chem. Soc.* **104**: 7174.

Ziegler, F.E. and Jaynes, B.H. (1987) *Tetrahedron Lett.* **28**: 2339.

Zimmer, H. and Walter, R. (1963) *Z. Naturforsch.* **18B**: 669.

Zschiesche, R., Grimm, E.L., and Reissig, H.U. (1986) *Angew. Chem. Int. Ed. Engl.* **98**: 1104.

Zurflüh, R., Wall, E.N., Siddall, J.B., and Edwards, J.A. (1968) *J. Am. Chem. Soc.* **90**: 6224.

Zutterman, F., De Wilde, H., Mijngheer, R., De Clercq, P.J.M., and Vandewalle, M.E.A. (1979) *Tetrahedron* **35**: 2389.

INDEX

Frequently used terms may not be indexed after their initial appearances in the text. These include polarity alternation, retrosynthetic analysis, synthon, etc.